D0405196

TK
5101
P334
2000

# The N...11

# Teleco...

## Second E...

DATE DUE

BOOKS MAY BE RECALLED
AFTER 10 DAYS

| | | |
|---|---|---|
| DEC 2 3 1998 | | |
| | | |
| JAN 08 2001 | | |
| MAR 0 8 2002 | | |
| | | |
| MAR 1 0 2002 | | |
| | | |
| | | |
| | | |
| | | |
| | | |

Demco, Inc. 38-293

# McGraw-Hill Telecommunications

# The *New* McGraw-Hill Telecom Factbook

## Second Edition

Joseph A. Pecar
David A. Garbin

**McGraw-Hill**

New York   San Francisco   Washington, D.C.   Auckland   Bogotá
Caracas   Lisbon   London   Madrid   Mexico City   Milan
Montreal   New Delhi   San Juan   Singapore
Sydney   Tokyo   Toronto

**Library of Congress Cataloging-in-Publication Data**

Pecar, Joseph A.
    The new McGraw-Hill telecom factbook / Joe Pecar, David Garbin.
      p. cm.
    ISBN 0-07-135163-9
    1. Telecommunication.  I. Garbin, David A.  II. Title

TK5101 P334 2000
384—dc21

00-055011

*McGraw-Hill*

*A Division of The McGraw-Hill Companies*

Copyright © 2000 by The McGraw-Hill Companies, Inc. All rights
reserved. Printed in the United States of America. Except as permitted
under the United States Copyright Act of 1976, no part of this publication
may be reproduced or distributed in any form or by any means, or
stored in a data base or retrieval system, without the prior written
permission of the publisher.

1 2 3 4 5 6 7 8 9 0  QM/QM  9 0 9 8 7 6 5 4 3 2 1 0 9

ISBN 0-07-135163-9

The sponsoring editor for this book was Steve Chapman and
the production manager was Pamela Pelton. It was set in
Vendome by Patricia Wallenburg.

Printed and bound by Quebecor/Martinsburg.

 This book is printed on recycled, acid-free paper containing a
minimum of 50% recycled, de-inked fiber.

McGraw-Hill books are available at special quantity discounts to use
as premiums and sales promotions, or for use in corporate training
programs. For more information, please write to the Director of Special
Sales, McGraw-Hill, Two Penn Plaza, New York, NY 10121-2298. Or
contact your local bookstore.

Information contained in this work has been obtained by The
McGraw-Hill Companies, Inc. ("McGraw-Hill") from sources
believed to be reliable. However, neither McGraw-Hill nor its
authors guarantee the accuracy or completeness of any information published herein and neither McGraw-Hill nor its authors
shall be responsible for any errors, omissions, or damages arising
out of use of this information. This work is published with the
understanding that McGraw-Hill and its authors are supplying
information but are not attempting to render engineering or
other professional services. If such services are required, the assistance of an appropriate professional should be sought.

"Science, for its progress, its understanding and so on, depends so much on truth and on clarity. And the question is: Are we living in a society where truth and clarity are an important element of our culture?"

*—Isador Iaisah Rabi,*
*Distinguished Nobel Prize winning physicist*

To our wives Linda Garbin and Nancy Ann Pecar

# CONTENTS

# PREFACE

You're reading the second edition of the best-selling *McGraw-Hill Telecom Factbook*. Shortly after its 1993 debut, the *Factbook* ranked sixth in sales among McGraw-Hill's broad selection of technical books. In more recent years, despite an unparalleled rate of technological advance and market development, the original *Factbook* has repeatedly captured the monthly "bestselling technical title" spot at Amazon.com, the highly successful Internet-based bookseller. Perhaps most significantly, McGraw-Hill has received more requests for update of the *Factbook* than for any of its other technical books.

Why such a high level of interest in the *Factbook?* There are two principal reasons. First, while most enterprises need telecommunications products and services, a rapidly growing number can't exist without them. Maintaining competitive edge and improving profitability are increasingly dependent upon telecommunications resources, thus making the topic of critical interest to enterprise decision-makers.

The second reason for the *Factbook's* popularity is that its originally stated objective—to present telecommunications technologies in a manner easily understood by business people with nonengineering backgrounds—has been met with a level of success rarely achieved in the technical publishing industry.

# Benefiting from the Telecommunications Revolution

The United States telecommunications industry, currently growing at 8 to 12 percent per year, reached $450 billion in 1999 and is projected to be worth $690 billion by 2004. As the established leader for new telecommunications technologies, products, and services, the United States accounts for nearly one half of the global market. Because businesses expend a large share of the dollars that create the market, business people have a vital need for knowledge that permits them to select the array of services providing the greatest benefit and competitive advantage at the lowest cost.

To succeed in today's workplace, people must be able to exchange information quickly and accurately. Job content is therefore increasingly influenced by telecommunications. Required business management skills now include knowledge of available telecommunications services, their application to changing organizational needs, the ability to work with technical professionals, and the expertise to acquire and use telecommunications services efficiently and cost-effectively. The pivotal role that telecommunications plays in private and business arenas is amply demonstrated by its spectacular growth-rate statistics, a few of which are mentioned below.

Over the past decade or so, U.S. telecommunications usage has grown at an unprecedented rate. From 92,000 cellular telephones in 1984, the number of wireless subscribers grew to 86 million by 2000, a growth rate of 60 percent per year! Complimenting these impressive U. S. statistics, by 1998 worldwide wireless subscribers totaled over 200 million with 50,000 new subscribers added per day.

Matching eye-popping wireless growth, starting with 80,000 in 1989, by 1998 the number of devices accessing the Internet reached 147 million, increasing by 67,000 per day. With 850 million Internet World Wide Web (Web) pages to peruse, in 1998 the average call duration was 30 minutes, an increase of 55 percent over 1997. Even more startling is a recent prediction, published by Motorola, that there will be more mobile devices than PCs connected to the Internet by the end of 2003.

To place current growth rates in perspective, consider that although it took a century to reach 700 million wired telephone subscribers, respected industry leaders predict the next 700 million will be added in just 15 years. And, in those same years, 700 million new wireless subscribers will also be added. To keep pace with this demand, between 1998 and 2000, over 1,000 new service providers have appeared. The evidence then is compelling. A telecommunications revolution of astonishing proportions is now underway, a revolution propelled by its own dynamics, and augmented by similar developments in the data processing industry.

What becomes evident in reading this book is that the revolution consists not just in exploding service, subscriber, network, and product statistics. Rather, these skyrocketing numbers presage an even more profound revolution, that is one in which information technologies are being applied in ways that radically alter how business advantage is created.

On the plus side, growth, rapid technological development, and sweeping regulatory changes are creating an astounding number of new opportunities for technology providers and consumers alike. On the down side, so many options are being created that merely remaining abreast of developments has become a major managerial challenge.

Even for telecommunications professionals, keeping up-to-date on the plethora of public and private offerings for the delivery of voice, data, imagery, video, and other telecommunications services is now a formidable task. Moreover, among engineers with such technical expertise, rarely does one find sufficient business acumen to, on their own, adapt information technologies in ways that lead to new and innovative business processes and capabilities.

By the same token, it is difficult for business people to articulate their needs in terms that telecommunications professionals can understand—and that suppliers can address with accurate, competitive proposals. What is required is a "knowledge bridge" between business needs and the growing spectrum of telecommunications offerings. *The New McGraw-Hill Telecom Factbook* is designed to satisfy that requirement.

## Objectives

The principal objective of this updated book remains to provide a comprehensive introduction and insightful perspectives into modern telecommunications services and their underlying technologies. A second objective—no less important than the first—is to employ a presentation style easily understood by government and commercial telecommunications planners, managers, users, and professionals who do not have the time to sift through multiple publications, complex formulae, and mathematics only to be forced to draw their own conclusions regarding technology, performance, and market alternatives.

All important telecommunications services and technologies are treated, but the quantity of information is limited to that needed for a complete understanding. In addition, rather than just treating topics individually, expert interpretations provide a valuable grasp of "bottom line" relationships among emerging services, technologies, and industry standards.

Simplicity of presentation style does not sacrifice the ability to familiarize readers with industry terminology and essential concepts—which is often the case with introductory material. To accomplish this, we systematically present basic definitions as part of explanations of larger concepts. This equips the reader not only with terminology, but also with rationale behind real-world applications, a tremendous advantage for thorough understanding and memory retention.

In this respect, perhaps the feature of the original *The McGraw-Hill Telecom Factbook* contributing most to its popularity and effectiveness is the rich array, quality and clarity of its over 160 figures. Using these figures the book goes far beyond simply identifying and defining panoplies of services and technologies. Inimitable technology taxonomy charts form the basis for uncommonly clear explications that fully apprise readers of essential differences and the practical advantages among alternative technology options.

# Plan of the Text

Although many of its topics are subjects of individually published textbooks, as noted our book's material has been carefully selected so that readers do not have to deal with more information than necessary to achieve learning objectives.

Under this approach, new material is placed into the context of material already presented, highlighting topic interrelationships while minimizing text length and complexity. Accordingly, the book begins by defining telecommunications and its essential terms. Telephony and the historical development of voice networks are treated next. This ordering is selected since the majority of U.S. network traffic is still voice, and the lessons-learned in achieving the impressive capacity, quality of service and reliability characteristics, so long associated with today's voice networks, apply to and must be embedded in tomorrow's integrated information networks. Furthermore—at least during some transition period—the next-generation networks will evolve from, or at least be required to "interwork" with existing voice networks.

Thus Part 1, Introduction, begins with terminology and background material that can be covered in several hours. Because government legislation and regulation have had such profound impacts on U.S. telecommunications, a historical review of the structure it has imposed is presented as a foundation for succeeding technical material.

Part 2, Telecommunications Fundamentals, describes and explains the primary telecommunications systems "building-blocks," that is the transmission, multiplexing, switching and advanced networking elements. If these terms are unfamiliar, you need not be concerned since Part 2 begins with an easily understood expository of basic concepts, techniques and devices. This knowledge enables readers, even those with no engineering background, to fully understand the operational principles and performance characteristics of all elements upon which telecommunications networks—and the services they render—are grounded.

Does that mean that reading this book prepares one to design "packet switched," "circuit switched," or "asynchronous transfer mode" networks? No it doesn't! But what the first edition proves it will do for even non-technical readers is impart a clear understanding of the purpose, operation, applications and advantages of all major telecommunications technologies.

With Part 3 the book's focus shifts from telecommunications fundamentals and building-block technologies to the truly vast number of telecommunications services of relevance to business and residential users. Parts 3 and 4 present voice and data services at the premises, metropolitan and wide area level, with supporting traffic engineering, service selection criteria and methods. In Part 5, all forms of terrestrial and satellite-based wireless service are identified, explained and compared. The terrestrial category treats cellular, PCS, paging, specialized mobile service, mobile data, and existig and emerging fixed-location wireless services. The satellite-based category addresses handheld or vehicular terminal, transportable terminal, and VSAT-based services.

Of course, writing about any rapidly developing technical subject is much like aiming at a moving target. While much of its content addresses telecommunications principles and terms of reference that are relatively unchanging, the updated book retains the first edition's modular structure, one designed to facilitate new editions to take into account ongoing U.S. and global telecommunications developments.

To help readers anticipate new developments and minimize "future shock," Part 6, Outlooks for the Future, presents summary conclusions and postulates likely trends and outcomes. As in the first edition, this part incisively examines past super-performers to determine what characteristics earned them "killer," or less violently, "key" technology/application status. The analysis not only facilitates a pedagogically effective summary, but in fact reveals a number of attributes that appear to be "common denominators" in many of the most successful developments to date. Finally, all parts and chapters of the book emphasize available telecommunications services, and corresponding business applications.

As you begin this book, please know that its authors have expended every effort to make it both a highly rewarding learning experience and as enjoyable and as easy to read as possible. Should this second edition be as popular as the first, there will no doubt be future updates. McGraw-Hill and the authors earnestly solicit your comments and suggestions. The authors e-mail addresses are appended to the bios located on the last page.

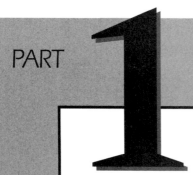

# Introduction

**Chapter 1**

# Definitions, Terminology, and Background

This chapter introduces telecommunications by defining its terms. Next it describes current and historical telecommunications structures in the United States—structures shaped by regulatory initiatives and judicial decisions. Telephony and the development of voice networks are described first since they were first to appear and the operational characteristics are familiar to most readers. Moreover, today's data networks use technologies that to a large extent evolved from voice networks. In fact, as shown later, voice and data networks which appear to users as separate and distinct, often share elements. Due to the enormous investment in installed plant, current networks will provide the basis for tomorrow's advanced "all-service" networks.

Because the Internet is still expanding at a near geometric rate, this chapter uses it as a model to describe telecommunications structures supporting emerging data, video, imagery, and other services. Again, reflecting global popularity and growth, this chapter closes with a description of telecommunications structures designed to support mobile subscribers, whether in vehicles or on foot, as well as mechanisms to interconnect mobile-subscriber telecommunications systems with those primarily intended for fixed-location subscribers.

The distinction between local, long-distance, and global telecommunications structures, and the impact that U.S. regulatory and international agreements have on those structures are presented in Chapter 2. Thus, Part 1, Chapters 1 and 2 supply the general background information and basic terminology definitions needed to establish a context for succeeding chapters. Topics introduced in these chapters are revisited later in greater detail, describing operational and technical characteristics of telecommunications technologies (Part 2) and the details of service offerings (Parts 3 through 5).

To make this book easy to use, discussions focus on telecommunications services that support business applications. We first provide readers with a working knowledge of telecommunications services and technologies, and then match them to business applications. Later sections furnish business telecommunications planners and users with guidance needed to develop technical specifications and request for proposal (RFP) packages, as well as to negotiate contracts with suppliers. In short, this book provides the means to apply advanced telecommunications capabilities to modern business needs, to enhance efficiency, profitability, and competitive advantage.

# Telecommunications Defined

What is telecommunications? The word is derived from the Greek *tele*, "far off," and the Latin *communicare*, "to share." Communications is the process of representing, transferring, interpreting, or processing information (data) among persons, places, or machines. The process implies a sender, a receiver, and a transmission medium over which the information flows. It is important that the meaning assigned to data be recoverable without degradation.

More specifically, *telecommunications* is any process that enables one or more users (people or machines) to pass to one or more other users information of any nature delivered in any usable form—by wire, radio, visual, or other electrical, electromagnetic, optical, acoustic, or mechanical means.

# Telecommunication Services versus Facilities

A *telecommunications service* is a specified set of information transfer, and information transfer supporting, capabilities delivered to a group of users by a telecommunications system. In this book, telecommunications services are treated in terms of voice services (Chapters 9 through 12), data services (Chapters 13 through 15), and wireless services (Chapters 16 and 17).

At the highest level, business applications are unique aggregations of telecommunications services that satisfy particular enterprise needs, e.g., medical/health care, hospitality, airline, retailing, etc. Most lower-level business applications, such as station-to-station calling within business premises, are enterprise-independent. All businesses use them.

It is necessary to distinguish between telecommunications services and the telecommunications systems/networks/equipment/components (facilities) by which the services are delivered, since one of the major decisions users must make is the extent to which they will satisfy their telecommunications needs via privately owned facilities rather than obtaining them from service providers, such as telephone companies, that retain ownership of facilities.

This distinction might appear trivial, but economic trade-offs among these options are extremely complex and are the subject of continuing controversy. Distinctions between options are often subtle. One example involves virtual private network services, provided by public telephone companies. Here the provider retains ownership but offers services intended to be indistinguishable (by customers or users) from those obtained when privately owned or dedicated-leased facilities are used. Customer or user options described below include obtaining services using shared public facilities, leased but dedicated private facilities, leased virtual private facilities, and privately owned facilities. Services delivered over facilities available to the general public are usually provided under tariffs, or at least subject to approval by regulatory agencies.

# Introduction to Voice and Circuit-Switched Systems, Networks, and Components

The following network definitions and descriptions are derived from those provided by AT&T in *Engineering and Operations in the Bell System*, Second Edition. A *telecommunications network* is a system of interconnected facilities designed to carry the traffic that results from a variety of telecommunications services. The network has two different but related aspects. In terms of its physical components, it is a facilities network. In terms of the variety of telecommunications services that it provides, it can support many traffic networks, each representing a particular interconnection of facilities. (The distinction will become more evident later.)

A *network* consists of nodes and links. Nodes represent switching offices, facility junction points, or both, and links are transmission facilities. *Traffic* is the flow of information within the network, among nodes, over links. Three characteristics influence the nature of a network. First, traffic must be carried over large geographic areas. Second, traffic is generated at virtually any time, although the duration of each information exchange may be short. Third, the ability to exchange information (i.e., connections) must be available with relatively short delays.

Figure 1.1 illustrates key aspects of a telecommunications network. Part A of Figure 1.1 shows a situation in which no switching is available and users at end points (business locations) are directly connected by transmission links. This example illustrates that a single site would need three sets of end-point equipment and three transmission links. Such a network is impractical since it requires separate user equipment and transmission links for every potential connection. The design is improved by including a simple switch at each end-point location, as shown in Part B of Figure 1.1, allowing each user to access any one of three transmission links and eliminating two of the three telephones needed in the Part A example. Part C shows the most efficient design with switching at a central point, which minimizes the number of switches, transmission paths, and telephones.

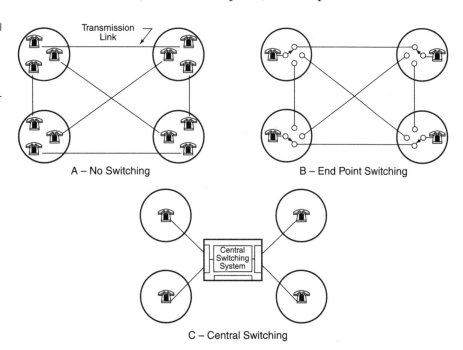

**Figure 1.1**
Transmission, equipment, and switching trade-offs.

In a network with no switches, if there are n locations, the number of independent transmission paths required is n•(n−1) divided by 2, or 6 for this example. The number of required telephones is n•(n−1), or 12 for this case. With central switching, the number of telephones and transmission paths required is reduced to n, or 4 in this case. This most economical network design is achieved at a performance price—the number of simultaneous connections through the network is limited. While the network in Figure 1.1A supports six simultaneous connections, those of Figure 1.1B or C support only two simultaneous connections.

A limited capacity for simultaneous calls is not normally a problem in well-engineered networks since the simultaneous use by all or even many users is unlikely. The simplified examples shown in Figure 1.1 illustrate one of the more important considerations in network design—the trade-off between network resource sizing and the probability that information exchange cannot be completed (indicated by busy signals in telephony systems). This performance parameter is referred to as *grade of service* (GOS) and is discussed further in Chapters 6 and 11.

The components of a facilities network are divided into three categories: switching, transmission facilities, and station equipment.

## Switching Systems

*Switching systems* interconnect transmission facilities at various locations and route traffic through a network.

## Transmission Facilities

*Transmission facilities* provide communication paths that carry user and network control information between nodes in a network. In general, transmission facilities consist of a medium (e.g., the atmosphere, copper wire cable, fiber optic cable) and various types of electronic equipment located at points along the medium. This equipment amplifies or regenerates signals (Chapter 3 defines analog and digital signals), provides termination functions at points where transmission facilities connect to switching systems, and can combine many separate sets of call information into a single "multiplexed" signal to enhance transmission efficiency.

## Station Equipment

*Station equipment* is generally located on the user's premises. Its function is to transmit and receive user information (traffic), and to exchange control information with the network to place calls and access services from networks. This information is conveyed in the form of electrical signals. Station equipment is one of several varieties of what is known as *customer premises equipment* (CPE).

Figure 1.2 illustrates the components a local telephone company must have to support local calling. The network shown comprises two switching central offices (COs), a term referring to a telephone company building in which network equipment is installed. The number of central-office switches in local telephone companies depends on the size of the area served, but in this example two switches are sufficient to illustrate the local network operation.

**Figure 1.2**

Components of a local telephone company network.

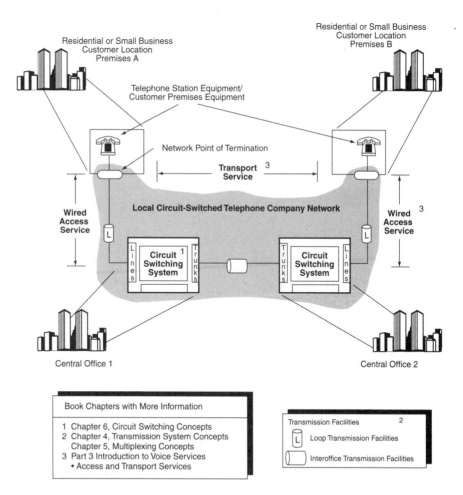

*Loop* transmission facilities connect switching systems to customer station equipment located in various business or residential premises throughout the served area. A *local loop* is a transmission path between a user/customer's premises and a central office. The most common form of loop, a twisted pair of copper wires, is called a *line* (see Chapter 4 for details). Telephone companies that employ copper wire subscriber loop facilities are referred to as *wireline* carriers as opposed to telephone companies using radio loops. The connection between the telephone company's network and CPE is formally called the *network point of termination* (POT) or, alternatively, the *network interface* (NI).

Interoffice transmission facilities connect the telephone company's switching systems. In a network, a *trunk* is a communication path connecting two switching systems. As can be seen, trunks are used to establish end-to-end connections between loops to customer station equipment.

As described above, station equipment, often furnished by the user, is used to obtain network services from local telephone companies, i.e., to complete calls to or from any other user's station equipment. More specifically, the network services needed to complete a call include *originating access service*, to connect station equipment to the local telephone company's central-office switch serving the user's area; *transport service*, to route the call through the network to the central office serving the called party; and connection to the called station equipment via *terminating access service*, again via the local loop. This segmentation of telecommunications network services into access and transport services applies to all network configurations and is used (and expanded upon) throughout this book.

Switching and transmission systems provide for signaling, a major network function that is rapidly advancing in standards and techniques. *Signaling* is the process of generating and exchanging information between components of a telecommunications system to establish, monitor, or release connections (call-handling functions) and to control related network and system operations (other functions). Put more simply, among other functions, signaling is the capability that responds to a caller's "touch-tone" or rotary dialing instructions, sets up connections to, and rings called parties. Today, as in the past, voice networks typically employ *circuit switching*, a process that establishes connections on demand and permits the exclusive use of those connections until they are released. Chapter 6 presents signaling and circuit switching details.

# Interconnection and Hierarchies of Networks

In most cases, concatenations of "facilities networks" are required to complete calls between station equipment. For example, long-distance calls usually involve two local telephone company networks, one long-distance network, and customer premises equipment (comprising inside wiring, station, and other customer-owned equipment). Thus, end-user-to-end-user connections include station equipment, transmission facilities, and switching systems residing in user, local, and long-distance networks. Chapter 2 illustrates how long-distance networks interconnect and relate to local telephone company networks.

With many in-building or campus "customer" networks connected to local telephone switches, and multiple local telephone networks connected to long-distance networks, the ensemble forms a "hierarchy of networks". This hierarchical structure of customer, local, and long-distance networks has characterized the *Public Switched Telephone Network* (PSTN) since the inception of long-distance service in the United States. (Station equipment and building or campus CPE is normally owned by users, so it is not included in either local or long-distance public telephone company network segments.)

In the United States during the early part of the twentieth century, American Telephone and Telegraph (AT&T) designed, owned and operated virtually all the facilities-based long-distance networks, most of the local telephone company networks, and even much of the CPE. "Interoperability" and overall or end-to-end network performance was essentially "designed in" and assured by AT&T. As explained in Chapter 2, ownership and operation of local and long-distance networks in the United States today is shared among many companies with end-to-end operational integrity and performance largely accomplished through "open-system" standards and interface specification agreements among owners of interconnected networks.

Increased competition in the United States, privatization of previously government-owned foreign telephone systems, and the advent of mobile or cellular telephone service throughout the United States and worldwide, means that most calls, originating in one network, are completed in other networks. In addition to the problem of maintaining technical performance integrity across networks, there is the added complexity of collecting the separate charges applied by each network provider and rendering a bill to the customer for each call.

To arrive at these charges on a monthly basis, various telephone companies must *settle* accounts among themselves. The magnitude of the administrative complexity of interconnecting national and world-wide networks is exemplified by the fact that AT&T operates a global clearinghouse that settles its accounts with operators of over 220 different foreign-country networks.

The next two subsections focus on data, mobile-subscriber, and other service telecommunications structures. Data and mobile-subscriber telecommunications structures involve interconnections of even larger numbers of networks than traditional voice networks. Furthermore, as will be shown, most data and mobile-user networks either share facilities with or use voice networks for originating or terminating access service.

In summary, although terms such as "the U.S." or "the global" telecommunications network occur in the literature, these abstractions do not refer to any single, monolithic entity. Rather, they are agglomerations of interconnected local, regional, national, and international networks, owned and operated by literally thousands of small, medium, and large commercial or government service providers. Whereas interconnection of local, regional, national, and international networks has always characterized telecommunications, what has changed drastically over the last two decades is the number of enterprises and the number and variety of network types and underlying technologies. The rest of this book is devoted to explaining these developments—using official, industry-accepted terminology in ways both business-people and engineers can relate to—with no compromise of the validity of either the conceptual or technical operating principles presented.

# Introduction to Packet-Switched Systems, Networks, and Components

In circuit-switched voice networks, switching and transmission systems respond to user dialing instructions to establish end-to-end connections (circuits) between originating and terminating instruments. As noted above, circuits may include switching and transmission facilities in multiple networks. Once a connection is established, a *full duplex* circuit (i.e., a transmission path capable of transmitting signals in both directions simultaneously) is held or dedicated to that call until it is released.

Although full duplex circuits make it possible for two or more parties to speak at the same time, normally that does not happen. In the absence of special network provisions, when full duplex circuits are used for voice, one direction's capacity is nearly always idle or wasted. Moreover, even when a person is speaking, inter-syllable, inter-word or inter-speaker pauses mean that end-to-end resources may be conveying information only about 50 percent of the time.

To enhance efficiency, where transmission resources are scarce or expensive, (transoceanic circuits for example), techniques such as *Time Assigned Speech Interpolation* (TASI), for analog voice signals, or *Digital Speech Interpolation* (DSI) for digital voice signals, are used to detect speech pauses and switch circuits to serve active speakers. TASI and DSI and how they apply to analog and digital voice signals are explained further in Chapter 3. Unfortunately, TASI/DSI techniques apply only to some network circumstances, and only to voice traffic.

The advent of data information exchange over circuit-switched networks led to even greater inefficiencies. In on-line or real-time interactive sessions for example, human operators use terminals to view information from and manually correspond with remote computers or other terminals or resources. Because human operators do not generate information constantly, in these circumstances, dedicated, circuit-switched transmission paths are actually used less than 20 percent of the time.

To solve this problem, a new type of switching, *packet switching* was invented. One of the great advantages of packet switching is that it provides a universal means to allocate network resources to users (devices) *only* when they have information to transmit, and makes those resources available to other users or "uses" otherwise. Best of all, it applies to any type or mix of traffic—voice, data, video, imagery, etc. Modern data communications networks route traffic among nodes and transmission links using packet switches (the type of packet switch used in the Internet, the world's most widely known packet switching network, is called a *router*).

A *packet* is a quantity of data that is transmitted and switched as a composite whole. A packet consists of user data (for example, some portion of the contents of an electronic mail message) and control information, the latter including the network address of both originating (sending) and terminating (receiving) equipment. Since the quantity of data packets are able to carry is relatively small, the contents of large messages must be segmented and inserted in multiple packets, in a process referred to as packet assembly, often in devices known as *packet assembler-disassemblers* (PADs).

Control information contained in each individual packet is sufficient to allow network packet switches to route it from originating to terminating equipment. In many networks, multiple packets containing information from a single message may traverse different paths, incur different delays, and thus arrive "out of order." In such instances, PADs or equivalent computer-based processing restore messages to original compositions.

To be routed through packet-switched networks, information in analog signal format must first be converted to a digital format (requiring analog-to-digital or A/D converters), and the resulting digital signals assembled into packets as just described. Details of analog and digital signal characteristics, conversion processes, the structure of packets, and packet switching concepts are found in Chapters 3 and 7 respectively. For now, this thumbnail sketch provides a basis for (1) introducing top-level packet-switched network features and capabilities; (2) contrasting them with those of circuit-switched networks discussed above; and, (3) illustrating ways in which today's circuit-switched networks provide access to the Internet for most residential and small-business users and otherwise play important roles in the delivery of packet-switched network services.

Figure 1.3 illustrates essential components of a packet-switched network and its relation to a local circuit-switched telephone company network. Note that the top portion of the figure depicting a circuit-switched telephone network is functionally identical to that shown in Figure 1.2, except customer locations are shown to include both personal computer and telephone equipment. What is implied in the figure is that the customer locations at Premises A and B are served by ordinary analog, wired subscriber loops or lines—such as those described above in relation to Figure 1.2. A further implication is that the PC and telephone equipment share a single line so that use of the line for voice telephone traffic makes it unavailable for PC traffic, and vice versa. An alternate example might allocate separate lines, one for voice and one for data traffic.

Since PCs generate digital signals and analog subscriber lines do not support digital signal traffic, the PCs must use modems to transform the digital signals to analog signal formats. Modems and their attributes are presented in Chapter 3.

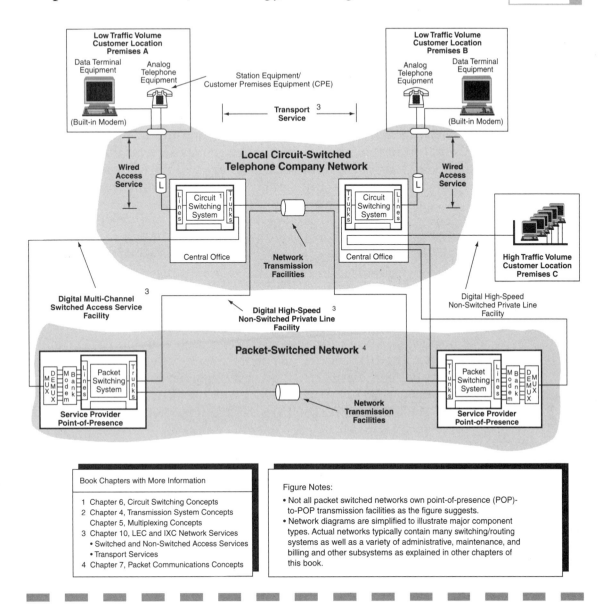

**Figure 1.3** Components of a packet-switched network.

The vast majority of the nearly 150 million devices accessing the Internet in 1999 are owned by individuals in private residences or small businesses. In either case, the majority of those owners obtain access to both public telephone and Internet service via today's analog subscriber lines, as illustrated in Figure 1.3. There is a growing trend to *digital subscriber lines* (DSL—see Chapters 3 and 4) and these may one day replace analog access. Whether analog line or DSL, individuals and small businesses remain likely to obtain access to both voice and packet networks via local telephone companies, whether they be one of today's incumbent companies or one of the emerging competitive class defined in Chapter 2.

The bottom portion of Figure 1.3 delineates key packet-switched network components and shows how access to such networks is accomplished using local telephone company facilities. For simplicity, only two of potentially many packet switching/routing systems normally residing in most networks are shown in the illustration. For the purpose of an example, we can consider that the packet switching/routing systems shown are owned by Internet Service Providers (ISPs). Under this assumption we can now examine the process by which subscribers establish PC-to-ISP connections, in what ways that process is similar to, and in what ways it differs from, placing voice telephone calls.

ISPs maintain ordinary "10-digit" telephone numbers for the sole purpose of receiving packetized data from "dial-up" customers. In establishing a dial-up connection between a PC and an ISP, or in the vernacular to use switched telephone company (telco) service to log on to an ISP, a subscriber typically employs communications software residing in the PC.

When invoked, this software activates a modem that automatically dials the ISP number. The subscriber's modem establishes a standard voice connection with a modem at the ISP through telco circuit switches. The ISP maintains a bank of modems for the express purpose of receiving calls from subscriber PCs. Once physical circuits are established between PC and ISP modem pairs, modem-to-modem synchronization takes place enabling ISP modems to restore or reproduce the packetized, digital signals generated by PCs.

Beyond PC-to-ISP connections is the need to establish communications sessions that allow users to browse the 850 million Web pages, to send or receive electronic mail (e-mail), to download files, or to conduct any other Internet-based transactions. That process is discussed next.

An IP (*Internet Protocol*) address is an identifier for computers or other devices connected to the Internet. On the Internet, IP addresses mimic the role telephone numbers play in allowing circuit switches to establish connections between calling and called-party telephone instruments. *Protocols* define strict procedures for initiation, maintenance, and termination of data communications, and IP, in particular, defines packet formats and packet addressing. Thus, IP address information contained in packets is precisely the information packet switches need to route packets from source to destination.

The packet switching/routing systems in Figure 1.3 examine each packet for source and destination address information and route them through networks accordingly. In addition to these decoding and processing capabilities, packet switches/routers contain input and output buffering, i.e., memory elements to store packets temporarily, and internal switching connecting input and output buffers. Routing actions are based on a multiplicity of factors including address information, network traffic loading or congestion, data priority, etc.

A router has all the attributes of a packet switch, but in addition has the ability to interconnect networks of differing design. Unlike packet switches that are designed to function in monolithic network environments, routers are equipped to perform such essential *network-edge* functions as address translation and, in general, to implement more complex network routing and control decisions. Since the Internet comprises an unstructured interconnection of hundreds of separate networks, it employs routers. Remember that it is for simplicity only that Figure 1.3 shows only two interconnected packet switches/routers. In wide-area networks, packet switches/routers are typically richly interconnected.

Figure 1.3 reveals two possible ways of provisioning inter-packet switch transmission facilities. First, these high-speed, digital, non-switched private lines may be obtained from incumbent (or newly emerging competitive) telephone companies under some sort of leasing agreement. The other alternative is that packet switched network owners construct, operate, and maintain private facilities. Companies providing both circuit and packet switched service obviously opt to integrate circuit and packet-switched transmission since that takes best advantage of the enormous installed telco plant and existing cross-country rights of way, and offers economies of scale. It represents yet another example of the interconnection between circuit and packet-switched networks that, at first, may appear to be totally separate and distinct entities.

The last aspect of Figure 1.3 that needs to be treated is the mechanism for connecting high traffic-volume customers to packet-switched networks. This situation is depicted at "Premises C" in the right-hand portion of the figure. Premises C could be a business housing perhaps hundreds or thousands of PCs or other DTE (*data terminal equipment*—any device that can send data, receive data or perform both functions), most of which require access to local and wide-area data networks.

In most cases, such premises employ *local area networks* (LANs) connected to on-premises packet switches or routers. See Chapters 7 and 13 for popular LAN architecture and configuration descriptions. In the Figure 1.3 example, a router interconnects a LAN with the Internet.

The amount of traffic generated and addressed to the ISP in the Premises C example justifies use of a high-speed, digital, non-switched facility, available from local telcos. As the figure shows, these facilities pass through the same telco wire centers that accommodate analog subscriber lines, but are *not* switched by telco circuit switches. In many urban areas, telco competitors, such as Metropolitan Fiber Systems, Inc., offer alternative means for obtaining high traffic-volume access to packet-switched networks.

In the preceding section, we have described how packets are routed from the originating subscriber to the destination terminal using the destination's IP address on every packet. But how does the originating subscriber know the IP addresses of the millions of computers on the Internet? The good news is that he doesn't need to know. From the PC user's point of view, software in the computer and within the Internet greatly simplifies establishing connections and exchanging information.

To "humanize" or make addressing and session establishment more user-friendly, the Internet uses recognizable names that identify computers. These names are arranged in a hierarchy, going from the general group (or domain as groups are known on the Internet) to which a computer belongs, to a more specific group, and to the computer itself. Within the Internet, domains are defined by IP address, and all devices sharing a common part of the IP address are said to be in the same domain. It should be noted that although common names are what subscribers use to establish session connections, those names are converted to 32-bit numerical IP addresses that are much like public telephone numbers. The process is not unlike name-based "speed dialing" in telephone systems.

Search engines help users find desired Web site/addresses using English word queries. And bookmarks make return visits to favorite Web sites even easier. Other automation that simplifies Internet use includes the process of converting letters, messages, or files to or from packet format, i.e., assembling and disassembling packets. Here again, this task is accomplished in PCs using readily available software. Once subscribers select a domain name, the IP address information is automatically inserted into each packet.

As Internet popularity skyrockets, means are fast appearing to use it for the transmission of messages, voice, video, high-fidelity stereo music, high-resolution graphics and imagery, and other services. Of particular interest is the possibility of using the Internet for ordinary voice telecommunications, referred to as *Voice over IP*, or VoIP. Before any significant shift from plain old telephone service (POTs) can take place, instruments as simple as today's telephones, but with capabilities to access the Internet, and the means to exchange signaling among circuit- and packet-switched networks must be developed.

At this time, efforts to define *Session Initiated Protocols* (SIP) by the Internet Engineering Task Force (IETF) and the International Telecommunication Union—Telecommunication Sector's (ITU-T) work on the H.323 series of standards recommendations for multimedia communications over IP networks, indicate that solutions are being sought from both Internet and traditional telephony camps. Already top-level architectures suggest the use of gatekeeper equipment as a means to exchange signaling information between circuit- and packet-switched networks to enable voice calls to be placed and terminated easily on either network. While an ultimately ideal solution remains in the distant future, what this progress points to is an eventual possibility, at least from the users' vantage point, of seamless, highly integrated, multi-service networks—a day when a single subscriber line may satisfy all our voice, data, video, and other service needs simultaneously and economically.

Chapter 7 addresses all of the above topics with substantial detail on packet switching, International Organization for Standardization (ISO), and Internet TCP/IP protocols, LAN, frame relay, and asynchronous transfer mode (ATM) protocols. Also addressed in Chapters 7 and 8 are performance trade-offs among state-of-the-art switching techniques. While this introduction to packet switching emphasizes its salient advantages, inserting enough control information into each packet so that it can be routed as an entity through networks uses transmission capacity that could carry user information. As will be shown, thorough comparisons of switching technologies involves many factors such as packet size, frame size, switching speed, and the susceptibility of classes of user information (voice, data, video, imagery) to fixed and variable network delays.

# Introduction to Mobile-Subscriber Systems, Networks, and Components

In acquiring a broad understanding of modern telecommunications, as many lessons can be learned by studying the course of events that have shaped the evolution of *mobile-subscriber networks* (MSN), as by studying technical and operational principles of today's systems. Both perspectives are covered below, beginning with summary network design and capability descriptions.

Figure 1.4 illustrates essential components of a terrestrial-based mobile-subscriber telecommunications network and its relation to a local circuit-switched telco network. As in Figure 1.3, note that the top portion of the figure depicting a wired-access, circuit-switched telephone network is functionally identical to that shown in Figure 1.2.

The bottom portion of Figure 1.4 identifies key mobile-subscriber network components. Included is a radio network segment comprising wireless mobile-subscriber *station equipment* (e.g., radio-based telephones and pagers) and *cell site base transceiver stations.* Cell site base stations are fixed-location transmitter-receiver (transceiver) sites that communicate with mobile-subscriber station equipment. While nearly all large telecommunications networks employ radio or electromagnetic communications facilities (for example, the terrestrial microwave, satellite, and other transmission systems described in Chapter 4), what makes mobile-subscriber radio networks different is that unlike the fixed-location nodes treated in Chapter 4, customers carrying hand-held station equipment, or in vehicles in which such devices are mounted, are free to move about or roam at will.

As shown in Figure 1.4, stationary equipment segments of mobile-subscriber networks consist of *mobile telephone switching offices* (MTSOs, also referred to as *mobile switching centers* in International Telecommunications Union standards) and transmission facilities interconnecting stationary or fixed nodes.

Since the objective here is to focus on top-level operations and telco interconnections, only two cell site base stations are depicted. For large urban areas, depending upon traffic density, cell sites may be spaced at less than ten-mile intervals. In 1998, for example, there were approximately 3,000 cellular systems and 70,000 base stations in the United States. In some newer Personal Communications Systems

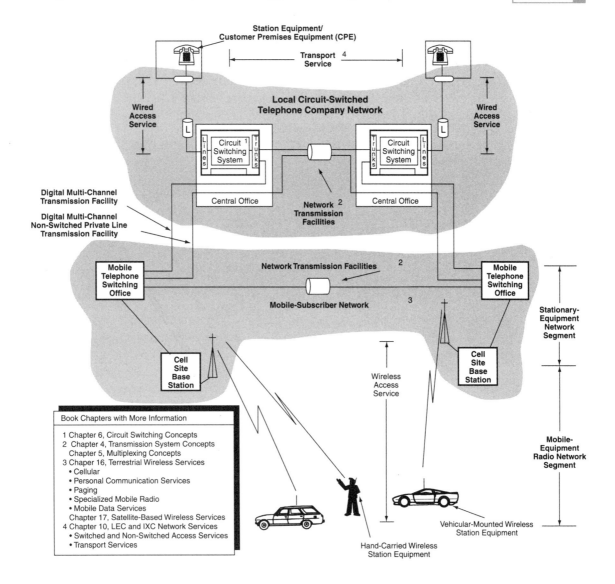

**Figure 1.4** Components of a mobile-subscriber communications network.

(PCSs), base stations may be just city blocks apart. Tokyo's Personal Handyphone Systems (PHS), for instance, use over 200,000 base stations spaced every 300 to 900 feet.

The simple idea behind cellular service, which makes wireless service feasible, is that the same frequency channels (see Chapter 3 for definitions of channel, frequency, and related terms) are systematically reused. Unlike previous radiotelephone designs using a single set of frequencies to cover an entire city, severely limiting the maximum number of simultaneous users, cellular service uses low-powered base-station transmitters covering relatively small geographic cells. Because cell designs operate on a non-interfering basis, today's systems are able to serve over 90 million U.S. subscribers. This is a startling growth from just 92,000 subscribers in 1984.

The purpose of mobile telephone switching offices (MTSOs) is to establish connections and route traffic originating from or terminating on mobile station equipment. For traffic originating and terminating at station equipment and base stations served by a single MTSO, that MTSO alone establishes connections between originating and terminating equipment.

For traffic originating or terminating in telcos other than those operating *mobile-subscriber networks* (MSNs), MTSOs are connected via digital multi-channel transmission facilities to wired-access local telco switches. For calls originating and terminating within a single MSN, but among subscribers served by different MTSOs, the inter-MTSO call traffic may be carried via transmission facilities provided by incumbent local telcos, or via private, MSN-owned or competitive transmission facility providers, as shown in Figure 1.4.

Network design and capacity planning efforts are based on either estimated or measured traffic loading and connection patterns. Figure 1.5 depicts nationwide cellular call volume as well as originating and terminating traffic patterns in U.S. MSNs.

**Figure 1.5**
Mobile subscriber network traffic patterns.

| Traffic Pattern | Statistic |
|---|---|
| Total MSN Originating/Terminating Calls per Day Traffic | 150 million |
| Traffic Originating and Terminating in an MSN | 6% |
| Traffic Originating in an MSN and Terminating in a Wired-access Telco | 57% |
| Traffic Originating in a Wired-access Telco and Terminating in an MSN | 37% |

Another way in which Figure 1.4 is simplified is that, like previous figures, it reveals no network management and control (M&C) components. Omission of M&C components within figures and in the commentary to this point in the book should not be interpreted as minimizing M&C importance. In fact, because most extant networks exhibit otherwise equally high technical performance, the ultimate competitive service provider discriminators now reside in M&C capabilities. Moreover, it is in cellular systems that M&C challenges reach their apex since each service provider must be prepared to accommodate, authenticate, manage, control, and render billings for not only its own customers, but for the "roaming" subscribers of other service providers, with service areas located perhaps half a world away.

But keeping Figure 1.4 simple not only facilitates explanation of essential MSN technical and operational characteristics; it allows a single figure to represent the wide variety of different mobile-subscriber cellular telephone, PCS, paging, Universal Mobile Telephone Service (UMTS—Europe), Personal Handyphone Systems (PHS—Japan), specialized mobile radio (SMR), mobile data, and other services, all discussed in more detail in Chapter 16.

To place in context the details of Chapter 16 and MSN evolution, (or perhaps more accurately—revolution) that spawned phenomenal mobile-subscriber growth, consider the following. From the beginning of the twentieth century until the mid-1980s, voice service was delivered to users almost exclusively via the twisted-pair copper wires noted above. Just after World War II, television burst onto the scene and the mechanism chosen to deliver TV service to consumers was radio-wave (a generic term) transmission. TV station operators generally found the highest spot in a city, erected tall antennas, and thus, almost overnight, were able to serve millions of viewers in metropolitan areas.

While at that time people possessing no engineering knowledge may not have completely understood technical terms like channel capacity, bandwidth, or frequency spectrum (see Chapter 3 for definitions and explanations), there developed a common perception that telephone loops had limited capacities, suitable for narrow-bandwidth voice service signals. The corollary view was that the atmosphere (or as some called it—the "ether") has a very large bandwidth capacity—suitable for broad-bandwidth TV signals.

In fact, a single broadcast TV signal does occupy about the same amount of the atmosphere's ability to support radio-wave signals as 1,000 voice signals—similarly broadcast. Moreover, most telephone subscriber loops in that venue were tuned or optimized to minimize loss (a process that made long multi-mile loops possible) for single-channel voice traffic, a design that deservedly earned them the title narrow-bandwidth channels. As a result, most people held the notion that wires or cables exhibited limited capacities and that the capacity of the atmosphere was very large and maybe inexhaustible.

In practice, exactly the opposite situation is true. Once a high-power TV station is allocated a broadcast frequency in a city, no other TV station can use the same allocation without interference and, unfortunately, the total spectrum of useable frequencies is limited and not inexhaustible. Radio-wave spectrum is in such demand globally, that a special organization, the *World Administrative Radio Conference* (WARC—a part of the 154-member International Telecommunications Union mentioned earlier) exists to resolve or prevent interference. The Federal Communications Commission (see Chapter 2) represents the U.S. at WARC and administers frequency allocation in the United States.

By contrast, the capacity of both metallic wires/cables and newer fiber optic cables (Chapter 4) is constantly expanding due to the development of new materials, sophisticated termination equipment, and signal design and fabrication technologies. Coaxial cable TV systems introduced in the 1970s now deliver hundreds of channels. Optical cable research has produced a practical single-fiber cable with the capacity to carry the simultaneous peak voice traffic loads of AT&T, Sprint and MCI combined! Of course, if still more capacity is needed, the option of using multiple cables is available.

In light of these facts one might be tempted to ask why knowledgeable engineers took what appears to be such a short-sighted approach. The answer is that when TV was introduced, AT&T and other telcos had already invested over $60 billion and half a century in copper, twisted-pair loop plant, a medium then unable to carry even a single TV channel. To invest in a cable replacement of like or greater value for a yet-untested TV market could never have been justified in any rational business case. Then too, installing new high-capacity cable plant may very well have delayed TV's rapid expansion and the advertising revenues that made its incredible growth rate possible. Finally, at that time, fiber optic cable technology was just emerging from laboratories.

By the early 1990s, however, some experts began to see and predict a communications transmission system paradigm change of truly profound proportions. In his book, *Being Digital*, Nicholas Negroponte,

Professor of Media Technology at the Massachusetts Institute of Technology (MIT) and founding director of its Media Lab, summarized his prediction as follows: "The idea, which I first discussed and illustrated in a Northern Telecom meeting at which George Gilder and I were speakers, simply says that information currently coming through the ground (read, wires) will come in the future through the ether, and the reverse. Namely, what is in the air will go into the ground, and what is in the ground will go into the air. I called it 'trading places.' Gilder called it the 'Negroponte Switch.' The name stuck.

"The reason I considered this trading of places self-evident is that bandwidth in the ground is infinite and the ether is not. We have one "ether" and an unlimited number of fibers. While we can be cleverer and cleverer with how we use the ether, in the end we ought to save all the spectrum we have for communication with things that move, which cannot be tethered, like a plane, boat, car, briefcase, or wristwatch."[1]

Not all prognosticators were as incisive as Negroponte. A late-1970s market-research study commissioned by Bell Labs, which invented cellular telephone technology, predicted a mobile-subscriber base of only 800,000 by 2000, and AT&T chose then not to pursue the business vigorously. In what history may judge to be one of the most fortuitous yet enlightened and invaluable regulatory actions ever taken, the FCC broke with its traditional stance that telecommunications services in the United States are best rendered by regulated monopolies.

Instead, the FCC decided to grant two separate cellular telephone licenses in each urban area, one to incumbent telcos, and the other to independent enterprises—originally by lottery. The new players being highly competitive and not shackled by monopolistic corporate culture must be credited, to a large extent, with creating the climate for today's worldwide 200 million mobile-subscriber market, and generating annual U.S. revenues of over $20 billion.

Cordless telephones are another manifestation of how the industry is creatively responding to customer needs for "untethered" service. Cordless telephones now operate in what are known as unlicensed Industrial, Scientific, Medical (ISM) radio spectrum bands. With nearly $5 billion in 1999 sales (about 80 million units) for models connecting portable cordless telephones to inexpensive base stations attached directly to wired-access telco subscriber loops, there are significant U.S.

---

1 Negroponte, Nicholas, *Being Digital*, Alfred A. Knopf, Inc.

and overseas efforts to extend basic "private" cordless operations to what has become known as *Universal Mobile Telephone System* (UMTS) service. With UMTS, single handheld telephones communicate with either base stations connected to telco subscriber loops, or with PCS or other cellular MSN service provider base stations.

European UMTS efforts include standards work for telephones supporting both DECT and GSM cellular standards. (Cellular service conforms to GSM standards in 85 countries. Because U.S. cellular telephone sales reached large numbers prior to GSM standards availability, other standards dominate North American operations. Chapter 16 describes popular historical and extant mobile communications standards.)

In the U.S., UMTS efforts are linked and complementary to *wireless local loop* (WLL) developments. WLLs are radio-based alternatives to conventional twisted-pair, copper wire loops. Wireless or radio-based alternatives have been possible and in limited use for decades. What has changed is that low-cost PCS and cellular base station technologies have made wireless access less expensive in some cases than wired access. Today the average cost of a 1,500-foot wired local loop ranges from $1,000 to $1,500 per subscriber. WLL solutions cost a quarter to a third as much. Chapter 4 describes historical and current manifestations of fixed wireless systems in more detail.

What the new breed of local telephone company competitors hope is that the convenience of a single wireless phone for use in and outside homes or business premises, combined with lower rates for in-premises use, will cause customers to abandon wired connections. Chapter 2 discusses other-than-cellular or PCS local telco competitors and related regulatory issues. The remainder of the book discloses additional technical and business factors that, at the end of this millennium, combine to make mobile-subscriber service the fastest growing telecommunications sector.

## Summary Introductory Remarks

Terminology, network, switching, component, and organizational definitions and concepts presented in this first chapter establish a solid technical and operational framework for acquiring a comprehensive, practical grasp of all important telecommunications technologies and services, and for coping with higher levels of detail presented in the rest of the book. What becomes evident as one progresses through the book is that the "revolutionary" growth and capability advancements outlined exist not just in exploding service, subscriber, network, product, and performance statistics. Rather, these skyrocketing numbers presage an even more profound revolution, one in which information technologies are applied in ways that radically alter how business or personal advantage is created.

It is axiomatic that potential business and personal benefits do not accrue from the simple existence of high technology. As one pundit puts it, "the music is never in a high technology nor any other kind of piano." History demonstrates that engineers who can make startling new discoveries, and possess exquisite knowledge of complex technologies, are rarely the people able to conceptualize new business applications and processes that put those technologies to work enhancing productivity, reducing cost, and increasing profits.

Most often, this type of pioneering is accomplished by people already armed with business acumen who acquire a working knowledge of telecommunications technologies. Seeing the "bridge" between the two, they often produce innovative business ideas or processes that enable companies not only to survive but take leading positions in fiercely competitive world markets.

Some benefits are relatively mundane. For example, few businesses today could be conducted without modern telephone and fax services. But, nearly every day, publications like the *Wall Street Journal, Business Week,* and even local newspapers report on new start-up businesses or ways of streamlining existing business made possible only by the application of Internet or some other telecommunications-using technology. Many experts credit the size and length of the economic boom of the 90s to the fact that over the last decade or so, U.S. companies have been extremely successful in applying new technologies to enhance productivity and increase business volume and revenues—all with fewer human or material resources, inventories, etc.

Correlation between telecommunications capabilities and benefits to enterprises should not be viewed as a recent development or the product of recent Internet or mobile-subscriber network popularity. Figure 1.6 shows a straight-line relation between plain-old voice telephone densities and a country's per capita gross national product (GNP) compiled in the early 1990s. While the figure does not disclose whether large GNPs enable countries to afford high telephone densities, or whether having high telephone densities produces large GNPs, the correlation is irrefutable. Moreover, it is safe to state that countries with leading GNP indicators could not sustain them without modern telecommunications resources.

**Figure 1.6**
Correlation between telephone densities and gross national product per capita.

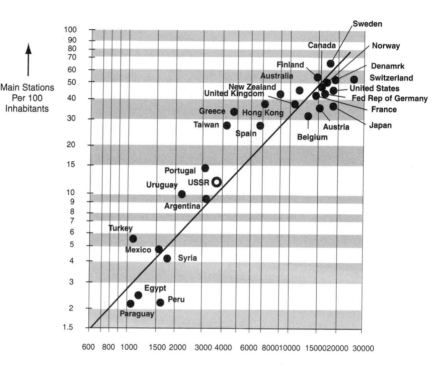

Source: "World Communication," A Gaston Lionel Franco Publication

There are also important personal benefits in addition to the salutary effects of telecommunications on companies and nations. In fact, telecommunications is the technology most directly related to man's highest faculty and his most sublime activities: namely, his acquisition of information, the thoughts by which he interprets and postulates new ideas and conclusions, and the exchange of his thoughts with others to refine and iterate the process. There is no question that telecommunications greatly augments both the potential quality and the timeliness of these powerful and uniquely human abilities.

Our nineteenth century's agricultural and transportation technologies nourished the burgeoning cityscapes created by the Industrial Age. Similarly, our twentieth century is viewed as the Age of Communications, marked by the birth and explosive proliferation of radio, television, and computers—technologies designed to distribute food for thought.

As we enter the twenty-first century—the Information Age—enabling technologies are advancing at the speed of light, beckoning to us with possibilities both exciting and perhaps unbounded. Read on—the fun is just beginning!

CHAPTER 2

# Regulatory and Service-Provision Structures

In the United States, federal and state governments are authorized to intervene in the telecommunications marketplace when free enterprise is deemed inadequate to ensure the economical supply and distribution of products and services. Telecommunications regulation at the federal level involves the *Communications Act of 1934* (as amended) administered by rulings of the *Federal Communications Commission* (FCC). State regulatory bodies, such as *public utility commissions* (PUCs), handle state-level regulation.

Regulatory impact has been so profound that an historical review of the structure it has imposed on telecommunications in the United States is a prerequisite to further discussions.

# Voice Service Provision in the U.S.

The story of voice services in the United States has been marked by the transition from service provision by a single regulated monopoly to a competitive marketplace with many sources of supply. This transition began with the regulatory breakup of the Bell System in 1984 and the opening of the long distance voice services market to competition.

## Historical Developments

The American Telephone and Telegraph (AT&T) company dominated both local and long-distance markets until 1982, by which time it had reached $155 billion in assets and over 1 million employees. AT&T, the parent company of an entity known as the Bell System, served over 144 million telephones through Bell operating telephone companies. The remaining 36 million telephones were served by some 1,450 independently operating telephone companies.

Since its incorporation in 1885, AT&T had been the subject of recurrent Department of Justice antitrust actions. Following burgeoning growth in the 1950s and 1960s, AT&T's market dominance prompted the 1975 Justice Department suit. After seven years in the courts, AT&T finally accepted a restructuring agreement known as the *Modification of Final Judgment* (MFJ), which was approved by U.S. District Court Judge Harold Greene in August 1983, and became effective January 1, 1984.

The divestiture of AT&T resulted in the creation of 22 *Bell operating companies* (BOCs) organized into seven regional Bell holding companies, commonly referred to as *regional Bell operating companies* (RBOCs), as shown in Figure 2.1. The purpose of this divestiture was to separate the increasingly competitive long distance market from the local services market, where incumbent telephone companies faced no competition. A regulatory mechanism had to be formulated to prevent the resources of the monopoly local telephone companies from subsidizing long-distance service and hindering the growth of competition. Hence, the MFJ called for the creation of BOC service areas, called *Local Access and Transport Areas* (LATAs), to distinguish local from long-distance calling markets. A LATA is a geographic area within each BOC's territory that has been established in accordance with the provisions of the MFJ, which defines the area in which a BOC may offer its telecommunications services. In 1989, there were 198 LATAs in the United States. Independent telephone companies (ITCs), the non-Bell exchange carriers, continued to provide telecommunications services within their franchised areas without LATA restrictions. *Intra-LATA* is a term used to describe services, revenues, functions, etc. that relate to telecommunications originating and terminating within a single LATA. *Inter-LATA* is a term used to describe services, revenues, functions, etc. that relate to telecommunications originating in one LATA and terminating outside that LATA.

**Figure 2.1**
The 22 Bell Operating companies (BOCs) organized into seven regional Bell Holding companies (RBOCs) that were the result of the MFJ.

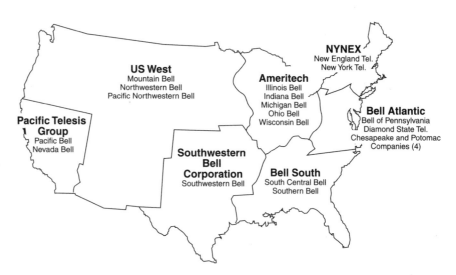

An exchange carrier (or *local exchange carrier*, LEC) is any company that provides switched telecommunications services within a defined area. In this book, the term LEC will be used when referring to either BOC or independent local exchange carriers. In order to understand the provisions of the Telecommunications Act of 1996, described later in this chapter, two additional categories of LECs must be defined. These categories are *incumbent local exchange carriers* (ILECs) and *competitive local exchange carriers* (CLECs). An ILEC is defined, with respect to an area, as the LEC that provided local exchange service to that area on the date of enactment of the Act. CLECs are companies that began providing alternative exchange service to an area in competition with an ILEC. Figure 2.2 shows the relationships among the categories of LECs.

**Figure 2.2**
Classification of local
exchange carriers.

| LECs – All Local Exchange Carriers | |
| --- | --- |
| ILECs – Incumbent local exchange carriers (providers of exchange service as of the date of enactment of the Telecom Act). | CLECs – Competitive local exchange carriers (companies which began providing alternative exchange service to an area in competition with an ILEC). |
| BOCs – Bell Operating Companies (22 specific telephone companies created from the divestiture by AT&T of the Bell System in 1984). | ITCs – Independent (non-Bell) telephone companies (some were incumbent LECs with their own franchised territory at the passage of the Act). Competitive LECs also fall into this category, but the term CLEC is most often used to describe the new companies. |

An *interexchange carrier* (IXC) is a company that provides telecommunications services between LATAs. Although the term "inter-LATA carrier" would be more precise, this book will use the MFJ term IXC for these carriers. IXCs other than AT&T were referred to as *other common carriers* (OCCs). This term is largely obsolete today.

The following sections provide insight into how the MFJ rules promoted competition in the interexchange market.

# LEC/IXC Operations, Responsibilities, and Restrictions

The best way to describe post-MFJ telecommunications facilities, service responsibilities, and restrictions is to trace an inter-LATA call via local and interexchange Public Switched Telephone Networks (PSTNs). The PSTN includes those portions of LEC and IXC networks that provide public switched telephone services. Figure 2.3 is used to analyze an inter-LATA call and to introduce additional terms arising from the MFJ.

**Figure 2.3**
LEC/IXC facilities and services used to completer inter-LATA calls.

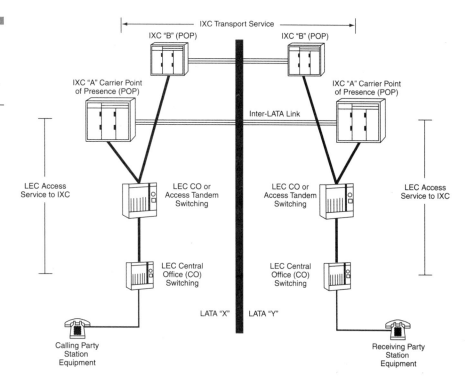

For this example, a calling party in LATA X places a call to a receiving party in LATA Y. The calling party is provided with dial tone via access services from his LEC (types of access services and their signaling, protocol, and operating characteristics are described in Chapter 10). LEC access services are used for both intra-LATA and inter-LATA calls, with the *central office* (CO) switches interpreting the digits dialed by the calling party and routing the call accordingly.

A CO is a BOC or an ITC switching system within a LATA where loops to customer stations are terminated for purposes of interconnection with each other and to trunks. For our example inter-LATA call, the LEC routes the call through its intra-LATA network from the originating switch to one of its access tandems, which connects to trunks terminating at an IXC *point-of-presence* (POP). An *access tandem* is a LEC switching system that provides a traffic concentration and distribution function for inter-LATA traffic originating or terminating within a LATA. The access tandem provides the IXC with access to more than one CO within a LATA, although more than one access tandem may be required to provide access to all COs within a LATA. In a LATA served by more than one IXC, the IXC to be used is selected (on a permanent or a call-by-call basis) by the calling customer.

A POP is a physical location within a LATA at which an IXC establishes itself for the purpose of obtaining LATA access and from which the LEC provides access services to its intra-LATA networks. An IXC may have more than one POP within a LATA, and the POP may support public and private, switched and non-switched services.

Once the call is accepted by the IXC, the IXC provides transport service through its network to its POP in LATA Y, for our example. From that point the call is completed via the LATA Y LEC network to the called party's station equipment, in a fashion similar to that described for the call originating in LATA X.

The MFJ restricts the lines of business in which the BOCs and AT&T may engage. For example, under the terms of the MFJ, the divested BOCs are allowed to:

- Offer local exchange service within specified geographical operating areas
- Provide equal access to BOC facilities for all IXCs
- Provide Yellow Pages directory publications
- Sell but not manufacture customer premises equipment
- Provide cellular mobile communications services
- Provide voice storage retrieval, voice messaging, and electronic mail services

Under MFJ terms, BOCs may not:

- Provide interexchange services
- Provide information services

- Provide any product or service that is not tariffed or otherwise regulated

The MFJ imposes no obligation on independent telephone companies, nor does it restrict the lines of business in which they may engage or the types of service they may provide. However, GTE is bound by a consent decree similar but not identical to the MFJ.

## Interexchange Carrier Access to LEC Networks

One of the more significant directives of the MFJ ordered all BOCs to provide IXCs with "equal access" to BOC networks. Switched IXC access to BOC networks means that, on a call-by-call basis, calls are routed through the LEC PSTN to the IXC POP. Because equal-access service is based on dialing, signaling, routing, and transmission plans that require particular CO capabilities, access services other than equal access were offered to IXCs for customers served by nonconforming COs. Initially, most other common carrier (OCC) access was provided using non-equal access facilities. The transition to equal access COs took almost seven years to complete.

LEC access services were provided as four feature groups, characterized by line-side access or trunk-side access at the CO switching system. Feature Group A access to the LEC networks is through a two-wire, line-side connection to the CO. This service exhibits the poorest quality, and, because it does not support automatic number identification (ANI), users must first dial an IXC local number and then, after receiving a second dial tone or a recorded message, enter up to 14 more digits (for caller identification and billing purposes) in addition to the dialed long-distance number. The advantage of Feature Group A for small IXCs was that all telephones in the local calling area were accessible through a single trunk group. However, it is now rarely used for access by IXCs.

Feature Group B access is provided via higher-quality four-wire trunk-side connections. Like Feature Group A dialing, Feature Group B permits users to access an IXC by dialing a special prefix and a *carrier identification code* (CIC) assigned by the *North American Numbering Plan Administrator* (NANPA). Because of the extra dialing requirement, Feature Group B is not widely used today for originating access. It is popular for terminating access trunk groups because it costs less than the other trunk-side access alternatives.

Feature Group C, an access arrangement available only to AT&T, offers "1 Plus" dialing, using four-wire trunk-side connections and ANI at most locations. Feature Group C was largely in place prior to divestiture. Direct-access line trunks run directly from COs to AT&T POPs, further enhancing quality and speed of service. As upgrades to equal access are completed, COs and AT&T will convert from Feature Group C to Feature Group D.

Feature Group D represents the long-term solution for equal access and, except for the fact that calls are generally routed through the LEC PSTNs rather than directly from COs, offers service comparable to Feature Group C.

To equalize the competitive positions among new IXCs and AT&T following divestiture, the BOCs were ordered to charge lower origination and termination charges for access arrangements providing lower-quality service. These charges are paid by IXCs to LECs and constitute a large percentage of the long-distance charges that IXCs charge their customers (roughly half of an IXC's revenue was paid to the LECs in access charges in the years following divestiture). A significant consequence of the lower access charges for new IXCs during the transition process to equal access was the ability of these IXCs to offer lower prices to their customers and gain market share. As equal access conversion was completed, the large user cost differences among IXCs disappeared.

One of the consequences of equal access and the use of access tandem switches as points of traffic concentration and distribution for all IXCs serving a LATA, is that the access tandems can represent single points of failure for all long distance service, even if a customer elects to use multiple IXCs. The consequences of this reliability vulnerability were graphically illustrated several years ago in the Hinsdale, Illinois, fire, which interrupted long-distance service completely for a large number of users in the Chicago area.

## LEC/IXC Tariffs

IXCs other than AT&T were originally regulated by the FCC. In the Competitive Carrier proceeding (Docket 70-252), the FCC decided to forbear (i.e., desist) from regulating new IXCs. As the dominant IXC, AT&T was still regulated and required to justify prices for its services as filed in tariffs. A *tariff* fixes the allowed rate (price) for a specific telecommunications service, equipment, or facility, and constitutes a contract between the user and the telecommunications carrier.

A formal process in which carriers submit filings for government regulatory review, possible amendment, and approval establishes tariffed services and rates. Tariffs, therefore, contain the most complete and precise descriptions of carrier offerings. Further, tariffed offerings cannot be dropped or changed without government approval. As competition took hold in the IXC market, price levels in tariffs were set by the market rather than regulation, even for AT&T.

LEC tariffs are under the jurisdiction of state authorities, with one notable exception. LEC tariffs for access facilities used to originate or terminate interstate traffic are under the jurisdiction of the FCC. Chapters 9 and 12 describe tariffs and network design and engineering techniques used to compare and evaluate the extent to which tariffed carrier offerings satisfy business applications.

## Major Carriers and Market Share

History has judged the deregulation of the long-distance industry to be a success by almost any measure. The number of long-distance carriers serving presubscribed customers has tripled in the last ten years to a total of 621 in 1997. Altogether, if one counts carriers, resellers, rebillers, service aggregators, subaggregators, and agents, there are over 1,500 long-distance businesses in the United States today. Predictably, the market share among carriers of the once dominant AT&T fell steadily from 82 percent in 1986 to 43 percent in 1998. The other major IXCs include MCI WorldCom and Sprint. Figure 2.4 shows the market shares of the three largest interexchange carriers from 1984 to 1998. In 1998, the second and fourth largest carriers merged to create MCI WorldCom and seriously challenge AT&T for the top spot among IXCs (Figure 2.4 combines MCI and WorldCom revenue for years prior to 1998). This process continued with the announcement in late 1999 of the merger of MCI Worldcom and Sprint. As of this writing, this merger is still under consideration by the FCC.

The increased competition has brought benefits to the consumer as well. The average cost per minute of all interstate switched voice traffic has dropped from 21 cents in 1986 to 15 cents in 1992 to 10 cents in 1998.

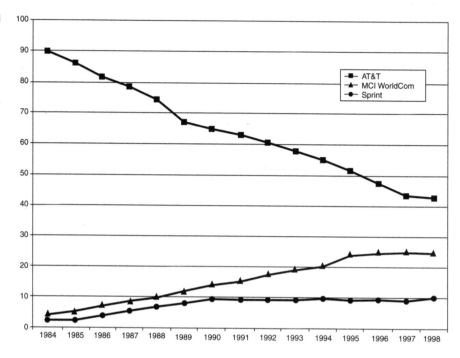

**Figure 2.4**
Long-distance market
share of the three
largest interexchange
carriers.

Source: FCC Industry Analysis Division

# The Telecommunications Act of 1996

The realization of a truly competitive long-distance marketplace set the stage for a bold move toward a complete overhaul of U.S. telecommunications policy. On February 1, 1996, Congress passed the Telecommunications Act of 1996 (the "Act"), the first comprehensive rewrite of the Communications Act of 1934. The Act dramatically changed the ground rules for competition and regulation in virtually all sectors of the communications industry. The Act's provisions fall into the following major areas, identified by the Title in the Act that covers the area:

- Title I—Telecommunications Services
- Title II—Broadcast Services
- Title III—Cable Services
- Title IV—Regulatory Reform

- Title V—Obscenity and Violence
- Title VI—Effect on Other Laws
- Title VII—Miscellaneous Provisions

This section focuses on provisions of Title I of the Act, Telecommunications Services. However, it should be noted that Title VI, Effect on Other Laws, superseded the AT&T Consent Decree that split up the Bell System and prohibited the BOCs from certain lines of business. In its place, the Act codifies the principles that will be used to remove the statutory and court-ordered barriers to competition between segments of the telecommunications industry.

## Provisions

Title I of the Act attempts to create opportunities for competition within the local exchange market (local loop) while allowing BOCs to enter the competitive inter-LATA market. It requires the BOCs to share their networks with would-be competitors in return for that entry. The following sections treat the major aspects of the telecommunications provisions.

### PROMOTING COMPETITION IN THE LOCAL LOOP

As noted above, the monopoly position of the BOCs required the MFJ to mandate that they open their local networks to the IXCs. Like the MFJ, the primary mechanism the Act uses to achieve competition is to require incumbent LECs to share their local networks with competitors.

Sections 251 and 252 of Title I establish the core ground rules for the sharing of local networks. Special provisions apply only to incumbent LECs because of their advantages as monopolies in the local loop. These are:

- To provide interconnection at any technically feasible point within their networks

- To provide unbundled access to network elements in a manner that allows the competitor to combine elements to provide telecommunications services

- To offer for resale at wholesale rates any telecommunications service that the incumbent LEC provides at retail rates to noncarrier subscribers

- To provide for physical collocation of equipment necessary for interconnection or access at the premises of the incumbent LEC.

All the above requirements are to be provided on a nondiscriminatory basis and on rates, terms, and conditions that are just, reasonable, and nondiscriminatory.

Compliance with these requirements is achieved through two means. First, compliance is part of a checklist of preconditions for BOC entry into the in-region inter-LATA market (see below). Second, Section 252 creates mechanisms, administered at the state level, to promote and ensure agreements between ILECs and their competitors. These mechanisms include:

- Voluntary agreements between incumbent LECs and their competitors
- Compulsory arbitration, which must be resolved consistent with the above network sharing requirements
- BOC filing of a statement of generally available terms and conditions for all carriers, in compliance with the above sharing requirements.

The terms and conditions of all arbitrated or negotiated agreements must be made available to other interconnecting carriers on the same terms and conditions. Challenges to the state decisions regarding these agreements are reviewable only in federal court. If a state fails to carry out its responsibilities under Section 252, the FCC is required to preempt state authority.

### ALLOWING NEW COMPETITION IN TELECOMMUNICATIONS MARKETS

Title I immediately bars state and local requirements that prohibit an entity from providing telecommunications services. It also establishes timetables and rules for BOC entry into the inter-LATA and other previously prohibited markets. The most important of these provisions are the rules for BOC entry into the inter-LATA market within its own service region. This is the most lucrative market for a BOC and is also the market where a BOC can exert the most undue influence due to its monopoly position.

Section 271 provides safeguards against this undue influence by specifying a state-by-state approval process conducted by the FCC. The FCC must find that a set list of preconditions has been fulfilled in a state before the BOC may offer in-region inter-LATA service there. These preconditions include:

- The FCC must consult with the Department of Justice and give substantial weight to its recommendation.
- The FCC must determine that entry is consistent with the public interest, convenience, and necessity.
- The BOC must have a facilities-based competitor operating under a state-approved interconnection agreement or not have received an interconnection request.

The most important precondition is that the BOC must have complied with a 14-point competitive checklist (similar to, but more expansive than the network sharing requirements of Sections 251 and 252). The BOC must have either filed an approved statement of generally available terms and conditions or entered into one or more approved interconnection agreements that include each of the following checklist requirements:

1. Interconnection in accordance with the requirements for incumbent LECs in Sections 251 and 252
2. Nondiscriminatory access to network elements in accordance with the requirements for incumbent LECs in Sections 251 and 252
3. Nondiscriminatory access to poles, ducts, conduits, and rights of way owned or controlled by the BOC at just and reasonable rates in accordance with the requirements of Section 224 of the 1934 Act
4. Local loop transmission from the central office to the customer's premises, unbundled from local switching or other services
5. Local transport from the trunk side of a wireline switch unbundled from switching or other services
6. Local switching unbundled from transport, local loop transmission, or other services
7. Nondiscriminatory access to 911 and E911, directory assistance, and operator call completion services
8. White pages directory listings for customers of other carriers' telephone exchange service
9. Nondiscriminatory access to telephone numbers for assignment to the other carriers' customers
10. Nondiscriminatory access to databases and associated signaling necessary for call routing and completion
11. Interim number portability through remote call forwarding, direct inward dialing trunks, or other comparable arrangements

with as little impairment of functioning, quality, reliability, and convenience as possible

12. Nondiscriminatory access to such services or information necessary to allow the requesting carrier to implement local dialing parity in accordance with the requirements for all LECs

13. Reciprocal compensation arrangements in accordance with the requirements for incumbent LECs in Section 252

14. Telecommunications services available for resale in accordance with the requirements for incumbent LECs.

### CONSUMER PROTECTION

Title I provides a series of measures to protect consumers' continued access to telecommunications services. The major provisions represent an overhaul of the universal service requirements and subsidies. These provisions are designed to evolve with technological change.

Section 254 directs the FCC to define the services to be supported by the federal universal support mechanism based on the recommendations of a federal-state Joint Board. Subsidies are to be explicit rather than hidden. For the first time, the concept of universal service is allowed to extend beyond the most basic telephone service. Services will be included within universal service to the extent that they are essential to education, health, or safety; have been subscribed to by a substantial majority of residential customers as a result of market choice; are being deployed in public telecommunications networks by carriers; and are consistent with the public interest.

Section 254 also provides direction on several pricing issues. It specifically requires rate averaging (i.e., the IXC charges to residential customers in rural and high cost areas must be no higher than those charged by the same carrier in urban areas).

## Implementation by the FCC

Following passage of the Act, the FCC initiated proceedings to implement its provisions. In particular, the public's attention was dominated by the FCC's actions in three major areas: Local Competition, Universal Service, and Advanced Telecommunications Incentives.

In August 1996, the FCC issued the landmark Local Competition Order, which included rules implementing Sections 251 and 252 of the Act. In particular, this order defined what constitutes a network element and what network elements must be offered on an unbundled basis. These network elements are, at a minimum:

- Local loops
- Switching capability
- Interoffice transmission facilities
- Databases and signaling systems
- Operations support systems.

Figure 2.5 illustrates the location of these unbundled elements in the LEC network. Note that competitors can gain access using dedicated facilities to incumbent LEC tandem switching offices as well as local switching offices. Connections to tandem switches can then use the incumbent LEC transport element to access many end offices. The Order went on to specify the methods of obtaining interconnection and access to unbundled network elements, including standards for collocation at the incumbent LEC facilities. Finally, the order addressed two critical pricing issues: the pricing of interconnection and unbundled elements and the pricing of incumbent LEC services on a wholesale basis for resale.

**Figure 2.5**
Unbundled network elements in the local exchange network.

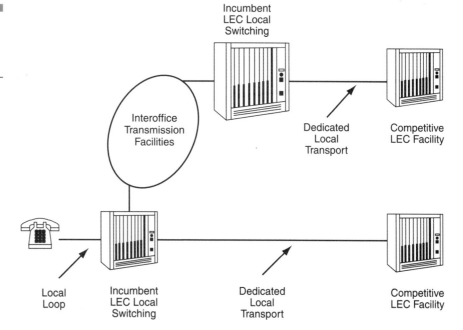

While it falls to state commissions ultimately to set the price levels for unbundled network elements, the FCC mandated a uniform methodology for the states to apply in determining these rates. This methodology was based on forward-looking economic costs (as opposed to a consideration of embedded costs). This was seen as the best method of encouraging the new entrants to make decisions without a distortion of the underlying economics. To aid the states in determining the rate levels for the unbundled elements, the FCC instituted the development of an accepted cost proxy model to determine the *Total Element Long Run Incremental Cost* (TELRIC) of each element in each area of the country.

It is generally accepted that facilities-based competitors using unbundled network elements provide the best chance for long-term deregulation of local telecommunications. However, in the interim the FCC also provided for the resale of services provided by the incumbent LECs. It mandated that these services be provided to competitors at wholesale rates equal to the incumbent LEC retail rates minus the cost of marketing, billing, collection, and other costs that will be avoided by the local exchange carrier.

In March 1996, a Notice of Proposed Rulemaking and Order established a Federal—State Joint Board to consider universal service issues. The Universal Service Order, adopted in May 1997, outlined a plan to revise universal service support consistent with Section 254 of the Act. The Order defined four distinct areas requiring support:

- Schools and libraries
- Rural health care providers
- Low-income consumers
- High-cost service areas

Each of these areas will now be supported by a separate Universal Service Fund. Money flows into the funds through contributions from carriers providing interstate services. The level of contribution is determined by the carrier's gross end-user revenue. The carriers are not being altruistic in their contributions; their costs are being recovered through surcharges on customer bills under various names (Universal Service Fund [USF] charges, National Access Fee, etc.).

The first two areas are new with the Act and are the subject of ongoing political debate. The fund essentially provides for discounts for access to the Internet for eligible recipients in these categories. Low-income customers traditionally received reduced monthly charges through Lifeline programs. The Order revised these programs to make them available in every state and territory and expanded the services that must be provided.

The last area, support for customers who are costly to serve because of where they are located (typically in rural or insular areas), was also traditionally supported before the Act. Incumbent LECs provided service to these customers at reasonable rates and received compensation through a previous universal service fund. Realizing that the current subsidy system could not work in a competitive environment, the FCC adopted a plan to replace implicit subsidies with support based on forward-looking costs of providing the services. A carrier will only be eligible for support when its costs as determined by a cost proxy model exceed a benchmark value. In both the interconnection and the universal service arena, the FCC consistently is moving to cost-based mechanisms that are essential in a competitive environment.

In February 1999, the FCC released its report on the deployment of advanced telecommunications capability, as required by Section 706(b) of the Act. The Act defined advanced telecommunications capability as high-speed, switched, broadband telecommunications capability that enables users to originate and receive high-quality voice, data, graphics, and video telecommunications using any technology. The overall conclusion of the report was that broadband capability was being deployed in a reasonable and timely manner to all Americans. Hence, no specific actions were required at that time by the FCC.

## Status

Despite the high hopes for rapid deregulation and a rise of competition in local area telecommunications, little progress was made during the first three years after passage of the Act. Since the Act touched on the most basic issues in regulatory law, including the fundamental principle of states' rights versus the authority of the federal government represented by the FCC, litigation followed every action taken by the FCC. Numerous appeals of the Local Competition Order were finally consolidated before the Eighth Circuit Court. The court held that the FCC's jurisdiction was limited to interstate services, that the FCC had no authority to adopt rules governing pricing, dialing parity, exemptions for rural LECs, and dispute resolution. It also struck down the FCC's requirement that incumbent LECs provide combined network elements on a bundled basis as well as on an unbundled basis.

Finally, the Eighth Circuit rejected the FCC's "pick and choose" rule that would have permitted requesting carriers to obtain particular provisions of any other existing interconnection agreement negotiated by an incumbent LEC. This case was ultimately settled in the U.S. Supreme Court in January 1999; the Eighth Circuit was reversed on the key issues and the authority of the FCC to implement the Act was reaffirmed. However, two and half years had passed where key issues involved in the process of deregulation remained ambiguous. While progress was clearly made in some areas, as indicated by the number of interconnection agreements negotiated by new competitive LECs, the actual amount of local service provided by competitive carriers remained small. Three years after the Act's passage, no BOC had passed the 14-point checklist to enter the long-distance market. Bell Atlantic successfully passed the checklist in New York in the fall of 1999 and was authorized to provide long-distance service there, but the authorization was immediately challenged in federal court.

Figure 2.6 illustrates the trends in local competition and brings home the fact that, although the current level of competition is low, the stage is being set for a robust competitive market. Local service revenue for competitive carriers represented only 3.6 percent of all revenue in 1998, but was almost doubling each year. While competitive LECs use, in total, less than 2 percent of incumbent LEC switched lines, there is potential for significant gains in usage. The figure indicates that, at year-end 1998, competitive LECs had operational collocation arrangements in switching centers from which incumbent LECs serve about 42 percent of their switched lines to residential customers and about 58 percent of their switched lines to business and government customers. In addition, a local service competitor that owns a telephone switch must acquire a numbering code (Central Office code or CO code or "NXX" code) for that switch before commencing operation as a facilities-based competitive LEC providing mass-market switched telephone service. By the second quarter of 1999, competitive LECs had acquired 20 percent of the numbering codes assigned.

## Major LECs and Market Share

Market share in the local services arena is largely determined by the monopoly service area of the incumbent LECs (as seen above, only about 2 percent of the local loops are provided by competitive carriers in an incumbent LECs service area). Nevertheless, the landscape has

**Figure 2.6**
Local service
competition
summary.

| | As of December 31, 1998 | As of December 31, 1997 |
|---|---|---|
| Share of Local Service Revenues | | |
| – Incumbent LECs | 96.5% | 97.7% |
| – Local Competitors | 3.5% | 2.3% |
| Lines Provided to CLECs for Resale | 1.7% | 1.1% |
| Lines Provided to CLECs as UNEs | 0.2% | 0.1% |
| ILEC Lines where CLEC has Collocation Agreement | | |
| – Residential Lines | 42.2% | 23.3% |
| – Business Line | 58.3% | 41.4% |

Source: FCC Industry Analysis Division

changed since the passage of the Telecommunications Act due to merg-ers of major players and the prospect of even more consolidation among the ILECs. Two of the major RBOCs formed at divestiture have merged with other RBOCs. Southwestern Bell and Pacific Telesis became SBC Communications and Bell Atlantic and NYNEX became Bell Atlantic Corporation. Figure 2.7 shows the distribution of local loops by holding company as of December 1998. At this writing, SBC and Ameritech have also merged and Bell Atlantic and GTE are in the final stages of having their merger approved by the FCC. All such mergers of regulated entities are scrutinized by the Justice Department for possible antitrust issues. The competing forces of new competitive entries on the one hand and the creation of fewer larger entities through mergers on the other will continue to play out over the next decade.

# Data Service Provision in the U.S.

The evolution of data service provision in the U.S. has taken a totally different path from that of voice service. The technology of packet

**Figure 2.7**
Distribution of local service revenue and local loops by holding company.

| Holding Company | Revenue | Lines |
|---|---|---|
| Bell Atlantic | $25,563,794.00 | 48,978,374 |
| SBC Comm | $21,715,508.00 | 46,520,010 |
| Bell South | $15,772,356.00 | 28,405,719 |
| Ameritech | $12,313,875.00 | 25,449,161 |
| US West | $10,668,772.00 | 23,355,703 |
| GTE | $13,935,796.00 | 21,251,644 |
| Sprint | $4,869,542.00 | 8,393,679 |
| Cincinnati Bell | $712,125.00 | 1,119,285 |
| Alltel | $649,546.00 | 982,321 |
| Frontier | $335,847.00 | 910,753 |
| Alliant | $216,352.00 | 367,335 |

*Source: FCC Industry Analysis Division*

switching is a relatively recent development; the first prototype networks were rolled out in the late 1960s. The ubiquitous provision of service by a monopoly service provider was not a starting point for the new technology. Instead, relatively small private networks came into being serving specific interest groups and organizations. The early packet switches were limited in size, so sharing among many organizations was not feasible and the switches were generally located on a user's premises. As technology advanced and higher-performance switches became available, public data services began to emerge. The data-service market is now divided into two major segments: the provision of private or virtual private network services to businesses and the phenomenon known as the Internet (with a capital I). The technical details of these services will be discussed in later chapters. The sections below describe the major market players in this industry segment and how they interrelate to provide the data services we enjoy today.

## Business Data Services

Data communications between two users requires much more standardization of equipment than does voice communications. Within a private network serving one organization, this is accomplished by using one manufacturer's equipment throughout the network. For example, thousands of private networks based on IBM's Systems Network Architecture (SNA) were implemented to interconnect IBM and compatible mainframe computers. The adoption of standards by national and international standards bodies was the first step in allowing various manufacturers' equipment to work together and public data services to come into existence. For business services, the key event was the adoption of the X.25 and related standards by the CCITT, an international standards body. National data transmission services based on X.25 were provided by organizations such as Telenet and Tymnet during the 1970s and 1980s. Due to technology limitations, the transmission speeds supported by these networks were relatively low (less than 7,000 characters per second) and there were no guarantees on end-to-end service performance. The situation began to change radically in the early 1990s as technological advances made very high-speed packet switches feasible and the standards bodies responded with new service standards that could take advantage of the increased capability. Wide area network services such as frame relay (see Chapter 14) became immensely popular among businesses and drove the demand for high-capacity public networks to provide these services. Fortunately, the technological advances made high-capacity switches a reality and the traditional exchange carriers began implementing data networks into their public network infrastructure. Today, the major suppliers of wide area data services to businesses are the same IXCs and incumbent LECs described above. The services are generally provided on an unregulated basis, but the same comments on industry structure made above also apply to the corresponding data service entities. More detail on how business services are structured and how to select the best services for a given organization is given in Chapters 12 and 15.

## Services of the Internet

Of all the concepts presented in this book, the Internet may be the most enigmatic. At this point in history, almost everyone has heard of it and very few have any idea what it really is. It cannot be denied that the Internet is one of the most revolutionary advances in our technological history. Its effect on our day-to-day lives and the way the world conducts business rival those of the telephone and the transistor. Having made that assertion, one would be hard pressed to point to the single idea or discovery that made all the difference. The story of the Internet is an excellent example of the triumph of common sense and the evolution of an extremely complex entity from building blocks of simple entities. Understanding the Internet today requires retracing these evolutionary steps and understanding the significance of key decisions made along the way.

## EARLY HISTORY

The Internet began with the packet-switching research projects sponsored by the Advanced Research Project Agency (ARPA) in the 1960s. The experimental ARPANET grew in the 1970s to support many organizations in the U.S. Department of Defense as well as many university and research organizations. The Transmission Control Protocol/Internet Protocol (TCP/IP, see Chapter 7 for a discussion of computer protocols and TCP/IP in particular) was developed to allow connections over a wide variety of network types. It was included in a popular release of the operating system that was freely distributed throughout the university community. This loose development of technology was distributed among many entities with common goals and research results were freely shared within the community.

In 1985, the National Science Foundation (NSF) funded several national supercomputer centers. The NSF desired to make these supercomputer centers available to the research community in universities around the country. Hence, the NSF funded a network linking the centers and offered to let any of the university computer centers that could reach this network physically connect to it. This was the "seed" of the Internet network, as we know it today.

The universities soon found that the network was useful for other things such as electronic mail, file transfer, and newsgroups. The traffic on the network began to grow dramatically. In 1987, the NSF awarded a contract to Merit Network, Inc. in partnership with IBM, MCI, and the state of Michigan to upgrade and operate the NSF backbone. This was the largest data networking project ever undertaken up to that time. By 1990, the backbone was expanded to 16 sites and the transmission speeds were upgraded again. The major function of the NSF network was to interconnect networks, and the term "Internet" was first defined as having connectivity to this backbone.

## NAPS AND COMMERCIALIZATION OF THE INTERNET

As a growing amount of Internet traffic became commercial in nature, it became increasingly difficult to justify the funding of the NSF backbone by the government as the major Internet transport mechanism. At a minimum, the government was essentially competing with private companies who were in the business of data communications transport. A number of private commercial backbone operators emerged, but they still relied on the NSF network to interconnect them. In May 1993, the NSF bowed out of the backbone business and radically altered the architecture of the Internet. Instead of a government-run backbone network, the NSF announced the formation of *network access points* (NAPs). Private commercial backbone operators could exchange traffic with each other at these single points and become, by definition, part of the Internet. The NSF awarded four contracts for NAPs in San Francisco, Chicago, New Jersey, and Washington, DC. On April 30, 1995 the NSF backbone was shut down and the NAP architecture became the Internet. Backbone and regional providers connected to the NAPs and began selling capacity to each other through peering arrangements. As this concept became more popular, the need for interconnect capacity grew. Metropolitan Fiber Systems currently operates five *Metropolitan Area Exchanges* (MAEs) which serve as de facto NAPs. This series of NAPs could be considered the top level of the Internet. In addition, large backbone operators interconnect with each other whenever and wherever it becomes feasible to do so. This structure allows the Internet to be dynamic and adapt to the explosive growth in demand that is occurring. Figure 2.8 shows the location of the major U.S. interconnect points on the Internet.

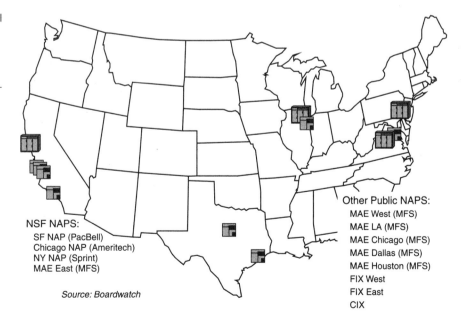

**Figure 2.8**
Major Internet connection points in the USA.

NSF NAPS:
SF NAP (PacBell)
Chicago NAP (Ameritech)
NY NAP (Sprint)
MAE East (MFS)

*Source: Boardwatch*

Other Public NAPS:
MAE West (MFS)
MAE LA (MFS)
MAE Chicago (MFS)
MAE Dallas (MFS)
MAE Houston (MFS)
FIX West
FIX East
CIX

## APPLICATIONS AND THE NATURE OF DEMAND

In order to understand the structure of the Internet today, one must understand the changing nature of the demand for Internet services. The initial network connected very specific points on the network (the supercomputer centers). As the more general research community was connected, the number of end-points multiplied. However, the archaic user interface limited the users to university students who were skilled in computer science. In order to make research opportunities more accessible even within the university community, a computer application and network protocol was developed that would change the face of history. The NSF, the European Laboratory for Particle Physics (CERN), and the National Center for Supercomputer Applications (NCSA) in the United States developed a graphical application that allowed users to browse and access data without specialized knowledge of the underlying computer technology. The network of users and content providers became known as the World Wide Web (Web) and the software for accessing the content was called a Web browser. The initial browser developed at NCSA was called MOSAIC and was distributed free to users of various computer systems. The developers of MOSAIC founded a commercial company dedicated to the evolution and distribution of browser technology called Netscape Communications. This technology, coupled with the rapidly decreasing price of personal computers, finally put the power of the Internet into the hands of the average person at home as well at work. With almost ubiquitous connectivity now available, businesses and consumers are being connected at an ever-increasing rate over the Internet with no end in sight. This phenomenon has driven the architecture of Internet access at the lower network levels.

## CATEGORIES OF INTERNET PROVIDERS

We have discussed the interconnect NAPs as the "top" level of the Internet and backbone *national service providers* (NSPs) as those carriers providing high-speed nodes and transmission links across the country interconnecting the major cities. Below this level, regional networks operate within a state or among several adjoining states. Most of these are remnants of the original NSF regional entities that connected universities. They typically connect to one or more national backbone operators or to a single NAP. They very effectively extend the Internet to smaller cities and towns in their area.

The extension of the Internet into the consumer market drove the need for a new level of entity, the *Internet Service Provider* (ISP). While offering dedicated connections to the Internet, ISPs predominantly thrive on dial-up customer accounts. They operate modem banks at ISP points-of-presence that offer dial-up connections to customers in the calling area of those points. In general, ISPs don't run their own backbone, but lease connections to a national backbone provider. There are approximately 6,000 local ISPs in the United States with combined revenue of $7 billion dollars in 1998.

## STATISTICS AND MARKET DATA

Reliable statistics regarding Internet and other data network usage are hard to come by. The lack of any overseeing regulatory body, such as the FCC for voice, means that accurate traffic and revenue data are not collected and put in the public domain as is required for regulation. Most carriers consider traffic and market data proprietary information. Nevertheless, general estimates can be made about the industry segment as a whole. By all estimates, traffic on the Internet is continuing to double every year. Figure 2.9 illustrates this growth using two different measures, the number of host computers on the Internet and the traffic carried by the Internet backbone networks. The first measure, shown in Figure 2.9 (a), shows a steady doubling each year since 1985. The second, shown in Figure 2.9 (b), reveals the effect of a sharp paradigm shift. The rather drastic increase in the traffic measure between 1994 and 1996 coincided with the development of the World Wide Web, the Web browser, and the introduction of the Internet to the general population. This increase in popularity coupled with the greatly increased traffic requirements of graphical applications like browsers, caused a quantum jump in traffic. However, since that time, the growth in traffic is following the now familiar pattern of doubling each year.

The significance of this growth and its effect on the national backbone providers should not be underestimated. While ISPs can react to growth by multiplying their numbers, the relatively small number of facility-based backbone providers must continually upgrade the capacity of their networks to keep up with this demand growth. Given the time required to deploy transmission lines and equipment, it is a daunting task to keep ahead of the demand curve. Most major NSPs doubled capacity between the middle of 1996 and the end of 1997. The two largest NSPs, MCI and UUNET, increased their backbone capacity six to eight times during this period.

With its acquisitions, WorldCom (now MCI WorldCom, Inc.) became the largest NSP in the Internet, subsuming the national backbones of

**Figure 2.9**
Growth of the Internet: (a) Number of connected computers, (b) Number of terrabytes/month of traffic.

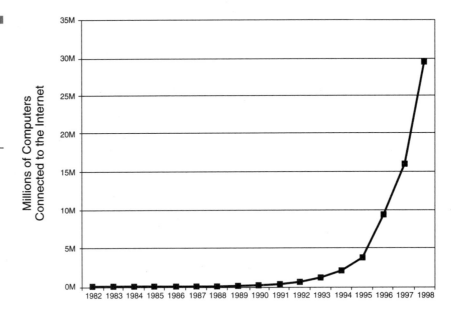

(a) Number of connected computers.

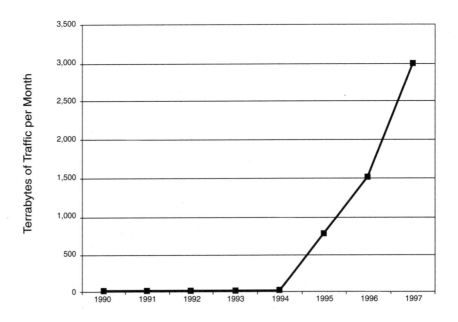

(b) Number of terrabytes/month of traffic.

UUNET, and ANSnet. MCI was required to divest itself of its national backbone as part of the Justice Department's approval of its merger with WorldCom. MCI WorldCom now controls just under 50 percent of the Internet backbone. Sprint is the next largest NSP with approximately 10 percent of the backbone capacity. As the next chapters will show, leaps in transmission and switching technology must continually be made to keep up with the ever-increasing demand for Internet capacity.

# International Telecommunications Service Provision

Under traditional arrangements, U.S. and foreign carriers jointly provide international services. The foreign carriers are often agencies of foreign governments, generally referred to as post, telephone and telegraph administrations or PTTs. Privatization in many countries is occurring, however, and in some cases the U.S. carrier may be dealing with its own foreign affiliates. Agreements on the rules and regulations governing international communications carriers must be among governments; U.S. carriers are bound by the regulatory authority of the FCC regarding their permissible actions and the agreements they may enter into with foreign carriers. The following sections describe the market structure and current developments for the provision of telephone service and other services.

## Market Structure

Through 1985, AT&T provided most of the international telephone service between the continental U.S. and foreign markets. In 1985, other carriers began to provide international telephone service on a pure resale basis. In order to provide facilities-based service to international points, carriers had to obtain facilities, execute operating agreements with foreign carriers, and then attract customers. An operating agreement is a contract between a U.S. carrier and its foreign correspondent under which two carriers agree jointly to provide service between two countries. The agreement contains an accounting rate which governs how much carriers compensate each other for handling each other's traffic. The carrier that bills the customer pays half a specified amount to its correspondent for each minute of jointly provided service. Under FCC's International Settlement Policy, the accounting rate is symmetrical, meaning that foreign carriers pay U.S. carriers the same amount per minute (the "settlement rate") when the U.S. carrier completes a call for a foreign carrier.

In the past, some accounting rates were so high that the U.S. carrier would pay out more to foreign carriers than it collected from customers. Those high settlement rates, however, meant that return traffic was very profitable. New carriers could not compete unless they received a share of the return traffic. The FCC adopted a proportionate return policy so that U.S. carriers would each get roughly the same share of foreign-billed traffic for any particular country as they had for U.S. billed traffic.

Because U.S. customers place far more calls than they receive, U.S. carriers make net settlement payments to most foreign carriers (i.e., they pay out more settlement payments than they receive). The total net settlement payments for all U.S. carriers grew from $0.4 billion in 1980 to $5.6 billion in 1996. Total U.S. carrier revenue for international calling, after settlement payments, grew from $2.2 billion to $9.2 billion over the same period.

The accounting rate agreements between U.S. carriers and foreign carriers also specify the currency units in which payments are made. Many accounting rates are defined in monetary units other than U.S. dollars, such as special drawing rights and gold francs. In such cases, changes in the value of the dollar can affect the net settlement amounts. Accounting rates may vary by time of day, by service classification, by the volume of minutes, or even by the locations in which calls originate and terminate.

U.S. carriers now offer services, such as call-back arrangements, that allow customers in foreign points to place calls through the United States to another foreign point. These services allow customers abroad to take advantage of relatively low U.S. service rates. Call-back service can be used to terminate calls in the United States or another international point. A call-back arrangement allows a customer in a foreign country to use foreign facilities to dial a preassigned telephone number in the United States. That call is not completed, but the presence of signaling information triggers a call back to the customer, who receives a dial tone from the U.S. carrier's switch. The customer can then place a call via the U.S. carrier's outbound switched service either to a point in the United States or to another international point. U.S.

call-back carriers typically operate as pure resale carriers. The FCC has determined that call-back using uncompleted call signaling does not violate either U.S. domestic or international law. Some U.S. carriers now offer "hubbing" or "reorigination" services that allow a foreign carrier to complete its customers' calls to countries with which it does not have an accounting rate agreement. Commission rules permit U.S. carriers to provide hubbing to foreign carriers in specified circumstances. Some countries impose value-added taxes on call-back and other countries have banned call-back entirely. The Commission ruled, however, that U.S. carriers are not authorized to provide call-back using uncompleted call signaling in countries which have expressly declared that call-back is illegal.

## Resale

The Commission has a long-standing policy of requiring carriers to permit the resale of their services. This has been critical to the development of international competition because it allowed carriers such as MCI and Sprint to provide service via resale while they negotiated operating agreements with foreign administrations. Carriers provide pure resale service by routing calls to an underlying carrier, which carries the traffic over its own facilities and pays settlements on that traffic to the carrier in the country of destination. Under the Commission's proportionate return policy, the underlying carrier gets the lucrative return traffic associated with the resale minutes.

U.S. carriers are now allowed to carry calls between the United States and some selected countries using interconnection arrangements that are outside the normal settlement process. This traffic is referred to as International Simple Resale because carriers initially had to offer the services over resold private facilities. Carriers are now permitted to carry International Simple Resale traffic over circuits that they own. The FCC first approved these arrangements for a handful of countries that offered competitive opportunities to U.S. carriers that were equivalent to the opportunities that the United States afforded foreign carriers. Determining whether a country met this standard was referred to as the Equivalent Competitive Opportunities test. The FCC now approves International Simple Resale arrangements between the United States and other *World Trade Organization* (WTO) members if it finds 50 percent of the U.S.-billed traffic for the country was settled at or below the benchmark settlement rate for that country adopted by the FCC. Where the benchmark settlement rate condition is not met, the FCC will authorize the provision of International Simple Resale where the WTO member provides resale opportunities that are equivalent to those available under U.S. law. For non-WTO countries, the Commission will authorize International Simple Resale only of both the settlement benchmark and resale opportunities conditions are met.

As of July 1998, the Commission had authorized International Simple Resale between the United States and Canada, the United Kingdom, Sweden, New Zealand, Australia, the Netherlands, Luxembourg, Norway, Denmark, France, Germany, Belgium, Austria, Switzerland, and Japan. The International Simple Resale arrangements have helped drive down both accounting rates and prices for calls to approved countries. In 1996, International Simple Resale telephone traffic accounted for $75 million compared with $14.1 billion for facilities-based telephone traffic.

## International Private Lines and Data Services

While telephone traffic still accounts for the vast majority of international revenue, there is also an increasing market for international non-switched circuits (commonly known as private lines) and for data services. Since the market for these services has been small, complex rules such as accounting rates, settlement rates, and proportional return do not apply. Each carrier bills customers for its portion of the service provided. While this appears to be simple and fair, there are undesirable side effects for customers. Consider the example of an international private line from the U.S. to France, ordered by a customer in the U.S. The price listed for the circuit by the U.S. carrier is only for a "half circuit." The cost to the customer for the other half circuit is the price charged by the French carrier, which is not listed in any U.S. tariff. Furthermore, the bill for the French half-circuit is rendered in French francs. While the U.S. carrier can act as a Coordinating Carrier in dealing with the foreign carrier, all terms and conditions of foreign tariffs apply. Given these variables, coupled with month-to-month currency fluctuations, the U.S. customer cannot enter into any kind of stable agreement with a carrier for this service. The same situation exists for data services. However, changes for the better are on the way.

As competition is allowed in more and more countries, alliances and joint ventures are being made between U.S. and foreign carriers. One result of these ventures is the end-to-end provisioning of international private lines and data services on a full channel basis. This means that U.S. carriers provide fixed prices and stable terms and conditions for these services regardless of changes in the underlying facilities provided by the foreign partner. The resulting competition among U.S. carriers for services provided in this manner will result in lower prices for the customers.

## Market Share

Today, the three largest toll carriers in the United States are also the three largest providers of international service between the United States and international points. Figure 2.10 illustrates that international service represents almost a fifth of the toll revenue billed by the largest toll carriers.

**Figure 2.10**
International toll revenues of U.S. carriers (1996).

| Carrier | Total Toll Revenue ($ Billions) | International Toll Revenue ($ Billions) |
|---------|--------------------------------|------------------------------------------|
| AT&T | $39.3 | $8.9 |
| MCI WorldCom | $16.4 | $3.8 |
| Sprint | $7.9 | $1.6 |

*Source: FCC Industry Analysis Division*

As deregulation occurs in more and more countries around the world, a new class of international service provider is emerging. These new companies are establishing points-of-presence in several countries, acquiring international transmission capacity through membership in new fiber optic cable consortia (in particular, the undersea cable consortia), and are selling a wide variety of voice and data services. These carriers, such as Carrier1 and Facilicom, are not encumbered by decades of restrictive regulation and outmoded physical plant. While their combined revenues are small, they represent an alternative to traditional national carriers for telecommunications users.

# 2

# Telecommunications Fundamentals

**Chapter 8**

# CHAPTER 3

# Basic Concepts, Techniques, and Devices

This chapter presents explanations of concepts and terminology crucial to anyone with the need to intelligently select or use of the wide array of telecommunications systems and services now vital to the success of nearly all business, professional, and personal endeavors. The material is consistent with and builds on the high-level network and organizational definitions and descriptions just presented in Chapters 1 and 2. It establishes a framework permitting readers, even those with little or no engineering background, to grasp and fully comprehend pivotal attributes of advanced, and sometimes complex, technologies described in the remaining parts of this book.

Many introductory books do not attempt to furnish nontechnical readers with a level of knowledge that allows them to converse effectively with engineers or to understand the practical implications of technical proposals and solutions. While this was no easy task, the first edition of this book proved it could be accomplished by systematically presenting basic definitions as parts of explanations of larger concepts. This equips readers not only with terminology, but also with the rationale behind real-world applications, a tremendous advantage for thorough understanding and memory retention.

Beyond terminology familiarization, this chapter provides an invaluable, up-to-date summary of all basic concepts, techniques, and devices relevant to today's advanced technology telecommunications systems.

# Analog Electrical Signals

In telecommunications networks, information is transferred in the form of signals. A *signal* is usually a time-dependent value attached to an energy-propagating phenomenon used to convey information. For example, an audio (sound) signal is one in which the information is characterized by loudness and pitch. Until the early 1960s, the PSTN evolved as an analog network. The meaning of the term "analog" in networks is illustrated in Figure 3.1. Figure 3.1A is a pictorial representation of a voice sound wave, i.e., how the compression and expansion of air would look if sketched on paper as a function of time. A microphone or transducer (the telephone handset in the figure) intercepting the sound wave converts differences in acoustic pressure to "analogous" differences in electrical signal amplitudes, as shown in Figure 3.1B. Once converted to electrical format, these signals can be transported through networks to loudspeakers (other transducers), which convert the electrical signals back to sound waves. Time sketches or representations such as those shown in Figure 3.1A and B are referred to as *waveforms*.

**Figure 3.1**
Analog voice acoustic and electrical signals.

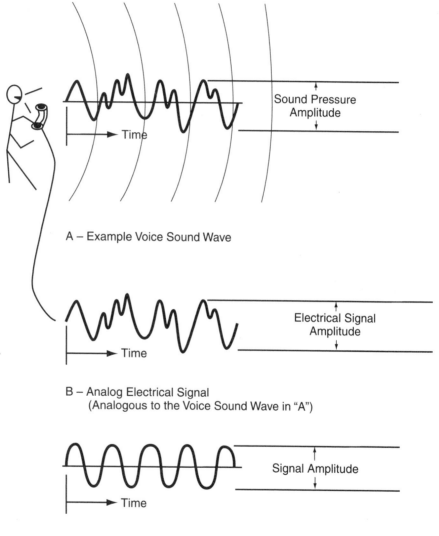

A – Example Voice Sound Wave

Sound Pressure Amplitude

Electrical Signal Amplitude

B – Analog Electrical Signal
(Analogous to the Voice Sound Wave in "A")

Signal Amplitude

C – Single Frequency Analog Electrical Signal

An *analog* signal is a continuous signal that varies in some direct correlation with an impressed phenomenon, stimulus, or event that typically bears intelligence. Sound waves and their electrical analogs are characterized by loudness (a quantity proportional to amplitude) and pitch. Analog signals can assume any of an infinite number of amplitude values or states within a specified range, in accordance with (analogous to) an impressed stimulus. Pitch refers to how many times per second the signal swings between high and low amplitudes.

Simple sound waves and signals can be made up of only a single tone, like a single note on a piano, as shown in Figure 3.1C. In this case the waveform consists of repeating identical cycles and is said to be of a single frequency, measured as the number of cycles that occur in one second of time. In communications, *frequency* was traditionally expressed in *cycles per second* (cps), but is now expressed in *hertz* (Hz), still equal to one cycle per second. Thus, 1,000 cycles per second is equal to 1,000 hertz, or one kilohertz (kHz).

Complex waveforms, such as those representing voice, are made up of combinations of many different single-frequency signals, each with a potentially different amplitude. *Bandwidth* is a range of frequencies, usually specified as the number of hertz of the band or the upper and lower limiting frequencies. One can speak of the bandwidth or *spectrum* that various signals occupy, or the bandwidth or spectral frequency response capabilities of transmission systems, high-fidelity amplifiers or other apparatus. For transmission channels or apparatus, a flat spectral response, or one in which all frequencies in a specified band are treated equally, is a desirable characteristic.

Since, in the past, transmission costs were related to bandwidth, analog signals used to provide commercially acceptable quality (toll quality) for telephone communications were normally limited to the range of frequencies in which most of the voice energy occurs—that is, the spectrum between 200 Hz and 3.5 kHz.

The amplitude of an electrical signal is measured in *volts* and referred to as voltage. Within an electrical circuit, the flow of electrons, called electric current, is measured in *amperes*. *Resistance* is a measure of opposition to electrical current by materials or free space, when a potential difference (*voltage*) is applied between two points. According to Ohm's Law, when a constant voltage (non-time-varying) of one volt amplitude is applied to a resistor with a resistance value of one ohm, it produces a steady or constant current of one ampere. For time-varying electrical signals, similar—though more complicated—relationships exist between electrical current, voltage, resistance, and in particular, quantities dubbed *inductive* and *capacitive reactance*. In these cases, opposition to the flow of alternating current (i.e., resistance and reactance) is called *impedance* and assigned the symbol "Z". In the remainder of this book, these simplified definitions and general relationships are sufficient.

Electrical power, measured in watts, is proportional to the product of voltage and current. For non-time-varying voltage and current examples, one watt is dissipated in a one-ohm resistor to which one volt is applied and through which one ampere flows (or more generally, watts = volts × amperes). As a familiar practical example, a music system audio amplifier capable of 100 watts, can produce sounds ten times as loud as a 10-watt amplifier. As will become obvious, power is one of the factors that enters into the determination of overall telecommunications channel or device quality and serviceability.

# Digital Electrical Signals

A *digital* signal is an electrical signal in which information is carried in a limited number of different (two or more) discrete states. The most fundamental and widely used form of digital signals is binary, in which one amplitude condition represents a binary digit 1, and another amplitude condition represents a binary digit 0. Thus a binary *digit*, or *bit*, is one of the members of a set of two in a numeration system that has two—and only two—possible different values or states.

Figure 3.2 shows an example time profile of a binary signal waveform (also known as a *bitstream*), and illustrates the two-level bit structure. The signal corresponds to information generated by *data terminal equipment* (DTE), any device that can act as a data source (transmitter), a data sink (receiver), or both.

Many DTEs format information into 8-bit bytes. (A *byte* is an 8-bit quantity of information also generally referred to as an octet or character.) The number of different bit patterns possible in a byte is 256. Assignment of byte patterns (e.g., various 8-bit sequences of 1s and 0s) to represent specific alphanumeric characters, punctuation marks, control signals, or other signs and symbols enables transmission of information, e.g., the English language, via a process called *binary coding*. The assignment of bit patterns to English letters and other symbols must be agreed upon by both the data source and sink DTEs. The *American Standard Code for Information Interchange* (ASCII), for example, uses 8-bit bytes (7-bit bytes for data encoding and an eighth parity bit for error detection, a process explained later in this chapter).

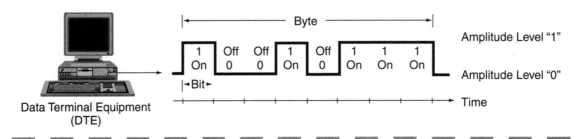

**Figure 3.2** *Digital signal example.*

# Analog Signal-to-Digital Signal Conversion: Encoding and Decoding

Virtually any analog signal can be converted to a digital signal using a codec. A *codec* (contraction of coder and decoder) is a device that transforms (encodes) analog signals into digital signals for transmission through a network in digital format and decodes received digital signals, transforming them back to analog signals. The motivation for analog-to-digital conversion is that digital transmission generally reduces costs and improves communications quality. The analog-to-digital codec process is illustrated in Figure 3.3.

Analog signal-to-digital signal (A-to-D) conversion involves *sampling*, *quantizing*, and *digitizing*. Figure 3.3B shows the process of sampling the amplitude of the analog signal at regular intervals. Figure 3.3B also illustrates quantizing, the replacement of the actual amplitude of the samples with the nearest value from a finite set of specific amplitudes. The samples are then digitized, completing the encoding process.

In binary coding, a binary digital signal is generated with sequences of bits that represent the quantized values of the analog signal amplitude samples. Figure 3.3C shows an example binary number code and the relationship to the decimal equivalents for the sample amplitudes shown in Figure 3.3B. As with the decimal numbering system, the least significant digit is farthest to the right, i.e., bit 1.

In binary format, the number 00000001 is equivalent to the decimal system number 1. The binary equivalent to the decimal number 2 is 00000010. The binary equivalent to the decimal number 4 is 00000100. As can be seen, each movement of digits to the left corresponds to a multiplication by a factor of 2 rather than 10, as in the decimal system. Thus the binary number system is said to be of base 2 whereas the decimal system is said to be of base 10.

**Figure 3.3**
Analog signal-to-digital signal and digital signal-to-analog signal concersion (ADC/DAC).

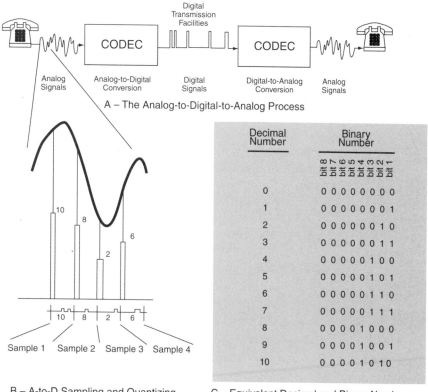

A – The Analog-to-Digital-to-Analog Process

B – A-to-D Sampling and Quantizing

| Decimal Number | Binary Number |
|---|---|
| | bit 8 bit 7 bit 6 bit 5 bit 4 bit 3 bit 2 bit 1 |
| 0 | 0 0 0 0 0 0 0 0 |
| 1 | 0 0 0 0 0 0 0 1 |
| 2 | 0 0 0 0 0 0 1 0 |
| 3 | 0 0 0 0 0 0 1 1 |
| 4 | 0 0 0 0 0 1 0 0 |
| 5 | 0 0 0 0 0 1 0 1 |
| 6 | 0 0 0 0 0 1 1 0 |
| 7 | 0 0 0 0 0 1 1 1 |
| 8 | 0 0 0 0 1 0 0 0 |
| 9 | 0 0 0 0 1 0 0 1 |
| 10 | 0 0 0 0 1 0 1 0 |

C – Equivalent Decimal and Binary Numbers

A-to-D conversion is essential to *pulse code modulation* (PCM). *Modulation* is the process of varying certain parameters of a carrier signal— i.e., a signal suitable for modulation by an information signal—by means of another signal (the modulating or information-bearing signal). For instance, in AM broadcast radio, a radio frequency (RF) carrier signal (at a frequency assigned by the FCC to a particular station, e.g., 630 kHz) is amplitude modulated by an analog voice or music audio electrical signal. In this way, the information in the audio signal can be carried via radio wave propagation from radio transmitters to radio receivers where it is reproduced for listening. Additional description of modulation and carrier transmission systems is provided in Chapter 4.

In our PCM example, the analog modulation signal is sampled, and the sample is quantized and digitized into a defined number of equal-duration binary-coded pulses (bits) representing the quantized amplitude samples of the analog signal. At a receiving point in a telecommunications system, the process is reversed and the original analog signal reconstructed.

Binary-coded numbers of a fixed length (i.e., the quantity of bits—*word length*—used to represent binary numbers) can only represent an analog signal amplitude with a limited degree of precision. This means that no finite PCM process can ever be totally without quantizing noise, the difference between the converted binary value and the actual analog signal's amplitude.

For voice applications, quantizing noise is controllable, and toll-quality voice can be produced for speech signals limited to less than 4 kHz bandwidth, if a sampling rate of 8,000 samples per second and 8 bits per sample (256 quantizing levels) are used. The impact of quantizing noise is further controlled in networks by preventing more than two or three successive PCM encoding/decoding processes from occurring on end-to-end connections.

# Transmission Channels, Circuits, and Capabilities

A *channel* is a single communications path in a transmission medium connecting two or more points in a network, with each path separated by some means; e.g., physical or multiplexed separation, such as frequency or time division multiplexing. (Multiplexing is defined and discussed in Chapter 5.) Channel and circuit are often used interchangeably; however, *circuit* can also describe a physical configuration of equipment that provides a network transmission capability for multiple channels. The characteristics of channels and circuits are determined by the network equipment and media used to support them.

Channels and circuits refer to unidirectional (one-way) paths or bidirectional (two-way) paths between communicating points. A *simplex* circuit is a transmission path capable of transmitting signals in one direction only (e.g., broadcast radio). A *half-duplex* circuit is a bidirectional transmission path capable of transmitting signals in both directions but only in one direction at a time (e.g., citizen band radio). A *full-duplex* circuit is a bidirectional transmission path capable of transmitting signals in both directions simultaneously (e.g., most telephone voice links).

Circuits are classified as either two-wire or four-wire, regardless of whether they use fiber or metallic cable, terrestrial or satellite radio links, infrared or other optical transmission, in atmospheric or free space media. (A description of various transmission media is provided in Chapter 4.)

A *four-wire* circuit uses two sets of one-way transmission paths, one for each direction of transmission. It may be two pairs (four wires) of metallic conductors or equivalent four-wire as in multichannel transmission systems. (Multichannel transmission systems are described in Chapter 5.) Four-wire circuits, or equivalents, are normally used for toll-quality long-distance transmission facilities, where, for technical reasons, unidirectional amplifiers are required. *Two-wire* circuits are normally used for local loops to subscribers due to the enormous wire and cable investment required to serve hundreds of millions of telephones.

Channels and circuits are designed to support either analog or digital signals. Digital channels must be designed so that the sending and receiving terminals interpret, in an identical way, each transmitted bit of information. For example, the receiving terminal must know whether the first bit of a byte received corresponds to the most significant bit or the least significant bit. Two mechanisms have been designed to provide this type of synchronization.

In *asynchronous* transmission, each byte or character is marked with distinctive START and STOP bits so that each of the bits between the START and STOP bits can be interpreted identically by the sending and receiving terminals. In *synchronous* transmission, a means is provided to match up sending and receiving terminals so that for a continuous stream of bytes, characters, or words, the significance of each bit is agreed to on a continuing basis. Synchronous transmission systems eliminate the need to include START and STOP bits in each transmitted byte and, hence, are significantly more efficient. Unfortunately, the words asynchronous and synchronous apply to several different time and frequency domain characteristics of telecommunications systems as explained in Chapter 5. Nevertheless, the interpretation just given, as well as others presented in Chapter 5, are all valid.

## Transmission Capacity

Within an analog network, *bandwidth* is the fundamental signal capacity characteristic that specifies the rate at which information can be exchanged between two points. A voiceband channel is defined as having a 4 kHz bandwidth—although, as noted earlier, in voice-grade analog channels, the speech signal is typically limited to a range of frequencies from 200 Hz to 3.5 kHz. The additional bandwidth allows for a guard band on either side of the speech signal to lessen interference between channels in some multichannel transmission systems. Analog circuits are available with larger bandwidths to support either multiple voiceband channels or special services such as broadcast program or video material.

The capacity characteristic associated with digital signals is *channel rate* or *bit rate*, that is, the number of bits (or bytes) per second that a channel or circuit will support. For example, a transmission facility that can support data exchange at the rate of 1 megabit per second (1 Mb/s or 1,000,000 bits per second), delivers the same quantity of information, i.e., throughput, as a 1 kilobit per second (kb/s or 1,000 bits per second) facility, but in only 1/1000 of the time. Bit-per-second capacity is often referred to as circuit, channel or product operating *speed*. Thus, manufacturers of the newer V.90 modems, discussed later in this chapter, typically describe their products as providing "high-speed" access to the Internet via voice-grade telephone subscriber lines.

Greater bandwidth and higher bit rates use more network resources and, historically, cost more. However, telecommunication service charges are also proportional to the length of time required to complete a voice call, or in the case of data communications the time required to complete an information exchange transaction. Unfortunately, the cost trade-offs between bandwidth, bit rate, and call holding times are neither linear nor simple; and, with the introduction of ever wider bandwidth fiber optic circuits, the relationships are constantly changing. As a result, matching business requirements with telecommunication service options is not straightforward, but involves the use of sophisticated modeling and estimation techniques.

Voice coding schemes that can pack more voice channels into a digital channel of a given bit rate capacity (without noticeably degrading intelligibility or speaker recognition), and compression techniques that remove data redundancy prior to transmission, will generally be more economical. Both capabilities are explained further following a brief discussion of device, transmission and other external factors that limit or degrade quality.

# Transmission and Device Quality

Telecommunications signals are subject to: 1) transmission impairments or degradation caused by practical limitations of channels (e.g., signal-level loss or attenuation, echo, various types of signal distortion, etc.); 2) signal-handling device imperfections; and, 3) interference from outside sources (such as power-line hum or interference from heavy electrical machinery).

Specification of allowable impairment levels, measurement of actual impairment levels in delivered systems, and measured performance in the presence of various types of interference are factors essential to the success of any telecommunications system or service procurement. Signal-to-noise ratio, percent distortion, frequency response, and echo are measurements that relate to impairments most noticeable to users in analog voice systems. Distortion refers to the inability of analog systems to react to weak signals in exactly the same proportion as strong signals. Similarly, frequency response refers to the ability to handle low- and high-frequency signal components equally. Amplitude or frequency-response imperfections corrupt communications and limit the ability to faithfully reproduce transmitted signals. Other signal impairments such as dispersion, cross, and intermodulation distortion are addressed in later sections or defined in the Glossary.

Errors in digital telecommunications systems, and noise or unwanted signal interference in analog systems or devices, are caused by a variety of different phenomena. In electrical circuits, one such ever-present phenomenon, *thermal noise*, comes from electron, atom and molecule movement. Sometimes referred to as *random* or *white noise*, in most communications channels or devices internally generated thermal noise produces a "flat" frequency-spectrum. That is, in a band-limited channel (a channel with finite or limited frequency bandwidth capability), the noise power level at any frequency within the band, on average, is the same as the level at any other frequency in that band. Thermal noise gets its name from the fact that its noise level or power is proportional to temperature.

External interference, especially when it results from the summation or aggregation of many different sources may appear as random or white noise and thus resemble thermal noise. In analog systems, when random noise begins to dominate, the effect can be described as what one hears when tuning a standard AM radio to weak or distant stations. More frequently, external noise is other than random and takes on what is often called a *deterministic* character. One familiar example of deterministic interference occurs on telephone channels when cross-talk among channels makes other, unwanted, conversations audible. As mentioned above, unwanted power-line hum, single-frequency tones, or tonal harmonics are also examples of deterministic interference.

Natural or man-made impulse noise results from high peak-power signal bursts interspersed among longer interference-free intervals. Impulse noise comes from disturbances like lightning, power system spikes, diathermy machines, etc.

In analog or digital systems performance is ultimately limited or determined by the ratio of desired signal power to undesired naturally occurring or man-made noise or interference power with which receivers or signal regenerators must contend. *Signal-to-noise* ratios (SNRs) are most often stated in decibels, or dBs, that is, ten times the logarithm of the ratio of desired signal power to undesired noise power. If the mathematical term "logarithm" is unfamiliar, there is no need for concern. In practical systems SNRs of 10 dB are generally regarded as quite good (at 10 dB, signal power is ten times as large as noise power), 100 dB incredibly good, and 1,000 dB, almost unheard of. Chapter 17 contains examples of how *distance* and *transmission frequency* effects the loss of signal power between radio transmitters and receivers.

In digital systems, *bit error rate* (BER) is the ratio of the number of bits received with errors to the total number of bits transmitted. BER and the average number of error-free seconds are dominant impairment measures for digital channels. A bit error or sequence of errors in channels supporting digitized voice may only be perceived as a "pop" or burst of noise to the listener. Thus, for digital voice applications, a BER of one part in one thousand ($10^{-3}$) yields acceptable results. However, with data transmission relating to financial or other information, even single errors can be catastrophic (e.g., a million dollars erroneously changed to a billion dollars). It is for this reason that error-detection and correction schemes, described in the next subsection, have been designed to enable virtually error-free (at least no undetected errors) data information exchange.

The ability to specify impairment limits and test procedures to assure compliance, as a condition of new-system acceptance, is essential when developing telecommunications system and service request for proposal (RFPs) and contract documents. This subject will be treated in more detail in Chapters 11, 12, and 15. For now, it is sufficient to understand that one reason for making the transition from analog to digital facilities is that analog signals are more susceptible to transmission impairments than are digital signals. Moreover, analog transmission facilities require more-expensive repeaters and amplifiers, more-precise tuning and adjustment, and greater levels of maintenance.

Similar cost savings and performance improvements result from the use of digital switching in lieu of older electromechanical and even electronic analog designs. While 20 years ago, the per-channel cost of PCM conversion (the process that makes possible conversion of network implementation from analog to digital) was high, today it is accomplished with inexpensive, integrated circuits. Therefore, the trend in modern telecommunications systems is to minimize use of analog facilities. Note that the transition to digital switching and transmission facilities within the public switched telephone network is being accomplished in a manner transparent to users and either preserves or improves the integrity and performance of voice-band channels while lowering costs.

Although the conversion to digital technologies was undertaken to improve voice service economic and technical performance, these technologies also facilitate emerging data and other nonvoice services, and the sharing of facilities by those services.

# Error Detection and Correction

Designers have developed two basic strategies for dealing with errors. One scheme appends redundant data (check bits) to original blocks of data in a manner that allows receivers to detect errors when they occur. As noted above, the ASCII uses 8-bit bytes, 7-bit bytes for data encoding and an eighth parity bit for error detection. The *parity bit* is set at the data source so that the total number of binary 1s is odd (odd parity) or even (even parity). Receivers (data sinks) check each received byte for even or odd numbers of binary 1s. By knowing whether the data source is using even or odd parity encoding, the data sink is able to detect all single bit-per-byte errors.

Note that because parity checking produces identical results for either "no actual errors" or some combinations of multiple errors, it only reliably detects single errors. By adding additional redundancy (more check bits), more-complex error detection schemes are able to detect multiple errors. An n-bit unit containing m data or message bits, and r check bits is often referred to as an $n = ( m + r )$ bit codeword. Ensembles of all legitimate or allowable codewords used to enable receivers to deduce that error(s) have occurred, are called *error-detecting codes*.

Error-detecting codes are typically used in systems where *Automatic Retransmission reQuest* (ARQ) messages are sent to data-sources by receivers that detect errors in received codewords. In many cases, receivers return *acknowledgement* (ACK) messages when no errors are detected, and *negative acknowledgement* (NAK) messages when errors are detected. Of course, such arrangements mean that in order for data sources to be able to retransmit codewords arriving at receivers with errors, they must store transmitted data until ACK messages are received.

By adding still more redundancy, error-detection and correcting codes can be created. Using error-correcting codes, receivers are able not only to detect errors but to deduce what the transmitted code must have been, eliminating or greatly reducing the need for retransmission. Such codes are also described as *forward error-detection and correction* (FEDAC) codes.

The use of either error-detecting or correcting coding involves a two-fold penalty. First, for fixed transmission channel or device speeds, adding redundant check-bits reduces effective user-data throughput rates. Second, error-coding techniques require *digital signal processing* (DSP) which adds complexity to both source and sink *data terminal equipment* (DTE).

Since low-cost, high-speed DSP large-scale integrated circuits largely eliminate complexity concerns, in most applications user-data throughput bit rates dominate error-coding technique selection criteria. Shedding light on such trade-offs, in an historical 1948 paper, Claude Shannon proved that the maximum error-free bit-rate that a bandwidth-limited channel can sustain is completely defined in terms of signal power-to-white noise power ratio, its bandwidth, and no other factors. What this means is that in otherwise identical channels, those with higher SNR ratios are able to support higher data-transfer rates. Now known as *Shannon's Law*, this landmark contribution still dominates modern information theory, and its error detection and correction disciplines.

Interestingly, maximum error-free user-data throughput rates are often not obtained from input signals with the lowest "raw" or uncorrected error rates, indicating that selecting optimum error detection and correction techniques can only be accomplished as an integral part of total systems engineering efforts. Other factors such as transmission impairment and noise characteristics also weigh in heavily.

For example, if data are transmitted in blocks of 1,000 bits each, and random noise causes one error in one thousand bits (a $10^{-3}$ bit-error rate), on average, every block of data will contain a single error. Consider next instead of random noise that impulse noise produces bursts of errors 1,000 at a time, but producing the same $10^{-3}$ average bit error rate. In this case, it can be expected that 500 error-free data blocks will be followed by one or two heavily damaged blocks. These two situations illustrate the need for radically different error-mitigation approaches that take into account the type of information and the impact that uncorrected errors have on specific organizational objectives and activities.

While error coding is one of the most arcane technical disciplines encountered in the study of information systems, fortunately, what is key to business and other telecommunications users has just been presented. To recapitulate, what is most important to users is, 1) the knowledge that errors in practical systems are unavoidable, 2) nearly universal operational requirements exist to detect, purge, correct, or otherwise account for errors, 3) recognition of the need for vendor/service provider competence in this arena, and, 4) a rudimentary understanding of the terms introduced and the techniques outlined above as a basis for evaluating alternatives and conversing with the designers and engineers proposing solutions.

## Modems for Transmitting Digital Signals Over Voice Networks

*Modems* (modulator/demodulators) are devices that convert digital signals generated by data terminal equipment (DTEs) to analog signal formats suitable for transmission through the extensive, worldwide connectivity of public and private, switched (dial-up) and non-switched telephone voice networks. Modems are designed to overcome analog network limitations and to support data communications over virtually any channel capable of delivering ordinary telephone service.

Figure 3.4 illustrates the use of modems in public and private voice networks. The pair of modems used on any given connection incorporates compatible modulation/demodulation designs. Figure 3.4 identifies some of the past and current Bell System and ITU (International Telecommunications Union, an international telecommunications standards-setting group) standards supported by modem manufacturers. Appendix A identifies and discusses U.S. and international standards-setting groups.

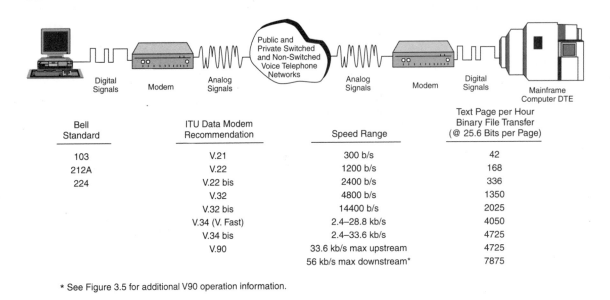

| Bell Standard | ITU Data Modem Recommendation | Speed Range | Text Page per Hour Binary File Transfer (@ 25.6 Bits per Page) |
|---|---|---|---|
| 103 | V.21 | 300 b/s | 42 |
| 212A | V.22 | 1200 b/s | 168 |
| 224 | V.22 bis | 2400 b/s | 336 |
| | V.32 | 4800 b/s | 1350 |
| | V.32 bis | 14400 b/s | 2025 |
| | V.34 (V. Fast) | 2.4–28.8 kb/s | 4050 |
| | V.34 bis | 2.4–33.6 kb/s | 4725 |
| | V.90 | 33.6 kb/s max upstream | 4725 |
| | | 56 kb/s max downstream* | 7875 |

\* See Figure 3.5 for additional V90 operation information.

**Figure 3.4**   *Data communications through voice networks using modems.*

Today's modems not only support a maximum rate in accordance with an international standard, e.g., V.22 bis (bis means the second iteration of the standard) at 2,400 bps, but also work with modems that support only lower-rate standards. Modems can be procured as standalone single-channel devices, as plug-in cards for personal computers, as credit card-sized PCMCIA (Personal Computer Memory Card International Association) cards, or in rack-mounted, multichannel configurations suitable for "modem pooling."

Advances in both design and large-scale integrated-circuit integration have simultaneously increased speed, as evidenced in Figure 3.4, and reduced modem cost. Modems built to the V.90 standard are intended to take advantage of the fact that except for analog subscriber lines to residential or small-business locations, telephone networks already incorporate digital technology on an end-to-end basis. Thus, for example, most Internet service providers (ISPs) connect to telephone networks via high-speed digital local loops. As illustrated in Figure 3.5, in many cases these ISPs use digital V.90 modems to communicate with their subscribers. As a consequence, analog transmission facilities are only encountered at the subscriber-to-telco link, and only one telco *analog-to-digital codec* (ADC) is needed.

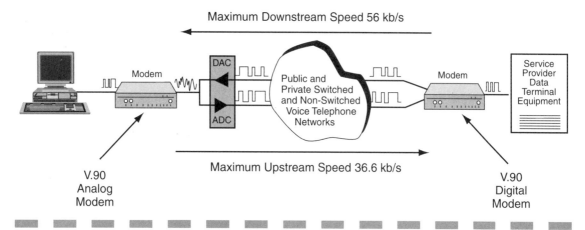

**Figure 3.5** *V.90 Asymmetrical Operation.*

What this means is that performance-limiting quantizing noise only occurs at the telco ADC, that is the ADC in the path from subscriber to ISP—or as it is called, the upstream path. Using the North American analog-to-digital conversion codecs described in the next section yields a maximum signal-to-noise ratio of 38 to 39 dB (the "noise floor"), which limits the upstream path to about V.34 speeds, or 33.6 kb/s. Because the downstream path incurs no analog-to-digital conversion, it can support 56 kb/s speeds.

Interface between standalone modems and DTEs is often via 25-pin connectors specified in the Electronic Industry Association's (EIA) RS-232-C or EIA-232-D standards. The telephone network interface is via RJ-11C or other standard modular telephone plug and jack connections.

Smart modems, used in conjunction with communications software installed in personal computers or other DTEs, execute a wide range of performance-improving and convenience functions. Examples include error control, automatic answering, called-party number storage and retrieval, auto-dialing, modem/terminal/network diagnostics, and data compression.

# Digital Subscriber Line Modems

Perhaps the greatest advantage of modems designed to convert digital signals to analog signals suitable for transmission through telephone channels is that they facilitate end-user-to-end-user data communications *with absolutely no modification* to interconnecting voice networks. There may be no other factor more important to the large-scale and rapid growth of the Internet than this ability to access it from anywhere on the globe with voice network connectivity. And yet, the timeliness and ubiquity of using voice channels for data traffic is not without penalties.

As noted in Chapter 1, because much Internet traffic is bursty in nature, it is rare that more than 20 percent of a voice channel's available capacity is used. Moreover,. data calls tend to be much longer than voice calls. It is not unusual for Internet subscribers to remain connected to local Internet service providers (ISPs) for hours. So although local exchange carriers (LECs) are able to carry data-modem traffic with no channel-design modification, LEC switch and transport facility loading increases significantly. Under fixed, "unlimited-in-time" service tariffs, per-call revenue decreases dramatically.

A second penalty associated with voice channel modem use affects Internet users and ISPs more than LECs. As amazing as it may be that inexpensive V.90 modems can squeeze up to 56 kb/s out of voice channels, even that rate is terribly inadequate for high speed streaming-data voice, video and other Internet traffic now becoming increasingly popular.

*Digital subscriber lines* (DSLs) is a generic name given to a class of digital services that enable users to connect data terminal equipment directly to LEC subscriber lines without voice channel modems. Of course this means that LECs must provide special central-office data service access termination equipment. Ultimately this can eliminate the need to "dial up" data connections. As long as user DTE equipment is powered on, it can transmit or receive data at any time. Another advantage of DSL service is that by eliminating loading coils and branch taps (installed to improve voice-only service) from subscriber loops, DSL modems may offer data rates hundreds and even thousands of times faster than the best V.90 voice-channel modems.

The sophisticated modulation and digital signal processing techniques and hardware technologies that make this "magic" possible are treated in Chapter 4.

# ▮▮ ▮▮ Data-Compression

*Data-compression* is a form of processing that, among other things, 1) eliminates or minimizes irrelevant information and redundant data, and 2) uses simpler characters, symbols, codes, or patterns to represent more frequently occurring data elements than those used for infrequently occurring data elements. More broadly its purpose is to minimize:

- Data storage size requirements
- The speed or bandwidth of communications facilities needed to move data from one location to another
- Complexity or cost of generating and processing data or information.

Material printed on blank business forms is a classic example of irrelevant information that should not needlessly be stored in computer databases or transmitted over communications channels. Instead, only unique or unknown data, perhaps originally entered manually in blank form spaces, should be stored or transmitted. All that is required to obtain such storage and speed requirement reductions is some means to associate data used to "fill in the blanks" with specified data fields in databases or communications signal structures. Of course, upon data retrieval or signal reception, completed forms (presenting both unique and unchanging form data) can be displayed on computer screens or printed.

In other developments, *on-line analytical processing* and modern *data warehousing* technologies are rapidly eliminating redundant data commonly encountered in early database and data communications systems. With low-cost, high-speed processing, it is no longer necessary or cost-effective to store or transmit computed data, along with data used to make the computations.

To illustrate this point, consider this simplified example. In the past, mainframes may have transmitted arrays of length, width, and area (i.e., length times width) data to nonintelligent terminals. Today, only length and width data need be transmitted and if corresponding area data are required, database retrieval or communications receiver system analytical processing capabilities handily generate the results—often with no perceptible delay. In more complex, real-world business management applications, potential savings are huge, as attested to by the large number of research, development, marketing, and sales activities addressing these technologies.

Techniques exploiting the fact that certain data representations occur more frequently than others (e.g., the letters "a" and "e" occur much more frequently in the English language than the letters "x" and "z"), are traceable to David Huffman's 1952 data compression patent. There Huffman discloses how, by assigning shorter bit-length symbols to frequently occurring characters and longer symbols to infrequently occurring characters, data storage and transmission rate requirements can be substantially reduced.

Analogous advantages are possible in still-shot imagery and video data. In large numbers of pictorials, sizable image segments correspond to repetitive representations, for example, walls or background areas of identical color and texture. Instead of transmitting or storing millions of identical picture elements (or *pixels*), advanced algorithms now, in effect, substitute a simple command that until further notice, every pixel is the same as the last one transmitted.

From Huffman's discovery forward, dramatic progress has been made in terms of what can best be described as application-independent lossless data compression technology. By lossless is meant that data recovery or reception processes, while producing storage and transmission economies, are able to reconstruct—without loss of detail or fidelity—all aspects of original, unprocessed stored or transmitted data. (Note that not all useful compression techniques need to be, or are, lossless. For example, less-than-perfect voice reproduction fidelity may be preferable to "blocked calls" under busy network conditions.)

Of course, overall progress must take into account both advantages and disadvantages. For data compression this means balancing economies against increased latency (added delay in data retrieval and transmission), error propagation, and data expansion. By error propagation is meant the exaggerated impact that system errors have on compressed data versus non data-compressed data. To mitigate this, error-detecting and correcting codes are used in conjunction with data compression.

For real-time voice and video transmission, excessive transmission delays may be intolerable. Latency is directly proportional to data compression algorithm and protocol complexity, and inversely proportional to processing speed. As a consequence, for real-time applications, efficient data compression algorithms and high-speed processing are crucial.

Data expansion, as opposed to data compression, can result if, for example, long symbols are erroneously substituted for the most frequently occurring data representations, or in general, when data-compression algorithms are mismatched to actual data. To circumvent this difficulty, a class of self-learning algorithms is emerging with abilities automatically to provide high-performance data compression for a variety of data types and representation statistics.

ITU's V.42bis standard specifies a state-of-the-art algorithm and is implemented in nearly all new modems. On average, it compresses data by a factor of 3.5 to 1. It avoids data expansion by continuously monitoring compression performance and switching it off if it degrades rather than increases speed. V.42bis also eliminates the danger of error propagation by retransmitting data damaged in transit. When it is used with V.34 modems designed to support 28.8 kb/s without compression, effective transmission speeds of 115.6 kb/s can normally be realized. V.42bis also supports Microcom's MNP 4 error correction and MNP 5 data-compression methods and other de facto industry standards used prior to V.42's approval. Since high-speed modems with data compression are now priced below $100, and the cost of voice-grade channels is independent of which modem is being used, the higher throughput and faster response time benefits justify the investment in nearly all applications.

Current microprocessors meet V.42-like algorithm requirements for serial text traffic at data rates corresponding to high-speed voice modems, that is, rates in the 56 to 128 kb/s range, but they are not adequate for the great variety of Mb/s to Gbps broadband applications and transmission channels found in today's networks. The next section of this chapter describes popular speech and audio processing alternatives designed to optimize fidelity while minimizing file size, transmitted data rates, and algorithm and processing complexity. The final section covers the same territory for image and video applications.

# Alternative Voice Coding, Speech, and Audio Processing Techniques

The voice coding PCM technique described above produces 64 kb/s digital signals—i.e., 8,000 samples per second times 8 bits per sample equals 64,000 bits per second. The specific implementation used in North America and Japan has been designated as μ-law (μ=255) PCM, and uses a process called companding to enhance the signal-to-quantizing noise, particularly for analog signals at small amplitudes.

In Europe, the European Conference of Posts and Telecommunications (CEPT) has standardized on a 64-kb/s PCM technique that uses a slightly different companding method and has been designated as A-law PCM. Both techniques result in high-quality speech and are widely used in telecommunications systems throughout the world. Experience has proved PCM-based voice and data (modem) traffic quality superior to that of most analog techniques. This is essential to the growth of digital networks, since, to date, neither LEC nor IXC facilities are designed to detect whether voice or data traffic is being carried, precluding the possibility of special treatment for data traffic.

From a private-line user's point of view, economics often dictates the use of lower than 64-kb/s bit-rate voice processing techniques. Low bit-rate or narrowband voice processing permits more voice channels to be transmitted over digital channels of given bit rate or throughput capacity. Narrowband voice is accomplished by using microcomputer or specialized hardware-based processing power to remove redundancies inherent in speech signals, thereby producing digitized voice signals at less than 64-kb/s.

One such technique that has achieved American National Standards Institute (ANSI) standard status is *adaptive differential PCM* (ADPCM). ADPCM requires only 32 kb/s to digitize analog voice signals and produces voice quality that is not perceptibly lower than that of 64 kb/s PCM digitization.

The widespread application of ADPCM within public networks has yet to occur, as it requires special administration because of link compatibility, signaling provisions, data limitations, restrictions on the number of tandem coding/decoding points, and other problems.

*Digital speech interpolation* (DSI) can result in further savings when used in conjunction with either PCM 64 kb/s or ADPCM 32 kb/s codecs. DSI takes advantage of the fact that in normal conversation, "quiet periods" permit nearly doubling the number of voice signals that can be accommodated if the equipment serving a large number of channels allocates channels only when a talker is active. DSI is a proven technology. An analog form called *time-assigned speech interpolation* (TASI) was used for years to enhance the efficiency of analog transoceanic voice communications.

Other voice processing techniques have been developed that produce intelligible, but not toll-quality, voice at rates as low as 2,400 bps. PCM and ADPCM are best described as waveform-encoding speech processing techniques. Although most narrowband voice processing techniques are referred to as voice-coding techniques (vocoding for short), often the term voice coding will be used to describe any process that converts analog voice signals to digital formats.

Narrowband voice coders typically use analysis and synthesis processes that model a talker's pharynx, larynx, mouth, and tongue (the voice tract). Transmitters then need only send slowly-varying model parameters, whether short "talk-spurts" (one to two-tenths of a second speech segments) contain unvoiced sounds (fricatives, i.e., consonants like "f," or "s"), voiced sounds (pitch-driven, e.g., vowels), and—if voiced—pitch information.

At receiving stations, electronic filters modeling the talker's voice tract are "excited" with locally generated signals. For example, during voiced talk-spurt intervals, short impulses at a rate dictated by the pitch information are used to excite the filter. During unvoiced intervals, the filter is excited with flat-spectrum or noise-like signals. More recent and more complex voice processors combine waveform encoding with voice encoding to produce near toll quality, but at output data rates substantially below PCM or ADPCM rates.

The National Security Agency has developed and produced several hundred thousand STU-III secure telephones that protect the highest levels of classified government information. STU-IIIs use a narrowband voice processing algorithm known as *linear predictive coding* (LPC) and produce digitized voice at 2,400 b/s. Newer models use a technique known as *code excited linear predictive coding* (CELP), which greatly increases voice quality and produces digitized voice at 4,800 b/s. The STU-IIIs include modems so that they can be used over virtually any channel capable of supporting ordinary voice service.

LPC and CELP can be used in nonsecure environments to achieve cost savings beyond what is possible with ADPCM in applications where good intelligibility without toll quality is sufficient. Although the $2,000 cost per unit for STU-IIIs is nearly an order of magnitude less than its $20,000 predecessors, it still represents a sizable cost differential relative to non-secure telephone equipment. Even so, for non-secure applications, systems built to "stuff" 24 narrowband voice signals in a single channel intended to carry just one 64-kb/s PCM digital voice signal can produce enormous savings. Note that well-designed arrangements would not require replacement of all station equipment but rather would only require STU-IIIs for trunks, a reduction by a factor of 10 or more relative to the number of stations.

The need to protect sensitive technological and economic information (versus military secrets, as distinguished in the 1984 Computer Security Act) might yet generate a large requirement for secure telephones in both nonmilitary government agencies and industry. Should this occur, there might be an opportunity to achieve information security and cost savings simultaneously.

Figure 3.6 identifies popular speech processing techniques, corresponding output bit rates, and as appropriate, applicable standards. The top three techniques are of the waveform-encoding variety. The rest all incorporate some form of voice-encoding or hybrids. Chapter 16 demonstrates how low and variable-rate voice encoding techniques can increase the maximum numbers of users that a single CDMS-based (IS-95 standards conforming) cellular telephone system can support—by 100 percent. For now it is important to understand that various techniques exist, why and for what applications they are needed, and some practical implications of their use.

**Figure 3.6**
Comparison of speech processing/voice coding techniques.

| Speech Processing Technique | Acronym Exposition | Codec Output Bit Rate | Standard/ Standards Organization* |
|---|---|---|---|
| PCM | Pulse Code Modulation | 64 kb/s | G.711/ITU |
| ADPCM | Adaptive Differential PCM | 32 kb/s | G.721/ITU |
| CVSD | Continuously Variable Slope Delta Modulation | 32 or 16 kb/s | — |
| LD-CELP | Low Delay Code-Excited Linear Prediction | 16 kb/s | — |
| RPE-LTP | Regular Pulse-Excited LPC with Long Term Predictor | 13 kb/s | GSM |
| VSELP | Vector-Sum-Excited Linear Prediction | 8 kb/s | Interim IS-54/ EIA-TIA-CTIA |
| CS-ACELP | Conjugate-Structure Algebraic-Code-Excited Linear Prediction | 8 kb/s | G.729/ITU |
| CELP | Code-Excited Linear Prediction | 4.8 kb/s | FS 1016/US DoD |
| QCELP | Qualcomm Code-Excited Linear Prediction | 4 kb/s | IS-96/TIA |
| LPC | Linear Predictive Coding | 2.4 kb/s | LPC-10/US Government |

*See Glossary for Acronym and Abbreviation Exposition*

Algorithm complexity and processing power are normally inversely proportional to output bit rate. PCM codecs, for instance, are relatively simple and can be implemented on inexpensive single integrated circuit chips but produce relatively high output bit rates. In contrast, very low bit-rate codecs demand processors with capabilities in excess of one million instructions per second (MIP).

All Figure 3.6 techniques are designed for voice telephony, as opposed to music or other audio applications. For voice telephony, speaker recognition and intelligibility are key quality performance factors. With music, key performance factors include high-fidelity sound reproduction, the ability to handle very soft and very loud sounds without distortion (large dynamic range), and from two to five simultaneous and independent channels for stereo and surround-sound systems.

Setting a very high-quality standard, the digital audio coding used on compact discs (16-bit PCM) yields a total range of 96 dB, accommodating loud sounds four billion times louder than the quietest sound or the noise floor. Taking 16-bit samples 44,100 times per second creates an output digital rate of 705.6 kb/s for each channel, an amount of data often too large to store or transmit economically, especially when multiple channels are required. As a result, for music-like applications, new forms of digital audio coding—often known as perceptual coding—are designed to permit the use of lower data rates with a minimum of perceived degradation of sound quality.

MP3, which stands for *Motion Pictures Expert Group* (MPEG), audio layer 3, is one such coding scheme. MP3 uses perceptual audio coding compression to remove superfluous information (that is, the redundant and irrelevant parts of sound signals the human ear doesn't hear anyway), and, in general, to take advantage of psycho-acoustic phenomena known as acoustic masking. The practical result is that MP3 shrinks the original sound data from a CD (with a bit rate of 1411.2 kilobits per second of stereo music) by a factor of 12 (down to 112—128 kb/s) without sacrificing sound quality.

Popular Internet music-quality audio processing and file formats include Audio Format (AU), Waveform Audio (WAV), and Audio Interchange File Format (AIFF). Also, broadcast High Definition Television (HDTV) is slated to adopt a variant of MPEG-2 for video compression and Dolby AC-3, a five-channel voice-compression capability.

# Alternative Video, Graphics, and Image Processing Techniques

MPEG, an acronym for a family of hardware and software compression standards and file formats, is designed to reduce digital video storage and transmission speed requirements. The acronym stands for Motion Picture Experts Group, pronounced "m-peg." MPEG is a working group of the International Organization for Standardization (ISO). MPEG achieves high compression rates by storing only changes from one frame to another, instead of each entire frame. Since some data are removed, MPEG is a lossy compression technique, but picture-quality degradation is generally imperceptible to the human eye.

MPEG-1 generates video frames with a resolution of 352-by-240 pixels at 30 frames per second (fps), quality slightly beneath conventional VCR levels. Resolution refers to image sharpness and clarity and signifies the number of distinct dots (pixels, or picture elements) that computer monitors, printers, scanners, or other devices are able to display, print, or detect.

MPEG-2, a newer standard, offers 720-by-480 and 1280-by-720 pixel resolution at 60 fps, with full CD-quality audio, performance levels meeting all major TV standards including NTSC (National Television System Committee), and as noted, even HDTV. Video for Windows (Microsoft) and QuickTime (Apple Computer) are competing "software-only" video compression schemes. *Digital Video Interactive* (DVI) is a "hardware-only" technique developed by General Electric. *Indeo*, a software-only product from Intel for computer-based video, emulates DVI. Currently ISO is working on MPEG-4 (there is no MPEG-3) that will be based on QuickTime's file format.

JPEG, for *Joint Photographic Experts Group*, and pronounced "jay-peg," is a compression technique for color images. Although it reduces file sizes by up to 95 percent, some detail may be lost. GIF, which stands for *Graphics Interchange Format*, is a lossless method of compression. Since lossless algorithms do not discard non-redundant image elements, maximum compression ratios depend on the amount of redundancy or repetition contained in an image. Whereas consistent color areas may compress to as little as 10 percent of original, for complex, non-repetitive images, GIF may only produce 20 percent savings.

Explosive Internet subscriber growth, and voice, image, and video-augmented Web site popularity, has created unprecedented interest in and need for both ultra-high-quality and ultra-high-efficiency compression technologies. Exacting requirements such as those for remote medical image interpretation collaboration and downloading CD-quality music, are already driving the need for ultra-high image resolution and audio quality file transfers. Fortunately these requirements can usually be met with file transfer rates in the order of minutes or tens of minutes.

At the other end of the spectrum are requirements where lower quality must be accepted to attain real-time voice and video performance. Example applications include voice-over-Internet two-party or multiparty conferencing voice communications, as well as n-party voice and video broadcast and/or teleconferencing, to support seminar or other distance-learning needs.

CHAPTER 4

# Transmission
# System Concepts

Building on definitions and basic concepts presented thus far, the next four chapters treat transmission, multiplexing, circuit-switching, and packet-switching concepts in substantial detail. As noted in Chapter 1, transmission facilities provide communication paths that carry user and network control information between nodes in a network. In general, transmission facilities consist of a medium and various types of electronic equipment located at points along transmission routes. This equipment amplifies or regenerates signals, provides termination functions at points where transmission facilities connect to switching systems, and often includes means to combine many separate sets of call information into single multiplexed signals to enhance transmission efficiency.

# Transmission Media

A *transmission medium* is any material substance or free space, (i.e., a vacuum) that can be used for the propagation of suitable signals, usually in the form of electromagnetic (including lightwaves) or acoustic waves, from one point to another; unguided in the case of free space or gaseous media, or guided by a boundary of material substance. *Guided media,* including paired metallic wire cable, coaxial cable, and fiber optic cable, constrain electromagnetic or acoustical waves within boundaries established by their physical construction. *Unguided media* are those in which boundary effects between free space and material substances are absent. The free space medium may include a gas or vapor. Unguided media including the atmosphere and outer space support terrestrial and satellite radio and optical transmission. This section provides brief descriptions of and defines terms associated with prevalent media. Later sections discuss how each medium relates to and supports various telecommunications services.

## Guided Media

For many decades, the most common medium supporting voice applications within residential and business premises (i.e., within the local loop) has been copper *unshielded twisted pair* (UTP) cables. UTP, illustrated in Figure 4.1, is basically two wood-pulp or plastic-insulated copper wires (conductors), twisted together into a pair. The *twists,* or *lays,* are varied in length to reduce the potential for signal interference between pairs. Wire sizes range from 26 to 19 AWG (American Wire Gauge, i.e., 0.016 to 0.036 inch in diameter), and are manufactured in cables consisting of 2 to 3,600 pairs. A *cable* is a group of metallic conductors or optical fibers that are bound together, with a protective sheath, a strength member and insulation between individual conductors/fibers, and contained within a jacket for the entire group.

**Figure 4.1**
Unshielded twisted pair (UTP) cable.

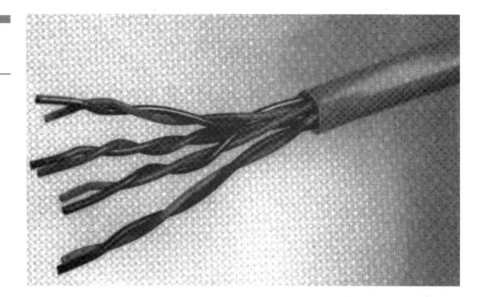

In electrical circuits, a *conductor* is any material that readily permits a flow of electrons (electrical current). In twisted pair wire, the electrical signal wave propagates from the sending end to the receiving end in the dielectric material (insulation) between the two conductors. Due to the finite conductivity of copper, the medium for guided wave transmission is fundamentally dispersive. With *dispersion,* complex signals are distorted because the various frequency components that make up the signals are affected differently by a medium's propagation characteristics and paths.

Dispersion limits the upper bit rate that a medium can support by distorting signal waveforms to the extent that transitions from one information state to another cannot be reliably detected by receiving equipment (e.g., logical 1 to logical 0 value changes). Graphical illustrations of the effects of dispersion are included in the optical fiber medium discussion below. The impact of dispersion on signals is similar in all media.

*Shielded twisted pair* (STP) cable is similar to UTP, but the twisted pairs are surrounded by an additional metallic sheath before being clad with an insulating jacket.

*Coaxial cable,* shown in Figure 4.2, consists of an insulated central copper or aluminum conductor surrounded by an outer metallic sheath that is clad with an insulating jacket. The outer sheath consists of copper tubing or braid. Coaxial cable with solid metallic outer sheaths reduces leakage of signals relative to braid-type designs. Because of its strength characteristics, cable-television distribution systems normally use aluminum coaxial cable with solid outer sheaths to minimize radio frequency interference (RFI) with aircraft navigation and other life-safety systems.

**Figure 4.2**
Coaxial cable.

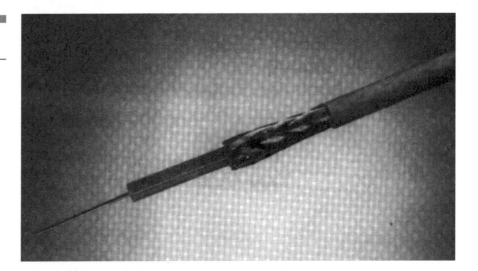

Cables and other media differ in the following ways:

- Bandwidth capabilities (for example the number of voice conversations that can be supported per circuit).
- Susceptibility to electrical interference from other communications circuits or from unrelated electrical machinery or natural sources such as lightning.
- The ability to handle either analog or digital signals.
- Cost.

With UTP, one pair is commonly used for each voice conversation. However, with multiplexing schemes to be described later, UTP has been adapted to support 24 or more voice-grade analog signals per pair. Coaxial cable, on the other hand, exhibits useful bandwidths in the hundreds of MHz (1 MHz = 1,000,000 hertz or cycles per second = one megahertz), and certain metallic cables are able to transmit several thousand voice channels. Because it is considerably more expensive per foot than UTP, applications for coaxial cable are generally those requiring large bandwidth such as multiple-channel voice, cable television, image transfer, and high-speed data networks.

Until the *Institute of Electrical and Electronic Engineers* (IEEE) established its *10BaseT* specification variant to its IEEE 802.3 local area network (LAN) standard (see Chapters 7, 9, and 13), UTP had not played a major role in the support of digital signals or data communications. Now UTP cabling that complies with 10BaseT can support error-free transmission at rates over 100 Mb/s over distances up to 300 feet. This accomplishment represents a breakthrough for premises wiring in that a single type of low-cost cabling and modular connectors now supports voice, data, and some video/imagery applications. Chapter 9 describes modern approaches and standards applying to premises wiring systems.

*Optical fibers* are composed of concentric cylinders made of dielectric materials (i.e., nonmetallic materials that do not conduct electricity). At the center is a core comprising the glass or plastic strand or fiber in which lightwaves travel. Cladding surrounds the core and is itself enclosed in a light-absorbing jacket that prevents interference among multi-fiber cables. Figure 4.3 illustrates multi-fiber cable that can be purchased with between two and 144 fibers.

Today, using optical fiber media requires that electrical signals be converted to light signals for transmission through hair-thin strands of glass or plastic to light-sensitive receivers, where light signals are converted back to electrical signals. *Light emitting diodes* (LEDs) and *light amplification by stimulated emission of radiation* devices (lasers) are two solid-state or semiconductor conversion/transmitter technologies used with fiber optic cables. *PIN diodes* and *avalanche photo diodes* (APDs—see the Glossary for definitions) receive or detect lightwave signals and convert them to electrical signals. APDs are capable of amplifying light and are used in receivers requiring high light sensitivity. Lasers and APDs are used in long-length cable runs and broad bandwidth applications.

Optical fibers are either single mode or multimode. A *single-mode, step index* fiber optic cable is depicted in Figure 4.4. The center portion of the figure shows a cable segment with perhaps an electrical signal-to-light signal converter/transmitter on the left, and a receiver/detector, perhaps containing an APD that also functions as a light-to-electrical signal converter/amplifier, on the right.

**Figure 4.3**
Fiber optic cable.

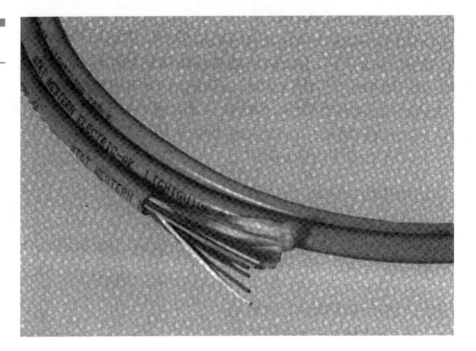

Single-mode fibers have sufficiently small core diameters that electromagnetic waves (lightwaves) are constrained to travel in only one transverse path from transmitter to receiver. This requires the utmost in angular alignment of light-emitting devices at points where light enters the fiber, and it results in higher transmitter/receiver costs than for multimode fiber systems. The figure shows an example single-mode cable consisting of a 9-micrometer (a micrometer or micron is one millionth of a meter and a meter is equal to 39.37 inches) diameter fiber core surrounded by a 125 micrometer cladding. Micrometer is often written as μmeter or simply μm and referred to as micron.

The word "index" is shorthand for index of refraction which is defined as the ratio of the speed of light in a vacuum to the speed of light in a material. At the boundary between two materials, for example air and water, differences in the index of refraction of the two materials cause incident lightwaves to bend or be reflected at the boundary. Fiber cables can be manufactured to exhibit a constant index of refraction from core centers to core extremities, or treated to be variable. A *step index* optical cable exhibits a uniform refractive index across the core and a sharp decrease at the core-cladding boundary as shown in Figure 4.4.

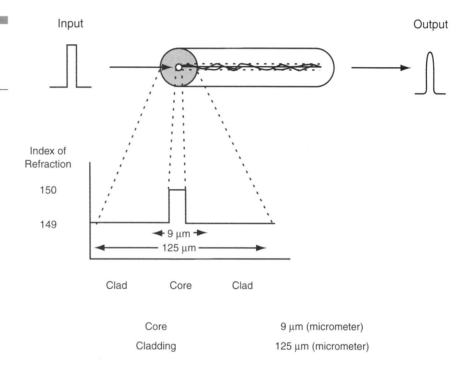

**Figure 4.4**
Step index single-mode fiber optic cable.

*Multimode cables,* with much wider fiber cores, allow lightwaves to enter at various angles, and reflect off core-cladding boundaries as light propagates from transmitter to receiver. A step-index, multi-mode fiber optic cable is shown in Figure 4.5. This figure shows the construction a typical multimode cable to consist of a 100-µm fiber core surrounded by a 140-µm cladding.

In this figure, angular lines within the core represent various paths (modes) that lightwaves may take while traversing from end to end. With a step index design, the speed of light is uniform throughout the core. Note that light rays traversing cores straight down the axis travel shorter distances than those that bounce off core-clad boundaries. As a consequence, at the receiving end, several replicas of the input signal arrive in time-delayed sequence, the axial component arriving first.

As noted in Chapter 3, complex signals, like the impulse function shown in Figure 4.5, are composed of many frequency components. In multimode cable, when the highest frequency components arrive (via different modes and therefore different paths lengths) with differential delays equal to one-half wavelength[1], (that is completely out of phase) they tend to cancel each other out. When this occurs, the high-frequency signal components that create steep or rapid rise-time pulse leading and trailing edges (as shown in the figure), are attenuated and output pulse waveforms become rounded and flattened. In fiber optic cable this phenomenon is called *mode dispersion* but the effects of dispersion in any media previously mentioned produce similar effects.

**Figure 4.5**
Step index multimode fiber optic cable.

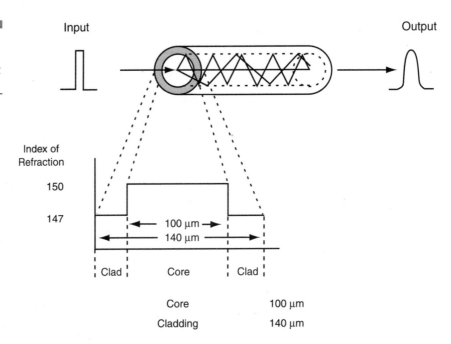

*Chromatic* or *material dispersion* in fiber cables produces the same pulse-flattening/bandwidth-limiting degradation, but is caused when the cable materials affect input signal frequencies or wavelengths differently. Finally, *structural dispersion* in fiber cables is attributable to unequal fiber-material structural alterations caused by different signal wavelength components. Important to the broadband and long-distance networks being deployed today is the fact that single-mode fibers, which are free of mode dispersion, can be designed so that

---

1 Chapter 5 defines wavelength and the relationship between frequency and wavelength.

structural and chromatic dispersion cancel each other out. Such designs are commonly referred to as *dispersion shifted.*

As noted, dispersion in any medium reduces the maximum bit rate at which receiving pulse detectors are able to distinguish digital-signal state transitions, that is changes between logical "1" and "0" signal representations. In other words, dispersion limits maximum useful upper-channel bandwidth and/or practical cable-run lengths.

To minimize dispersive effects in multimode fiber cable, cores are manufactured with *graded* rather than *step* index properties. Figure 4.6 illustrates such a cable. Here the index of refraction, as one moves outward from the axis, is varied so that the speed of light is greatest at the outer extremity. For this case, axial light rays travel slower than those that bounce off core-clad boundaries. As a result, although direct mode paths are shorter, multimode path delays tend to be more equal and the equivalent bandwidth or high bit-rate cable capacity greater than similarly constructed step-index fibers.

**Figure 4.6**
Graded index multimode fiber optic cable.

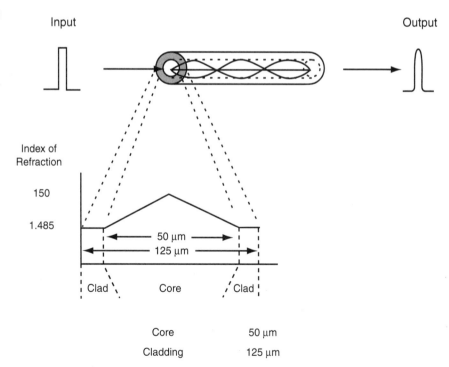

Another approach mitigating dispersion effects employs specially designed signal pulses called solitons. Soliton transmitters generate pulses with shape and power characteristics to compensate for fiber dispersion—thus greatly extending the distance between amplifiers beyond what would otherwise be needed. Solitons are considered an advanced fiber transmission technology but the theory behind them was first noted by John Scott Russell in 1834 while he was observing barges plying the Edinburg-Glasgow canal in Scotland. By combining both amplitude modulation (AM) and frequency modulation (FM), significantly higher digital data rates can be achieved. Combined with dense wavelength division multiplexing, this yields incredible single-fiber traffic capacities.

The principal advantages of lightwave transmission using fiber optic cable (also referred to as lightguide) are ultra-wide bandwidth, small size and weight, low attenuation relative to comparable metallic media (attenuation is signal level or amplitude loss per foot of length), and virtual immunity to interference from electrical machinery and man-made or natural atmospheric electrical disturbances. In particular, fiber cables employing *dense wavelength division multiplexing* (DWDM—discussed in the Chapter 5) far exceed metallic cable channel capacity, often by many orders of magnitude (multiples of 10).

Disadvantages include the added cost for electro-optical transmitters and detectors, higher termination costs (largely manpower for making physical connections and splices), and overall higher installed cost in short-distance applications such as premises wiring systems.

Advantages outweigh the disadvantages, however, and fiber optic cable is rapidly becoming the transmission medium of choice in applications such as high-capacity multichannel metropolitan and wide area interoffice trunks, transoceanic cables, and high-speed data communications. In addition, decreasing costs have resulted in fiber optic penetration into the realm of premises wiring. This is occurring in enterprises that require wide bandwidth communications among campus buildings, and/or directly to desktops for specialized applications, such as image file transfer. Reflecting this development, Lucent offers a composite cable that includes eight metallic twisted pairs (UTP) and two fiber strands within a common outer sheath. Such composite or hybrid cable is intended for information outlets within offices serving desktop voice and data terminals.

Moreover, although fiber installation and connection costs were initially greater, for most multichannel applications the reverse is now true. For example, coaxial transoceanic cables installed in the 1950s exhibited a $40,000.00, 20-year cost per circuit. Today's fiber transoceanic cables cost less than $40.00 per circuit and could soon be less than $4.00 per circuit.

One factor contributing to such economies is that viable photonic technologies are emerging that permit cost-competitive signal processing, filtering, switching, and multiplexing to occur in optical-versus-electrical signal domains (see Chapter 8). One such technology is the *erbium-doped fiber amplifier* (EDFA). As noted, signal loss (i.e., attenuation) and dispersion set limits on the maximum distance between transmitting and receiving network nodes. To extend these distances, signal amplifiers or regenerators, sometimes called repeaters, are required. In early fiber cable installations, these repeaters converted optical signals to electrical signals for amplification and back to optical for retransmission.

Today EDFAs provide in-line amplification of signals in optical form and can amplify several independent (wavelength division multiplexed) signals with virtually no mutual interference and negligible crosstalk. Whereas electrical signal amplifiers require electrical power, EDFA amplifiers consist of 10 to 100 meters of optical fiber with erbium-doped cores (erbium is a rare earth element), and optical *pumping* signals coupled to erbium doped segments. Pumping causes erbium atoms to transfer energy through stimulated emission and, in the process, amplifies traffic signals. Besides being amazingly simple, EDFA amplifiers can typically be spaced 10 to 100 times as far apart as conventional repeaters.

EDFAs were installed in the America One and Columbus II transoceanic cables in 1994. TAT 12/13 and TCP-5 followed with erbium-doped fiber amplifiers, now the dominant technology in both long-haul continental U.S. and transoceanic optical transmission systems. Driven in all markets by "bandwidth-hungry" new Internet and intranet voice, data, video, imagery, and other applications, trans-Atlantic traffic alone is expected to grow at about 80 percent per year.

Traditionally, transoceanic cables were the purview of monopolistic telcos that financed undersea cables for their own use. However, responding to market projections and profit potential, Fiber Line Around the Globe (FLAG), an independent submarine cable venture, began offering service to multiple carriers in 1997. Currently FLAG provides 10 Gbps of capacity to Sprint, AT&T, Cable and Wireless, China Telecom, Deutsche Telekom, MCI WorldCom, and others. Global Crossing and CTR Group (Project Oxygen Network) are two other similar ventures, which in combination with FLAG constitute about $14 billion in recent transoceanic cable investment providing several hundred landing points in hundreds of countries.

At the end of 1998, in the U.S. alone, interexchange carriers had installed 3.6 million miles of long-haul fiber transmission. Incumbent local exchange carriers had 16 million miles of fiber in the ground and their competitive local exchange counterparts had installed 3 million miles of fiber. The optical equipment market was $100 million in 1999 and some analysts expect it to top $5 billion by 2003.

With DWDM, single fibers can carry traffic equivalent to peak voice loads of AT&T, MCI, and Sprint networks combined and may soon support an incredible 80 Gbps. It is difficult to overstate the importance of fiber optic and related developments. They are watershed technologies and, in the opinion of some experts, as important to modern information systems as the invention of the transistor.

## Unguided Media

*Unguided media,* the atmosphere and outer space, are used in terrestrial microwave radio transmission systems, satellite, mobile telephone, and personal communications systems. *Terrestrial microwave radio transmission systems* consist of at least two radio transmitter/receivers (transceivers) connected to high-gain antennas (directional antennas that concentrate electromagnetic or radio wave energy in narrow beams) focused in pairs on each other, as illustrated in Figure 4.7. The operation is point to point; that is, communications can be established between only two installations. This is contrasted to point-to-multipoint systems, such as broadcast or citizen band radio.

In long-distance carrier applications, terrestrial microwave is an alternative to guided metallic or fiber optic cable transmission media. For this application, antennas are normally mounted on towers and require an unobstructed or line-of-sight path between the antennas, which typically can be separated by up to 30 miles. Strings of intermediate or relay towers, each with at least two antennas and repeater/transceivers to detect and amplify signals, interconnect switching centers in different metropolitan locations. By 1980, AT&T had installed 500,000 miles of terrestrial microwave facilities. While not much growth in mileage has occurred since then, significant progress has been made in conversion from analog to digital microwave transmission technologies.

The term microwave is often taken, somewhat arbitrarily, to mean frequencies above 890 MHz. The telecommunications industry use of microwave radio propagation usually refers to operating frequencies above 2.56 GHz. The FCC has assigned frequencies in the 4-, 6-, and 11-GHz bands for long-haul telecommunications common-carrier use. Microwave radio, operating in the FCC-allocated 18- and 23-GHz bands, is now a popular short-haul transmission medium within private networks. Private microwave terrestrial applications include high-capacity transmission between privately owned switches and direct multichannel access to exchange carrier networks.

**Figure 4.7**
Terrestrial
microwave.

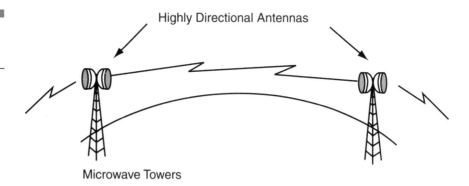

Highly Directional Antennas

Microwave Towers

In the latter application, private microwave transmission is used to connect CPE directly to carrier facilities in lieu of LEC-provided alternatives. *Digital termination systems* (DTS) provided by some carriers permits operators of private networks to use digital microwave equipment to access carrier networks. Private microwave equipment constitutes a fast-growing industry segment. Depending upon capacity and other capabilities, private point-to-point microwave sites range in cost from $15,000 to $150,000.

One advantage of terrestrial microwave is that it obviates the need to acquire right of way (except for towers) or to bury cable or construct aerial facilities. In areas where the cost of right of way is very high, or, as is true in some rural areas, where cable installation costs might be prohibitive, significant savings can be realized with microwave. However, as a transmission medium, the earth's atmosphere creates problems not encountered with other media. For example, trees, obstructions, heavy ground fog, rain, and very cold air over warm terrain can cause significant attenuation or signal power loss. Decades of experience have led to conservative tower spacing, space and frequency diversity (alternate physical paths and frequencies), and other engineering practices to offset these difficulties and achieve reliable operation. Another problem associated with microwave transmission is the possible inability to obtain operating licenses due to dense usage and the associated frequency congestion in a number of metropolitan areas.

Commercial *satellite communications* entails microwave radio, line-of-sight propagation from a transmitting earth terminal (usually ground based, but also shipborne or airborne) through free space (the atmosphere and outer space) media to a satellite, and back again to earthbound receiving terminals. In essence, satellites are equivalent to orbiting microwave repeaters. In addition to signal repeating functions, some communications satellites provide signal processing and switching capabilities.

Terminals, sometimes termed earth stations, consist of antennas and electronics necessary to:

- Interface satellite equipment with terrestrial systems.
- Modulate and demodulate radio frequency (RF) carrier signals with multiple (multiplexed) voice and data signals.
- Transmit and receive RF carrier signals to and from satellites.
- Otherwise establish, support, and control communications among earth terminals.

Figure 4.8 illustrates the satellite network topology.

**Figure 4.8**
Satellite network topoligy.

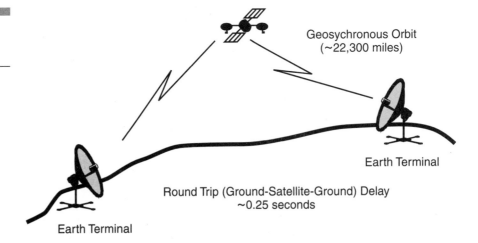

Geosychronous Orbit
(~22,300 miles)

Earth Terminal

Round Trip (Ground-Satellite-Ground) Delay
~0.25 seconds

Earth Terminal

The first satellites were Russian Sputniks 1 and 2 launched respectively on October 3 and November 4, 1957. The first U.S. satellite was Explorer 01 launched February 1, 1958. The first telecommunications satellite was the U.S. Score, launched December 18, 1958. The four principal satellite mission areas are scientific, earth observation, telecommunications, and navigation. Satellites are also classified as either geosynchronous, low earth-orbiting (LEO), medium earth orbiting, or elliptical, in accordance with their orbital trajectories and distance from the earth.

Most commercial satellites operate in three frequency bands: C band (4/6 GHz), Ku band (11/14 GHz), and Ka band (20/30 GHz). In each case, the lower-frequency band is used for downlink (satellite-to-earth terminal) transmission and the higher-frequency band for uplink transmission. GE Americom's GE-4 satellite has twenty-four 36-MHz, 110-watt Ku-band transponders, four extended 72-MHz Ku-band transponders, and twenty-four 36-MHz C-band transponders. A geosynchronous satellite located at 110° western latitude and employing fully switchable transponders, GE-4 provides 50-state, Canadian, Mexican, Caribbean, and Central American two-way communications coverage. For high-capacity C-band communications, antennas are typically parabolic in shape and 10 meters (about 33 feet) in diameter. Ku- and Ka-band antennas can be as small as 1.5 and 1.0 meters (5 and 3 feet), respectively.

Since 1964, approximately 1,000 telecommunications satellites have been placed in geosynchronous orbit (a circular orbit 22,300 miles above the equator). Satellites in *geosynchronous orbits* appear stationary from any point on earth. In these orbits, normally only three satellites are required to provide nearly global earth coverage (the exceptions are the mostly uninhabited polar areas). Even though radio signals travel at the speed of light, the time required for them to travel from the earth terminal to the satellite is approximately 0.12 second. For a single "hop"—that is, a connection involving only one uplink and one downlink—the round-trip delay is about a half a second. This is noticeable and sometimes annoying during voice conversations. It should also be noted that the long, round-trip transit time demands special techniques to permit efficient data communications.

The *very small aperture terminal* (VSAT) earth terminal market grew to nearly $1 billion by the early 1990s, concentrated mostly in America. However, beginning in the 1990s, following privatization and liberalization of telecommunications markets, over 20,000 VSATs have been installed in Europe. By 1998, more than 300,000 interactive VSAT terminals were installed worldwide. This C- or Ku-band technology uses small antennas (1.5 to 6 feet in diameter) and is now extremely popular within financial, banking, brokerage, retail, gas station/convenience store, restaurant, pharmacy, automobile, and hospitality industries.

Small sites use two-hop links and transmit to public hub stations, which relay signals to other corporate locations, as shown in Figure 4.9. Installed costs per terminal are in the less-than-$10,000 range, and VSAT services are generally significantly less expensive than tariffs associated with equivalent, concatenated LEC intra-LATA and IXC inter-LATA services. Service quality and reliability is excellent. On average 99.9 percent of the time there is less than 1 error in 1,000,000 characters sent by satellite, and VSAT remote equipment fails only once every 10 years.

**Figure 4.9**
Very small aperture terminal (VSAT) satellite network topology.

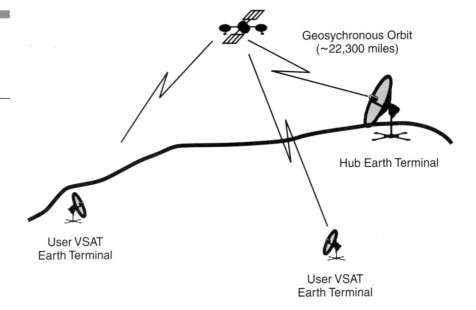

An example VSAT application is the Ford North American VSAT Network (FORDSTAR) with more than 6,500 receiver sites that Ford uses to distribute video and data services to affiliate dealers throughout Canada, Mexico, and the United States. In August 1999, GE Americom announced that FORDSTAR will be transitioned to its GE-4 satellite, described briefly above.

*Low earth orbit* (LEO) satellites are typically placed in circular orbits a few hundred miles from earth. At those altitudes, it takes from 90 minutes to several hours to encircle the globe. As a consequence, the time over which any particular satellite remains in view from fixed points on earth can be measured in minutes. Therefore, when LEOs are used for telecommunications missions, in order to provide continuous service, a fleet of satellites is needed. For example, backers of

Iridium, the satellite-based mobile telephone system, launched nearly 90 satellites between 1997 and 1999. The entire system operates much like terrestrial-based wireless cellular telephone systems introduced in Chapter 1 except that its transceiver "base stations" are moving and located in space rather than on the ground. With a sufficient number of satellites, such systems can provide coverage essentially anywhere in the world. Chapter 17 has more details.

*Direct broadcast satellite* (DBS) is another popular telecommunications satellite service. DBS permits small and relatively inexpensive dish antennas, installed at individual user residences, to directly receive high-power, satellite-television signals. DBS systems were analog until 1994 when Hughes Electronic Corporation launched the first digital system, with Primestar following soon thereafter. Since that development, the number of subscribers reached 10.7 million by October 1999, with a 1998 to 1999 growth rate of 34 percent.

Though DBS was originally intended only for broadcast TV, most long distance carriers have either purchased, or have considered purchasing, stakes in DBS companies, based on the promise that the medium is well suited to transmit video magazines, video clips, and voice, as well as ordinary text. Hughes Network Services and America Online have formed a strategic alliance that now offers an Internet access service, called DirecPC, that uses ordinary telephone lines for user-to-Internet traffic, but substitutes 400-kb/s satellite links for download traffic.

Current DBS companies include Canadian Satellite Communications Inc. (CANCOM); DirecTV Inc. (a subsidiary of Hughes Electronic Corporation); Echo Star Communications Corporation; ExpressVu Inc.; Primestar Partners; U.S. Satellite Broadcasting Co.; and WorldSpace Inc.

Lest readers, after learning of the spectacular growth in fiber-based media and the phenomenal technical capabilities, should jump to the conclusion that alternative media may be abandoned, according to a recent Business Communications Company prediction, the worldwide space manufacturing and services industry is expected to grow 30 percent per year to more than $250 billion by 2003. World LEO and geosynchronous commercial space market revenues are expected to grow from nearly $21 billion in 1968 to $65 billion in 2003, an annual growth rate of 26 percent.

Other portions of this book describe in greater detail how media are incorporated into transmission systems and how they support specific voice and data telecommunications services.

# Voice Frequency Transmission Systems

Two broad categories of transmission systems are *voice frequency* (VF) transmission systems and *carrier transmission* systems. *VF transmission* supports the ubiquitous local loops connecting business and residential customer premises with serving central offices. The median loop length is 1.7 miles, with 95 percent of all loops less than 5.2 miles. VF transmission is also used for short interoffice connections. The medium most often used for VF transmission is copper twisted-pair, multipair cable. In local loops, VF transmission includes LEC *main, branch, distribution,* and *drop cables,* collectively known as outside plant, with inside wiring completing connections to on-premises telephones and other CPE devices.

In 1980, the pre-divestiture Bell System investment in transmission facilities was $60 billion. Sixty percent of this investment was in local-loop plant, with the interoffice 40% being split 26% for metropolitan areas, 5 percent rural, and 9 percent long distance. In the future, fiber optic cable can be expected to replace copper twisted-pair cable in the local loop environment as it proves economical. Since today's copper cable provides power for telephones, with the advent of fiber, some alternative mechanism will have to be devised if telephone operation independent of other electrical power availability is to be maintained. It is likely that normal AC convenience outlets, with battery back-up will power tomorrow's fiber-connected telephones. Of course, wireless telephone service in any of the developing forms, i.e., cellular, PCS, or universal mobile telephone system (UMTS) handsets provide an additional back-up mechanism for corded systems disabled by electrical power outages.

# Carrier Transmission and Modulation Systems

A *carrier system* is a transmission system in which one or more channels of information are processed and converted to a form suitable for the transmission medium used by the system. When such a system is used with multiplexers described in the next chapter, the key objective is to transmit multiple information channels over a common medium. Another carrier system objective is to use a medium in a way that augments its ability to propagate or carry signals. A prime example of the latter application is the use of modulation and coding to increase a medium's useful bandwidth or bit-rate handling capacity.

Some form of modulation and demodulation process is fundamental to all carrier systems. *Modulation* is a process of varying certain parameters of a carrier signal—i.e., a signal suitable for modulation by an information signal—in accordance with one of the characteristics of another signal (a modulating and usually an information-bearing signal). In most cases modulating signals are *baseband* signals, that is signals in original form, unchanged by any modulation process. Analog voice signals generated by microphones and digital signals generated by PCs or other DTE, described in Chapter 3, are examples of baseband signals.

## Modulation and Demodulation

Figure 4.10 shows the relationship between digital baseband and sinusoidal carrier (single frequency) modulator input signals, and resulting modulator output signals for three different modulation processes. The top part of the figure presents "time profiles" for the input digital signal and modulator output waveforms. Beneath the digital sequence and baseband signal profile rows is an example of a modulator output waveform that corresponds to *amplitude shift key* modulation (ASK), one of several possible forms of amplitude modulation.

In this example, digital baseband signal levels corresponding to logical "1" values produce modulator output bursts of carrier signal sinusoidal cycles. Baseband signal logical "0s" inhibit modulator output signals altogether. Although not shown in the figure, the carrier input signal, for all the modulation examples shown, is simply a steady, unperturbed sinusoidal signal, i.e., what the ASK modulator output would look like if the baseband signal consisted of nothing but "1s".

The next modulation scheme example is called *phase shift key* (PSK). In this process, baseband 1s and 0s produce diametrically opposed—or phase-reversed versions—of the input carrier signal, as modulator output. Since each carrier signal cycle is said to consist of 360°, the illustrated PSK modulator example shifts the input carrier phase by 180° for each baseband signal's transition from logical 1 to logical 0 states.

The last time-profile modulation example in Figure 4.10 is dubbed *frequency shift key* (FSK). Here, modulators produces two different

**Figure 4.10**
Example modulation
types and time and
frequency spectrum
profiles.

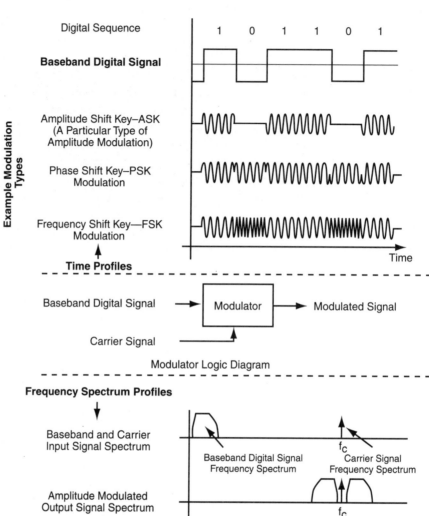

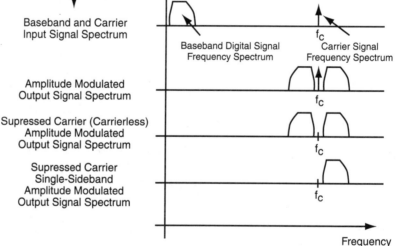

output frequencies, one corresponding to logical "1" input baseband signal states, and the other corresponding to logical "0s".

The lower part of the figure illustrates the relationship between modulator input and output signal spectral densities. As explained in Chapter 3, complex waveforms, such as those representing voice or data, are made up of combinations of many different single-frequency signals, each with potentially different amplitude. *Bandwidth* is a range of frequencies, usually specified as the number of hertz of the band or upper and lower limiting frequencies. *Spectral density* specifies power levels for each frequency, or spectral component, of a signal.

The shape of the baseband signal density (shown in the "baseband and carrier input signal spectrum" row in the figure) is not unlike that for typical voice signals. Note that the spectral density for a carrier signal is depicted as a single line since, theoretically, all its power is concentrated at the single or "point" frequency—$f_c$ in this case. If a baseband signal with the indicated spectral density is used to amplitude modulate a carrier signal $f_c$, then the modulator output signal has the spectral density shown in the next row. That is, it exhibits a single frequency component at frequency $f_c$, and two sidebands, each (ideally) with shapes identical to the input baseband signal, with the lower sideband being inverted or reversed.

The next two rows in the figure depict modulator output signal spectra for the case of *suppressed carrier* (carrierless) and *suppressed carrier-single sideband* amplitude modulation processes, respectively. One reason why suppressed carrier modulation is used is that carrier signals contain no modulating or information-bearing signal content. Despite this, transmitting them "usurps" up to 50 percent of a transmitter's total power. When carriers are suppressed, transmitter power that would have been dissipated in "informationless" carrier signals is used instead to increase information-bearing signal component signal-to-noise ratio.

Whereas suppressing carriers conserves power, suppressing sidebands—that is transmitting only a single sideband, conserves bandwidth. By suppressing one of the sidebands, all a transmitter's output power is concentrated in half the bandwidth; if implemented properly, this process sacrifices no modulating signal information content.

The penalty paid for power and spectrum efficiency is that demodulators are more complex than those required for standard AM. The fact that standard AM demodulators are so simple is the reason why for nearly a century, AM radios were so popular and inexpensive. Today, with low-cost advanced solid-state componentry, especially where bandwidth is fixed by unalterable physical constraints (as it is for example, in subscriber loops discussed below), the use of sophisticated modulation and coding schemes to optimize performance is commonplace.

The examples shown in Figure 4.10 portray just a few of a large set of possible modulation processes used in telecommunications systems. For example, the illustrated PSK example is more accurately described as binary phase shift key, since only two possible modulator output phases are possible. That is, the modulator output signal is either in-phase or out of phase or 180° phase-shifted with the input carrier signal phase. Higher-order phase modulation schemes are possible.

For instance, baseband signal bits can be grouped as two-bit bytes. Since there are exactly four possible "1" and "0" patterns in two-bit bytes, each possible pattern can be related to four possible modulator output phase values (for example, in-phase or 0°, 90°, 180°, and 270°). Such four-phase modulators are called *4-ary* and higher-order systems m-ary. In an analogous way, *m-ary* discrete amplitude modulators are possible, as are m-ary combinations of both phase and amplitude modulation.

Modulation-coding details are beyond the scope of this book but what is crucial is an appreciation of why this technology is important. Engineers are constantly challenged to determine and specify the best combination of modulation and coding (see Chapter 3) processes for each medium and traffic type. Success relates directly to telecommunications service quality and economy. Assume that a particular modulation-coding combination permits spacing repeaters twice as far apart, or allows a particular fiber span to support twice as many user information channels—at the same performance quality level. Either way there are enormous benefits in terms of reducing provisioning costs and increasing revenue and profit.

One final signal modulation/demodulation-detection characteristic that is a powerful performance-enhancing factor is coherence. Demodulation is a process inverse to the modulation process just described. In other words, a signal modulated to be efficiently carried by a particular medium circuit is transmitted to a remote facility where it is fed to demodulation equipment—along with a local replica of the carrier signal used to create the transmitted modulated signal. With these two inputs, demodulators should recover high-quality version or replica of the original baseband signal.

In *coherent demodulation*, the local version of the carrier signal is ideally in perfect phase and frequency agreement with the carrier signal used in the modulation process. At a minimum, coherent demodulation produces output signal-to-noise ratios 3 dB better than those of non-coherent demodulation. This has the same impact as doubling transmitter power and again, in some cases, may double traffic-handling capacity on otherwise identical transmission systems. In advanced optical transmission systems, frequency, phase, and timing coherence can increase receiver performance by 10 to 15 dB relative to simpler intensity (amplitude) modulation/direct detection; it facilitates improved multi-channel wavelength division multiplexing (WDM), described in the next chapter.

## Digital Subscriber Line (DSL)— Modulation/Coding Enhancements for UTP Local Loops

While business customers with large numbers of subscribers or other broadband service requirements have been able to take advantage of high-speed transmission technologies, cost has kept high-speed access out of the reach of most small-business and residential users. High transfer rates typically require expensive equipment and construction to bring special transmission media such as fiber optic cable to customer locations. Telephone companies are willing to make the investment where high traffic concentration ensures a reasonable return on investment. However, the nearly one quarter of a trillion dollar cost of rewiring access (including outside plant and inside or in-building wiring) to over 120 million households and small businesses precludes that alternative.

Accordingly, a significant amount of research has focused on how to use existing copper loops to support high-speed data transmission. The result is a whole new approach to carrying high-speed digital signals over copper wires that promises to bring fast Internet access to small businesses and households across the country. The approach consists of two major elements. The first involves a relatively simple modification to UTP loops, namely the removal of *loading coils* and *bridging taps* at multiple distribution points, originally installed to optimize voice service on long runs and make loop reassignments easier, respectively. The second is the application of inexpensive but advanced adaptive *digital signal processing* (DSP)-implemented modulation and encoding techniques—in transceivers providing reliable, nearly error-free transmission at rates up to 25 Mb/s.

The following sections present technical details behind this new technology, known as *digital subscriber line* (DSL—or xDSL to accommodate the many emerging varieties), as well as advantages and issues associated with its use. Basic frequency spectrum, bandwidth, modulation, coding, digital signal processing, and other concepts presented above and in Chapter 3, are applied here in explaining DSL.

## BACKGROUND AND BASIC DSL TECHNICAL CONCEPTS

The highest transmission speed that can be achieved over voice networks, using voice-channel modems discussed in Chapter 3, is 56 kb/s. This restriction is primarily due to limited circuit-switched network voice channel bandwidths (approximately 3400 Hz). However UTP copper local loops are not themselves intrinsically limited to this nominal voice channel bandwidth. If data could be stripped off at the central-office end of UTP local loops and connected directly to data networks, much higher bit rates could be supported over local loops, and on an end-user-to-end-user basis over those data networks. To understand current local-loop digital transmission limitations, a review of historical design practices is helpful.

In most carrier service areas, some subscribers are invariably located at greater distances from central office switches than others, and therefore require very long local loops. One problem with very long copper loops is electrical signal energy loss. In the past, telephone companies dealt with this problem primarily in two ways.

First, loading coils were added to modify loop electrical characteristics resulting in high-quality voice transmission over extended distances (typically greater than 18,000 feet). Loading coils are placed at about 6,000-foot intervals, so some runs involve multiple coils.

Second, remote *subscriber loop carrier* (SLC), *optical networking unit* (ONU), and other equipment is often installed closer to subscriber locations to ameliorate line loss problems or to increase the number of subscribers that can be served (pair gain systems). Thus, SLCs and ONUs terminate shorter copper loop runs, aggregate or multiplex individual circuits, and transmit those signals back to central offices using higher-capacity and less "lossy" circuits. This approach allows more subscribers to be served at greater distances from central office switching equipment.

While these solutions are practical for voice applications, loops using them present special obstacles to widespread DSL deployment. Loading coils are completely incompatible with the higher frequencies needed for high-speed digital transmission and must be removed on loops using DSL modulation and coding.

For loops connected to SLC/ONU remote terminals, DSL transceiver equipment must often be installed in unattended outdoor cabinets where limited space, power, and other support facilities necessitate compact, low-power designs. Also, some existing SLC/ONU terminal-to-central office facilities may have to be upgraded to accommodate higher DSL bit rates. Both issues are economic; they represent higher implementation cost that affects telco DSL business cases, selection of serving area segments for upgrade, and deployment rate decisions.

Assuming loading coil and bridge tap removal and the availability of compact, low-power and cost-effective transceivers, two significant technical DSL deployment challenges remain. The first is insertion loss or signal level attenuation, which, even without loading coils, increases with frequency for fixed-length UTP loops. Figure 4.11 illustrates signal attenuation loss as a function of frequency for a UTP copper loop, where the frequency parameter in the figure is proportional to the bandwidth in which DSLs must operate.

Since bit rates, bit-error rates, bandwidth, and signal-to-noise ratios are all related, it follows that because signal power loss varies with both frequency and loop length, the maximum bit rate supported by DSL transceiver designs must vary with, and decrease with, loop length. The implication is that in large carrier-service areas, using a given DSL technology, telcos are unable to offer the same maximum bit-rate service to distant subscribers that they can offer to closer-in subscribers. To achieve target bit rates over conventional loop lengths (less than 18,000 feet), efficient modulation and encoding techniques able to transmit in excess of one bit per second per Hz of bandwidth are needed.

The concept of selecting modulation schemes to conserve bandwidth was introduced earlier in this chapter. Recall that single-sideband modulation transmits all necessary information-bearing signal components, but uses up only half the bandwidth that ordinary dual-sideband amplitude modulation demands. When we are dealing with digital signals, both modulation and digital signal encoding techniques can be applied to "squeeze" information bearing signals into bandwidth limited media—such as UTP subscriber loops now being discussed.

For digital signals, a measure of bandwidth conservation success can be expressed as the number bits per second that can be transmitted in a single Hertz of bandwidth. Not too many years ago, the Bell Standard 103 (V.21) modem delivered only 300 b/s using nominal 3 kHz telco voice channels. As noted in Chapter 3, today's V.90 modems, selling at a fraction of inflation-adjusted dollars deliver up to 56 kb/s. Specific instances of efficient modulation and encoding schemes are presented below in this chapter and throughout the book. The technology is key to UTP, wireless, and any other telecommunications applications where bandwidth is at a premium.

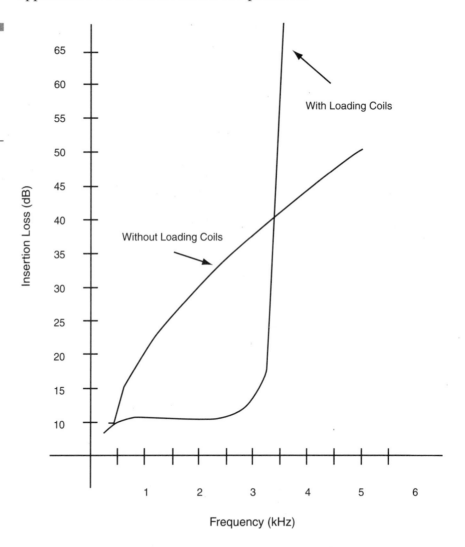

**Figure 4.11**
Insertion loss (signal attenuation) versus frequency with and without loading coils on a 24-gauge copper UTP loop.

The second significant technical challenge facing DSL designers is coping with crosstalk. Crosstalk is caused by undesired signal energy radiated from one UTP pair, coupling to and interfering with desired signals carried in adjacent copper loops in the same cable bundle.

Whereas in most transmission links, thermal noise (see Chapter 3) limits performance, in multi-pair cable bundles, the cumulative effect of large numbers of interfering signals can exceed thermal noise and dictate performance thresholds.

Because cable bundles leaving central offices contain large numbers of loops, crosstalk can be a major factor in the performance of DSL systems. The amount of interference depends on the number of loops in a bundle using signals in the same frequency range as DSL signals (i.e., other digital services). This dependence on present and future service mix in cable bundles makes engineering DSL services a complex process.

At central offices where all local loops terminate, main feeder cables carry more wires than either branch feeder or distribution cables. Hence, crosstalk is greatest at central offices and interferes more with weak signals entering offices than with strong signals leaving them. The net result is that higher bit rates can be transmitted to users (*downstream*) than from users (*upstream*). This general phenomenon is referred to as *near-end crosstalk* (NEXT), a limitation of multi-circuit cables. NEXT, and other performance parameters such as attenuation per foot, are typically measured by manufacturers and specified for each of their multi-circuit cable products. Chapter 9 describes this and other performance-limiting cable characteristics.

Fortunately, this phenomenon matches most Internet and other data user requirements. In general, the quantity of information sent from network servers to users far exceeds what users send to network servers (mostly "query" information). As a result, most of the DSL installations to date provide what is called *asymmetric digital subscriber line* (ADSL) service wherein downstream bit rates are typically much higher than upstream rates. Figure 4.12 shows the frequencies used by a typical ADSL system, and a lower baseband-spectrum portion for sharing loops with traditional *plain old telephone service* (POTS i.e., voice, facsimile, voice-modem).

### DSL MODULATION AND CODING TECHNOLOGIES

Spectacular Internet user growth, heightened interest in broadband multimedia Internet services, and extraordinary new revenue opportunities are spurring manufacturers, LECs, CLECs and ISPs alike to conduct "pull-out-all-the-stops" efforts to capture markets. This section summarizes salient DSL design approaches and characteristics. In the early 1980s, a modulation-coding scheme that carries two bits of information per Hz of analog bandwidth was defined, primarily as a candidate for supporting 144-kb/s ISDN *Basic Rate Interface* (BRI) signals (see Chapter 8). The line code used, is designated *2B1Q* (2 Binary, 1 Quaternary), because successive pairs of user data bits are mapped into one of four (quaternary or 4-ary) symbols as shown:

| First Bit | Second Bit | Quaternary Symbol |
|-----------|------------|-------------------|
| 1 | 0 | +3 |
| 1 | 1 | +1 |
| 0 | 1 | −1 |
| 0 | 0 | −3 |

In ISDN DSL applications (referred to in some literature as *IDSL*), and first generation *high bit-rate* DSL (HDSL) applications, 2B1Q signaling employs four-level pulse amplitude modulation (PAM), uses frequencies from 0 to 80,000 Hz, and transmits up to 160-kb/s rates over 18,000 foot loops.

**Figure 4.12**
Frequencies used by a typical ADSL system.

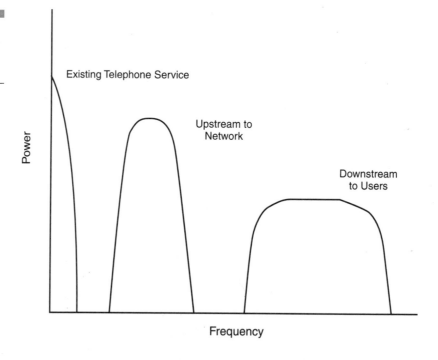

At the same time, another modulation-coding scheme, called *Carrierless Amplitude and Phase* (CAP) modulation was introduced by AT&T;

it is able to transmit from 2 to 9 bits per Hz of analog bandwidth. This enables transceivers to transmit the same amount of information in a smaller and lower range of frequencies than 2B1Q, which equates to longer loop reaches. Figure 4.13 contrasts 2B1Q and CAP frequency spectra. As implied in the modulation discussion above, maximum DSL loop transmission rates are determined by loop signal-to-noise ratio and the number of bits per Hz the modulation scheme supports. For given noise levels, longer loops result in lower signal-to-noise ratios and lower transmission rates. Figure 4.14 presents, by way of a practical example, the relative advantages among techniques by contrasting 2B1Q and CAP loop length with bit-rate performance.

**Figure 4.13**
Frequencies used by CAP and 2BIQ DSL modulation schemes.

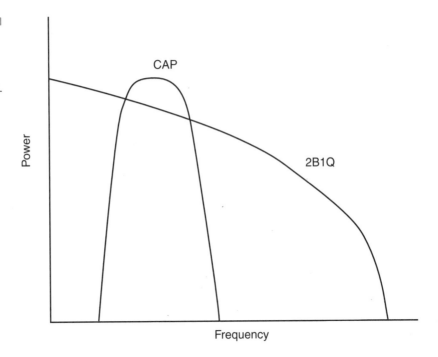

Another modulation technique, known as *Discrete Multitone* (DMT), has been selected by the *American National Standards Institute* (ANSI), *European Telecom Standards Institute* (ETSI), and the *International Telecommunications Union* as a standard for ADSL. DMT's ANSI T1.413 standard divides the available bandwidth into 256 subchannels with subcarriers that can be modulated from zero to a maximum of 15 bits/second/Hz, as depicted in Figure 4.15. This permits up to 60 kb/s data rates per subcarrier.

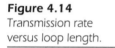

**Figure 4.14**
Transmission rate
versus loop length.

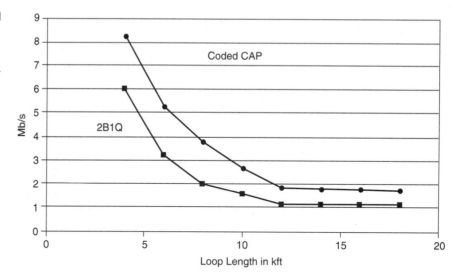

At lower frequencies where copper twisted-pair attenuation is low and signal-to-noise or interference ratios are good, adaptation between transmitter/encoders and receiver/decoders typically results in high data-rate (greater than 10 bits/second/Hz) operation. In less favorable conditions, the operation drops back to lower rates (e.g., 4 bits/second/Hz), or a particular subchannel may not be used at all.

Both DMT and CAP support what has been coined *rate adaptive* DSL (RADSL). With DMT, adaptive rate adjustment steps may be as small as 32 kb/s, which is a very fine gradation for payload bandwidths in the order of 6-8 Mb/s. CAP provides 340-kb/s adaptive steps with the lowest rate alternative to no service at all being 640 kb/s.

DMT designs exhibit higher latency (the time it takes to get information through a network) and require greater amounts of digital signal processing. Interactive voice and video applications are latency sensitive and current DMT designs may breach some service specifications. Low-cost and powerful high-speed digital signal processors now mitigate processing requirement disadvantages that once excluded DMT from consideration for some applications.

Beyond loop length and upstream and downstream bit rates, other DSL technical design and application trade-off factors include: one- or two-pair (UTP) requirements; combined POTS and data traffic capability (is a separate loop for voice service needed?); tolerance to mixed-loop wire gauge and bridge taps; mixed-service crosstalk tolerance; latency (cumulative network information delay); and bit-rate adaptation to loop conditions. With respect to one- or two-pair (UTP)

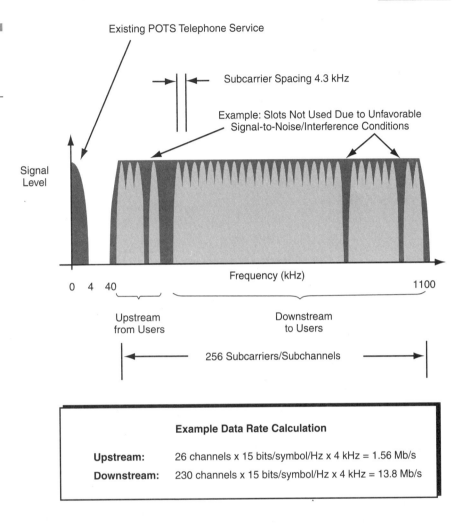

**Figure 4.15**
Discrete multitone modulation design for ADSL.

Existing POTS Telephone Service

Subcarrier Spacing 4.3 kHz

Example: Slots Not Used Due to Unfavorable Signal-to-Noise/Interference Conditions

Signal Level

Frequency (kHz)

0   4   40                                                          1100

Upstream from Users

Downstream to Users

256 Subcarriers/Subchannels

**Example Data Rate Calculation**

**Upstream:**    26 channels x 15 bits/symbol/Hz x 4 kHz = 1.56 Mb/s

**Downstream:**  230 channels x 15 bits/symbol/Hz x 4 kHz = 13.8 Mb/s

requirements, the ANSI/T1/E1 technical committee is currently developing an HDSL2 single UTP standard to supplement its two-pair HDSL standard. Although initially SDSL stood for symmetric DSL, in some literature it is now used to denote single-pair DSL.

Accommodating voice traffic in DSL loops can be accomplished either by ensuring that DSL spectra lie outside the normal voice band (0-3400 Hz) or by digitizing voice signals and multiplexing them along with other DSL traffic. Figure 4.16 depicts an ADSL and POTS single-UTP line subscriber location arrangement in which POTS traffic occupies the 0—3400 Hz spectrum and up and downstream high bit-rate data traffic use two separate higher-frequency spectrum allocations.

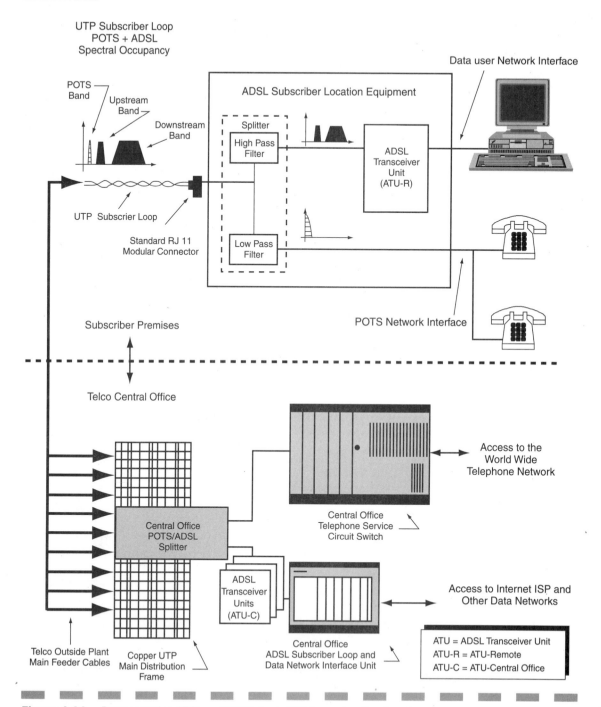

**Figure 4.16** Representative ADSL and POTS single UTP line subscriber location arrangement.

Tolerance to mixed-loop wire gauge and bridge taps, mixed-service crosstalk tolerance, and latency are all performance characteristics directly related to specific modulation-coding techniques. For example, HDSL2 exhibits significantly improved crosstalk immunity and compatibility and reduced latency, (less than 500 milliseconds) than its predecessor HDSL.

Since the quality or signal handling condition of any loop depends upon both fixed or static physical as well as time-varying or pseudorandom factors (such as crosstalk among same- or mixed-traffic sources), manufacturers now produce products offering manual (set up at installation) and dynamic abilities to adapt to local conditions. Rate-adaptive techniques work with both asymmetric and symmetric DSL designs. Current models now support up to 7 Mb/s downstream rates.

Very high bit-rate digital subscriber line (VDSL), the latest development and asymmetric in its initial form, is now appearing in products that support 25.92 Mb/s downstream rates over cable lengths in excess of 3,500 feet. Original goals targeted SONET and SDH base rates (51.84 and 155.52 Mb/s, described in the next chapter) over loops shorter than 1,000 feet. To attain these bit rates, implementation must avoid high crosstalk levels encountered at central offices. Accordingly, an attractive application for VDSL loops is on short runs from subscriber CPE to ONUs.

Successful concatenated operation of hybrid VDSL UTP subscriber CPE-to-ONU loops and fiber-based ONU-to-central office links may be the first practical, and therefore widely implemented, appearance of "fiber-to-the-curb" provisioning of universal broadband access.

### CURRENT STATUS

As with most recent emerging technologies, frenetic vendor/service provider developments are outpacing related standards activities. Consequently, most available products employ proprietary designs and are not interoperable with DSL products from other manufacturers. This situation is being addressed vigorously by international and national standards bodies such as ITU, ANSI's T1E1.4 committee, ETSI (the European Telecommunications Standards Institute), DAVIC (the Digital Audio Video Council), and the ATM Forum.

The first standard for consumer-oriented DSL, known as *G.Lite*, has been adopted by the ITU and may provide a market thrust that only standards-based implementation produces. With standards-based products, it is conceivable that low-cost DSL modems shipped with PCs may obviate at least a portion of expensive telco-specific installations, particularly evident in most current high bit-rate telco facilities. Recent trends already point toward a shift in emphasis from telco central office to CPE market segments.

Moreover, DSL sales are likely to benefit directly from applications of HDSL2 in carrier T1 service markets. Deployment of T1, a 1.544-Mb/s service introduced in the 1960s Bell System (T1 multiplexers and service are described in the next chapter), exceeds 120 million miles in the U.S., and 250,000 new lines are expected to be installed by the end of 1999. Much of the new demand is driven by Internet ISPs using T1 service to connect with telco central offices which in turn connect to mushrooming numbers of Internet users via voice modem-equipped subscriber loops as described in Chapter 1.

Even without most of the technological advancements alluded to above, DSL is already the most sought after high-speed service offered by LECs. Cost for service varies with bit rate, but at consumer-oriented bit rates (640 kb/s downstream), prices are typically under $50 per month. For about twice the price of a second telephone line, customers enjoy downstream bit rates ten times faster than today's best POTS-compatible modems. The limiting factors to the widespread rollout of DSL remain loop plant topography and loading coils and remote terminals in loop plant. By the third quarter of 1999, approximately 275,000 DSL lines had been deployed in the U.S., a number that doubled every quarter last year. From a worldwide hardware perspective, ASDL modem shipments grew 74 percent sequentially in 3Q99 and may top 1.2 million by year end.[2]

Overall, the prognosis appears to justify exponential growth predictions. This conclusion rests upon demand for broadband access—driven by the popularity of multimedia telecommunications services—and startling new capabilities that companies combining advanced digital signal processing and integrated circuit technologies have demonstrated they can produce. Broadcom, a company founded in 1991 (now making more than 95 percent of the chips used in set-top cable modems discussed in the next section) and a leading contender in SDSL markets, has grown so fast that its founder, Henry Samueli, has a unique problem of managing over 350 stock-owning millionaire employees.

Significant efforts at Lucent Microelectronics Inc., Analog Devices, Rockwell Semiconductor Systems, Texas Instruments, 3Com, Pairgain,

---

2   Cahners In-Stat Group November 23, 1999 press release.

Copper Mountain Networks, Cisco, Alcatel, and many others including some heavy hitters like Nortel, Siemens, US West, and Microsoft, all seem to confirm that DSL will play increasingly important roles over the next decade.

## Cable Modems—Adapting Cable TV Networks for Internet and other Data Communications Services

Cable modems provide high-speed Internet access via cable television networks. Cable modems compete with voice-band modem and DSL telco subscriber loop-based, as well as direct-broadcast satellite approaches discussed earlier in this chapter. Because today's cable TV networks employ broadband coaxial cable or hybrid coaxial cable-fiber optic media capable of supporting hundreds of TV signals, cable modems are promoted principally on the promise of delivering downstream bit rates 10 to 1000 times as fast as voice band modems. Part of the business case rests with the fact that homeowners in over 60 million residences now wired for cable TV entertainment service are prime and ready customers for the access service part of burgeoning Internet markets.

Figure 4.17 highlights major residential and cable TV head end-located components making Internet access via cable TV networks possible. Head end is a term describing a cable TV company's central facility for aggregating video and audio program material (and now Internet and other data) and re-broadcasting it to service subscribers. Cable Modem Termination Systems (CMTSs), added to traditional entertainment-only head-end sites to add data service, provide interfaces between ISP and existing cable TV networks. As shown on the figure, a CMTS contains down- and upstream modulators and other processor-based communications interface and management and control components. Figure 4.17 is simplified in the sense that single cable networks may serve hundreds of thousands of users and the intermediate service nodes and facilities required to provision that many subscribers with data service are not shown.

As with DSL transceivers, cable modems in residences include sophisticated, bandwidth-efficient modulator and demodulator equipment, made practical by the application of advanced digital signal processing and *application-specific integrated circuit* (ASIC) technologies. First-generation cable modems used proprietary designs prohibiting the use of cable modems made by different manufacturers from being used on the same network and creating an impediment to lower-cost, mass-produced hardware. Today, most equipment conforms to MCNS/DODIS (multimedia cable network system/data over cable service interface specifications) specifications in the U.S. and DVB/DAVIC (digital video broadcasting/digital audio video council) specifications in Europe.[3]

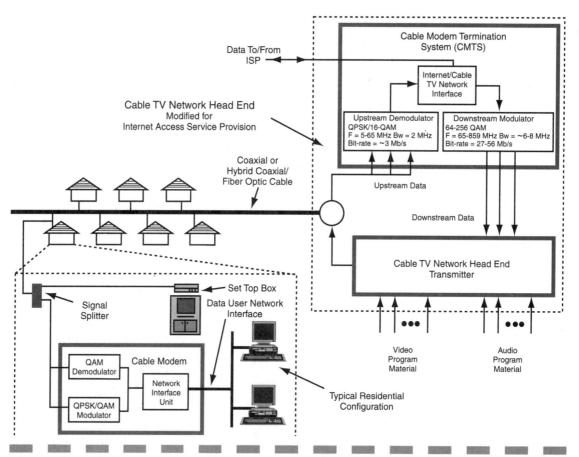

**Figure 4.17**     Cable modem cable TV network head end and residential Internet access arrangements.

---

3 Multimedia Cable Network System Partners Limited leads the DODIS project in the U.S. The Digital Video Broadcasting Group is a European organization that publishes its work through ETSI. The Digital Audio Video Council, a voluntary special-interest group, promotes international audio video application and service standards.

Figure 4.17 reveals that 64-256 QAM in downstream and QPSK/16 QAM in upstream are the modulation techniques used in cable modems. The nomenclature M-QAM is defined as quadrature or quaternary amplitude modulation (QAM) combined with m-ary order phase modulation, modulation schemes explained earlier in this chapter.

The figure also reveals how signal splitters feed both "set-top" boxes or cable-ready TV sets, and cable modems connected to PC or other residential user Internet equipment. Note that whereas voice-band and DSL modems use dedicated loops or channels to telco central offices, cable modem users share common channels with neighbors. The method of sharing is similar to that used in Ethernet local area networks (LANs describer in Chapters 7 and 13).

Because of this LAN-type operation, some users have been able to log onto other neighbors' PCs and rummage through private files. This possibility is made more likely by the fact that with cable modems, user PC to CMTS links are full period; that is, it is not necessary for PCs to make dial-up connections as is common with voice band modem service. With cable modems, unless users intentionally terminate sessions, if your PC is on, you have a connection to a head-end CMTS. As a consequence, for those users who leave their PC equipment on continuously, whatever vulnerability exists, exists full time.

While encryption and other firewall measures can mitigate privacy or security vulnerability, the use of shared LAN-type channels makes it possible for data-hungry large file-streaming users to monopolize limited shared-channel capacity. Again, cable modem data access service providers can devise methods to prevent single users from unfairly usurping capacity, but this too represents a service feature that users should consider when comparing Internet service providers.

Another consideration that to date appears to be unique to cable-modem and satellite-based ISPs is the commingling of what traditionally has been separate decisions when selecting access and ISP service. Incumbent LECs are under a mandate to open their networks to competitors (see Chapter 2's discussion of the Telecommunications Act of 1996). This includes sharing local loops with DSL service competitors.

As a result, whether voiceband or DSL modems are used, subscribers are free to choose any ISP. In contrast, cable and satellite-based service providers appear to be free from any legal constraint to enter into exclusive joint ventures with single ISPs, that in effect forces subscribers to deal with partnering ISPs. While some local jurisdictions have mandated that franchised cable providers support connections to multiple ISPs, these mandates have not faced judicial review and the eventual outcome is unclear. In the end, it may turn out that Internet access choice may be better determined by bundled content and higher-level ISP services than by the merits of any access technology.

# Carrier Systems for Multichannel Telco Transmission Systems

Whereas DSL and cable-modem carrier system technologies are designed principally for subscriber-access applications, historically the first carrier systems were used to support telco inter-switch, multi-channel trunking requirements. These carrier systems are classified as either analog or digital carrier systems.

### ANALOG INTER-SWITCH CARRIER SYSTEMS

Analog carrier systems use repeaters that correct for medium impairment characteristics, and produce at their outputs linear scaled versions of the input signals. Analog carrier systems can carry speech, data, video, and supervisory signals, although they are best suited for speech signals. Analog carrier systems operating over multipair UTP cable, N-carrier, coaxial cable, L carrier, and radio systems have largely been replaced by digital facilities.

An economic comparison of analog versus digital carrier facilities must include whether or not interconnected switching systems are analog or digital. With the current transition to digital switching, increasing emphasis is on digital carrier systems.

### DIGITAL INTER-SWITCH CARRIER SYSTEMS

Digital carrier systems are designed to transmit digital signals, using regenerative versus linear repeaters, and time division multiplexing. Multiplexing is a technique that enables a number of communications channels to be combined into a single broadband signal and transmitted over a single circuit. At the receiving terminal, demultiplexing of the broadband signal separates and recovers the original channels. The primary purpose of multiplexing is to make efficient use of transmission capacity to achieve a low per-channel cost. Multiplexing methods used in telecommunications systems are the topic of the next chapter.

# Fixed Wireless Systems

Until the mid-1980s, nearly all 180 million telephones then installed in the United States gained access to telcos via wired local loops. In 1988, the FCC authorized a class of service allowing telephone companies to use digital radios instead of copper wires when cost effective. These were called Basic Exchange Telecommunications Radio Service, although analog "rural radios" had existed for some time; it took BETRS's digital technologies to reach viable operational and cost performance thresholds. The decision was historic in the sense that the FCC, for the first time, recognized the equivalence of wireless and wireline telco access technologies for fixed, and not just mobile subscribers.

By 1990, over 60 BETRS installations served several thousand U.S. customers. Most installations use frequency allocations in the lower UHF band (ultra-high frequency band, 300 MHz to 3 GHz), although some designs operate in VHF (very high frequency band, 30 MHz to 300 MHz) and 800 MHz spectra. These early designs employ time division multiplexing (see Chapter 5) and air interfaces similar to those used in today's digital cellular radios (see Chapter 16).

Broaching the fixed wireless topic might be easier if it involved completely separate and distinct systems and technologies. Beyond the multiple access technique similarities just mentioned, modern fixed wireless approaches often incorporate technologies identical to those used in mobile wireless systems—with some telcos even using common facilities to provision fixed and mobile service. Clearly, cellular mobile subscribers are free to discontinue fixed wireline access service, using instead mobile telephones in residences or businesses locations. In fact, subscribers often use both. Aggressive and highly competitive bulk long-distance mobile telephone carrier plans have already enticed many to limit use of wired telephone service to lengthy local voice and Internet calls.

Originally it might have been accurate to define fixed wireless as radio-based telco access facilities used in lieu of ubiquitous copper twisted-pair loops or more modern fiber guided-media alternatives, to connect to fixed user terminals. But, as we have seen above and will see in ensuing discussions, depending upon network and network interface unit design and placement, wireless facilities originally intended for wired or corded telephones may eventually support cordless or corded, fixed, or mobile telephones or other CPE station equipment.

As a specific instance, the ETSI is unifying Digital European Cordless Telephone (DECT), Global System for Mobile Communication (GSM) cellular, and other existing standards leading to Universal Mobile Telecommunications Systems (UMTS). UMTS is a member of the International Mobile Telecommunications-2000 (IMT-2000) "family of systems" formerly known as the Future Public Land Mobile Telecommunications System (FPLMTS), an ITU project. With UMTS, single portable telephones and integrated telco networks may satisfy indoor, outdoor, pedestrian, and vehicular communications requirements.

Further evidence of the broad scope of services under the fixed wireless umbrella is the fact that some important fixed wireless systems involve neither incumbent nor competitive local exchange carriers. For example, both terrestrial and satellite-based service providers now offer wireless point-to-multipoint video distribution, directly competing with traditional cable TV offerings. The point is that there are so many new interrelated wireless standards, service, technology, and business-area developments under way that to place them in context, it is best to first define categorical radio environment, design and application attributes, and then selectively present the more important individual initiatives.

To that end Figure 4.18 depicts all possible wireless radio environments in terms of cell size. Products falling into the smallest (150 to 1500 feet) home cell category include cordless telephones that communicate with at least one customer-owned base station, which itself is wire-connected to LEC local loop facilities. This category also includes wireless data local area networks (LANs—introduced in Chapter 1 and discussed in detail in Chapters 7 and 13) such as Apple's "Airport" product that interconnects multiple personal computers or computer peripherals at 10 Mb/s rates. Wireless data LANs for home or business are sometimes portrayed as products that do for PC data traffic, what cordless telephones accomplish for voice traffic.

Next are picocells that find application in larger buildings. Products in this category are similar to the cordless telephone and wireless LANs just described. They differ principally in that large-building system designs contain a sufficient number of strategically located base stations to ensure high quality and drop-free call connections, regardless of where users toting portable station equipment may roam. In most large buildings both wired and wireless telephone and data LAN facilities are necessary. In those cases, wired telephone systems (private branch exchanges [PBXs] or Centrex service discussed in Chapter 6) and wired LANs (described in Chapters 7 and 13) must be seamlessly integrated with wireless components to guarantee satisfactory per-

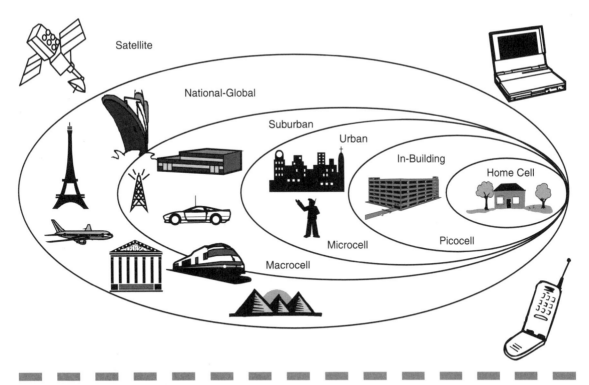

**Figure 4.18** Radio environment for wireless communications.

formance. Indeed, a key advantage claimed by PBX and Centrex equipment manufacturers offering both wired and wireless products is designed-in wired-to-wireless interoperability.

Macrocells, cells serving urban and suburban areas, may use base stations separated by a few thousand feet in high-traffic areas, to 15-20 miles in less densely populated suburbs. The cellular telephone systems that spawned explosive subscriber growth since the mid 1980s use macrocells, the exact size of any cell being determined by detailed traffic engineering and other technical performance analyses. When first introduced, cellular telephones were too heavy and bulky to be carried by pedestrians. As a consequence, peak communications traffic loads tended to occur during rush hours, generated by telephones mounted in vehicles trapped on clogged roadways.

With hand-carried station equipment, more mobile user sets may be "packed into" any given urban space than vehicles. The basic method by which cell-based wireless systems are able to accommodate greater and greater numbers of users within a given area is to subdivide those areas into smaller cells. Over the past decade, rising numbers of subscribers in densely populated urban areas have forced designers to what are referred to as microcells. The newer personal communications systems (PCSs) and Handyphone systems, mentioned in Chapter 1 and elaborated on in Chapter 16, incorporate city-block sized (300—900 feet) microcells wherever high user-traffic conditions exist.

Finally, satellites, able to provide fixed or mobile wireless service to national and even global sized cells, are treated in Chapter 17.

Beyond cell size, Figure 4.19 identifies three additional attributes by which the growing number of existing and future wireless systems can be further classified. These attributes, portrayed along three orthogonal axes, are *terminal mobility, supported data rate,* and the *frequency band* in which systems operate.

Terminal mobility is a measure of how fast terminals change position relative to base stations, if we understand that in satellite-based wireless systems both user terminals and satellite base stations may be moving. As a wireless system discriminator, what is important is the question of how much mobility a system can tolerate since nearly all traffic and synchronization performance parameters are affected by both relative velocity and acceleration.

Then too, hand-off mechanisms, the complex means by which wireless control systems maintain communications as mobile terminals frequently and rapidly cross cell boundaries, may be virtually nonexistent (or simple "one-time" manual set-up procedures) where user terminals remain in a single cell or are only moved infrequently. In Figure 4.19 hand-carried cordless or cellular telephones are classified as slow mobile whereas terminals used in automobiles or aircraft qualify as fast mobile. Wireless LANs typically support terminals in the lower moveable mobility category (e.g., PCs and workstations) but may also have to serve slow mobile terminals.

Illustrating the impact that cell size and mobility have on viable wireless systems, IMT-2000 standards specify the following minimum user data rates:

| Environment | Minimum Data Rate |
| --- | --- |
| Vehicular | 144 kb/s |
| Pedestrian | 384 kb/s |
| Indoor Office | 2.048 Mb/s |
| Satellite | 9.6 kb/s |

**Figure 4.19**
Three important
wireless system
attributes.

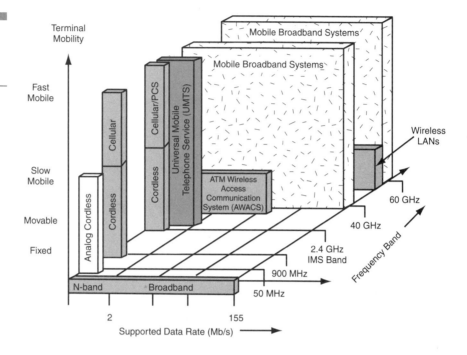

While mobility tends to limit data rates, another important wireless technology trend, the transition from single to multimedia information (voice, data, video, imagery) capabilities, generally leads to "bandwidth-hungry" applications and higher data rates. One instance of this trend is the direct broadcast satellite (DBS) example mentioned earlier in this chapter. Originally built for video TV traffic, DBSs now support broadband downlink Internet data traffic, which may itself have multimedia content. Likewise, most cellular telephones, networks, and standards are being adapted to, or are already capable of carrying, Internet and other data communications traffic.

Thus, data rate capability is an important attribute for categorizing emerging wireless systems. Figure 4.19, for example, while not showing details, indicates that existing cellular systems, when adapted, will support data communications at rates up to 115.2 kb/s. By comparison, UMTS objectives target full mobility service rates up to 2 Mb/s. The European Advanced Communications Technologies and Services (ACTS) program's wireless LAN projects (Median and Magic Wand) promise rates to 155 Mb/s.

Finally, Figure 4.19 identifies frequency band allocation as the third attribute by which wireless systems are classified. The figure shows some historical, current and future cordless and mobile telephone frequency allocations. For example, the first analog cordless telephones operate at 50 MHz. As digital modulation and control signal designs appeared, most manufacturers opted to move to the 900-MHz band. The most recent models are migrating to the Industrial Medical Scientific (ISM) 2.4 GHz band. In those bands, licenses are not required for each telephone although the FCC imposes design restrictions to minimize interference among telephones and with other radio equipment. More information regarding performance characteristics and design rationale motivating mobile telephone spectral allocation and selection is found in Chapter 16.

Standards groups around the world currently envision the following frequency allocations for newer broadband applications:

| Application Area | Frequency Band |
| --- | --- |
| Wireless LAN | 2, 5, 17, 19, 40, and 60 GHz |
| Mobile Broadband Systems | 40 and 60 GHz |
| Broadcasting/Distribution | 11, 12, 14, 18, 22, 28, and 40 GHz |
| Mobile Satellite | 11, 14, 20, 30, and 40—50 GHz |

Mobile Broadband Systems (MBS) is a forward-looking concept, the development of which is sponsored by Europeans but with global participation, advocating wireless systems exhibiting both broadband (2 to at least 100 Mb/s data rates) and mobile attributes. Its architecture encompasses all application areas described above and numerous others such as *digital video broadcasting* (DVB), *microwave video distribution system* (MVDS—38-42 GHz), *local multipoint distribution service* (LMDS—26-32 GHz), and *microwave multipoint distribution systems* (MMDS—a 2 GHz mid-70s technology).

These last examples are sometimes referred to by the oxymoron "wireless cable TV" with most installations now providing broadcast TV. Like traditional wired cable TV, however, wireless counterparts are increasingly adding multimedia service. Even without multimedia services the market is huge. Currently wired cable TV reaches over 60 million American homes, with wireless (mostly DBS) penetration now topping 10 million homes. Figure 4.20 shows a terrestrial wireless multipoint broadcast TV distribution system. Once radio signals are received and video TV signals recovered, distribution to individual TV sets in multi-unit dwellings is identical to that used for wired cable TV. As such, the only technical difference between wired and wireless cable TV systems is the transport from service provider head ends to customer premises network interface units.

**Figure 4.20**
Terrestrial wireless
broadcast TV
example.

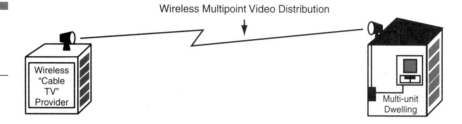

Figure 4.21 illustrates components of fixed-wireless local loop (WLL) facilities and shows their relationship to wired local loop, mobile telephone, switching and other telco facilities. The top portion of the figure credits this illustrated telco with using modern fiber-based digital loop carrier feeder and distribution local loop segments. In the past, copper twisted-pair runs extended from central office wire centers all the way to customer CPE. As alluded to previously, the technology to implement fiber-to-the-home (FTTH) or fiber-to-the-curb (FTTC) is well in hand and waiting only for a business case that justifies the investment.

Inside customer premises, depending on size and complexity, individual circuits are terminated on wire or fiber distribution frames (see Chapter 9) and/or network interface units. From there, inside-premises distribution systems complete the connection to user CPE. The figure implies that in large installations, connections between "outside" wiring and user CPE may involve on-premises circuit switching (PBXs) or LANs. Of course, just as with wired local loops, customers may opt to use cordless telephones or wireless LANs within premises, as alternatives to building wiring.

The middle part of the figure essentially substitutes WLL for wired facilities. Like mobile telephone system facilities, WLLs contain a sufficient number of base stations and radio links to handle engineered levels of traffic and geographically dispersed customer locations. Again, except for radio transceivers, customer-premises facilities and CPE can be identical to those served by wired local loops.

The bottom part of the figure depicts telco-owned mobile telephone system components and their connection to other telco facilities. As observed above, nothing stops subscribers from using cellular mobile terminals to satisfy their fixed indoor telephone requirements. Recall that the philosophy behind UMTS is that single instruments be designed to satisfy both indoor cordless and outdoor mobile telephone requirements.

**Figure 4.21**
Telco fixed-wireless local-loop components and options.

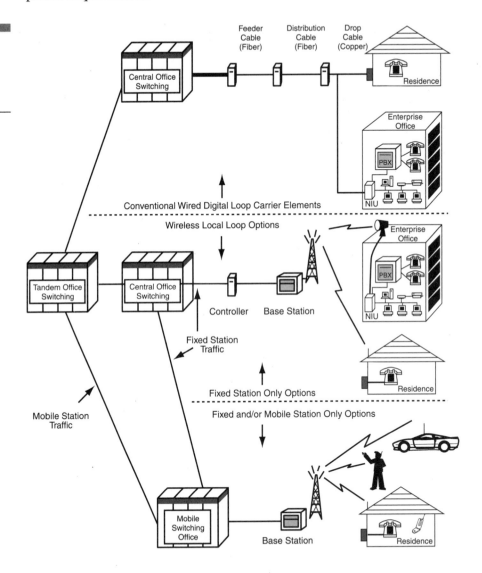

The major switch manufacturers have already brought to market full complements of equipment and software products to implement Figure 4.21 capabilities. As described in Chapter 6, one of the advantages of modern city-wide digital Centrex is that it delivers an extensive set of intelligent-network features to subscriber telephones. Nortel's Reunion broadband wireless product line offers the capability of "twinning" mobile and wired Centrex supported telephones such that either can be used and both respond to the same listed directory telephone number. Similarly, Lucent's FLEXENT Wireless Local Loop platform allows telcos to integrate wired, WLL, and mobile facilities so that customers can use existing standard telephone devices (such as cordless and corded phones, fax machines, PCs with modems, or answering machines) or any of the new breed just discussed.

# CHAPTER 5

# Multiplexing Concepts

The previous chapter covers how all types of guided and unguided media are arranged to provide circuits between telecommunications network nodes and devices. This chapter amplifies earlier references to *multiplexing*, that is, techniques that at transmitting nodes combine a number of individual communications channels into a common frequency band or a common bit stream for transmission, usually over single circuits. At receiving terminals, demultiplexing equipment or processes separate and recover the individual channel components of multiplexed signals. Multiplexing makes more efficient use of transmission capacity to achieve a low per-channel cost.

In theory, the dimensions of *space, time, frequency* or *code division*, separately or in combination, can be employed to keep individual channel signals from interfering with each other. Accordingly, these are the dimensions upon which alternative multiplexer designs are based. Although purists would maintain that space division may legitimately be used to describe a form of multiplexing, in essence with space division, a separate wire, fiber optic strand, or radio frequency link (that is a separate circuit) is assigned to each channel or signal. Practically speaking then, space division multiplexing means no multiplexing at all.

As a consequence, there are three basic multiplexing methods used in telecommunications systems. These are *frequency division multiplexing* (FDM), *time division multiplexing* (TDM), and *code division multiplexing*, all of which are treated in this chapter. As explained further below, *wavelength division multiplexing* (WDM), introduced in the fiber optic media discussion, is a lightwave version of FDM.

# Frequency Division Multiplexing

The simplest and oldest type of FDM divides a transmission circuit's frequency spectrum into subbands, each supporting single, full-time communications channels on a non-interfering basis with other, similarly multiplexed, channels. FDM multiplexing is suitable for use with analog carrier transmission systems. Standard AM and FM broadcast radio is an example of FDM where different stations occupy FCC-assigned portions of the standard broadcast band. Cable TV is another example where different stations are assigned frequency bands on a single-cable medium, and are selected by appropriate frequency conversion equipment using either standalone converter boxes or cable-ready TV set tuners.

Recently, more sophisticated frequency-hopping techniques, in which input signals are assigned a unique *frequency-hopping* pattern, as opposed to being an assigned to fixed subbands, have been fielded. In frequency-hopping multiplexing or multiple access systems, each individual input signal (or user) is assigned a hopping pattern in a manner that prevents or at least minimizes interference with other user signals. Frequency hopping has two principal advantages. First, it minimizes the effects of narrow-band interference by the ratio of the total bandwidth over which frequency hopping occurs to the interference bandwidth. Secondly, it tends to "hide" transmissions from unauthorized eavesdropping and it makes it much more difficult, in military situations, for adversaries to intentionally jam or degrade friendly communications.

# Asynchronous Time Division Multiplexing

The quality of signals undergoing frequency division multiplexing and demultiplexing processes is dependent upon the design and precision of analog FDM equipment components. Because TDM multiplex equipment incorporates digital techniques, it possesses the digital versus analog advantage over FDM multiplexing equipment.

In TDM, a transmission facility is shared in time rather than frequency. Figure 5.1 illustrates how this is accomplished using D-type TDM channel bank equipment. *D-type channel bank* is terminal equipment used for combining (multiplexing) individual voice-channel signals on a time division basis. D-type channel banks provide interfaces for *n* analog signal inputs. Each input signal is directed to a codec for conversion to PCM samples. Channel bank equipment and a repeated digital line together comprise a digital carrier system.

For the example in Figure 5.1, the codecs use the 64-kb/s North American standard previously described. Each individual codec produces an 8-bit sample at the rate of 8,000 samples per second (8 kHz), or one 8-bit sample every 125 microseconds. (A *microsecond* is one-millionth of a second.)

Each of these samples is interleaved in time sequence into a single high-speed digital signal, as shown on the figure, beginning with one signal designated as channel 1, and continuing through channel 24.

The process is then repeated for successive 8-bit samples from each channel. Following the 24th channel's sample a *framing bit (bit BF)* is added by the transmitting equipment to permit receiving equipment to derive frame synchronization and thus identify the time correlation between 8-bit information segments and each of the input analog channels. The central importance of maintaining transmitter-multiplexer and receiver-demultiplexer bit, byte, and frame synchronism is presented later in this chapter.

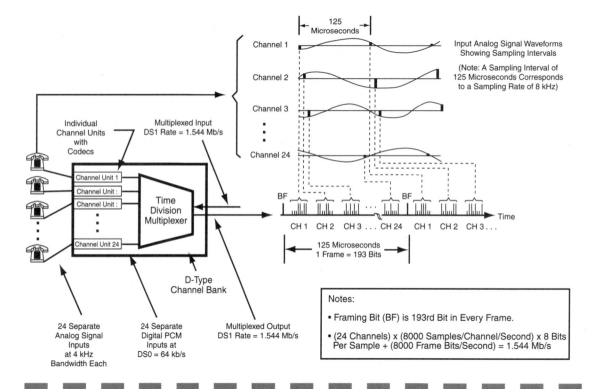

**Figure 5.1**   *Time division multiplexing (TDM).*

In a TDM system, a *frame* is a sequence of time slots, each containing a sample from one of the channels served by the multiplex system. The frame is repeated at the sampling rate, and each channel occupies the same sequence position in successive frames. The TDM process can be described as assigning a *time slot* in each frame on a high-speed TDM *bus* (also referred to as a *highway*), to each of *n* input channel signals. Here *bus* is defined as one or more conductors (or some medium) that connect a related group of devices.

In our example, the frame rate is 8 kHz, the sampling and frame interval is 125 microseconds, and each frame consists of 24 time slots. The TDM bus bit rate (and the TDM output bit rate) is 1.544 Mb/s, which is the product of 24 times 64 kb/s for each PCM input signal, plus frame bits occurring at 8 kb/s ($24 \times 64,000 = 1.536$ Mb/s and 1.536 Mb/s + 0.008 Mb/s = 1.544 Mb/s).

The example TDM process concentrates 24 voice channels into one equivalent four-wire, full-duplex circuit. Transmission facilities of this type are referred to as *T1-type digital carrier* (or simply *T carrier*) facilities. Other transmission facilities have capacities greater than 1.544 Mb/s. For example, the AT&T FT3 lightwave fiber transmission system supports 44.736 Mb/s, or 672 individual 64 kb/s, PCM channels. To permit higher levels of multiplexing concentration, a multilevel TDM digital signal hierarchy has been developed.

The *DS1 level* in the hierarchy corresponds to the 1.544-Mb/s TDM signal already described. Although not formally a member of the hierarchy, the *DS0 level* refers to the individual time slot digital signals at channel rates of 64 kb/s. Two DS1 signals are digitally multiplexed to produce a *DS1C level* signal containing 48 DS0 channels and they require a transmission facility that supports 3.088 Mb/s. Four DS1 signals comprise a *DS2 level* signal containing 96 DS0 channels, requiring a 6.312-Mb/s transmission facility. A *DS3 level* signal results from the digital multiplexing of 7 DS2 signals, supports 672 DS0 channels, and requires a 44.736-Mb/s transmission facility. Finally, a *DS4 level* signal supports 6 DS3 level signals (4,032 DS0 level signals) and requires a 274.176-Mb/s transmission facility.

The DS designation refers to the signal-level hierarchy and is independent of the type of carrier facility, except of course for compatibility requirements. For example, T1-types carry DS1 signals and FT3 lightwave systems carry DS3 signals. Often T1 and DS1 are used interchangeably, but strictly speaking, DS1 refers to the rate and the format of the first level of the hierarchy, while T1 refers to a particular equipment/cable arrangement used to transport DS1s among network nodes.

Figure 5.2 summarizes the *asynchronous digital signal hierarchy* (ADH) in the United States. Associated capacities and multiplexer designators defined originally by AT&T are today established by ANSI (American National Standards Institute). Note that in North America the level-4 line rate is 139.264 Mb/s rather than the rate originally defined.

**Figure 5.2**
The U.S. TDM asynchronous digital signal hierarchy (ADH).

| Level | Digital Transmission Hierarchy Level Designator | Number of DS0 Channels | Line Bit Rate | Multiplexing/Demultiplexing (MULDEM) Equipment Designators |
|-------|------------------------------------------------|------------------------|---------------|-------------------------------------------------------------|
| Level "0" | DS0 | 1 | 64 kb/s | |
| Level 1 | DS1 | 24 | 1.544 Mb/s | T1 or D"n" 24:1 |
| Level 1 | DS1C | 48 | 3.088 Mb/s | M1C 2:1 |
| Level 2 | DS2 | 96 | 6.312 Mb/s | M12 4:1 |
| Level 3 | DS3 | 672 | 44.736 Mb/s | M23 7:1  M13/MX3 28:1 |
| Level 4 | DS4* (Not Used) | 4032 | 274.176 Mb/s | M34 6:1 |

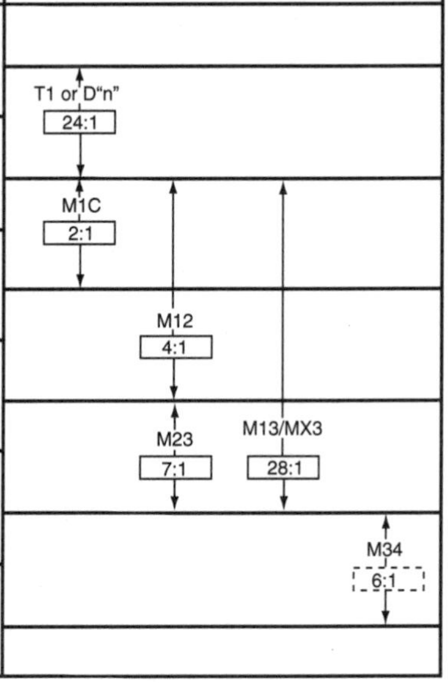

*In North America, the Level 4 lines rate is 139.264 Mb/s.*
*Note: Line rates are not multiples of 1.544 Mb/s, e.g., 28 x 1.544 = 43.232 Mb/s.*

In a network, a *clock* is a device that generates a signal that provides a timing reference. Clocks are used to control functions such as setting sampling intervals, establishing signal rates (bps), and timing of the duration of signal elements, such as bit intervals. In a completely synchronous TDM network, all participating nodes must use timing provided by a single master network clock. This does not happen today in most TDM networks.

Note in the digital signal hierarchy that line rates at level 2 and above are higher than the sum of constituent member rates. The added "overhead" bits are necessary to maintain synchronization in an inherently asynchronous TDM operation resulting from less-than-perfect network timing devices spread throughout networks. A synchronous digital signal hierarchy (SDH) developed to capitalize on broadband and highly stable fiber media is discussed following a summary of T1 and other ADH multiplexing operations and equipment. Complete definitions for the words asynchronous and synchronous along with more detailed multiplexing and transmission performance implications are offered later in this chapter.

In addition to the multiplexing functions just described, channel units in channel bank equipment provide proper voice frequency and signaling interfaces for central-office trunk, loop, or other assigned circuits. Thus, from analog inputs to a multiplexer, to analog outputs from a remote demultiplexer, TDM equipment is essentially transparent to both user information and signaling associated with various forms of telephone service. (Signaling associated with voice services is discussed in Chapter 6.)

## T1 Carrier Systems

Historically, T1 digital carrier systems have played a major role in the transition from analog to digital facilities. As one of the fastest growing LEC and IXC facility segments in the early 1990s, T carrier became commonplace in large private networks, and, due to declining costs and wider service options, was increasingly found in moderate-sized networks, growing to over 120 million miles in the North American continent. At the end of the decade, Bell Atlantic forecast a 35 percent growth rate for T1 service.

Introduced in 1962 within the old Bell System, T1 carrier supports 24 full-duplex voice channels using just two pairs of unshielded twisted pair (UTP) 19 AWG cable, constituting a low cost and reliable alternative to single, analog channel-per-twisted-pair voice frequency or other analog frequency division multiplexing (FDM) carrier systems. Because of its importance, this section expands on the preceding definitions to describe representative DS1 facilities, T1 carrier system signal structures, and operational capabilities.

Figure 5.3 illustrates a typical installation providing 24 analog voice-grade channel end-to-end connectivity between two business premises, via LEC and/or IXC networks. Within the business premises, 24-channel DS1 signals are generated by customer premises equipment (CPE), for example, the D-type channel bank equipment described above or one of the other types of T1 multiplexers described below. *Channel service units* (CSUs), at the least, and possibly *data service units* (DSUs) are required to connect CPE to DS1 service. As shown in the figure, CSUs terminate telephone company digital circuits, and protect networks from harmful signals. In fact, CSUs or equipment incorporating CSU functions must be designed in accordance with Part 68 of the FCC Rules and Regulations, entitled "Connection of Terminal Equipment to the Telephone Company."

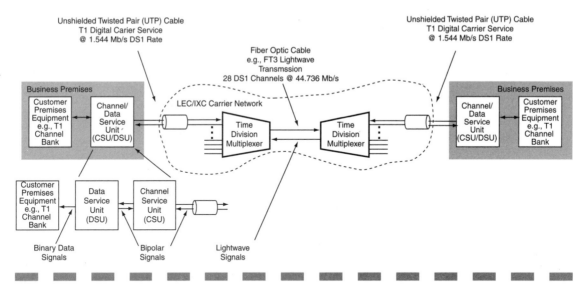

**Figure 5.3**    *DSI/T1 service connection arrangement example.*

Other CSU functions include line conditioning and equalization, error control, and the ability to respond to local and network "loop-back" circuit testing commands. In a telephone company, *line conditioning and equalization* is the spacing and operation of amplifiers so that gain provided by the amplifiers for each transmission frequency compensates for line signal loss at the same frequency.

DSUs, meanwhile, provide transmit and receive control logic, frame synchronization and timing recovery across T1 and other digital circuits (when these functions are not implemented in other CPE). DSUs also convert ordinary binary signals generated by CPE to special *bipolar signals.* Bipolar signals, described below, are designed specifically to facilitate transmission at 1.544-Mb/s rates over UTP cable, a medium originally intended for 3-kHz voiceband signals. As indicated in Figure 5.3, the trend today is for manufacturers to combine CSU and DSU functions in a single device for interface with DS1, DS3, and the new digital services described in Chapters 14 and 16. Representative CSU/DSU manufacturers include ADC Kentrox, Lucent, Coastcom, Timeplex, and Verilink. Some telcos and other service providers supply equipment from various manufacturers in delivering digital services.

Figure 5.3 depicts DS1 signals undergoing higher levels of multiplexing in long-haul end-to-end networks. An FT3 fiber optic transmission segment supporting 28 DS1 signals at a combined DS3 rate of 44.736

Mb/s is the example in the figure. Note that binary, bipolar, and light-wave signal formats appear at various points in the end-to-end connection, but that these conversions remain transparent to user CPE.

## DS1 Bipolar Signal Format

As described above, T1 carrier systems carry 24 channels on two pairs of copper twisted-pair cable (one pair for each transmission direction). Signals are applied directly to cable pairs in bipolar format in which positive and negative pulses, always alternating, represent one binary signal state (for example the state corresponding to a binary bit value of "1"). Under the *Alternate Mark Inversion* (AMI) convention, pulses correspond to binary "1"s and the absence of pulses corresponds to binary "0" states. With *Alternate Space Inversion* (ASI), pulses correspond to binary "0"s. Figure 5.4 illustrates the relationship between a binary pattern of "1"s and "0"s and signals in bipolar AMI format. The figure shows how "1"s are transmitted, alternatively as positive or negative pulses, and how "0"s are conveyed by the absence of a pulse. The T1 line's bipolar signal can be transmitted approximately one mile over 19/22 AWG twisted pair before requiring a *repeater*. A repeater, which is normally DC-powered over the line, regenerates signals; that is, it detects "1"s and "0"s, recovers symbol timing, and generates reconstructed versions of the received signal for transmission down the next segment (span) of lines.

The bipolar format has significant advantages, particularly when used with twisted-pair transmission media. One advantage is that single-bit errors can be easily detected at any point along the transmission path. For example, if at any instant in time, interference causes a pulse to be detected when in fact a no-pulse "0" condition is correct, a bipolar violation occurs. That is, instead of the polarity reversal (AMI) that should normally occur between two successive "1"s, two successive pulses of the same polarity will be generated. Bipolar violations also occur when interference causes the annihilation of positive or negative pulses, causing a transmitted "1" state to be falsely detected as a "0". Detection of bipolar violations permits error correction and improves the quality of end-to-end transmission.

In T1 carrier systems, receivers and repeaters must synchronize to their incoming signals by detecting the timing between positive and/or negative pulses. Since transmitted "0s" suppress the transmission of pulses, random binary information signals that generate long sequences of "0"s degrade the ability of receivers to "lock onto" and track the timing of incoming pulse signals. The original design in 1962 dictates that to maintain proper operation, on the average there must be at least one "1" in 15 bits, and at least three "1"s in 24 bits.

**Figure 5.4**
Binary signal and bipolar signal format relationship.

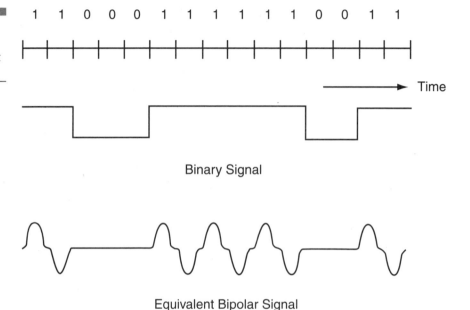

Various zero-suppression coding techniques have been implemented to prevent degradation due to long strings of "0"s, a characteristic of nonvoice signals. Some early techniques limit DS0 channel capacity to 56 kb/s. These techniques are acceptable for voice transmission but cannot support error-free 64-kb/s data communications. The technique known as *bipolar eight-zero substitution* (B8ZS) replaces a block of eight consecutive "0"s in the user data stream with a code containing bipolar violations in the fourth and seventh bits. When eight "0"s occur they are replaced with the B8ZS code before being multiplexed onto the T1 line. At the receiver, detection of the bipolar violations permits B8ZS codes to be converted back to 8-bit strings of "0"s, allowing the full 64 kb/s use of DS0 channels for data applications. Unfortunately, B8ZS is not compatible with early T1 facilities, and consequently carriers (particularly LECs) do not always have B8ZS facilities available. Telecommunications users must therefore be cautious when procuring digital transmission services for data applications.

# T1 Superframe (SF) Signal Format

Over the years a number of T1 framing formats were defined by AT&T. By far the most popular is the D3/Mode 3 D4 format for framing and channelization. Shown in Figure 5.5, the D3/Mode 3 D4 bit stream is organized into superframes, each consisting of 12 frames. Note at the top of the figure that every other framing bit BF is determined and coded to produce patterns that are easily recognizable by receiving DSUs. BFs marking odd frames (i.e., $BF_t$—or terminal framing bits) produce a sequence of alternating "1"s and "0"s, whereas BF(s) marking even-numbered frames (i.e., $BF_s$—or signaling framing bits) produce a sequence of alternating groups of three "0"s, followed by three "1"s.

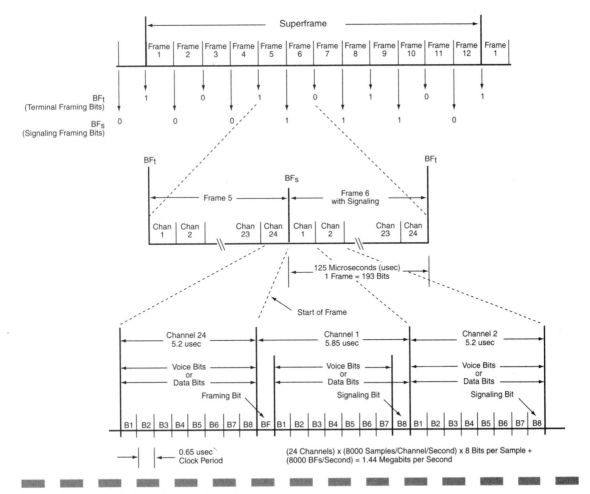

**Figure 5.5**  DSI/T1 signal framing format.

The bottom of Figure 5.5 shows that the framing bits (the 193rd and last bit in each frame) are inserted between the 24th and 1st channel words. Each channel word consists of 8 bits (named B1 through B8). Channel words represent eight-bit samples, taken at the rate of 8,000 samples per second, and correspond to 24 different sources of voice or data information. In the D-type channel bank example described, the channel words represent digital versions of 24 analog voice signals.

For voice transmission, signaling information—or information exchanged between components of a telecommunications system to establish, monitor, or release connections—must be transmitted with the channel voice samples. This is accomplished by sharing the least-significant bit (B8) between voice and signaling as shown in the figure, a process termed *robbed bit signaling* (RBS). The B8 bits carry voice information for five frames, followed by one frame of signaling information. This pattern of B8 assignment to voice and signaling is repeated during each successive group of six frames, as shown in the figure.

## T1 Extended Superframe (ESF) Signal Format

Widespread popularity and use of T1 circuits—that is, carrier-supplied DS1 service—within corporate networks makes their reliability crucial to business success. Yet until ESF was introduced, both users and carriers were often unaware of gradual deterioration of digital circuits until they failed catastrophically. ESF takes advantage of the fact that with modern technology, not every framing bit needs to be used for framing and synchronization. In ESF, the superframe is extended in length from 12 to 24 frames. Of the 24 framing bits in an extended superframe, six bits are used for framing (synchronization), six bits are used for error checking, and the remaining 12 bits are used for a 4-kb/s *facility data link* (FDL), a communications link between channel service units (CSUs) and telephone company monitoring devices.

To permit error detection, the sending CSU station examines all the 4,608 data bits within an extended superframe, and, using an algorithm, generates a unique *cyclic redundancy check* (CRC) code. This CRC code is transmitted using six of each group of 24 framing bits. Using received data bits, a remote station employs the same algorithm to recalculate the CRC code. The remote station detects the CRC code sent by the transmitting station, and compares it with the locally generated CRC code. Generally, if the two CRC codes match, there are no errors. This technique detects 98.4 percent of all possible bit-error patterns. What's more, this information is stored and provides a historical record of performance over time so that degradation can be detected before total line outages occur.

The FDL can be used to exchange performance data among CSUs with ESF monitoring capabilities, and to remotely control CSU operational and test modes from central network management positions. For networks with multiple tandemed T1 bipolar UTP and fiber optic cable links (as in the Figure 5.3 example), when combined with individual link performance indicators like the number of bipolar violations and various failure/degradation alarm signals, the end-to-end CRC data provide powerful predictive and diagnostic tools for T1 network management, control and maintenance. The real motivation for, and advantage of, ESF lies in startling increases in T1 uptime and availability. Reports indicate that once installed, ESF can result in 70 percent fewer incidents of T1 line outages, and 30 percent reductions in T1 circuit downtime per incident. Although T1 carrier originally connoted a specific transmission medium, the term T1 has taken on a generic meaning in the industry and is often used to describe all manner of transmission services and equipment operating at 1.544 Mb/s.

## Digital Multiplexer/Digital Cross-Connect System Equipment Types

Multiplexers and switches are fundamental, related building blocks in telecommunications systems. As explained in Chapter 6, today's digital circuit switches employ time division multiplexing techniques, and emerging technologies (such as Asynchronous Transfer Mode [ATM] discussed in Chapters 7, 8, and 14) seamlessly combine both techniques. Currently, the simplest digital multiplexers are channel banks. Figure 5.6 illustrates the difference between simple channel bank multiplexers and the next level of sophistication, often referred to in the literature as *flexible* or *single aggregate* T1 multiplexers. Channel banks usually accommodate 24, 48, or 96 plug-in channel units and generate DS1 signals at 1.544 Mb/s, which correspond to 24 DS0, 64-kb/s signals, that in turn correspond to 24 analog input signals.

Flexible T1 multiplexers support a variety of digital input signals ranging from sub-rate (less than DS0 rate) to multiple DS0 rate digital signals. *Networking multiplexers* (not shown in the figure) represent a more sophisticated level of T1 multiplexers that combine capabilities previously described and permit simultaneous connection to multiple T1 transmission facilities for automatic or manual reconfiguration in response to circuit outages or changing traffic conditions.

**Figure 5.6**
Time division
multiplexer
configurations.

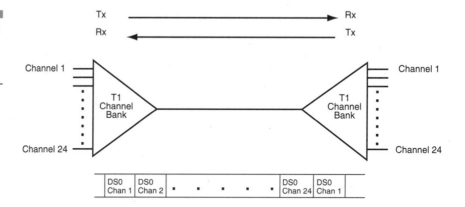

T1 Channel Bank Multiplexer

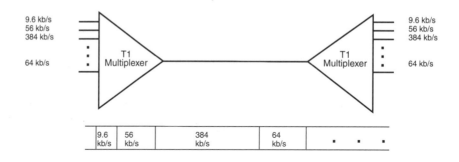

Flexible T1 Multiplexer

The *digital access and cross-connect system* (DACS) is a more recent generation of switching/multiplexing equipment that permits per-channel DS0 (64-kb/s) electronic cross-connection from one T1 transmission facility to another, directly from the constituent DS1 signals. Before DACS availability, if a DS0 signal arrived at a network switching node in one DS1 transmission facility and needed to be relayed to another DS1 transmission facility, an arrangement of back-to-back channel bank multiplexers and switches was necessary, as shown in the middle portion of Figure 5.7. That is, incoming composite 24-channel versions of the DS1 signals arriving on one DS1 transmission facility had to be

**Figure 5.7**
DCS, DS1, and DSO grooming; DSO concentration; DSO segregation operations.

*Photo courtesy of COASTCOM.*

16 Port DCS          8 Port DCS

COASTCOM DXC
DCS Products

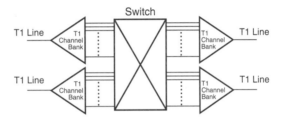

Conventional DS0 Channel
Switching Among DS1 Lines

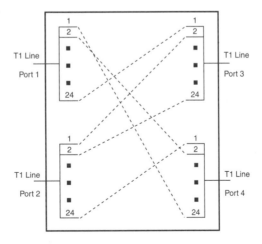

Digital Cross-Connect System
(DCS) DS0 Channel Switching Among
DS1 Lines

demultiplexed into 24 separate DS0 signals, each connected to a switch, which in turn had to be connected to other channel banks for remultiplexing, and finally connected to other DS1 transmission facilities.

In contrast, a DACS allows the 24 individual DS0 channels in a particular T1 line to be distributed among any of the other T1 lines connected to the DACS, as shown in the lower portion of Figure 5.7, with no requirement for channel banks or external switching. The cross-connect capability (referred to as a flexible add/drop capability) of DACS can be used to *groom* or segregate DS0 channels by type (e.g., voice, data, special services 4-wire/2-wire, etc.), increasing the fill of T1 lines and enabling more efficient use of resources.

The compact COASTCOM (a manufacturer of CPE T1 multiplex equipment) models shown in the figure are sized to support 8 and 16 T1 lines respectively, while some Lucent models terminate up to 127 lines. Moreover, connection paths within DACS are software controllable, which facilitates network management, and eliminates the need for manual connections. While DACS configuration changes are not made on a call-by-call, or circuit-switched basis, a DACS does implement channel switching—that is, the ability to reconfigure networks in response to outages, time-of-day traffic variations, or to accommodate growth and organizational changes.

Representative multiplexer manufacturers include Lucent, COASTCOM, Netrix, Timeplex, and others. DACS manufacturers include Lucent, Bytex Corp., COASTCOM, Frederick Engineering Inc., Rockwell Network Transmission Systems, Tellabs, and others. Lucent refers to its proprietary DACS products as *digital access and cross-connect systems* (DACCS).

Figure 5.8 illustrates the so-called *fractional T1* service that IXCs and some LECs are now able to offer using flexible multiplexing and DACS equipment. In the past, T1 service was available only in integral 24-channel DS0 increments. Business users with lesser requirements often had to resort to more expensive single-channel data services, such as Digital Data Service (DDS), or purchase a full T1 service and pay for the unused capacity. With fractional T1, users can obtain service in DS0 increments at per-channel rates that are almost always less expensive than other alternatives. The lower portion of Figure 5.8 illustrates cost comparisons of MCI WorldCom basic services for 2,500-mile circuits.

Fractional T1 access service is not offered at all transmission speeds by all LECs. Speeds of 128 kb/s and 384 kb/s are commonplace, but there are still areas where a full T1 access line must be purchased to gain access to IXC networks and their fractional T1 service offerings.

The cost justification for such instances must be established on an individual basis, often necessitating a thorough knowledge of carrier services and tariffs and computer-based network modeling and optimization tools. Applications of such tools to the design of voice and data networks are treated in Chapters 12 and 15, and, where appropriate, throughout the remainder of this book.

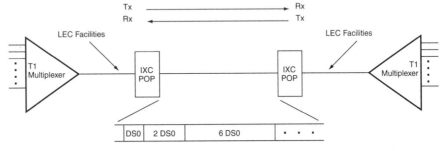

**Figure 5.8**
Fractional T1 service examples.

Fractional T1 Example

| MCI WorldCom Monthly Interoffice Rates (2500 miles) | | |
|---|---|---|
| **Service** | **Cost** | **Cost/DS0** |
| Fractonal T1 Services | | |
| 64 kb/s Digital Private Line | $ 1,272 | $ 1,272 |
| 128 kb/s Digital Private Line | $ 2,544 | $ 1,272 |
| 384 kb/s Digital Private Line | $ 7,632 | $ 1,272 |
| Full T1 Service | $16,460 | $ 686 |

# Higher-Order Multiplexing

Figure 5.9 illustrates how higher-order DSn asynchronous multiplexed signals are built up from lower-order tributaries. Here, *tributaries* are defined as low-rate inputs to multiplexers. The lowest order DS1 multiplexing employs a *D-Type* channel bank like the one described above. Among other characteristics, in original Bell System terminology, a D-Type multiplexer includes analog-to-digital coders and decoders and *sampling gates* to sample transmitter input signals and deliver receiver output samples to correct channels. Note that DS1-level multiplexing is byte-interleaved. That is, 8-bit (byte) encoder samples from one

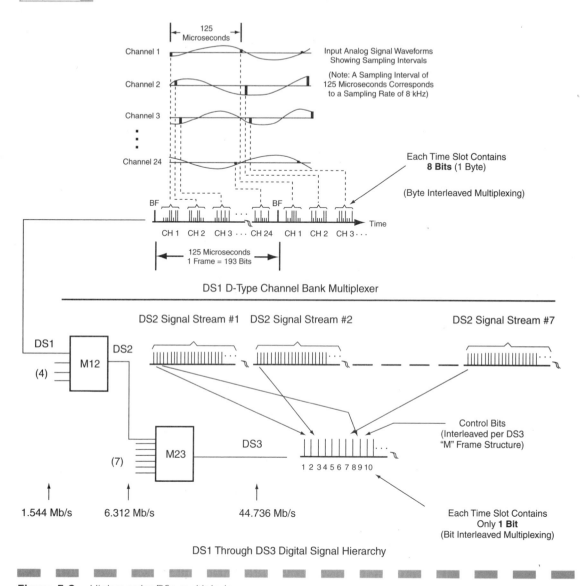

**Figure 5.9** *Higher-order DSn multiplexing.*

input channel are placed in assigned time slots, interleaved among other channel 8-bit samples.

The next level multiplexer combines four DS1 1.544 Mb/s signals into a 6.312-Mb/s DS2 signal, followed by a multiplexer that combines seven DS2 signals into a single DS3 44.736-Mb/s signal. As noted, multiplying four times 1.544 equals 6.176 Mb/s and not 6.312 Mb/s, the actual DS2 multiplexer output signal rate. The principal reason for this is that the four separate DS1 tributary signals are normally asynchronous with each other and with the clock used to generate the output DS2 signal. Similarly seven times the DS2 rate of 6.312 is 44.184 Mb/s, less than the DS3 rate of 44.736 Mb/s.

As explained below, one method of compensating for the possibility of "dropped" or "repeated" input signal bits caused by asynchronism is to allocate "extra" bits to each DS3 frame (i.e., more bits than would be required in synchronous digital multiplexing operations). To fit extra bits into a nominal frame period requires higher DSn nominal bit rates than would be expected from the ratio of number of channels each level supports. Please note that DS3 involves bit interleaving as opposed to DS1's byte interleaving process.

### SYNCHRONIZATION, FREQUENCY, AND TIME REFERENCE STANDARDS

Figure 5.10 shows the impact of asynchronous sampling. In the top half of the figure, it is evident that the sampling rate set by the receiver clock is too fast relative to the rate of the incoming bit stream. As a result, bits labeled "B," "E," and "H" are all repeated. The bottom half of the figures illustrates the opposite possibility, that is, a case where the sampling rate set by the receiving clock is too slow. In this case, incoming bits labeled "C," "G," and "K" are dropped. For voice traffic, an occasional added or missing bit (slips) might only cause an audible "pop" and be tolerable. In data traffic, however, missed or added bits might be disastrous, and that is why various mechanisms have been developed to preserve bit integrity in asynchronous systems.

In understanding and appreciating the operation and advantages of synchronous versus asynchronous telecommunications, a basic familiarity with how frequency and time reference standards are used to achieve network synchronization is indispensable. As explained in Chapter 3, bit timing and bit rate or frequency are inversely related. For example, for a signal with a rate or frequency of 100 bits per second, the time duration of each bit is a-hundredth of a second. In more general terms, bit duration is equal to the reciprocal of the frequency at which bits occur, or the number "1" divided by the bit rate.

Establishing accurate time and stable frequency reference standards for use in telecommunications networks involves three different but often related concepts of *date, interval* and *synchronization*. In some financial and military information systems, precise global date epoch agreement is crucial. Effective January 1, 1972, the *International Consul-*

*tative Committee for Radio* (CCIR) created a *Coordinated Universal Time* (UTC) to satisfy such requirements in telecommunications networks.

**Figure 5.10**
Impact of asynchronous sampling with frequency offsets.

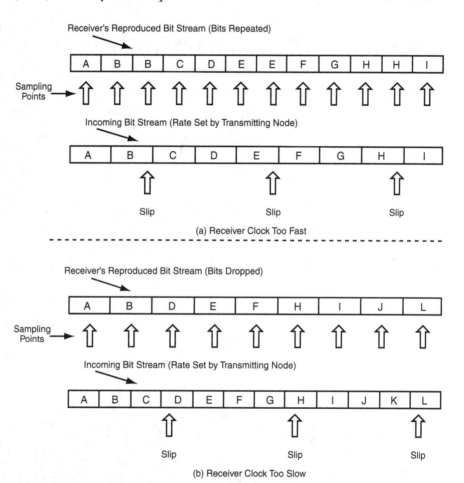

In other applications, accuracy in the measurement of the duration of phone calls, or multiplexer bit or frame intervals, may be far more important than absolute date-time epochs marking the beginning or the end of certain activities. Synchronization, the third important frequency-time concept, perhaps has more network relevance than either absolute date-epoch or interval. For example, it is not necessary for an orchestra to begin a concert at a precise hour, minute, or second of the day, but it is essential that orchestra members begin at the same instant and maintain the same tempo or pace. Similarly, in digital data transmission, the exact instant in time that marks the beginning of frames, bytes, characters, words, etc., is critical to error-free information recovery, whereas the precise hour, minute, or second of the day that marks such events is, in most cases, irrelevant.

Prior to 1924, signal synchronization in public switched telephone networks (PSTNs) was not required. In that era, voice signals were each sent separately in the frequency band in which they were generated (baseband), that is frequencies less than 4kHz. Although this eliminated the need for receiver decoding or demultiplexing equipment, it also meant that separate media circuits were required for each signal channel. As noted above, if baseband data is to be properly recovered, the introduction of more efficient carrier and multiplexing systems makes "matching" transmitter and receiver carrier frequencies, clock frequencies, and sampling and frame timing necessary.

Moreover, as higher levels of multiplexing evolve, so does the need for higher-accuracy frequency references. Figure 5.11 tracks the increasingly accurate frequency references that evolved as the number of multiplexed channels carried by single Bell System transmission systems grew.

**Figure 5.11**
Frequency reference accuracy progression related to Bell System UTP and coaxial cable carrier channel capacity.

| Channels/Circuit | Carrier/Multiplex System | Reference Frequency Accuracy (1 Part In) |
|---|---|---|
| 1 (Baseband) | None (<1924) | N/A |
| 12 | J, K | $10^6$ |
| 600 | L1 (1946) | $10^6$–$10^7$ |
| 1860 | L3 (1953) | $10^7$–$10^8$ |
| 3600 | L4 (1967) | $10^7$–$10^8$ |
| 10,800 | L5 (1978) | $10^9$–$10^{10}$ |

By the 1980s, the need for accurate frequency and stable timing signals to coordinate AT&T's analog and digital transmission was filled by the Bell System Reference Frequency Standard (BSRFS) and a nationwide distribution network (called the Bell System Carrier Synchronization Network). The center of this tree-like network in Hillsboro, Missouri, originally used three interlocked cesium frequency standards accurate to one part in $10^{11}$. (A 1,000 Hertz frequency source is accurate to one part in $10^3$ if its frequency never deviates more than one Hertz from its intended 1000 Hertz rate). A cesium standard is a *primary reference source* (PRS) frequency generating equipment in which a specified hyperfine transition of cesium-133 atoms is used to control its output frequency, the accuracy of which is intrinsic, i.e., achieved without calibration.

The ANSI T1.101 committee defines the four levels or "strata" of accuracy requirements for frequency generators (oscillators or clocks) used in North American networks, shown in Figure 5.12. The PRS, Stratum 1 level demands one part in $10^{11}$, an accuracy that can be met only by a cesium clock on site, or via the global positioning satellite system (GPS) or an equivalent remote source. The ITU in G.811 recommends a PRS clock at each international switching center, and use of the aforementioned Coordinated Universal Time (UTC) as its reference. With this approach, under normal conditions, "slip" rates on any 64-kb/s DS0 channel should be less than one in 70 days. The benefits of ultra-accurate clocks and the pivotal role they play in enabling synchronous telecommunications network operations, are amply demonstrated in the remaining parts of this book.

**Figure 5.12**
ANSI T1.101 Strata
accuracy
requirements for
network frequency
references.

| Stratum Level | Reference Frequency Accuracy (1 Part In) |
|---|---|
| Stratum 1 (Primary Reference Source) | $10^{11}$ |
| Stratum 2 | $1.6 \times 10^8$ (0.0025 Hz at 1.544 MHz) |
| Stratum 3 | $10^6$ (7 Hz at 1.544 MHz) |
| Stratum 4 | $10^6$ (50 Hz at 1.544 MHz) |

## SYNCHRONOUS, ASYNCHRONOUS, PLESIOCHRONOUS AND ISOCHRONOUS DEFINITIONS

The words synchronous and asynchronous may, for example, apply to all or most links in a network, but not to an overall network. Every transmitter-receiver supporting DS1 links, must be pair-wise synchronous or—in other words—synchronized. Yet, as stated above, the original and prevalent implementation of the digital signal multiplexing hierarchy makes it accurate to describe it as an Asynchronous Digital Hierarchy (ADH). What justifies this apparent contradiction is the fact that although each ADH link transmitter-receiver pair are perfectly synchronized, clock bit rates and/or framing among different pairs are not.

Moreover, when we refer to asynchronous transmission as bytes or characters marked with distinctive START and STOP bits (as in Chapter 3), we mean something entirely different from the lack of timing among various network link clocks. Since there are many possibilities for confusion, it is time to provide some strict definitions and clarify their usage, keeping in mind the date, interval, and rate aspects associated with time and frequency synchronicity.

Various telecommunications dictionaries define *synchronous* as an adjective relating to a condition; perhaps the best definition is "a condition among signals or signal components derived from the same frequency or timing reference source and hence identical in frequency or rate." Note that in becoming synchronized to a DS1 signal, a receiver may lock onto its bit-stream rate, before it locks onto or detects frame-marking components of the input signal. In this intermediate state, it can properly be said that the transmitter and receiver are bit synchronous but not frame synchronous. Later in its acquisition cycle, a receiver might achieve bit, byte, frame, and superframe synchronization—as implied in the DS1 extended superframe discussion earlier in this chapter.

Since the DS1 frame contains exactly 193 bits per frame (see Figure 5.1), the frame rate is precisely the bit rate (nominally 1.544 Mb/s) divided by 193. Most experts agree that although the bit and frame rates are numerically different, in a given signal they are perfectly synchronized and in fact synchronous.

Conversely, *asynchronous* is defined as "a condition among signals or signal components derived from different frequency or timing reference sources and hence having different frequencies or rates." *Plesiochronous* refers to a condition intermediate between synchronous and asynchronous. Plesiochronous is defined as, "a condition among signals or signal components derived from different frequency or timing reference sources, but of nominally the same frequency, within some stated degree of precision." For example, because they are not derived from the same atomic clocks, DS1s generated by AT&T and MCI are plesiochronous. Although they are arbitrarily close in frequency, over the long term "slips" can occur if one company transports the other's traffic without some means for tracking and correction.

*Isochronous,* the last term needing treatment here, generally refers to a class of traffic in which signal frequency or timing characteristics must be preserved if received information is to be intelligible or useful. Data traffic, e-mail for example, exhibits no intrinsic timing or rate dependence. Transmitting data traffic at a lower rate has no impact other than slowing delivery. However, voice and video traffic are intimately tied to transmission timing and rate. Slow the delivery rate and at best altos may sound like bassos, and at worst in digital voice systems, become totally unintelligible.

One of the advantages of circuit-switched and TDM service is that sampling and transmission rates intrinsically preserve sensitive isochronous traffic timing and frequency characteristics. Some packet-switching services discussed in Chapter 7 offer no such intrinsic guarantees. *Quality of service* (QoS) in voice applications normally encompasses a wide variety of parameters including loudness, signal-to-noise ratio, and others. When describing packet switched networks, QoS usually refers to the ability to meet isochronous traffic latency and delay requirements. As a rule of thumb, latency or end-to-end delay should be less than 150 msec with no more than 30 msec of delay variance of jitter in networks carrying voice traffic. QoS considerations arise throughout the remainder of the book.

With these definitions clarified, we can complete the higher-order ADH discussion.

Figure 5.13 depicts the DS3 frame structure. The top of the figure shows that DS3 "M" frames consists of seven subframes of 680 time slots each, for a total of 4,760 time slots total. In this part of the figure, the subframe structure of a single DS3 frame is laid out from left to right.

To present the structure of the subframe in more detail, the lower part of the figure arrays the DS3 frame in rows (corresponding to subframes) and columns (blocks). By convention, the first bits transmitted in a frame correspond to the top-most and left-most position (marked start for "start of frame"), with bits in adjacent columns in row one next, then successive rows following. Each subframe/block intersection contains user information and various types of control bits used to manage and transport user information through the network. In the array portion of the figure, the end of frame occurs at the bottom, right-most subframe/block intersection. The method of using arrays to depict complex signal structures is a standard industry practice and used in the remaining parts of this book.

In Figure 5.13, the "C" bits are bit-stuff control bits. The entire eighth block is reserved for actual stuff bits. Clearly, much of the DS3 structure is defined to accommodate asynchronicity among tributary signals and local DS3 node clocks. The total frame length is slightly less than the standard PCM 125 microsecond voice sampling interval to ensure that isochronous voice information can be transported without degradation.

Overall, digital multiplexing transmission services are key to the modernization of voice services, growth of video conferencing, and the development of metropolitan and wide-area data communications. In spite of the limitations of asynchronous and fixed bandwidth channelization operations, for the foreseeable future, ADH TDM systems, originally designed for voice service, will continue to play a role, be adapted for voice, data, video, and integrated applications, and constitute a major source of traffic for rapidly growing global, fiber optic media-based synchronous transmission facilities, discussed next.

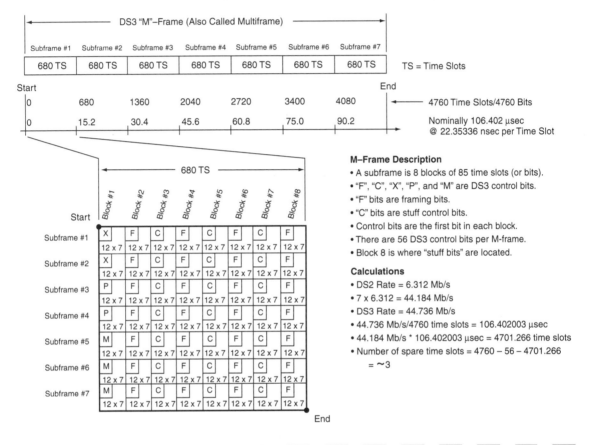

**Figure 5.13** Asynchronous DS3 signal structure.

# Synchronous Time Division Multiplexing and the Synchronous Optical Network (SONET)

Seeming to be an efficient means of transmission for optical networks, concepts for synchronous digital transmission first appeared in the early 1980s. On the international front, ITU's I.121 *Broadband Integrated Services Digital Network* (BISDN) (see Chapter 8) umbrella standard addresses switching, multiplexing, and transmission facilities able to meet expanding broadband, multimedia requirements. In particular, the G.707/708/709 recommendations define *synchronous transport module* (STM-n) signals as building blocks for a *synchronous digital hierarchy* (SDH). SDH is a digital transport structure that manages user information payloads and transports them through synchronous transmission networks.

In the same timeframe, SONET, an abbreviation for *Synchronous Optical Network*, was conceived by R. J. Boehm and Y. C. Ching of Bellcore[1]. By the end of 1984 the Bellcore work was submitted to ANSI's T1 Committee and in 1988 SONET interface standards, to be used in LEC/IXC optical networks, were approved. Like the ITU standards, SONET defines a hierarchy of *synchronous transport signals* (STS-n).

SONET defines standard optical signals, a synchronous frame structure for multiplexing digital traffic, and operations procedures so that fiber optic transmission systems from different manufacturers/ carriers can be interconnected. Figure 5.14 illustrates the relationship between ITU and SONET n-level electrical and optical carrier signals, along with line bit rates for each. Note that STS-n signal rates are all exact "n" multiples of the STS-1 rate of 51.84 Mb/s. STM-n rates are also integer multiples of the STM-1 rate.

The SDH is similar to the DSn TDM signal hierarchy just described. However, synchronous multiplexing allows component service signals to be combined into higher rates in a manner that simplifies multiplexing and demultiplexing, and offers easy access to SONET payloads. Figure 5.15 compares ADH and SDH multiplexing, revealing both similarities and differences. The figure's message is that synchronous multiplexing is a "one-step" process that makes network-wide add/drop and cross-connect simple and economical. By contrast, since asynchronous multiplexing is primarily designed for point-to-point multiplexing, add/drop and crossconnect at intermediate nodes is an unwieldy multi-step process. Thus, the enormous digital crossconnect benefits available at DS1 levels are not extensible to higher ADH levels.

---

1 Bell Communications Research (Bellcore) was formed by federal mandate in 1984 as a result of AT&T's divestiture to provide research and development services to the then seven RBOCs. It was purchased by Science Applications International Corporation (SAIC) in 1996 and is now called Telecordia.

**Figure 5.14**
Synchronous Digital
Hierarchy (SDH)
signals.

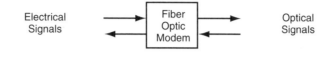

| STM-n Level | STS-n Level | OC-n Level | Line Rate (Mb/s) |
|---|---|---|---|
| | STS-1 | OC-1 | 51.84 |
| STM-1 | STS-3 | OC-3 | 155.52 |
| | STS-3c | OC-3 | 155.52 |
| | STS-9 | OC-9 | 466.52 |
| STM-4 | STS-12 | OC-12 | 622.08 |
| | STS-12c | OC-12 | 622.08 |
| | STS-18 | OC-18 | 933.12 |
| | STS-24 | OC-24 | 1244.16 |
| | STS-36 | OC-36 | 1866.24 |
| STM-16 | STS-48 | OC-48 | 2488.32 |
| STM-64 | STS-192 | OC-192 | 9953.28 |

STM-n = ITU–Synchronous Transport Module Level "n"

STS-n = SONET–Synchronous Transport Signal Level "n"

OC-n = SONET–Optical Carrier Level "n"

The bottom portion of the figure correctly implies that DS1 signals, or any ADH level signal can be directly multiplexed into or demultiplexed from high-rate (155.520 Mb/s and above) SDH signals. This capability results in significant equipment economies at switching center and signaling transfer points, and essentially extends today's programmable DS1-limited DACS benefits to all multiplexing levels.

It may appear strange to some readers for the first block diagram representation of an SDH synchronous multiplexer to portray an example in which all input signals are asynchronous ADH signals. In fact, standards for, and the technical feasibility of, networks in which most components operate synchronously won't make asynchronous equipment inventories disappear overnight. The good news is that SDH designs accommodate older equipment in ways that produce immediate technical and operational improvements and economies.

Figure 5.16 presents graphically some internal details of how STM-n multiplexers handle various U.S. and European ADH signals. In ITU terminology, for example, DS1 tributaries are first mapped into C11-containers. Attaching path overhead and assigning a pointer translates C11 containers to VC-11 virtual containers and TU-11 tributary units, respectively. Four TU-11s can be mapped (multiplexed) into a tributary unit group (TUG-2) and seven TUG-2s mapped into a higher order virtual container, VC-3. Lastly, three VC-3s, after being mapped into

**Figure 5.15**
ADH and SDH
multiplexing
compared.

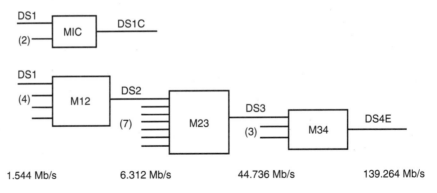

(a) U.S. DSn Asynchronous Digital Signal Hierarchy Multiplexing

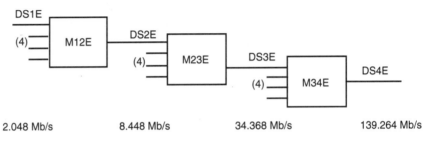

(b) European DSn E Asynchronous Digital Signal Hierarchy Multiplexing

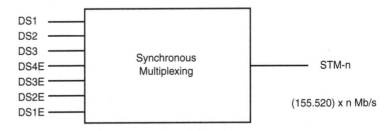

(c) Synchronous Digital Hierarchy Multiplexing

administrative units (AU-3s), are mapped to administrative unit groups (AUGs), "n" of which may be combined in an STM-n signal.

Although the figure's focus is on assembling broadband, high-capacity STM-n signals from lower rate tributary inputs, it must be emphasized that the process is entirely reversible. Furthermore, unlike ADH processes, once frame synchronized, an ingenious system of *pointers* permits direct extraction even the lowest order tributary signals.

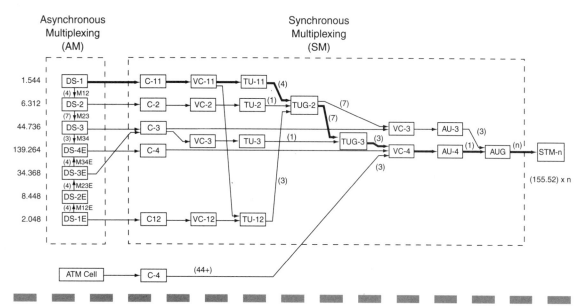

**Figure 5.16** *Synchronous Digital Hierarchy multiplexing structure.*

While not all readers require in-depth SDH multiplexing knowledge, what is essential is understanding its underlying principles of operation and its capabilities. The fact is that ITU and SONET SDH designs provide a solid basis for powerful, flexible, reliable, and cost-effective means for transporting both existing ADH and fast-growing packet and cell-based multimedia traffic over high-capacity fiber optic transmission systems.

The next two figures provide readers with a practical vision of how STM/STS-based SDH multiplexers accomplish their magic. Figure 5.17 depicts an STS-1 frame in a row-column array (similar to that used in Figure 5.13 to represent DS3 frames), with the start of frame at the upper left-most corner, and the end at the lower right corner.

STS-1 is 125 microseconds long and consists of two parts, a *section overhead* (SOH) portion occupying the first three columns and all nine frame rows; and, a *synchronous payload envelope* (SPE) into which user information is mapped. Each row/column intersection represents 8-bit bytes so that each frame comprises a 9-row by 90-column array of bytes. A single column is used to embed *path overhead* (POH) in the SPE providing communications between SPE assembly and disassembly points, that is nodes where user information is mapped to and from SPEs. Section and path overhead support link and network operations and maintenance, error control, pointer adjustment for timing offsets, and other relevant network functions. (*Pointers* are overhead bytes that provide a simple means of dynamically and flexibly phase-aligning STS payloads, thereby permitting ease of dropping, inserting, and cross-connecting payloads in networks. Transmission signal wander and jitter are also readily minimized with pointers.)

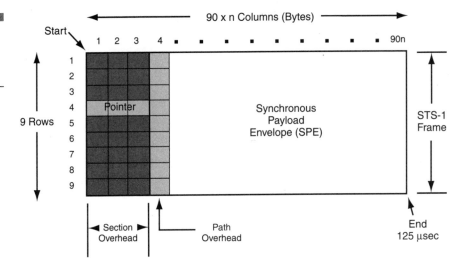

**Figure 5.17**
SONET synchronous transport signal (STS) frame structure.

To explain user information-to-SPE mapping, we will begin with a DS1/DS3 example. Because each DS1 frame is 24 bytes plus one frame bit long, and since SONET allocates 3 bytes of DS1 specific overhead, each DS1 frame [or in SONET terms virtual tributary (VT)] uses 27 SPE bytes. DS3 mapping places bytes from each of 28 DS1 VTs consecutively in adjacent columns (byte-interleaved), using a total of 84 columns (3 columns per DS1 × 28 DS1s per DS3). Since a single SPE contains 87 columns with one used for POH, the SONET DS3 mapping leaves 2 columns vacant, and normally "stuffs" them with all zeros. Because SONET standards define STS-1 mappings for ADH signals, it is possible to assemble and disassemble SPEs with equipment from different manufacturers.

STS-3 and higher-rate SONET signals are generated by byte interleaving STS-1 signals. Thus, STS-3 frames begin with 9 instead of 3 SOH columns and STS-3 SPEs contain 261 columns. Each STS-3 frame, therefore, comprises a 9-row by 270-column array of bytes. A single STS-3 can accommodate three DS3s or 2,016 DS0 channels. Figure 5.18 shows the STM-1 frame structure (with a bit-rate capability of 155.52 Mb/s, equal to that of STS-3), annotated using ITU terminology and designators.

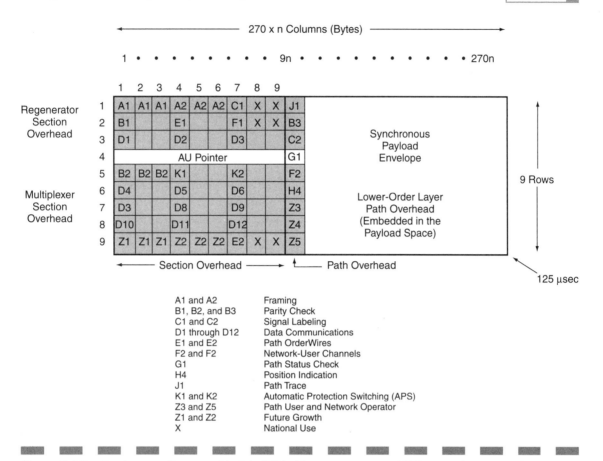

**Figure 5.18** Synchronous Transport Module (STM-n) frame structure and ITU annotation nomenclature.

SONET provides mappings for transporting services with bandwidths exceeding STS-1 capabilities. Signals associated with such services, dubbed concatenated, are treated as a single payload that does not comprise byte-interleaved lower-rate signals. SONET signals for transporting concatenated signals are designated by adding a "c" to the STS-n SDH nomenclature (e.g., the STS-3c and STS-12c entries in Figure 5.14). STS-nc uses a special section overhead pointer and only a single 9-row column for path overhead in the SPE.

At SONET nodes, STS-n output signals are generated using local timing. Since tributary or transiting signals may not be of identical frame timing or frequency, as mentioned earlier, SONET permits SPEs to "float" and uses pointers to keep track of actual SPE positions relative to locally defined frames. Frequency offsets are accounted for by incrementing pointer values one byte at a time. Figure 5.19 illustrates this floating SPE capability. Chapter 8 and Figure 8.7 explain how Asynchronous Transfer Mode (ATM) cells are mapped to SONET STS-nc payloads.

# Wavelength Division Multiplexing

Wavelength division multiplexing (WDM) increases the signal-carrying capacity of existing fiber spans (relative to spans operating with no WDM), by combining two or more optical signals of different wavelengths on a single fiber. WDM cannot be explained without a precise understanding of the term wavelength and its relation to frequency, so lets begin there.

*Wavelength* is a distance parameter associated with the motion of waves through or along a medium or free space. By wave motion, we mean the propagation of energy (or some kind of disturbance). One of the easiest types of wave motion to visualize is that produced by dropping a pebble in a pond of water. The waves radiating outward are familiar to all. An observer at a fixed distance from pebble impact would see water rise and fall in a smooth sinusoidal manner, resembling the single frequency analog electrical signal amplitude versus time signal representation shown in Figure 3.1, Part (c).

Looking down at a pond after a pebble has fallen, one can directly observe what is meant by "wavelength" (usually denoted by the Greek letter lambda [λ]). Wavelength is the distance between adjacent crests (or troughs) of waves radiating outward. In wave motion, propagation velocity is the speed with which a given value (e.g., a crest value) of the sine function (also referred to as a given phase of a sine function) travels away from the energy source or disturbance. This velocity (also termed phase velocity), usually represented in equations by the letter "v," is given by the product of frequency (the rate that water rises and falls at any fixed point from the disturbance) and wavelength—expressed as $v = f\lambda$. This fundamental relationship applies whether we are talking about water waves, vibrating strings, sound waves, or electromagnetic (that is radio or light) waves.

Since wavelength and frequency are directly related by propagation velocity in carrier media, wavelength division multiplexing could also be described as frequency division multiplexing. Industry practice, however, uses the term WDM for optical signal multiplexing. The ITU has defined 40 standard wavelengths that fiber optic WDM equipment manufacturers may use.

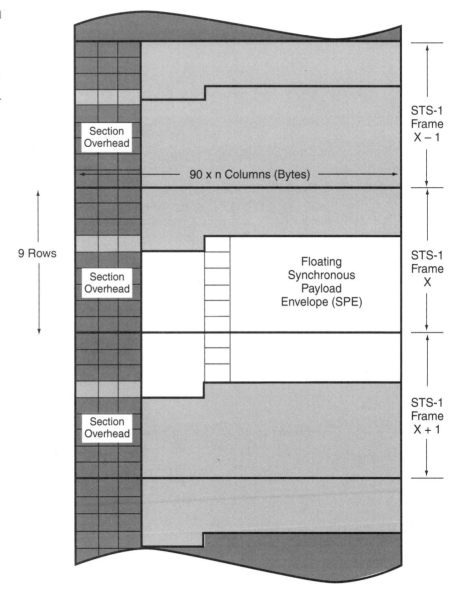

**Figure 5.19**
SONET offset
phase/frequency
floating SPE handling
example.

Figure 5.20 compares block diagram representations of fiber-based transmission (a) with no WDM; (b), with two-channel WDM; and (c) with eight-channel dense wavelength division multiplexing (DWDM). The "no WDM" Part (a) portion of the figure shows full-duplex electrical signals (conceivably already highly multiplexed signals—like DS3 signals) as inputs to and outputs from *fiber optic modems* (FOMs). In this case, FOMs transmit and receive optical carrier signals over two separate fibers that perhaps constitute a telco's current "outside transmission plant" connecting remote switching or multiplexing centers.

Part (b) could provide electrical signal-to-electrical signal performance exactly equivalent to the Part (a) configuration, but requires only a single "outside plant" fiber. In this case, a single fiber handles the same traffic that was split between two fibers in the Part (a) configuration. If, in fact, current outside plant consists of two separate fibers, as in Part (a), adding the indicated WDM-capable couplers offers a telco the possibility of doubling its outside plant's traffic-handling capacity. Herein lies WDM's economic and new service-provision response-time advantage. The large capital outlays and long lead times associated with deploying new fiber cable can often be deferred or avoided altogether. When we remember that LECs and IXCs now have installed over 20 million miles of fiber cable, the potential for savings, as capacity requirements continue to escalate, is enormous.

Note that in Part (b), couplers are arranged to provide *bidirectional WDM* (that is, east-to-west and west-to-east traffic handling capability). In *unidirectional WDM*, multiple wavelengths travel in the same direction on optical fibers. Specific requirements dictate the choice, but for two-channel installations, bidirectional WDM is usually preferred since it maintains the one-to-one relationship with electrical signal equipment input and output channels.

Part (c) of the figure depicts an eight-wavelength bidirectional dense wavelength division multiplexer, offering a potential eight-fold increase in capacity. Nortel offers a "short-reach" eight-wavelength OC-192 bidirectional product that uses the following wavelengths:

| Blue (λ2, 4, 6, 8) Wavelengths | Red (λ1, 3, 5, 7) Wavelengths |
| --- | --- |
| 1529, 1530 | 1550, 1552 |
| 1532, 1533 nm | 1555, 1557 nm |

"nm" stands for nanometer, which equals $10^{-9} = 0.0000000001$ meters. To permit extended-reach DWDM, Nortel and others offer multi-wavelength optical repeaters. With erbium-doped fiber amplifiers, spacing of tens of miles is possible. With each wavelength supporting OC-48, 2.5 Gbps rates, eight-wavelength DWDM installations can carry an aggregate of 20 Gbps of traffic. With 16-wavelength models in test beds and 40- and even 96-channel capabilities on the horizon, experts see no end to the growth of fiber's ever increasing traffic-handling capacity.

**Figure 5.20**
Comparison of no WDM, two-channel WDM, and eight-channel DWDM.

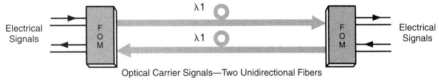

(a) Fiber Outside Plant—No Wavelength Division Multiplexing (WDM)

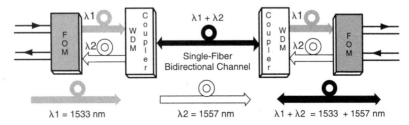

(b) Two-channel WDM—Single Bidirectional Fiber

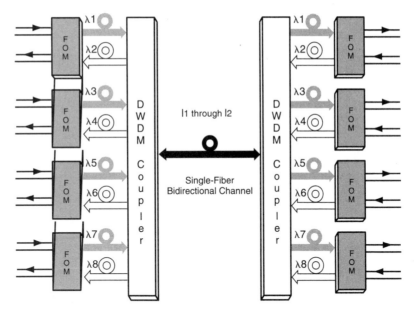

(c) Eight-Channel Dense Wavelength Division Multiplexing (DWDM)—Single Bidrectional Fiber

# Code Division Multiplexing

As explained above, FDM divides a transmission circuit's frequency spectrum into subbands, each supporting single, full-time communications channels on a noninterfering basis. In TDM, a transmission facility is shared in time rather than frequency. Unlike either FDM or TDM, with *code division multiplexing* (CDM), individual channel-signals are modulated with special, *orthogonal coding signals* in such a way that multiple signals can be transmitted in the same frequency band and at the same time without interfering with each other.

To recover each transmitted signal individually, receivers must be equipped with an identical version of the modulating orthogonal signals. A simple example illustrates CDM's operating principle. If three signals are represented as $S_1$, $S_2$, and $S_3$, and three modulating signals as $C_1$, $C_2$, and $C_3$, then the output signals from three separate modulators would be: "$S_1*C_1$", "$S_2*C_2$", and "$S_3*C_3$", where using the asterisk means "S1" multiplied by "$C_1$".

As a consequence, a receiver's composite input signal is of the form:

Receiver input signal          $= (S_1*C_1 + S_2*C_2 + S3*C3)$

Now to extract the $S_1$ signal, a receiver-decoder multiplies the receiver input signal by a "local version" of $S_1$'s encoding signal, that is "$C_1$". This produces the following $S_1$ decoder output:

S1's decoder output signal    $= (S_1*C_1 + S_2*C_2 + S_3*C_3)*C_1$
$= (S_1*C_1*C_1) + (S_2*C_2*C_1) + (S_3*C_3*C_1)$
$= S_1$

The orthogonal coding signals $C_1$, $C_2$, and $C_3$ are designed such that the cross product of different coding signals is "zero," whereas the cross product of the same encoding and decoding signals is "1". In practice, approaching this result requires a high degree of synchronism between transmitters and receivers and high quality of coding signals. By high-quality codes we mean a set of codes, which when multiplied together, yield a product that is very close to "0" for different codes, and close to "1" for the same code.

In real systems, to approach these ideal results, the bandwidth occupancy of coding signals is often hundreds of times as large as any of the input signals. Thus, the code division multiplexing process is often referred to as a spread spectrum modulation technique. Whenever wireline or fiber optic receiving equipment, satellites, or mobile base stations are designed to receive and handle multiple signals, they are said to exhibit *multiple access* capabilities. Accordingly, linking frequency, time, and code multiplexing to multiple access capabilities produces *frequency division multiple access* (FDMA), *time division multiple access* (TDMA), and *code division multiple access* (CDMA).

Most of the development and refinement of CDMA has taken place to produce military wartime communications with either anti-jam or covert characteristics. As it turns out, because orthogonal coding signals used in CDM/CDMA systems discriminate against all but local versions of themselves, in anti-jam applications they serve to recover desired signal power while minimizing the effects of intentional enemy jamming signals. Conversely, in covert communications applications, with sufficiently wide spread spectrum bandwidths, it is difficult for enemy receivers that do not possess encoding signals even to detect that friendly transmissions have taken place, much less have the ability to surreptitiously listen in on and recover the content of such signals.

Perhaps the most extensive commercial use of CDMA to date is in mobile telephones and base stations. In the United States, designs conforming to EIA/TIA's IS-95 standard used in Sprint's PCS service and others employ advanced CDMA and power control approaches. Additional details are found in Chapter 16.

# CHAPTER 6

# Circuit-Switching
# Concepts

Up to now, we have covered telecommunications fundamentals dealing with the generation and exchange of information in electronic form over various transmission networks. The fundamentals of *circuit switching*, presented below, center around the need to share telecommunications resources. For example, if a separate telephone and transmission line were required to connect you with every other person you needed to talk with, you probably couldn't fit into your office. The earliest telephone systems used just such an arrangement, however, which quickly threatened to bury communities in telephone poles and wiring.

The basic notion behind circuit switching was previously illustrated in Figure 1.1, showing how it does away with the need for dedicated transmission media running between each pair of communicators. Instead, the idea of *resource sharing* is introduced, wherein each person can access transmission resources as necessary—"just-in-time" communications, so to speak. More formally, *circuit switching is a process that establishes connections on demand and permits exclusive use of those connections until they are released.* In telecommunications networks, circuit switching establishes a sender/receiver voice or other information signal path in a manner analogous to an electrical switch's enabling current to flow from a power source to a light bulb.

The earliest switching arrangements revolved around telephone operators, who in the beginning were young men, often on roller skates, hurtling across central offices with plug-ended cords to connect telephone calls. These fellows tended to be rude to callers, however, and were soon replaced by better-mannered young ladies.

A major leap forward occurred with the advent of the *step-by-step* automatic telephone switch. Invented in 1889 by Almon T. Strowger, this switch was used, with only minor improvements, in telephone company COs up to the 1970s. An undertaker in wild and woolly Kansas City, MO of the mid-1880s, Mr. Strowger suspected his competitors of bribing central telephone operators to learn of the latest gunfights, and where the bodies could be found. His automatic switch was meant to foil their efforts. It did.

In modern packaging, circuit-switching techniques are still used to connect a greater number of telephones than there are transmission paths between them. "Switching" is the method by which particular transmission channels are connected between parties who wish to use them, disconnected following their use, and then reconnected between new parties.

# Circuit-Switching Fundamentals

*A switching system is an assembly of equipment arranged to establish connections between lines, between lines and trunks, or between trunks.* Included are all kinds of related functions, such as monitoring the status of circuits, translating addresses to routing instructions, testing circuits for busy conditions, detecting and recording troubles, sensing and recording calling information, etc. As indicated in Chapter 1, whereas circuit switching establishes connections on demand and permits exclusive use of those connections until released, packet switching establishes communications by sharing transmission channel capacity on a packet-by-packet basis. The details of packet switching are presented in Chapter 7.

The earliest circuit switch designs employed plug boards or electromechanical switch contacts, and to a large extent were controlled by human operators. Today, digital circuit switches are essentially specialized minicomputers. Every digital circuit switch has three essential elements: switching matrices, a central control computer, and interfaces, which among other functions, provide termination points for input and output signals, as illustrated in Figure 6.1.

## Switch Matrices

A *switch matrix* is the mechanism that provides electrical paths between input and output signal termination points. Modern matrices are electronic, using either time-division or space-division switching. A *time-division switch* employs the TDM process described in Chapter 5, in a *time-slot interchange* (TSI) arrangement. Figure 6.2 shows how TSI uses input and output TDM buses.

TSI can be thought of as having storage locations (buffers) associated with specific time slots in a TDM bit stream that corresponds to input signals, and with specific time slots in a TDM bit stream that correspond to output signals. As indicated in Figure 6.2, the switching operation can change the order of the time slots by changing the order of the information from input and output buffers.

In space division, a physical, electronic (spatial) link is established through the switch matrix. Where older space-division switches used electromechanical mechanisms with metallic contacts, modern space-division switches are implemented electronically using integrated circuits. (Chapter 8 describes emerging "photonic switching" technologies.

With photonic switches, light wave signals are switched directly as light waves rather than being converted to electronic signals for use with conventional switching mechanisms.)

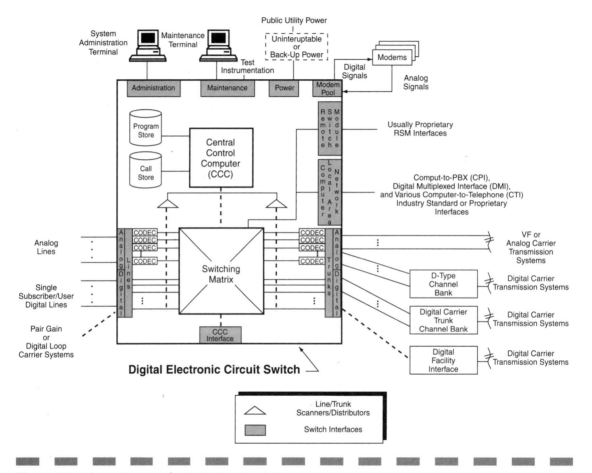

**Figure 6.1**    Block diagram of a digital circuit switch.

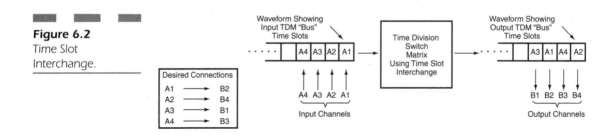

**Figure 6.2**
Time Slot
Interchange.

For small line/trunk sizes, switches can use time division alone. For large switches, combinations of time division and space division are required due to speed limitations over TDM buses. For example, to build a time-division switch that would accommodate 100,000 connections, each representing a 64-kb/s signal, would require a TDM bus speed of 100,000 times 64 kb/s or 6.4 billion bits per second. Consequently, popular large switch designs today use concatenations of both time division and space division switch modules, e.g., time-space-space-time division (TSST) or even TSSST approaches. For these large switches the space division module is referred to as a *time multiplexed switch* (TMS). A TMS is defined as an element of a time-division switching network that operates as a very high-speed space-division switch whose input-to-output paths can be changed for every time slot.

Although in Figure 6.1 the switch matrix is shown as a single component, most manufacturers implement the matrix using TSI and TMS modules. Small-capacity switches generally use one TSI module. Large switches may use several TMS modules and a quantity of TSI modules sufficient to accommodate the number of required lines and trunks. Modular designs are an economical way to accommodate a wide range of initial requirements and to add capacity for growth without the need to replace common switch components.

*Switching fabric* is a relatively new expression in the telecommunications lexicon. In this context, "fabric" refers to the physical structure of a switch or network. *It includes all the "interwoven" information-bearing components or channels that transport signals from input-to-output ports* (a *port* is defined as a physical device interface). In the above example, the ensemble of TSI and TMS modules in a particular switching product comprise its switching fabric. Switching fabrics do not encompass central control computers or adjunct system components described below, even though historically switch manufactures considered them to be integral parts of their switching products.

The term fabric applies to all input port-to-output port information-bearing physical network structures, independent of intervening switching (circuit, packet, frame or cell relay), multiplexing (time, space, frequency, or wavelength division), modulation, or transmission system techniques employed. Transmission, modulation, and multiplexing are described in Chapters 4 and 5. The various types of switching techniques and their characteristics are described in this chapter and in Chapters 7 and 8.

# Central Control Computer

In today's digital circuit switches, connections through matrices and other switching system operations are controlled by a central computer. Virtually all switches now use *stored program control* (SPC). With SPC, new services, special features, and configuration changes can be implemented as changes in a software program versus more costly changes in hardware. The basic components of an SPC system are illustrated in Figure 6.1. They are:

- One or more high-speed processors (central control computer) that sense input and output circuit conditions and execute stored program instructions.
- Semipermanent program store memory, containing operating system and applications software programs.
- Erasable, temporary call store memory to record and accumulate data during call processing.
- Scanners through which central control acquires input signal information such as on-hook, off-hook, etc.
- Distributors through which central control drives switch operations.

Such designs are made possible by digital computer technologies and permit the operation of a switching system to be altered by software program changes. To provide the reliability essential for telephone operations, SPC processor arrangements typically use full centralized control, with independent multiple processors that can load share, and functional multiprocessing in which different functions are allocated to different processors. Over 80 percent of an SPC switches functionality is implemented in software.

Whereas computers used in some data processing operations are acceptable with hour-per-month downtimes due to failures, well-designed circuit switches deliver availability performance that keeps downtime in the range of minutes per year. For example, the mean time between service-affecting failures for central office switches is one failure in 40 years. Because continuity of some business operations depends on available telecommunications, equipment and service reliability requirements constitute an important aspect of procurement specifications. It is interesting to note that in the past several years, most telecommunications failures significant enough to make national or regional newspaper headlines have been the result of software "glitches" and not hardware failures.

## Switch Interfaces

The diagram shown in Figure 6.1 applies to all types of circuit switches, i.e., central office switches, tandem switches, business premises-located private branch exchanges (PBXs), and key telephone systems (KTSs). Central office switches (also referred to as CENTRal EXchange [Centrex] systems), PBXs, and KTS systems are discussed in the next sections. Not all the interface types illustrated in Figure 6.1 and discussed below are available in all switches.

The most important circuit-switch interfaces support lines and trunks. Reflecting today's network status of mixed analog and digital facilities, switches may be configured with both analog and digital line/trunk cards. As shown in Figure 6.1, trunk connections between switches can be via VF analog carrier, or any of several digital carrier systems. The D-type channel banks can be used with older analog switches and provide input signals suitable for use with digital carrier facilities. Such arrangements permit the transition from analog to digital to occur at different rates and at different points for both switching and transmission systems.

Interfaces between digital switch trunk outputs and digital carrier facilities can occur directly—without analog-to-digital conversion in codecs—via multichannel built-in digital facility interfaces, or by means of digital channel bank equipment as shown in Figure 6.1. Direct interconnection with digital facility interfaces greatly simplifies and reduces the cost of interconnecting switching and transmission systems.

On the line side of the switch, either analog or digital interfaces can be used. Central office switches will support analog lines for the foreseeable future because of the large number of existing analog telephones. Illustrated in Figure 6.1 are clusters of codecs connected to analog line switch ports. Codecs are required to convert analog signals to and from digital formats suitable for digital matrix switching (*ports are interfaces through which signals or information gain entry or egress*).

On central-office switches, line ports accommodate analog and digital subscriber lines, Integrated Services Digital Network (ISDN lines described in Chapter 8), or lines to vendor-specific customer premises equipment that the CO is designed to accommodate.

Businesses that have purchased digital PBXs may elect to use analog telephones (and therefore analog line interfaces) because they are cheaper, in spite of the fact that they forfeit many features only available with digital telephones. All digital PBX manufacturers provide, and prefer to sell, digital subscriber line interface cards and the proprietary station equipment that goes with them.

Some switches provide line-side interfaces for multiplexed signals carried in digital loop carrier (DLC) systems. The types of services associated with various line/trunk facilities are described in more detail in Chapter 10.

Subscribers with analog telephones can selectively use lines between their offices and the switch for either voice or data modem traffic. One side effect of using digital technology is that once the decision is made to use digital telephones, only digital signals can be carried on station equipment-to-switch lines. That is, those lines can no longer support analog telephone or modem traffic.

*Modem pool* interfaces are provided in most digital switches as an alternative way of providing PCs and other data terminal equipment (DTEs) located in a user's office with PBX connectivity to other DTEs via PSTN or private voice networks. A modem pool is a group of modems connected to a digital switch. Each modem is provided with a digital input connection from a PC to the switch and an analog connection from the switch to the PSTN as shown in Figure 6.1. Each PC or DTE connection is associated with a switch extension number. In an office, the PC or DTE is connected to a data terminal connection provided on the digital telephone.

To use the service, an office user dials an extension corresponding to the modem pool's digital connection side (manually or automatically via the DTE), the switch selects a non-busy modem and connects the modem output through the switch, usually to the PSTN or private switched or non-switched voice networks.

One advantage of modem pools is that not every user needs to have a dedicated modem in his office. However, since modems can be procured in the $50-$150 cost range (less if PC plug-in modem cards are used), the upgrade to digital station equipment might not be justified unless there are reasons other than reducing the number of modems needed at a particular site.

As Figure 6.1 indicates, some switch designs are capable of supporting direct digital connections to collocated computers and local area networks (LANs). (LANs are discussed in Chapter 13.) Two switch-to-computer specifications advanced by the industry are known as the *computer-to-PBX interface* (CPI) and the *digital multiplexed interface* (DMI) specifications. Both specifications operate at DS1 rates of 1.544 Mb/s. A single CPI/DMI connection multiplexes into a single connection what would otherwise involve 24 separate switch-to-computer channels. Introduced by Northern Telecom Inc. (now Nortel Networks Corporation) and AT&T in the early 1990s, these interfaces have been eclipsed by more recent computer-industry driven computer telephone integration (CTI) initiatives to use modern computer hardware and software technologies to improve plain old telephone service (POTS).

The burgeoning new field of computer telephony integration has as its overall goal using computer intelligence to enhance making, receiving, and managing telephone calls. CTI connects computers to telephones or telephone switches so that information and commands may be shared. Events from computers can trigger events on telephone systems, and vice versa. CTI payoff is most dramatic in modern interactive voice response (IVR) and automatic call distribution (ACD) systems now employed by enterprises to streamline response to customer telephone requests for information, products, and services. The impact that CTI is having on PBX designs and markets is addressed in the next section. For this discussion of general circuit-switching operational principles, the key point is that CTI interfaces have assumed "center-stage" prominence.

Other switch interfaces support administration and maintenance terminals and processing systems. Most larger switches include built-in test and diagnostic capabilities to automatically perform basic switching equipment and loop transmission impairment tests. Maintenance interfaces also permit connection to standalone test instrumentation for more extensive testing.

Digital interfaces permit data communications directly between the switch central control computer and other processors used for support functions. For example, in campus installations, some manufacturers place a central switch in one building and use *remote switch modules* (RSMs) in other locations to concentrate traffic and simplify inter-building wiring. In such cases, data communications between the central switch control computer and RSM controllers provide identical service to all stations whether directly attached to the central switch or to the RSM(s).

The last major switch interface is with primary power. For certain premises systems, loss of telephone service during public utility power outages is not acceptable. Consequently, battery systems with back-up motor generators are often installed.

Among a variety of minor line and trunk interfaces supported by most switches are provisions for audio signals for music on hold, dial dictation, and loudspeaker paging systems.

# Private Branch Exchanges (PBXs)

A *PBX* is a circuit-switching system serving a commercial or government organization and usually located on that organization's premises. PBXs provide telecommunications services on the premises or campus, e.g., internal calling and other in-building or on-campus services, access to public switched telephone networks for outside calling, and access to private networks connecting various organizational sites.

## Evolution

Industry pundits define as many as eight generations of PBX development. For our purposes we see PBX development spanning four generations thus far. First-generation PBXs were operator-controlled plug-and-jack types. Second-generation PBXs performed switching mechanically or electronically but used analog designs. Third-generation PBXs are characterized by digital switching, stored program computer control; and digital and/or analog peripheral, loop, and trunk transmission interfaces. These systems support digital telephones, integrated voice and data terminals (IVDTs), and digital interface units for directly connecting PCs and other data terminals through the PBX. Telephones generally use embedded codecs and exchange digital signals.

Early IVDTs manufactured and offered by PBX vendors combined personal computer and voice telephone functions. However, PBX vendors, unable to compete in the rapid technological advancement and the price-driven PC industry, found their IDVT market has all but vanished. Moreover, because most PBX vendors do not use open-system, industry standard-based local area network designs for telephone/DTE to PBX or PC-to-PC connections, today's digital PBXs are used principally to switch voice and perhaps some digital modem or digital facsimile traffic.

Another third-generation PBX limitation is that most manufacturers use proprietary telephone-to-switch digital signal formats. To take advantage of the features offered, a user must buy the PBX manufacturer's proprietary digital telephones (thus forfeiting competitive bids). The only way to obtain competitive station equipment bids is to install analog station equipment, which might not be capable of taking advantage of all the features supported by the PBX.

Fourth-generation PBXs possess all third-generation capabilities and, in addition, support external computer-based control of call management and other switch functions using open architecture industry standard interfaces. Thus, fourth-generation designs permit one to "custom program" PBX-based networks within a single building or campus. Moreover, large companies can establish central computer-based control centers with the ability to manage national PBX-based networks from one or more (for reliability) locations. Clearly, such interfaces are key to advanced and highly effective computer telephone integration within enterprises with installed PBXs.

The advantages that such open architecture PBX interfaces yield are difficult to exaggerate. In truth they can be compared to the seminal capabilities that similar computer-to-LEC and IXC switches produced by making carrier-based *virtual private networks* (VPNs) possible. VPNs, now the flagship offerings of telephone companies, use stand-alone computers to remotely control public switching and transmission facilities and to deliver virtual private network services that are superior to and indistinguishable (by business customers) from those delivered by yesterday's facilities-based private networks. Chapter 10 contains the details. Figure 6.3 portrays an example fourth-generation PBX.

Although some labeling details are omitted, the figure shows PBX components to be the same as those of the generic digital circuit switch (Figure 6.1) described in the previous section. Interfaces for analog and digital station equipment are shown. The possibility of connecting data terminals or PCs directly to telephone instruments is illustrated. The more likely office scenario of connecting telephone sets to PBXs, and PCs to separate local area networks (LANs), is also illustrated. The number of desktops with LAN-connected PCs reached 60 million in 1997. The corresponding 1997 estimate of the number of business telephones was 85 million. As will be evident throughout the remainder of this book, the potential for PC-based CTI applications to augment and enhance telephone functions is already clearly established.

For now, envision the possibility of individuals with LAN-connected PCs beside telephones, having the ability to use Microsoft, Netware, or other third-party CTI application programs that permit PC-based directory, auto-dialing, coordinated voice-mail, paging, e-mail, conferencing, and other capabilities to streamline what is referred to as converged voice, data, video networking, and communications.

In Figure 6.3, digital telephones may be either single line or multiline. Multiline telephones generally incorporate visual displays to

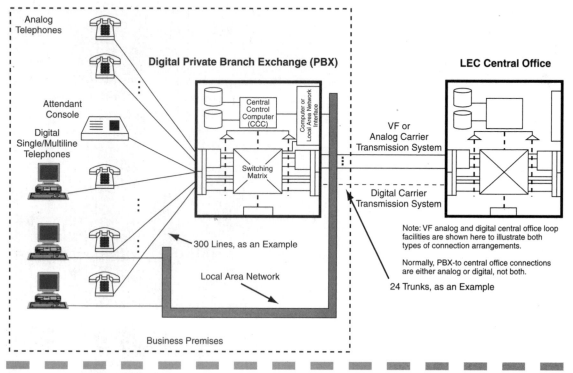

**Figure 6.3**   Fourth-generation private branch exchange.

indicate busy or ringing status of multiple extension numbers. Until the mid-1970s, multiline telephones were available only with key telephone systems.

Third- and fourth-generation PBXs support a variety of analog and digital carrier connections to local and long-distance network facilities. In the example in Figure 6.3, PBXs with DS1-type digital carrier service interfaces can support 24 voice or data channels using a single four-wire circuit, versus 24 separate two- or four-wire circuits if voice frequency (VF) analog carrier interfaces are used. Digital PBX trunk interfaces thereby offer the potential for significant savings in network connection and transmission costs. A photograph of Lucent's Definity Enterprise Communications System (PBX) is pictured in Figure 6.4. Definity is recommended for installations with as few as 200 stations but will support up to 25,000 stations and 4,000 trunks or a total of 29,000 ports. Lucent credits the Definity with an availability of 99.9 percent, which translates to less than eight hours of down-time per year. Nortel's Meridian 1 Communications System (PBX) is shown in Figure 6.5.

Depending upon the option selected, Meridian models support from 128 to 10,000 ports. Meridians can be equipped to support a wide range of applications such as CTI, IVR, and customer-controlled routing. Remote expansion cabinets up to two miles away can be connected via fiber optic links to main PBX modules for campus-like installations.

**Figure 6.4**
Lucent Definity
Enterprise
Communications
System (PBX).
Courtesy of Lucent
Technologies Bell
Labs Innovations.

## Services and Applications

In this book, PBX services comprise the category of voice services defined as *premises transport services.* A comprehensive listing of voice services is provided in Part 3's Introduction and Chapters 9 and 10, which describe voice services at the level of detail needed to prepare specifications for procurement packages. Chapter 12 also includes a section on how to go about selecting PBX and other voice systems.

**Figure 6.5**
Nortel Meridian 1 Communications System (PBX).

In the data processing world, an *application* is a software program that performs some useful task. PBX applications are defined in similar fashion. *Functional applications* are individual tasks or telecommunications services provided by the PBX, such as station-to-station calling. At the highest level, *business applications* are unique aggregations of telecommunications services that satisfy particular enterprise needs, e.g., hospitality, retailing, financial, etc.

Basic dial PBX functions have remained unchanged for decades. The most fundamental is station-to-station calling, using between three and five digits. Next is the ability to connect inside telephones (stations) with public and private external networks. To access these networks automatically, a station user dials an access code (such as "9"). This characteristic trunk-side connection has historically distinguished PBXs from key systems, which were directly connected to central office (CO) lines, and therefore did not require dialing an access code to place an external call. PBXs are also characterized by attendant service, whereby an on-premises operator connects two-way trunks carrying incoming external ("listed directory number") calls with internal stations. Attendants can also use two-way trunks to place external calls.

In addition to connecting with two-way CO trunks for attendant console operation, modern PBXs also support both *direct inward dialing* (DID) and *direct outward dialing* (DOD), using analog or digital facilities. DID allows incoming calls to a PBX to ring specific stations without attendant assistance. Similarly, DOD allows outgoing calls to be placed directly by PBX stations. DID greatly reduces the number of required console attendants, compared with systems in which all incoming external calls must be extended to PBX stations. Details on the operation of DID, DOD, and other network services furnished by LECs are provided in Chapters 10 and 12.

Other important PBX functional applications fall into the administrative category. These include the ability to produce reports such as *station message detail recording* (SMDR), which show calling and called numbers together with time, date, and duration of calls; station/feature moves, adds, and changes; maintenance support; voice messaging (also called *voice mail*); automatic call distribution (ACD), which directs large numbers of incoming calls to specific departments or attendants within an organization); message center/directory services; cable/inventory/energy/property management; and, of course, an ever growing number of new CTI-based capabilities.

For large-building, multi-building, or campus installations, premises telecommunications services can be extended from a central PBX switch in the main building to aforementioned remote switch modules (RSMs) in other buildings. Figure 6.6 shows a central PBX with a switching matrix that includes a high-speed time multiplexed switch (TMS) module. The central switch connects to a group of local TSI switch modules, each sized to support station and trunk requirements for the building in which it is located (e.g., Building A).

RSMs installed in other buildings are connected to the central PBX using high-speed transmission over fiber optic cable to support central TMS module-to-remote TSI module traffic, and a channel for control signals between the central PBX computer and remote TSI module controllers. Designs such as this provide identical services to all users regardless of their building location. RSMs can accommodate station and trunk capacity requirements on a building-by-building basis. Distances between the central PBX location and the RSMs can extend for miles, depending upon site equipment configuration.

**Figure 6.6**
*PBX remote switch
module connections.*

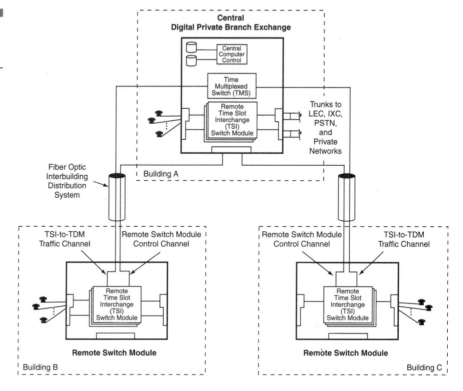

**Figure 6.6**
*PBX remote switch module connections.*

## Components

A modern PBX is comparable to a specialized computer, attached to and controlling a switch matrix. As such, it incorporates a control portion (*central processing unit,* or *CPU*), a data-manipulation portion (arithmetic unit), and a peripheral connection portion (input/output unit). The operating system software transforms this general-purpose computer into a PBX controller, and applications software define what functions the PBX can perform. Peripherals such as telephones, modems, and other items of voice terminal equipment are the "eyes and ears" of the PBX, furnishing information needed to set up, connect, supervise, and tear down calls.

### SWITCHING EQUIPMENT AND ARCHITECTURE

*Port interface capabilities.* Lines and trunks are the most elementary PBX input/output ports. Over time, the meaning of the term *line,* originally denoting an extension number appearing on a telephone, evolved also to mean *station*, the telephone itself. A population of lines, each carrying a certain amount of call traffic, is supported by a smaller number of *trunks*, which connect the PBX to network resources. The ratio of trunks to lines is generally on the order of 5–15 percent.

With the advent of digital PBXs, the distinction between lines and trunks began to blur. Most of the major PBX products are now port oriented, combining lines, trunks, attendant consoles, and utility circuits together as common sources of traffic demand. The PBX cabinets themselves are moving toward universal card slot design, where any printed circuit board (except common control) can be inserted into any card position.

*Centralized vs. distributed switching.* Early PBXs employed a centralized switching design—all telephones were connected to one switching entity that performed all PBX functions. As noted above, a number of manufacturers offer distributed PBX switching architectures. Users occupying a large building, campus, or office park can now elect to use multiple modular nodes (remote switch modules) that can be installed and linked by digital facilities, and when properly designed, seamlessly emulate a centralized switch. Distributed PBXs with low port densities per module are systems of choice in the *shared tenant service* market because initial installation costs are low and modular expansion is inherent in the system design.

### PERIPHERALS

The term peripherals, derived from the data processing industry, refers to items of equipment that send inputs to the system processor (PBX) and receive outputs from it. Telephones, data devices, attendant consoles, and other items of terminal equipment fall into this category.

The principal PBX peripheral is the telephone. Several milestone improvements have marked its evolution—the manual dial, the integrated handset, and dual tone multifrequency (DTMF) signaling models, known in the Bell System as "touch tone." Up until the early 1970s, the vast majority of telephones, both business and residential, were rotary-dial. It was not until the close of the decade that DTMF telephones, made economical by large-scale integration manufacturing techniques and the FCC's equipment registration program, became pervasive in the U.S. marketplace. Figure 6.7 illustrates a modern, multi-featured, programmable telephone manufactured by Lucent.

**Figure 6.7**
Lucent 4624 IP
programmable
telephone.
Courtesy of Lucent
Technologies Bell
Labs Innovations.

## Adjunct Systems

Adjunct systems are made up of multiple components and require systems engineering prior to installation. Some of the more popular adjunct systems alluded to above are described further below.

### VOICE MESSAGE PROCESSING

The voice message processing industry was founded on two separate business needs: comprehensive call coverage with more accurate message-taking, and information exchange on a non-real-time basis.

In its most basic form, *voice message processing*, often referred to as *voice mail*, allows callers to leave recorded messages. Instead of having to hang up after dialing to a ring-no-answer or busy signal, callers hear the telephone being answered, then a prerecorded greeting (generally in the called party's voice), followed by instructions on how to record, review, change, or erase a voice message to be left for the called party. Later, called parties retrieve messages and perform additional annotation, distribution, and delivery verification functions.

PBX voice message processing systems can be configured in numerous ways, although the most widespread usage is as a final call coverage point, where an incoming call is forwarded to a voice mailbox, as described above.

Ease of use and number of features of a voice message processing system are related to its degree of design integration with the PBX. Integrated systems support personalized greetings (in the called party's own voice), automatically reply with a voice messages to senders, redirect messages to other voice mailboxes, and can broadcast messages within voice mail communities.

Integrated systems do not require the caller to know the called party's extension number to leave a message or to perform other functions. Also, the called party is generally notified of messages via lamp indicators or text on a telephone visual display, rather than via muted ringing or "stutter" dial tone. Once messages are retrieved, the system signals the PBX to deactivate the message-waiting indicator.

Non-integrated voice message processing systems are unable to provide personalized greetings for each station. All calling parties hear the same standard message and must then identify the called party, either by speaking his name or by keying in his extension number (only from a DTMF telephone). Where the caller has spoken the called party's name, in the absence of voice-recognition capabilities, an operator must later transcribe and distribute the message.

In configuring voice message processing systems, it is important to ensure that callers can easily reach a live person when additional assistance is required. This is achieved either by direct selection using a dialed code or by a timeout feature, required for callers using rotary-dial telephones. Today, in large enterprise systems, voice message processing is typically just one of many functions included in call center operations available from both third-party and PBX vendors.

### SYSTEM MANAGEMENT

Even in the smallest configuration, a PBX nevertheless includes stations, trunks, and features, as well as telephones and associated wire and cable. As PBX size increases, a user organization is forced to devote resources to the task of monitoring and controlling PBX services, repairs, rearrangements, and configuration, including adjunct systems. This task is often referred to as system management.

At one time, all available system management functions were performed at the attendant console. Although the number of management operations was limited, the process was awkward because while in use for administration, the console could not be used to process calls. Later, separate *maintenance and administration panels* (MAAPS) were introduced, which were proprietary terminals designed specifically to interface with the particular PBX. Next, commercial off-the-shelf computer terminals were used in the interests of cost savings and ease of operation.

Today, the trend is toward PCs with industry-standard graphical user interfaces (GUIs) for maintenance, administration, and system management. Functions performed include:

- Long-distance call costing
- Billing reconciliation
- User cost allocation
- Software feature and physical moves, adds, and changes
- Automatic station relocation
- Traffic analysis
- Network administration and optimization
- Site administration
- Inventory control
- Directory services
- Cable system management
- Service orders
- Trouble handling

In the past, PBX system management encompassed more functions than those associated with LEC-provided Centrex service, described in the following section. Because a user-owned PBX switch is located on-premises, whereas a LEC-owned Centrex switch is located at the CO, PBX users traditionally exercised greater control over system management. Recently, however, LECs have begun to offer tariffed and special contract system management capabilities that virtually mirror those of the PBX. As the full effect of the 1996 Telecommunications Act unfolds, telco-based offerings will become even more competitive with PBXs.

## Operational Characteristics

Significant PBX product improvements occurred in the 1980s, in terms of increased cabinet capacities, enhanced performance specifications, more and better features, and reduced logistical and environmental demands. In short, today's PBX products deliver more "bang for the buck" than ever. Yet they have also become more commodity-like, with little to differentiate competitors, apart from price.

## System Capacities

Stations (lines) are the basic measure of PBX capacity. They correspond to electrical circuits on a printed circuit board (PCB), representing the ability to connect one telephone or data terminal. Stations may be analog or digital. Analog station PCBs tend to contain a higher circuit density than digital PCBs—often on the order of 2 to 1 (e.g., 16 analog stations vs. 8 digital stations). Digital telephones need a digital station PCB irrespective of whether voice is digitized at the telephone, or at the PCB.

PBX station traffic is "concentrated" before being connected by a trunk to an outside network. Concentration is achieved through contention of a particular quantity of stations for a lesser quantity of trunks. The concentration ratio typically ranges anywhere from 5 percent trunking to 15 percent. Governing factors are system size, the traffic-handling requirement and specified call-blocking probability (i.e., the chance for encountering a busy signal when all trunks are in use, a performance characteristic of networks described further below).

The presence of analog or digital data stations may have significant impact upon the total station quantity, the PBX's ability to handle traffic, total system size, and cost.

## System Performance

How well a PBX performs is largely determined by its ability to process calls based upon the division of traffic, i.e., internal versus external incoming and outgoing calls, and provisioning of trunk quantities adequately to handle external traffic. To ensure satisfactory installed performance, purchase requisitions and requests for proposals (RFPs) must include trunking, signaling, and transmission quality performance specifications. This section discusses these important parameters and their cost and performance implications.

*Grade of service.* A PBX significantly reduces the number of transmission channels required between the business premises and the LEC central office. The quantity of PBX-to-CO trunks is typically around 10 percent of the number of stations, as indicated in Figure 6.3, where 300 stations are served by 24 trunks. This reduces both the quantity of LEC transmission channels that a user must pay for and the number of PBX trunk interfaces that must be installed and maintained.

However, PBX users compete for a limited number of trunks; if all trunks are in use, the next outgoing call is *blocked* and the next incoming call receives a busy signal. An adequate quantity of trunks must be specified to ensure that station users receive acceptable grades of service.

*Blocking probability,* often referred to as *grade of service* (GOS), is an important measure of the adequacy of telecommunications networks. Other grade-of-service indicators include an estimate of customer satisfaction with a particular aspect of service, such as noise or echo. For example, the noise grade of service is said to be 95 percent if, for a specified distribution of noise, 95 percent of the people judge the service at or better than "good." GOS measurements apply to all aspects of telecommunications networks, not just PBXs. In many cases the literature equates GOS only with the probability of a blocked call. When used without further explanation, GOS generally refers to blocking probability.

In terms of call-blocking performance, GOS represents that portion of calls, usually during a busy hour, that cannot be completed due to limits in call-handling capabilities. For example a GOS of $P=0.001$ means that only one call in 1,000 would be blocked. GOS is a performance factor that merits understanding and careful interpretation. For instance, in the Figure 6.3 example, 300 trunks serving 300 stations would guarantee non-blocking access to the LEC network. Thus, a serving trunk would always be available for each station, so no contention or blocking could occur. However, if only three trunks were ordered for the Figure 6.3 example, the GOS would be such that most calls would be blocked.

In the non-blocking example, since most people don't use phones continuously, an organization would be paying for capacity that is seldom used. In the second case, all trunks would be used almost 100 percent of the time, but the GOS would be unacceptable. An optimal design is one that yields an acceptable GOS and yet maintains a reasonably high level of facilities utilization.

For design purposes, GOS can be estimated from total traffic intensity. Based on assumptions regarding the randomness of call arrivals, holding times, and other factors, tables are available that relate traffic intensity and the number of servers (trunks) to the probability that a call will be blocked. In planning a new system, the best source for estimates of traffic demand is call detail records (for example, SMDRs mentioned earlier) from an existing system. If no historical data are available, average industry estimates can be used.

*Centi-call seconds* (CCS), is the term used to quantify traffic intensity or demand. A CCS is 100 call seconds of traffic during 1 hour. Therefore, a single traffic source that generates traffic 100 percent of the time produces 36 CCS of traffic per hour, or 3,600 seconds of traffic every 3,600 seconds. An equivalent amount of traffic could also be generated by 10 sources that only generate traffic 10 percent of the time. That is, 10 sources of traffic generating 3.6 CCS, contribute the same total traffic as a single 36-CCS traffic source.

If the total traffic intensity generated by 300 subscriber stations is equal to 360 CCS (each of the 300 stations is used two minutes out of each hour), then the trunk utilization rates and the number of trunks needed to achieve GOS levels of 0.1, 0.01, and 0.001 are shown below:

| GRADE OF SERVICE (GOS) | P = 0.1 | P = 0.01 | P = 0.001 |
|---|---|---|---|
| NUMBER OF TRUNKS (REQUIRED TO ACHIEVE GOS) | 13 | 18 | 21 |
| PERCENTAGE UTILIZATION OF TRUNKS | 77.0% | 64.0% | 48.0% |

If the traffic generated by the 300 subscribers totaled 2,700 CCS, the results would change as follows:

| GRADE OF SERVICE (GOS) | P = 0.1 | P = 0.01 | P = 0.001 |
|---|---|---|---|
| NUMBER OF TRUNKS (REQUIRED TO ACHIEVE GOS) | 76 | 89 | 100 |
| PERCENTAGE UTILIZATION OF TRUNKS | 98.7% | 84% | 75% |

Two conclusions can be drawn from these calculations. First, for a given level of traffic, designing for better GOS results in poorer trunk utilization. The second is that there are economies of scale. That is, as one aggregates more traffic, greater resource efficiency is achieved while providing acceptable GOS performance. These phenomena form the basis for switching system and network design. Optimized designs can produce significant cost savings with essentially the same user quality of service.

Some telephone companies use a GOS objective of P=0.005. Smaller private networks are designed for GOS levels on the order of P=0.01. A typical large tandem switch, such as the Lucent 4ESS, can terminate

over 100,000 trunks with a blocking probability of P=0.005 and channel occupancy of 70 percent.

Chapter 11 revisits GOS as it applies to utilization efficiency of any
telecommunications switch or transmission resource, in terms of rudimentary traffic engineering principals. It also describes tariffs illustrating the relationship between performance and LEC and IXC service costs.

*Signaling and transmission quality.* Signaling and transmission quality
are important subjective yardsticks by which users measure the performance of PBXs and other types of switching systems. The term
subjective is used because even though standards exist governing these
parameters, quality is perceived "in the ear of the listener." Often,
users distinguish between good and bad service, without being able to
isolate individual evaluation factors.

Key variables are station line levels, noise, distortion, and crosstalk.
Line levels encompass several technical characteristics. The relative
loudness of sound on the line is one such variable. Dial tone, busy
tones, and DTMF tones might be too loud or too soft. The same goes
for being able to hear the distant party through the handset, and to
hear oneself (*sidetone*) in the handset.

Noise can be background static or hum, as well as intermittent loud
interruptions, caused by electrical impulses on the line. Sometimes,
feeding the PBX with poor-quality power or running wire and cable
near high-voltage sources will produce static and "60-cycle hum." Even
a poor plug-to-jack connection can cause annoying crackling sounds
in the handset.

Distortion is a signaling and transmission problem caused either by
the PBX components themselves, or by poor-quality network facilities
and connections. Distortion causes DTMF tones to fluctuate, producing a warbling effect. It also creates shifts in the quality of voice
reproduction due to frequency variations, sometimes making a voice
sound high or low pitched, or simply unrecognizable.

Crosstalk is caused by electrical coupling or transformer effect,
which superimposes one set of signals upon another, generally signals in
adjoining cable pairs as explained in Chapter 4. When interfering voices
are audible, the problem usually stems from faulty extension or trunk
printed circuit boards or from improperly installed wire and cable.

A common source of degraded signal and voice quality is a poorly
grounded PBX system, either because an improper ground was selected, or because the grounding design itself was faulty. Two major
sources of grounding problems are electrodes tied to plastic pipe, and

attempts to use electrical conduit or building steel as the system grounding electrode.

# Features

Sets of features transform the PBX from a simple port-to-port connection device into a powerful information transfer system. As described previously, features are software-driven routines that enable the PBX to perform certain repetitive functions.

PBX features are categorized as system, attendant, station, and management-related. *System features* are centered around processor-oriented functions, applicable to all categories of PBX users. *Attendant features* enable a console ("switchboard") operator to answer external calls, extend them to PBX stations, serve as a call-coverage point, and assist users in placing external calls. *Station features* help individual telephone users to communicate with people and to access other information resources more efficiently. *Management features* help the PBX administrators to review traffic information, change feature assignments, associate costs with premises and network services, and keep track of the system configuration.

Studies indicate that users richly equip PBXs with features—the top features being, in order of popularity:

- Toll restriction
- Automatic route selection
- T1 network interface
- Station message detail recording
- Voice mail capability
- Automatic call distribution
- System management
- Modem pooling

To this point, this chapter has presented the essential design, operational capabilities, and top-level features and characteristics of digital circuit switches, in general, and PBXs, in particular. Although the choice of either hardware or software implementation varies among manufacturers, in most designs, 80 percent of switch functions described are allocated to software. As noted, the historical practice of using embedded computer hardware and proprietary hardware and software designs to implement PBX functions is rapidly giving way to the use of high-performance general-purpose computer hardware and applications written to be executed by Microsoft Windows, NT, or other popular operating system-compatible software.

This development has already resulted in dramatic shifts in PBX design, the aforementioned open architecture or industry standard-based computer or LAN-to PBX interfaces, and PBX markets and marketing strategies. Another development contributing to these changes is the emergence and growing popularity of Voice over IP (VoIP) networks. Since the IP networks that provide VoIP service are all packet-switching networks, voice traffic collected and concentrated by PBXs can only be transported to or from those networks via special gateways. Of course VoIP traffic originating and terminating in PCs connected to LANs have no need at all for PBXs.

All these factors have led to dramatic PBX market changes. Traditional PBX vendors like Nortel Networks, Lucent, and Siemens are scrambling to protect their telephony turf by migrating from closed, proprietary designs to open architectures. To overcome the stigma of older rigid and inflexible designs, some are dropping PBX descriptors altogether. Lucent, for example, refers to its large-system PBX product line as Definity™ Enterprise Communications Servers (ECSs).

Meanwhile, Dialogic, Genesys Labs, Quintus, and other CTI specialty companies are vigorously marketing products that run under familiar Microsoft NT, IBM AIX, DEC UNIX, Sun Solaris, and other operating systems, can incorporate a variety of independent software vendor (ISV) application program packages, and interoperate with nearly all PBX vendor models. Some pundits predict a progression from "open PBXs" with many functions implemented in standalone computers, to "dumb PBXs" where most functions are implemented externally, to "UnPBXs," loosely defined as a PBX in a PC. Because many businesses depend on continuity of telephone service, the moderate view is that it is unlikely that ultra-reliable PBX hardware and software products developed by traditional vendors at cumulative costs measured in billions of dollars will be abandoned overnight.

From this chapter's point of view, the circuit switching/PBX functional, operational, and design principals presented are valid and apply regardless of by whom or how they are implemented—and will be so long as circuit switching remains a viable telecommunications option. Chapter 7 describes packet switching and applications in local, metropolitan, and wide-area network, Ethernet, frame relay and cell relay (Asynchronous Transfer Mode—ATM) settings. Chapters 13 and

14 present important and popular packet switched-based products and techniques for on-premises and within metropolitan and wide-area networks.

# Centrex (CENTRal EXchange) Systems

Centrex is a LEC service offering that delivers advanced PBX-type features without the need to purchase or lease switching equipment; it greatly reduces the need for premises floor space, commercial power, and heating, ventilation and air conditioning (HVAC) that would be required by PBXs. Centrex service can be provided from the same central office switch used for residential telephone service. Originally intended for customers with many stations, Centrex is now a candidate for users with just a few business lines. Figures 6.8 and 6.9 illustrate Lucent's 5ESS®, and Nortel Network's Supernode digital central office switches, both of which provide the Centrex services described below.

Figure 6.8 is a close-up view of Lucent 5ESS central office switch port cards. According to Lucent, a full-sized 5ESS supports up to 250,000 lines and 100,000 trunks and exhibits a "six nines" (99.9999%) availability. That translates to just 10 seconds of downtime per year. Worldwide, Lucent states that 5ESS switches in 52 countries now service over 108 million lines and 48 million trunks. Nortel's Supernode exhibits similar capabilities. Overall, most central offices are now equipped with digital switches. As an example of central office pricing, Lucent estimates that, for large configurations, the cost of a 5ESS is in the range of $300 per line.

## Evolution

Centrex was introduced by the Bell System in the 1960s as a CO-based premises service offering for medium-to-large-sized users. Within the Bell System, Centrex was first offered using analog, electromechanical switches (e.g., the No. 5 Crossbar, a relay-implemented class of CO switches that at one time served more than 28 million lines). Following that, Centrex debuted in electronic switching vehicles such as the 1ESS (1965), and its successor, the 1AESS (1976). Digital Centrex was intro-

**Figure 6.8**
Close-up view of
Lucent 5ESS central
office switch port
cards. Courtesy of
Lucent Technologies
Bell Labs
Innnovations.

duced in 1982 using AT&T's 5ESS, and in 1983 using the Northern
Telecom DMS-100.

Early Centrex offerings provided only basic business telephone
services. In the mid-1970s, PBXs outpaced Centrex in the application
of microprocessor technology, producing numerous advanced features, and capturing a majority share of the premises service market.
Following the breakup of AT&T in 1984, industry observers predicted
the demise of Centrex. Instead, the LECs have succeeded in employing
digital switching to re-establish Centrex as a viable PBX alternative,
not only retaining existing customers, but attracting new ones.

Figure 6.10 shows a CO implementation of digital Centrex circuit
switch and typical interconnections with business premises. Note that

**Figure 6.9**
Nortel DMS
Supernode central
office switch.

the logical components parallel those of the generic circuit switch, (Figure 6.1) and the PBX, (Figure 6.3). Centrex station lines are similar to ordinary telephone loops, except that the line loss is usually limited to values well below the maximum for an ordinary loop to maintain a good grade of service between users on the same premises who make heavy use of Centrex services and features. An internal office-to-office call might involve two multi-mile loops as opposed to shorter in-premises, inter-office wiring lengths encountered with PBX-based solutions. Note also that providing Centrex service to 300 subscribers necessitates 300 separate loop circuits, whereas the same number of subscribers can be served with only 24 trunks using a PBX, as described above.

Three examples of possible loop access implementations are shown. First there are the legacy copper twisted pairs, some of which have been installed for years. In Moscow, for example, some old paper-insulated wires installed at the turn of the century are still in use! Modern gel-filled plastic insulated cables may outlast earlier designs unless fiber or other technologies render them obsolete first.

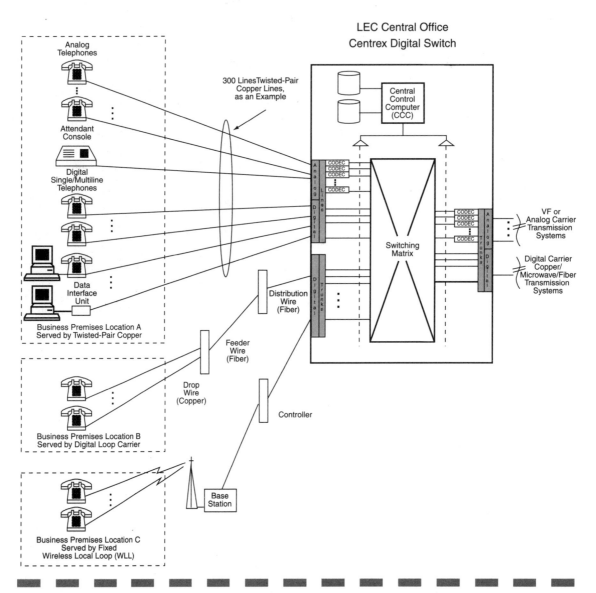

**Figure 6.10** *Digital Centrex.*

As a step toward bringing broadband fiber optic cable to homes and businesses, hybrid fiber cable had already debuted. The fiber distribution-feeder connections to copper premises "drop" segments shown in the figure now typify new installations. A number of older wire-based digital loop carrier installations remain in service.

Finally, as mentioned in Chapter 1, wireless local loops to fixed-station equipment represents a third option that is gaining popularity and is less expensive in some instances than cable-based alternatives. While this is still not as familiar as wired service options, Ameritech began its trial of "wireless Centrex" in February of 1993 at the University of Chicago. The potential for convergence of cordless telephone, cellular, PCS, and other wireless networks for fixed and mobile voice, data, and video users is introduced in Chapter 1 and treated further in Chapters 4, 9, Part 5 Introduction, Chapters 16 and 17.

## Components

*Switching equipment and architecture.* The CO switches delivering today's Centrex services incorporate modular designs, both in upper-level architecture (e.g., CPU/memory, switch matrix, administration, etc.) and in lower-level peripheral connections. For example, individual microprocessor/memory units control groups of line/trunk ports, with each group either slaved to a supervisory microprocessor in the distributed hierarchy (digital architecture), or managed directly by a master CPU (analog architecture).

The ability to add common control intelligence and peripheral ports in modular increments translates to lower rates for Centrex customers across the entire size spectrum. By avoiding large common control step functions, Centrex switching charges exhibit nearly linear cost characteristics as configurations grow.

System management requirements, however, particularly those involving CPE, such as cable management, call costing, voice processing, conferencing, and CPE repair/MAC (moves, adds, and changes) work, do create cost step functions as the work performed by the LEC, or its *fully separated subsidiary* (FSS), increases.

*Port interface capabilities.* Each Centrex mainstation number (unique telephone number) is normally supported by a twisted pair of copper wires in the LEC outside plant. These pairs are analog current-loop facilities powered by −48 VDC directly from the CO switch. For this reason, CO service is rarely interrupted by a commercial power failure, unless that failure is caused by physical damage that also affects the CO and/or local loop.

Trunking facilities for Centrex match and can connect to any of the analog or digital carrier PBX trunking facilities described above. Centrex trunking connections occur on the "back-side" of the CO switch, to access either tandem switches at IXC network gateways or other COs. Private-line facilities are also used with Centrex, and may require terminations at the premises POT and tieback to the CO.

*Centralized vs. distributed switching.* The ability of Centrex to provide uniform, cost-effective service to multiple locations is contingent upon switching equipment compatibility and availability of remote modules. Remote modules are smaller versions of the main Centrex switch, tied back to the host CO by digital transmission facilities, but designed to operate in standalone mode if the umbilical link is cut. Most CO switch suppliers offer several levels of remote module capability. An important issue is service degradation resulting from loss of the remote-to-host link.

If a mixture of Centrex and PBX service is used to implement a metropolitan area or city-wide network, different exchange codes for each location could be encountered. This creates a situation where station-to-station dialing within the network must be seven digits or more, in the absence of costly foreign exchange tariff arrangements. Also, the LEC may apply tariffed message unit charges to these calls.

Circumstances involving a CO "conversion" (upgrade) occurring at the time of large-scale Centrex project planning may afford the opportunity to negotiate with LECs for dedicated exchange codes. This allows exclusive use of up to 10,000 numbers, and network-wide four- or five-digit dialing, while avoiding message unit charges.

*Peripherals.* Figure 6.10 shows digital Centrex accommodating analog and digital, single and multiline station equipment. Until recently, multiline Centrex service necessitated use of key telephone system (KTS) CPE. Newer Centrex electronic key systems and digital single line telephones use Centrex lines over digital local loops. The CPE is proprietary to the manufacturer of the digital Centrex switch.

PCs and other data terminals can be connected via digital Centrex. Centrex lines used for digital single line and multiline station sets are terminated at the Centrex switch on special line cards. These lines require special administration and cannot be used with analog station equipment. Centrex, like a PBX, provides modem pool service for data communications via public and private voice networks. Centrex modem pooling, however, is more complex than that for a PBX, generally because of prohibitions against locating CPE (modem pooling equipment) with the LEC CO.

Since divestiture, the RBOCs have been prohibited from manufacturing CPE, requiring their LEC sales forces to inform a Centrex customer prospect of this constraint. In October 1985, however, Bell

Atlantic received a "Prime Contractor" waiver to the provisions of the MFJ. This waiver allows Bell Atlantic, as the parent RBOC company, to furnish an integrated systems proposal, which includes both Centrex switching and associated CPE. In such cases, the RBOC must ensure that the LEC maintains a neutral referral program, including other CPE suppliers, even to the extent of showing competitive equipment catalogs. Under provisions of the 1996 Telecommunications Act, manufacturing bans are lifted three years after an RBOC passes the "checklist" test as described in Chapter 2.

## Adjunct Systems

*Voice message processing.* Voice message processing provided with Centrex service is integrated either by connection with a voice-mail system collocated at the CO or at a remote location. The voice-mail system communicates with the Centrex switch over a high-speed digital data link. Where the system is located on the customer's premises or at a centralized voice mail provider's location, this data link runs back to the CO.

As in the case of PBX voice-message processing, today's Centrex voice-message processing is an integrated offering, providing the same ease of use and feature-richness previously associated only with a PBX-based system. The large scale of the system enables the LEC to tariff its service at a rate competitive with PBX-based voice mail.

One of the effects of the 1984 Modification of Final Judgement and divestiture was that the RBOCs began aggressive efforts to make Centrex and other LEC services more competitive with PBX-based alternatives. This was accomplished by offering customers "special contract" alternatives to usually more expensive tariffed services and modernizing switch and adjunct facilities. Reflecting this trend, voice-message processing and related interactive voice response and advanced call center capabilities are now integrated with Centrex service. For example, nearly a decade ago Bell Atlantic tariffed voice mail service using Lucent's (then AT&T's) 5ESS CO switches and Lucent's Audix system as CPE. Previously, Audix systems had been furnished only with Lucent PBX products.

*System management.* System management capability enables Centrex users to control and manage features, calling privileges, and restrictions. The system management function relies on computer equipment installed on the customer's premises, connected by data links to the CO switch providing Centrex service.

System management is valuable where the following circumstances exist:

- Frequent changes in feature assignments
- Frequent internal personnel moves
- Growth mode, with frequent assignment of new stations
- Requirement for internal control over Centrex
- Need for rapid accomplishment of MAC work
- Requirement for cost control

Again, powerful Centrex system management capabilities have been developed, rivaling those available with PBXs. Often called *operations support system* (OSS), or *operating system control* (OSC), these systems consist of standalone computers, generally powerful workstations, that not only connect with the CO switch for feature management but also control numerous other premises-related functions such as:

- Directory service
- Maintenance and repair scheduling
- Inventory control
- Cable management
- Moves, adds, and changes (MAC)
- Billing services

It is important to maintain *system management security* through frequent changing of computer passwords and other more elaborate "trusted-system" techniques. Changes made to the system are used to update the LEC's billing and service records, so accuracy and control are critical to system management operations.

Costs associated with Centrex system management can be equivalent to those of a PBX. This results from the addition of management functions that relate to CPE, rather than just the CO switch. Centrex system management hardware is therefore moving into the customer premises as part of an RBOC strategic approach to capture new tariff or contract revenues from CPE-based operations, maintenance, and management services.

In the past, certain changes to the switching configuration could only be performed at the CO, while others could be made on premises using a customer administration terminal. The terminal-based changes were not charged individually, but the CO-based changes generally were, which made frequent MAC work expensive.

Today, the trend is to enable the customer to make all necessary changes to the Centrex configuration using an RBOC-provided workstation. This lowers the cost of individual changes, which tends to reduce system operating expense. The initial capital cost of the management system, however, reflecting its new functionality, can approach that of a PBX.

## Operational Characteristics

The task of defining Centrex operational characteristics is complicated by issues relating to physical location and ownership of the switching equipment and the CPE. Users need to stay informed about these issues in order properly to plan and implement Centrex systems.

Decisions involving selection of terminal equipment and cable systems compatible with Centrex features and functions can be complex. Thus, the LEC needs to be brought into the planning process, and its inputs included, where appropriate, in specification and RFP structure.

Responsibilities created by location and ownership of equipment making up the Centrex system must be carefully identified. This simplifies not only the planning and acquisition process, but also ongoing system management. In general, the concepts described under the PBX section apply to Centrex as well. There are, however, some distinctions, which are pointed out below.

## System Capacities

Up to 100,000 lines of CO switch capacity means that an individual Centrex customer generally need not be as concerned with accurately forecasting growth requirements as in PBX installations. Service packages begin at several lines and can theoretically expand to tens of thousands. Type, size, and remaining capacity of the CO serving a particular location will, of course, determine actual growth capability.

Unlike the PBX, there is no concentration of station lines prior to connection with the Centrex switch. Each Centrex extension number is served by a dedicated pair running between the CO and the customer premises. Once terminated at the premises main distribution frame, pairs are cross-connected either to single line or multiline telephones or to separate key equipment.

Centrex is dependent upon the traditional single twisted-pair copper loop. In the past, attempts to increase digital voice and data throughput have been hampered by the bandwidth limitations of twisted pair, and constraints on flexibility imposed by analog outside plant design. Difficulties are encountered where high-speed digital services are run over the local loop, which may contain analog load coils, bridge taps, and multiple splice points—all of which mitigate against high-rate digital bitstreams. It is only in the last decade that advanced digital signal processing (DSP) techniques and the very large-scale integrated circuit (VLSI) chips that make them affordable have produced practical Integrated Services Digital Network (ISDN—see Chapter 8) solutions and the DSL breakthroughs described in Chapter 4.

Widespread use of broadband local loop fiber (that is fiber-to-the-home or fiber-to-the-curb) awaits a compelling business case that justifies abandoning the enormous investment sunk in copper-wire plant. Incremental upgrades using hybrid fiber-wire as noted above are paving the way to when the right "killer application" appears. Of course not even the most advanced Centrex switching fabrics are currently equipped to provide broadband video switching service on a scale that even approaches existing voice channel capabilities.

## System Performance

Both analog and digital Centrex switches are engineered by the LEC to carry anticipated amounts of offered traffic at various levels of blockage. The sheer size of most LEC switches usually means that initial or growth requirements of any single user are not likely to affect the blocking grade of service. However, although it might be highly unlikely in normal conditions, during a crisis or some popular event (such as the surge in calls for reservations to a rock concert, which, some years ago, resulted in the loss of dial-tone service in the Washington, DC, area), contention could begin on the line side of the switch, caused by a physical concentration ratio of incoming lines to paths through the switch's internal network. It is therefore important that a Centrex prospect with crucial communications needs query the LEC about what grade of service levels are being offered.

Note that calls placed within an exchange code provided by the switch do not require a trunk connection as they are station-to-station (intra-switch). This reduces blockage potential. Even where interexchange trunking is required, the large number of circuits connecting that switch to the PSN normally ensures that a trunk will be available when needed.

Where distributed Centrex arrangements are employed, involving use of remote switching modules and other peripherals, additional factors come into play. The distributed design must include sufficient node-to-host links to carry the offered traffic at the specified GOS. The remote module may be engineered to carry intra-module traffic only if the host link is lost. If network trunks are contained only at the host location, a link failure could cause loss of long-distance service, call accounting, and other centralized functions.

CO switches are often engineered to a lower overall traffic intensity number than are PBXs, e.g., 6.5 CCS/line (Centrex) versus 7.0 CCS/line (PBX), simply because residential subscribers tend to present a lower level of traffic than business customers. A modern CO is engineered for between 4 and 5 CCS per line standard. Above that standard, it might be possible to obtain a guaranteed GOS level under either tariff or special contract, but at extra cost.

The use of CPE data stations usually results in higher per-station traffic intensities than with telephones since holding times for data calls are longer than those for voice calls. CPE data stations tend to have less GOS impact in a Centrex versus a PBX environment because of the increased number of lines available in the CO switch. Information on all such connections should be furnished to the LEC as part of the service planning and selection process, however, if a minimum guaranteed GOS is required for either voice or data.

## Features

Centrex features are provided by both analog and digital switches and are dependent upon switching and software release versions. As previously noted, digital Centrex, where available, provides features comparable with those of fourth-generation PBXs.

The analog 1AESS switch, however, nearly 30 years after its introduction, is still in service although there are 20 times as many 5ESS and DMS-100 central offices as there are 1AESS central offices in operation today. Some RBOCs retaining 1AESS switches have added digital "front ends" to make them compatible with emerging end-to-end digital PSN.

In general, Centrex features are roughly equivalent to PBX features in terms of type and quantity. For marketing purposes, Centrex features are often categorized by the LECs as basic or enhanced.

# Key/Hybrid Telephone Systems

A *key telephone system* (KTS) is an arrangement of multiline telephones and associated equipment that permits station users to depress buttons (keys) to access CO or PBX lines and KTS features. Typical feature operations include answering or placing a call on a selected line, putting a call on hold, using an intercom path between phones at the same location, or activating an audible signal.

A KTS permits interconnection among on-premises stations without the need for central office or PBX switching systems. A multiline telephone incorporates visual indicators to show idle, busy, or ringing status of two or more lines (telephone numbers). KTSs are used by small businesses, or individual *community of interest* groups within larger businesses. Figure 6.11 illustrates a KTS and its relationship to other switching systems. Although not shown in the figure, KTSs behind Centrex service are arranged much as the PBX example.

**Figure 6.11**
Key telephone
systems.

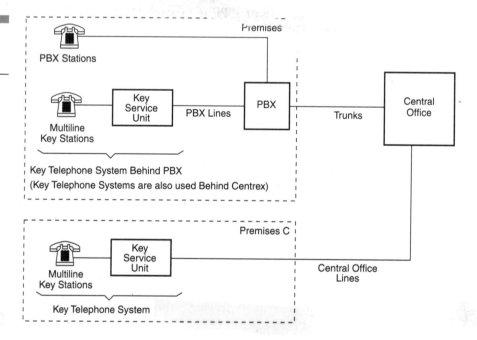

KTSs are directly connected to CO, Centrex, or business lines, so it is not necessary to dial an access digit to make an outside call. Station-to-station dialing in a KTS is accomplished by one- or two-digit intercom codes, with the number of codes determined by system size.

Attendant services are generally less sophisticated than those of a PBX, although hybrid and digital systems now support virtually all console functions.

## Evolution

In the mid-1920s, the Bell System first provided key telephone service using custom-engineered assemblies of lamps, keys, and wiring plans. These arrangements were complex and field labor intensive and, in the early 1930s, evolved into a packaged electromechanical hardware arrangement, the 1A KTS, which standardized wiring and greatly reduced on-site labor.

### THE 1A2 KTS

The 1A2 KTS, first offered in the early 1960s, was even easier to install and used prepackaged components, consisting of printed circuit boards (PCBs) and miniaturized relays. This resulted in a more compact unit, while adding popular features such as line exclusion, dial intercom, music on hold, and paging access. Typical button arrangements provided for push-button access to 5, 9, 19, and 29 lines per telephone. 1A2 installations are characterized by one or more 25-pair "key cables" from each telephone to the central *key service unit* (KSU).

### THE EKTS

In the 1970s, the widespread availability of low-cost microprocessors led to the development of *electronic key telephone systems* (EKTSs). These offered PBX-like features unavailable in electromechanical designs. They also eliminated the need for key cables by using PBX-like switching and control capabilities, which permitted a multiline telephone to be connected to the KSU via a "skinny-wire" one-to-six-pair cable. Thus, the distinction between KTSs and PBXs began to blur.

### HYBRID SYSTEMS

A *hybrid telephone system*, which may be of analog or digital design, incorporates both traditional KTS and PBX functions and features. A strict definition of a hybrid is difficult, since the mix of functions and features in a particular product is based solely on what the manufacturer believes will produce a competitive edge.

The FCC determines whether a system is registered as a KTS or a hybrid system based on whether a single-line station can access only a single CO loop line or trunk (KTS registration), or a pool of CO loops (hybrid registration). Some systems can be configured for either type of service, and are allowed dual registration.

### DIGITAL SYSTEMS

Digital KTS systems, all of which are classified as hybrids, incorporate digital PBX-type architectures. For example, the Nortel products incorporate a stored program control, pulse code modulation, time division multiplexed switch design, proprietary 2B+D (an ISDN-based channel and protocol design discussed in Chapter 8) intra-system signaling, and codecs embedded in telephones.

Figure 6.12 illustrates a digital KTS and its connection to a CO. Although the labeling details are omitted, the figure shows that the major components of the digital KTS are the same as those of generic circuit switches previously described. The figure also indicates that, like a PBX, the range of station equipment includes single and multi-line digital telephones, attendant consoles, and digital interface units.

**Figure 6.12**

Digital electronic key telephone systems.

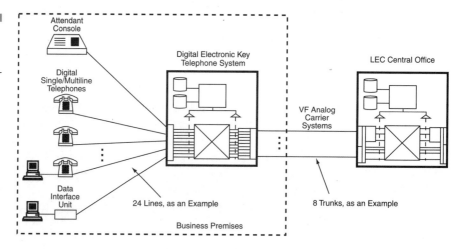

Hybrid and digital KTSs are capable of connecting to all types of analog trunks and tie trunks. *Tie trunk* is a term for non-switched, telco-provided, transmission service between PBXs and among PBXs, KTSs and telco switches. In the past, it generally connoted voice frequency or analog carrier service. Some newer high-end KTS systems are designed for direct T1, ISDN, and other digital carrier facility connection. Figures 6.13 and 14 are Nortel and Lucent model examples of modern key telephone systems.

**Figure 6.13**
Nortel Norstar
Integrated
Communications
System (KTS).

The Nortel Norstar Integrated Communications System (KTS) modular design accommodates many of the same CTI, IVR, and other features offered in Meridian PBXs. Equipped with Nortel's Companion line of wireless communication systems, Norstar users have a choice of using standard desktop telephones, or cordless handsets that allow users to roam throughout premises or campuses. The special pocket-sized Companion cordless telephones use the same listed directory number as assigned desk set numbers and afford users access to the same features available on desk sets. Lucent's Partner Advanced Communications System (PBX) offers similar capabilities and features.

## Components

*Port interface capabilities.* In traditional KTSs, KSU PCBs supported individual CO directory (telephone) numbers, which were then wired to multiline telephones. Thus each KSU PCB could support a number

**Figure 6.14**
Lucent Partner
Advanced
communications
System (KTS).
Courtesy of Lucent
Technologies Bell
Labs innovations.

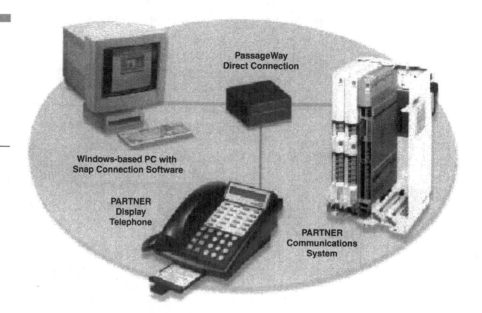

of telephones. This differed from the use of PBX PCBs which only supported individual telephones (stations). A traditional KTS line is equivalent to a PBX trunk, since both are used to connect the systems to a host switch.

Today's KTSs are more nearly like PBXs in that they impose limits on the quantity of stations and lines that can be accommodated. If a system is outgrown, telephones and internal PCBs are sometimes reusable, but the KSU cabinet and power supply generally have to be abandoned.

Traditional KTSs cannot connect to computers or LANs other than through a modem. Analog hybrids share this limitation. Today, vendors of higher-capability digital KTSs offer mechanisms for direct connection of PCs to their products, a connectivity that can generally be extended to other LAN-connected PCs in support of ever-increasing numbers of computer telephony integration operations. Thus, even small businesses are now able to "screen-pop" caller information before answering calls (and during them), set up conference calls and manage faxes and voice mail—with drag-and-drop ease, and much more. Nortel calls its KTS-to-PC product a Computer Telephony Adapter and Lucent refers to its as the PassageWay Direct Connection.

Interestingly, consistent with attempts to shed the stigma of older, limited capability "PBX" products by renaming them enterprise communications servers or communications systems, most vendors don't describe their KTS products as key telephone systems anymore. Nortel, for example has dubbed the Norstar KTSs Integrated Communications Systems. Likewise, Lucent refers to the Partner KTS as an Advanced Communications System.

*Centralized vs. distributed switching.* Traditional KTS and most hybrid/digital systems employ centralized switching designs. In this arrangement, each telephone is directly connected to a central KSU, which performs all required switching operations.

A recent, popular addition to the KTS industry is the "KSU-less KTS." These systems perform all common control functions using PCBs contained within each multiline telephone, and so do not require a KSU. They are designed for installation by the user, generally where premises wiring is already in place. Since there is no KSU, no communications closet space is required.

There is a limit to the quantity of CO lines and stations in a KSU-less system. Multiline telephones generally handle up to three telephone numbers, using between three- and six-pair horizontal wiring. Some telephones incorporate modular jacks for connecting ancillary equipment. At present, KSU-less systems are typically limited to 12 telephones.

*Wireless KTSs.* One of the most recent innovations is the wireless key system, such as the Siemens Gigaset 2420 shown in Figure 6.15. In this model, a single, small-footprint (about $10 \times 8 \times 4$ inches) corded base station/telephone station is attached to two local-loop telephone lines. Up to eight other cordless telephone sets transmit to and receive signals from the base station in the Industrial Scientific Medical (ISM) 2.4 Ghz band, a band that promises more capacity and roaming distance than earlier 50 Mhz and 900 Mhz cordless predecessors. The feature-laden, Gigaset provides caller ID/call waiting ID display, call transfer, call logging, built-in directories, distinctive call ringing and alerting, group calls, intercoms, handset paging, digital answering, base-station speakerphone, and other capabilities, most accessible from the remote cordless handsets.

Each cordless telephone is supplied with a desk stand that doubles as a battery charger. The system employs adaptive differential pulse code modulation (ADPCM) for high-quality voice and sophisticated frequency-hopping spread-spectrum modulation with excellent information security characteristics.

Competing with Siemens, Nortel offers its Companion wireless product that works with both its PBX and KTS systems. Companion places base stations throughout premises, at locations and intervals that guarantee high quality and minimize dropped calls due to marginal radio signal strength. Base stations are wired to a controller that interfaces with either the PBX or key system it supports. Again, remote cordless telephones are available in full-function models with all or most of the capabilities of wired counterparts.

**Figure 6.15**
Siemens 2420
Gigaset Wireless
Phone System, (KTS).
Courtesy of Siemens.

*Peripherals.* In traditional KTS applications, direct station selection/busy lamp fields (DSS/BLFs) are used in conjunction with a multiline telephone with line appearances for all telephone numbers in the system. Upon answering incoming calls, the attendants check the BLF. If the called party is not busy (indicator not lit), the attendant dials his intercom number by pressing a button on the DSS, and announces the call.

The call announcement may be received through the called party's handset, speakerphone, or over a call announcer, with either tone or voice alerting. To accept the call, the called party depresses the line button flashing on hold. If the call is not accepted, the attendant can take a message. Message waiting and electronic messaging features may also be used, and are particularly effective in situations where the called party does not answer.

Most KTS/hybrid telephones may be wall or desk mounted either in stock configuration or using a conversion kit. Some telephones incorporate additional features, such as long handset cord, cord swivel with strain relief, ringer volume control, and handset transmitter muting. Those with hearing aid compatibility include a control on the handset allowing line volume to be adjusted for each call. Some systems allow storing volume settings for future calls.

Speakerphones provide *full-duplex, handsfree operation*, allowing a user to place and receive calls without having to lift the hand set. A speakerphone also supports intra-office "roaming." The speaker is normally equipped with an on/off switch. Speakerphones also support *on-hook dialing* and *call progress monitoring*, which allow the user to hear dial tone, dial pulses or DTMF signaling, busy tone, ringing, intercept tones, and messages. Speakerphones can be used on both intercom and network calls.

Some key telephones assign individual basic features (transfer, add-on, and conference) to dedicated buttons for ease of use. Some also include dedicated "drop" buttons for removing parties from conference mode.

Dedicated hookswitch flash buttons are standard on behind-Centrex telephones to ensure that in-progress calls are not cut off by users attempting to invoke features.

*Soft keys* on proprietary instruments allow buttons to be programmed for either CO lines or features. Some systems support *macro keys,* which store up to six different keystrokes. This allows combining features such as speed dialing and specialized common-carrier service. Soft keys are generally assignable and programmable by station users.

## Adjunct Systems

*Voice message processing.* Traditional KTSs can be connected only to nonintegrated voice messaging systems, which do not deliver the capabilities of systems tightly coupled with call processing functions. Several hybrids and digital KTSs do incorporate integral voice messaging subsystems, which also perform automatic answering and allow direct internal dialing. Functions that would be housed in standalone equipment in PBX installations (such as auto attendant and voice mail) are often but not always embedded in KSUs. These adjunct systems are designed as PCB components and physically housed within the KSU.

Interactive voice response to callers can be provided by multiple voice response unit (VRU) PCBs, housed within a KSU. The attendant can activate VRUs as required, such as during busy periods, to invoke automated attendant service, or after hours to deliver recorded announcements. Tape units can be used to record calls and store them as messages.

*System management.* Traditional and other non-stored program control (SPC) KTSs do not offer maintenance and administrative capabilities since all features and functions are hardware oriented. Any additions or changes to these systems, as well as fault isolation and repair work, must be performed on site. No maintenance access panel (MAP) type terminal is available for these functions, apart from normal electronic test equipment. MAC work is accomplished largely through plug-in hardware module additions, while repairs involve module removal and replacement.

With SPC hybrids and digital KTSs, users can make certain feature changes at their own telephones. For more complex changes, the system administrator must activate a programming switch in the KSU and use a specific telephone in the system. Display telephones are commonly used for KTS system management functions since they are more economical than dedicated MAP-type terminals. As an option, vendors support administration and programming of KTS systems via local or remote PCs.

# Operational Characteristics

Like their larger PBX counterparts, KTS and hybrid systems have shrunk in size while growing in capability. From a feature and function standpoint, there is little to distinguish among these systems. Differentiation exists more in terms of cost and size than performance. It is therefore not surprising that KTS and hybrid systems suffer from being perceived in the same commodity-oriented perspective as PBX products.

# System Capacities

The KTS or hybrid system may either be *packaged*, with a certain capacity of lines and stations, or *port oriented*, in which case lines and stations can be intermixed to a maximum total capacity. Hybrids and digital KTSs are following the PBX path to port orientation, as it is more flexible and simplifies system expansion. System sizes vary widely, based upon the market segment targeted by the product.

In the traditional KTS, the number of stations (telephones) was unimportant, so long as it did not overtax the line lamp-illumination power supply. Modern systems impose limits on the number of telephones that may be supported. These limits vary based upon the type of system. Low-end space division technology KTSs may "max out" at six or eight stations. Several hybrid and digital KTSs can accommodate up to 120 stations.

Some of today's packaged hybrids and digital KTSs impose a line/trunk-to-station percentage on the system, generally on the order of 40 to 60 percent. This does not offer the flexibility of a port-oriented system, which allows lines and stations to be "mixed and matched." Note that in either case, however, once the upper limit of the system is exceeded, KSU upgrades are required.

## System Performance

*Grade of service.* In the small-systems world, GOS is largely measured—apart from voice quality—by the users' ability to access intercom paths and network connections. Early KTSs were constrained by a limited number of intercom talk paths or "links," analogous to station-to-station dialing paths in a PBX. Access to network resources, however, was non-blocking in line-per-station configurations, given the no-limit line capacity of 1A2 KTS systems, and the ability to terminate lines and trunks directly on multiline telephones.

Ironically, the later, more fully featured systems do not share this important capability. Because of system packaging, there are fewer lines than stations, which immediately introduces blocking probability. Even where lines are pooled, line-to-station contention can prevent placement of a network call.

Probability statistics indicate that for a given line-to-station ratio, blocking potential increases as the system size decreases. Thus, a four-line, six-station system can be expected to experience blocking more often than a 16-line, 24-station system. Many of today's hybrids and digital KTSs provide non-blocking service on intercom, but not on network connections.

## Features

KTS and hybrid features are categorized as system, attendant, station, and management related, some of which have been noted above. *Sys-*

*tem features* center around processor-oriented functions performed by the KTS or hybrid, applicable to all categories of users. *Attendant features* enable the attendant or receptionist to answer calls, announce them to intercom stations, and serve as a call-coverage point. *Station features* help the individual telephone user communicate with people and access other information resources more efficiently. *Management features* help the system administrator to access traffic information, change feature assignments, associate costs with premises and network services, and track the system configuration.

KTS and hybrid systems offer certain features not normally available in PBXs, e.g., intercom groups, call announcing, and music on hold. The average system delivers around 30 features, with most equipped as standard.

# Signaling System Fundamentals

Telephone exchange service was originally accomplished with manual switchboards and human operators. Prior to rotary-dial telephones, to make a call a user turned a crank that caused a lamp to flash on a panel at the central exchange. Responding to this *alerting* signal, an operator would then manually plug into the user's line (associated with the flashing lamp), and the user could verbally request connection with the called party. If the called party were served by a different exchange office, the operator used a similar process to alert a distant operator via trunks connecting the exchanges, and, again verbally, requested connection with the called party number.

The distant operator would check the called party's line for a "busy" condition. If it were not busy the distant operator would ring the called party's telephone. Both the calling and called party lamps remained lit during a call, providing the operators with indicators needed to *supervise* the use of customer lines and trunks. If the called party line were busy, the operator informed the caller.

Today, station equipment, switches, and transmission systems incorporate *call-handling* designs which generate and exchange signals to take the place of the above manual actions and verbal requests. The overall process is referred to as *signaling*.

Knowledge of signaling is important for several reasons. First, CPE acquired must be LEC and IXC compatible. Second, in private networks, CPE at different locations must be compatible. Third, telecommunications functionality increasingly depends on signaling capabilities, hence there is motivation to acquire business systems that ensure the ability to use new developments.

*Signaling* is the process of generating and exchanging information among components of a telecommunications system to establish, monitor, or release connections (call-handling functions) and to control related network and system operations (other functions).

As in prior years, telephone system users remain a part of the signaling process. They participate when they elect to use the service by going "off hook" (by lifting a handset), by dialing digits to access a service, by dialing telephone numbers, and by responding to various alerting signals such as audible dial tone, ringing signals, and recorded messages.

Signaling generates and transfers the functionally categorized signals described below.

Address signals convey destination information such as a dialed four-digit extension number, central office code, and when required, area code and serving IXC carrier code. These signals may be generated by telephone or other station equipment, or by the switching equipment itself.

Supervisory signals convey to a switching system or an operator the status of lines and trunks as follows:

- **Idle circuit**—Indicated by an "on-hook" signal and the absence of existing switching system connections on that line.

- **Busy circuit**—Indicated by an "off-hook" signal.

- **Seizure**—A request for service indicated by an "off-hook" signal in the absence of an existing switching system connection.

- **Disconnect**—Indicated by an "on-hook" signal subsequent to an established connection.

- **Wink-start**—Indicated by an "off-hook/on-hook" signal sequence on a trunk from a called office after a connect signal is sent from the calling office.

*On-hook* and *off-hook* are terms derived from the placement or removal of a handset at the telephone cradle, which closes or releases the *switchhook*, a plunger-activated switch built into the cradle. Although modern station equipment may use different switching arrangements, the on-hook/off-hook functions are still supported and the descriptive terminology is still in use. In LEC and IXC systems, supervisory signals are extended to billing and administration equipment for message accounting.

*Alerting signals* notify users, operators, or equipment of some occurrence, such as an incoming call. Included are ringing, flashing, recall, re-ring, and receiver-off-hook signals.

*Call progress signals* include dial tone, audible ringing tones, system-generated recorded announcements, and special-identification tones.

# Signaling Interfaces and Techniques

Signaling is described in terms of interfaces among network components, as well as techniques used to transmit signaling information between interfaces. *Signaling interfaces* exist between station equipment and transmission systems, between transmission systems and switching systems, and between transmission systems themselves. *Interfaces* are common boundary points between two systems or pieces of equipment.

## Loop Signaling Interfaces

An example of a *loop signaling interface,* one of three types of signaling interfaces, is shown in Figure 6.16. Although other signaling techniques can be associated with loop signaling interfaces, the technique illustrated in the figure is called *direct current (dc) signaling.*

**Figure 6.16**
Example of loop
signaling interfaces.

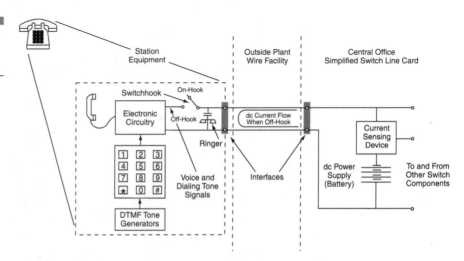

The left side of the figure shows the major parts of a telephone and its connection to a two-wire metallic VF transmission facility, which is connected on the other end to a line card in a CO switch. Loop signaling interfaces, indicated on the figure, derive their name from the metallic electrical loop formed by the line or trunk conductors (wires) and the circuits in terminating components.

Figure 6.17 illustrates signaling sequences for a typical call from one telephone to another through originating and terminating central offices. Event sequence is depicted by circled numbers. Figures 6.16 and 6.17 (as well as the text associated with them that is used to describe the principal signaling functions) are consistent with AT&T definitions and descriptions provided in AT&T's Engineering and Operations in the Bell Systems (see the Bibliography).

In Figure 6.17, the sequence begins with both calling and called-party telephones on-hook, or in an *idle* condition. A *seizure signal* is generated by the calling telephone when the caller removes the handset from the cradle (step 1 on Figure 6.17).

As Figure 6.16 shows, when idle, the telephone switchhook is open, and no dc electrical current can flow through the loop. Going off-hook closes the switchhook switch and permits current to flow through the line from the battery to a current sensing device, both at the central office. The current sensing device detects the off-hook status of the telephone and provides that information to other parts of the central office switch. This type of seizure signal is called *loop-start*, a supervisory signal generated by a telephone or a PBX in response to completing the loop current path.

An alternative seizure signal associated with loop signaling interfaces is *ground-start*, a supervisory signal generated by certain coin-operated telephones and PBXs by connecting one side of the line to ground (i.e., a point in an electrical circuit connected to earth).

*Wink-start* is yet another supervisory signal that consists of an off-hook followed by an on-hook signal, exchanged between two switching systems. The wink-start signal is generated by the called switch to indicate to the calling switch that it is ready to receive address signal digits.

Once the CO switch recognizes the telephone off-hook status, it generates and returns a *dial tone* over the loop (step 2 on Figure 6.17). Dial tone is a continuous tone formed by combining 350-Hz and 440-Hz tones.

Following receipt of dial tone, the user is free to dial the number of the party he or she wishes to call. In Figure 6.16, the caller does this by pressing appropriate dialing keypads, which generate a *dual-tone*

**Figure 6.17**
Signaling for a typical call connection.

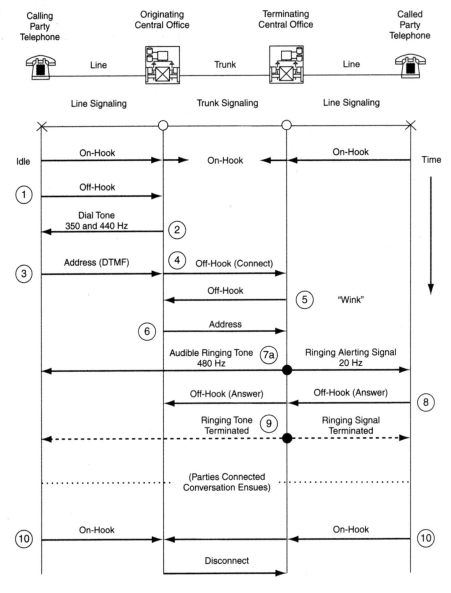

*multiple frequency* (DTMF) address signal (step 3 on Figure 6.17). DTMF signaling uses a simultaneous combination of one of a group of lower frequencies and one of a group of higher frequencies to represent a digit or character. Figure 6.18 illustrates the frequencies used to represent digits and characters.

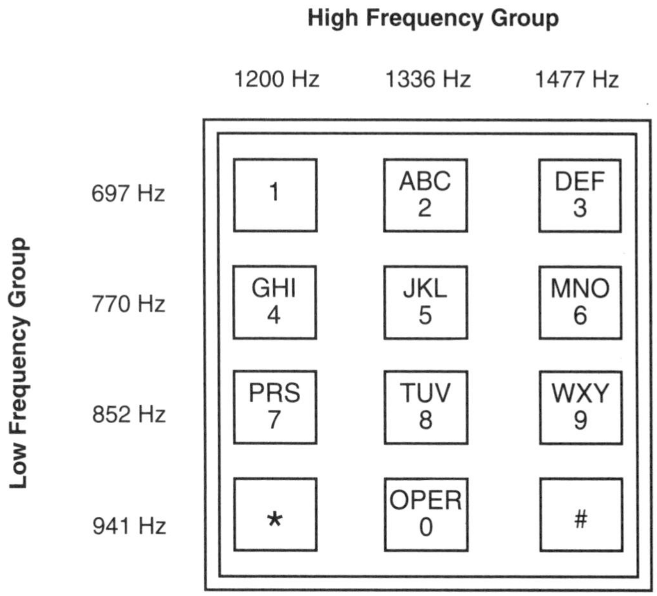

**Figure 6.18**

DTMF frequency groups.

**High Frequency Group**

1200 Hz      1336 Hz      1477 Hz

| Low Frequency Group | 1200 Hz | 1336 Hz | 1477 Hz |
|---|---|---|---|
| 697 Hz | 1 | ABC 2 | DEF 3 |
| 770 Hz | GHI 4 | JKL 5 | MNO 6 |
| 852 Hz | PRS 7 | TUV 8 | WXY 9 |
| 941 Hz | * | OPER 0 | # |

If we refer back to Figure 6.16, the telephone incorporates a DTMF tone generator, which under control of the dialing keys, sends the DTMF address signals over the same metallic loop conductors used for the supervisory signals. The originating CO receives the DTMF tones and stores them in a register for extension through the network via trunks to a terminating CO. The originating office seizes an idle interoffice trunk and sends an off-hook indication and a digit register request to the terminating office (step 4 on Figure 6.17).

The terminating CO sends a "wink" to indicate a register-ready status (step 6 on Figure 6.17) and the originating CO sends the address digits to the terminating CO (step 6 on Figure 6.17).

If a connection to the called-party telephone can be established (working line not busy), the terminating CO generates a *ringing signal,* a 20-Hz alerting signal that causes the called telephone to ring. The terminating CO also generates an audible ringing tone and returns it to the calling party's telephone (step 7a on Figure 6.17). This *call progress signal* is formed by combining 440-Hz and 480-Hz tones. When the called party is connected to the same central office as the calling party, the originating CO completes these actions.

When the called party answers (goes off-hook—step 8 on Figure 6.17), *ring tripping* occurs immediately so that he does not hear the 20-Hz alerting signal. The terminating central office removes the audible call progress ringing tone when it detects the called party's off-hook status (step 9 on Figure 6.17).

If the called party's line is busy, the central office sends a *busy signal* (another call progress signal) to the caller (step 7b, not shown on Figure 6.17). The busy signal is a combination of 480-Hz and 620-Hz tones, switched on for 0.5 second and off for 0.5 second. A *trunk-busy* signal is formed by the same tones but is repeated at a faster rate, such as 0.25 second on and 0.25 second off.

Once the call connection is completed, the parties can begin their conversation. In most cases, when either party hangs up (returns to on-hook status), the connection is released. This feature is called *first-party disconnect*, and without it, CO lines can get stuck in an in-use ("high and dry") condition, disabling both calling and called-party telephones.

PBX-to-central office (PBX-CO) trunks used to access public switched networks can use two-wire loops and loop signaling interfaces. PBX trunks and associated signaling may be arranged for rotary or DTMF service, as well as direct outward dialing (DOD), direct inward dialing (DID), or two-way operation. (Trunks arranged for two calling directions are called *two-way trunks*.)

Note that DOD trunks support only calls originating from PBX station equipment, and DID trunks support only calls terminating on PBX station equipment. (Trunks arranged to handle only one calling direction are called one-way trunks.) Once a connection is made, however, voice-message traffic is carried in both directions (note that one-way trunks do not imply one-way circuit paths).

Ground-start signaling is used on two-way PBX trunks to prevent simultaneous seizure from both ends, or *glare*. Glare is a condition that can occur if loop-start signaling is used on two-way trunks and can lead to conditions that take trunks out of service. Either ground-start or loop-start signaling can be used with DOD trunks, a choice driven by tariffs and overall premises requirements. Wink-start signaling is used with DID trunks.

## E&M Leads Signaling Interfaces

*E&M leads signaling interfaces*, used for connections between switches and transmission systems and between transmission systems themselves, support two-way operation. E&M leads is an interface in which the signaling information is transferred across the interface via two-state voltage conditions on two leads, each with a ground return, separate from the leads used for message information. The message and signaling information are combined and separated by means appropriate for the associated transmission facility.

E&M leads signaling is used in business telecommunications primarily for PBX tie trunks in private networks. Other types of tie trunks, such as one-way dial, were once common; however, FCC registration rules emphasize two-way E&M interfaces, which is consistent with most of today's analog equipment.

## Circuit-associated Signaling

In Figure 6.16, voice signals to and from the handset are electronically coupled to the same loop used for supervisory, call progress, and alerting signals. Thus, *circuit-associated signaling* is a technique that uses the same facility path for both voice and signaling traffic. Historically, this approach was selected to avoid the costs of separate channels for signaling and because the amount of traffic generated by signaling is small compared to voice, minimizing the chance for mutual interference. Circuit-associated signaling is contrasted with common-channel signaling that will be discussed later.

Circuit-associated signaling uses either in-band or out-of-band signaling techniques. In-band signaling uses not only the same channel path as the voice traffic, but also the same frequency range (band) used for the voice message. For example, in Figure 6.18, DTMF addressing signals use the same frequency band as the voice signals.

Out-of-band signaling uses the same channel path as the message but signaling is in a frequency band outside that used for voice traffic. Digital time division multiplexed and carrier system signaling is considered out-of-band. As previously shown in Figure 5.5, for one frame out of every six, the eighth bit in each channel of a DSO signal is "robbed" for the purpose of transferring signaling information.

In D-type channel banks, the channel units provide the means to convert switch loop closure, DTMF, and other analog signaling formats to equivalent digital formats. Available channel units for digital carrier support a full set of signaling capabilities including loop-start, ground-start, and E&M signaling for PBX tie trunk applications.

# Common-channel Signaling Interfaces

Signaling between switching offices may be provided on a per-trunk or a common-channel basis. In *per-trunk signaling*, signals pertaining to a particular call are transmitted over the same trunk that carries the call, as shown in Figure 6.19a. Interoffice signaling other than common-channel signaling falls into this category.

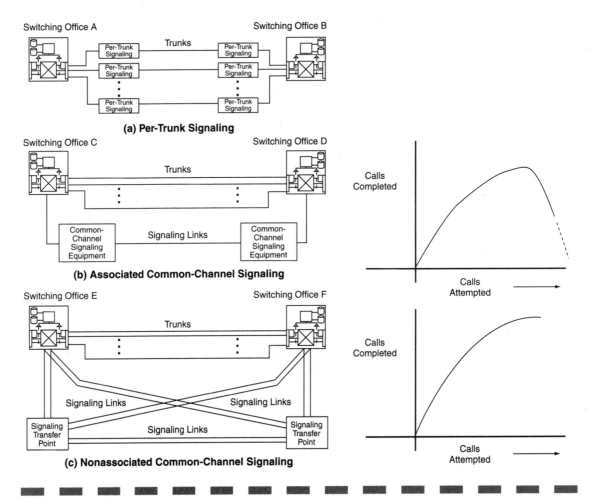

**Figure 6.19**   Per-trunk and CCS configurations.

*Common-channel signaling* (CCS) is a signaling system developed for use between stored program control switching systems, in which all of the signaling information for one or more trunk groups is transmitted over a dedicated signaling channel, separate from the user traffic-bearing channel. Figure 6.19b shows an associated CCS approach in which one signaling link per trunk group is routed between the switching offices terminating the trunk group, using the same transmission facility.

*Nonassociated CCS,* a more economical approach that greatly reduces call setup time and enhances total network routing flexibility, is shown in Figure 6.19c. Here signaling is routed through signaling transfer points over completely separate facilities so that the ability to complete a call on an end-to-end basis can be determined prior to the commitment of trunk and switch resources.

This is in contrast to circuit- and CCS-associated signaling, where traffic-bearing resources are already committed as the call progresses through a network. Should the final result be that a called party's telephone is busy and the call cannot be completed, with associated signaling, traffic resources have in effect been wasted during the call setup time. For greatly overloaded networks, as could occur during disasters or other high-volume call-causing events, circuits become blocked primarily from unsuccessful call attempts. This phenomenon is illustrated on the right side of Figure 6.19b, where at low levels of call-attempt incidence, all calls are completed by the network (except, of course, for those cases where the called-party line is busy or does not answer). During crises, such as the Three Mile Island nuclear accident, the number of call attempts saturates available circuits with call-attempt signaling traffic and the capacity for conversational traffic actually diminishes.

While nonassociated common-channel signaling prevents this failure (illustrated in the right portion of Figure 6.19c), it poses new reliability and availability problems. For example, if the CCS network fails, no calls can be completed even if all of the user-traffic channels are operative. To protect against such occurrences, CCS networks incorporate redundancy so that individual-link or equipment failures are not catastrophic. In recent years, however, we have witnessed several PSTN "crashes" caused by failures in CCS software.

In terms of the ability of U.S. telecommunications to withstand terrorist or military attack, it is not clear whether the relatively small number of signaling transfer points in today's public networks are adequate to sustain operations under determined conventional or nuclear military attacks. Whereas intentional enemy destruction of signaling transfer points is plausible, totally destroying associated signaling would require destruction of virtually all domestic telecommunications capabilities. National Security and Emergency Preparedness (NSEP) authorities constantly review and assess national telecommunications infrastructure vulnerabilities and develop plans, such as fall-back circuit-associated mechanisms to ensure that critical telecommunications can survive postulated threats.

# Signaling System No. 7 and Advanced Intelligent Networks

Common Channel Signaling System No. 7 (i.e., SS7 ) is a global telecommunications standard developed by ITU-T. The standard defines procedures and protocols by which public-switched telephone network (PSTN) elements exchange signaling information over digital packet-switched networks to effect wireless (cellular) and wireline call setup, routing, and control. The ITU definition of SS7 allows for national variants such as the ANSI standards in North America, and the ETSI standard in Europe.

The initial CCITT CCS standard, called *CCITT Signaling System No. 6* (or simply SS6), was supplanted by SS7. AT&T introduced and implemented a version of SS6 called *common-channel interoffice signaling* (CCIS) in 1976.

The AT&T and Bellcore versions of SS7 are commonly referred to as CCS 7 and SS 7 respectively. Although these versions are functionally similar, gateways are required between LECs and IXCs to account for differences—one of the penalties of divestiture and private telecommunications network ownership.

User interfaces with SS7 are addressed in the Q.930/931 part of ITU's General Recommendations on Telephone Switching and Signaling, a specification for *digital subscriber signaling systems* (DSS). DSS permits telecommunication users to capitalize on economies, service enhancements, and flexibilities offered by integrated (voice and data) digital networks. It also defines and sets standards for future business-system signaling requirements.

Using ITU terminology, Figure 6.20 depicts SS7 packet-switched network nodes and their relationship to service switching points (SSPs) which may be central-office, mobile-switching, or other subscriber traffic-bearing network circuit switches. SS7 networks support address, supervisory, alerting, and call progress signals and all of the nearly century-old calling/called party-driven call placement and take-down functions described above. They also provide generalized message-based communications among SS7 computers. Consequently, they are able to support a virtually unlimited set of enterprise-enhancing telco and customer-defined operations.

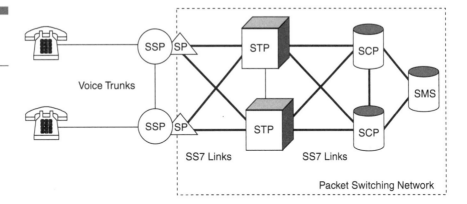

**Figure 6.20**
SS7 nodes and operation.

SS7 Signaling Nodes

- SSP (Service Switching Point)
  For example: a central offie, mobile switching office,
  access tandem, or other telecommunications switch.
- SP (Signaling Point)
- STP (Signal Transfer Point)
- SCP (Service Control Point)
- SMS (Service Management System)

For example, in addition to basic call management functions, SS7 networks and protocols are used for:

- Toll-free (800/888) and toll (900) wireline services.
- Call cost accounting associated with 800, 888 and 900 services.
- Accessing credit-card-calling data bases that permit special call screening and restriction.
- Enhanced call features such as call forwarding, calling party name/number display, and three-way calling.
- Local number portability (LNP).
- Wireless services such as personal communications services (PCS), wireless roaming, and mobile subscriber authentication.
- Operation, administration, and maintenance (OA&M) and customer billing and inter-company settlement activities.
- Efficient and secure worldwide telecommunications.

The above list only begins to reveal SS7 capabilities and uses. In fact the potential is so great that in the early 1990s, Bellcore and other similar groups and manufacturers around the world coined the term Advanced Intelligent Network (AIN) and began to define its capabilities—nearly all of which depend on SS7 facilities, protocols and standards. The notion behind AIN is that prior to connecting calls initiated by human subscribers or devices, databases are queried for special handling instructions. Databases may reside in telco networks or belong to customers—so long as they are compatibly interconnected.

Figure 6-20 highlights principal SS7 elements:

- **SP—Signaling Point.** A switch, or switch adjunct that endows the switch with SS7 capabilities.

- **STP—Signal Transfer Point.** A packet switch that routes messages to and from various nodes in an SS7 or a common channel interoffice signaling network.

- **SCP—Service Control Point.** The database or multiple database systems that provide enhanced SS7/AIN services, accessible from multiple locations through STPs.

- **SMS—Service Management System.** A nationwide or central database and processing system used to build, maintain and provide a master data source for SS7/AIN service. An SMS contains complete telco or customer-defined specifications including location data, numbering plan, features, screening actions, authorization codes, calling privileges, etc., upon which advanced telco services and/or unique business applications depend.

SS7 and AIN designs significantly reduce post-dial delay (call setup time); enable more efficient call routing methods; permit optimization of network traffic handling; facilitate network management; make worldwide telecommunications more efficient and secure; and reduce susceptibility to fraud. It is no exaggeration to state that AINs properly implemented on SS7 facilities are cornerstones of modern digital networks, with all the vulnerability that the term implies.

# Final Remarks

Because the scope of signaling concepts and technologies is so broad, and because successful, efficient, and reliable telecommunications service provision so crucially depends upon signaling system performance, we offer a short recapitulation to help readers catch their breath, and put in perspective the complex material just covered.

There are five major signaling techniques and three types of signaling interfaces. The signaling techniques are: direct current, in-band tone, out-of-band tone, digital, and CCS. These signaling techniques, also referred to as signaling systems, are classified as either *facility dependent* or *facility independent*. Direct current, out-of-band tone, and digital signaling techniques are facility dependent because they cannot operate separately from the transmission facilities used for the voice traffic that they control.

Inband tone (e.g., DTMF, SF, and MF) and CCS are *facility-independent* techniques in that their operation is independent of the facilities used to carry the voice traffic. *Single frequency (SF) signaling* is a method of conveying addressing and supervisory signals from one end of a trunk to the other, using the presence or absence of a single specified frequency, which exhibits significant susceptibility to fraud, and has all but vanished from the scene. A 2,600-Hz tone is commonly used. *Multiple frequency* (MF) is an interoffice address signaling method in which 10 decimal digits and five auxiliary signals are each represented by selecting two of the following group of frequencies: 700, 900, 1,100, 1,300, 1,500, and 1,700 Hz.

For networks built from a combination of tandem facilities, facility-independent, in-band signaling simplifies interfaces since signaling is automatically extended as part of the voice channel. That is, any facility that supports voice channels is also capable of transmitting signaling information. With the proliferation of digital facilities, however, CCS out-of-band signaling has become the dominant inter-switch signaling approach. SS7 is, therefore, a facility-dependent, out-of-band CCS signaling system.

The three types of signaling interfaces are: loop signaling, E&M leads, and CCS. From LEC and IXC points of view, combinations of interfaces and techniques are in three application-related realms—line signaling, interoffice trunk signaling, and special-services signaling.

Since station equipment and PBXs must be compatible with signaling arrangements of various LEC access services, line signaling is the realm of greatest importance to business telecommunications planners/users. Signaling associated with LEC access services is discussed in Chapter 10.

Telecommunications planners and users need to understand the characteristics of interoffice signaling to evaluate new services made possible by CCS signaling systems, and to select business systems best able to take advantage of digital access signaling, such as SS 7.

Special-services signaling refers to signaling used with special services, i.e., any of a variety of LEC and IXC switched, non-switched, or special-rate services that are either separate from public telephone service or contribute to certain aspects of public telephone service. Examples include PBX tie trunks, foreign exchange (FX), and private-line services (see Chapter 10). These services and their signaling arrangements are important to business telecommunication planners/users.

Finally, all signaling approaches can be classified as either *stimulus/ response* or *message-based*. As stored program control switches replace electromechanical switching systems, the trend is clearly towards SS7, message-based signaling. From a total network point of view, signaling systems are emerging as data communications subnetworks linking computers embedded in station, switching, and programmable transmission/multiplexing equipment.

In earlier parts of this book, switches, transmission media and components, and signaling are discussed on an individual basis. In fact, within any telecommunications network, all the switches taken collectively can be viewed as a *switching system*, and all the transmission components as a *transmission system*. Similarly, all the signaling components, distributed in these systems, comprise a *signaling system*. This systems-level notion is illustrated in Figure 6.21. While individual switches, transmission equipment, and other network components need to be specified, usually on a site-by-site basis, specification of network-wide switching, transmission, and signaling system performance is equally important in selecting public and private network services.

**Figure 6.21**
Notional system-level network switching, transmission, and signaling representation.

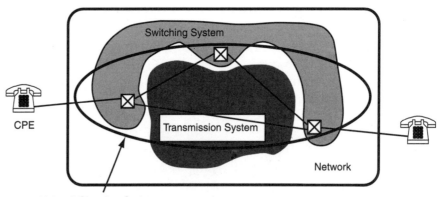

CPE

Switching System

Transmission System

Network

Network Signaling System
(Components in Both the Switching
and Tramsmission Subsystems)

# CHAPTER 7

# Packet Communications Concepts

This chapter presents fundamentals of data communications in preparation for a discussion of data services in forthcoming chapters. It relies on the introductory material presented earlier for terms of reference and background. In particular, it builds on the definitions and descriptions of digital electrical signals, binary bits, error detection and correction, data terminal equipment (DTE), digital carrier, time division multiplexing (TDM), and digital circuit switching presented in Chapters 1 through 6.

A significant difference between voice and data service is the extent to which human intervention is required to ensure end-to-end communications integrity, including diagnosis and recovery under failed or inadequate service conditions. For example, if an American places a telephone call to Japan that is answered by someone who cannot speak English, human intelligence is relied upon to seek an interpreter or to take alternative action. Similarly, if a call cannot be completed due to a network failure, a human determines the problem and takes corrective steps.

By contrast, data services are provided with minimal human intervention. As a consequence, more elaborate mechanisms are required to ensure that transmitting and receiving DTEs "speak the same language," and that service restoration actions are promptly taken under network failure conditions. This generally requires higher levels of hardware and software compatibility among DTEs and intervening data network elements than is required in voice networks.

For private data networks, it might be feasible to specify hardware and software from a single source, achieving compatibility through proprietary design. For public networks relying on universal connectivity supported by multiple vendors, standards and protocols defined by U.S. and worldwide organizations must be used. *Protocols* are strict procedures for the initiation, maintenance and termination of data communications, as described later in this chapter.

As we saw in Chapter 1, traffic characteristics impose different requirements on voice-versus-data network design. For circuit switched voice communications, a nominal post-dial delay (call setup) interval of several seconds is acceptable. However, data traffic often occurs in short bursts, resulting in long inactive periods interspersed with high-speed information exchange. So a dedicated non-switched channel would result in inefficient network utilization. In addition, setup time to establish a circuit-switched call would result in unacceptable response times for on-line data transactions, where terminal-operator requests for data must be responded to in a very few seconds.

Developed in the 1970s for long-distance data communications, packet switching, an alternative to circuit switching, drastically reduces or eliminates call setup time and inactive periods on circuits and is therefore well suited to bursty data traffic.

This chapter introduces basic packet switching principles and fundamental concepts underlying all protocols. It shows how different sets of protocols evolved as new technology changed the constraints under which the protocols operated. Finally, it describes the operation of major local and wide area network services in use today in terms of facility types and protocols.

# Packet Switching Fundamentals

A *packet* is a quantity of data that is transmitted and switched as a composite whole. A packet contains user data, destination and source information, control information, and error-detection bits, arranged in a particular format. A typical packet is shown in Figure 7.1. Packets are formed by segmenting user message information or data (which may be any number of bits or bytes) into packets of limited length by *packet assembler-disassemblers* (PADs), as shown in the figure. Packetization is used in virtually all data communications systems.

**Figure 7.1**
Typical packet format.

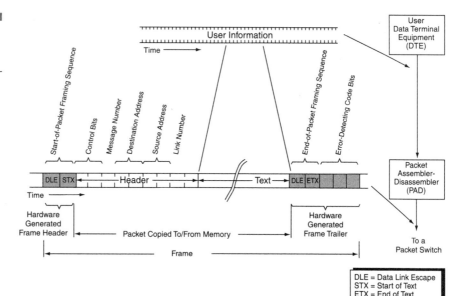

Sufficient information is embedded in packets to enable packet switches to route them through networks. A *packet header*, which precedes user data, may contain destination address, source address, link numbers, packet numbers, and other information. Specifically, a header is control information appended to a segment of user data for synchronization, routing, and sequencing of a transmitted data packet. Among adjacent and connected switching nodes, packets are encapsulated in *frames* which themselves include headers and trailers (codes), usually hardware generated, to indicate start-of-message and end-of-message events. The glossary explains in more detail several legitimate meanings for the word "frame" when used in telecommunications contexts.

A packet-switching network is designed to switch and transport information in packet form. Figure 7.2 illustrates how packet switching works and the differences between packet and circuit switching. In the figure, user messages, represented by the rectangles labeled A, B, and C are shown as DTE inputs and outputs. Message length is indicated by the length of the rectangles.

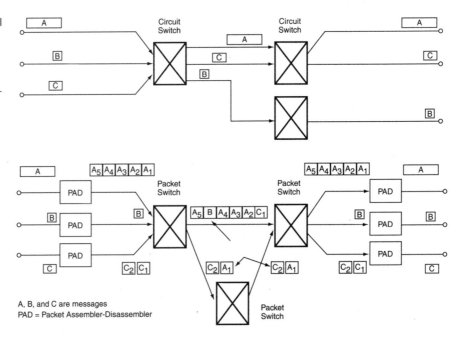

**Figure 7.2**

Example of circuit and packet-switched connections.

For circuit switching, illustrated in the upper half of Figure 7.2, channels between switches are used exclusively for individual message transmissions A, B, and C, assigned on a first-come, first-served basis. The circuit switches establish connections (data calls) between remote DTEs in a manner similar to that for voice traffic between two telephones. As with voice, channels remain occupied (out of service for additional calls) until released. As already noted, for interactive message traffic generated by human keyboard operators, actual information transfer may only occur in bursts, interspersed with long periods of inactivity. In this case, circuit switching makes inefficient use of potentially expensive transmission resources. Once all links are busy, new messages, even short ones, may experience unacceptable delay waiting for circuits to clear. In this example, three circuits are required between the switches to handle the information transfer.

In packet switching, messages A, B, and C are segmented into packets by PADs prior to being offered to packet switches. This operation is illustrated in the lower half of Figure 7.2 where packets corresponding to messages A, B, and C are processed by packet switches and interleaved on transmission links between the switches. Packet-switched networks provide more efficient data transport than circuit-switched networks because the connections through the network are used only while data is being transmitted. As a result, many different connections can share the same circuit.

Each packet switch is connected to one or more remote packet switches. In the transmission of message A, note in the figure that some packets from message A are delayed more than others. This occurs when packets from message A wait in a queue at the switch while packets from other messages are transmitted over the same circuit. Because of the random nature of packet arrivals from different sources, this situation occurs during normal operations. Packet switches must provide storage space for packets that are waiting, in the form of buffers. Because of this phenomenon, packet switching operations produce variable end-to-end message delays. For data applications this is normally not a problem, but it can degrade voice and video communications. For this and other reasons, packet switching has traditionally been used exclusively for data communications. Advanced, fast packet-switching technologies supporting voice, data, video, and other services overcome these limitations.

# Packet-Switch Functions and Capabilities

A packet switch consists of the following functional entities:

- Input and output buffering (memory elements to temporarily store packets).
- Processing for decoding header address, routing, and other information; error detection; and switch and network control.
- Internal switching to connect input and output buffers. The transmission of packets through a network requires three types of packet control procedures:
  - Routing control to determine the routes over which packets are transmitted.
  - Flow control to prevent congestion in the network and lock-ups or traffic jams.
  - Error control to deal with any transmission errors that occur.

In contrast to circuit-switched voice networks where signaling is invoked once to establish call connections for the duration of the entire call or transaction, in packet networks, each packet is examined for source and destination address information and acted upon accordingly. While this operation does result in efficient utilization of transmission resources, overall routing, error, and flow control impose significant processing requirements on packet switches. In fact, the throughput of packet networks is limited primarily by the processing capabilities of the packet switches.

In the Internet backbones of the major Network Service Providers (NSPs), the switches have a capacity exceeding 10 million packets per second. This dramatic improvement to the 300-packets-per-second performance of the switches in the 1970s networks has made possible the public data networks we depend upon today.

# Access and Transport Services

As noted above, once a call is established in circuit-switched networks, a dedicated, physical connection is established between telephones or other user station equipment, to be torn down at the end of the call.

Analogously, a connection-oriented packet-switched network transport service establishes logical connections in response to station equipment (DTE) requests. All packets entering the networks are delivered to terminating DTEs in the order in which they were received. As with voice calls, *connection-oriented* data services use separate procedures for connection establishment and end-to-end information transfer (connection establishment must take place prior to information transfer).

This service is referred to as *virtual circuit service* since, in the absence of degradation, message routing is logically identical to routing over circuit-switched facilities (i.e., all packets for a given logical connection follow an identical path through the network). Note, however, that circuit-switching inefficiencies are avoided since packets from multiple sources can be interleaved over the same physical transmission paths.

A *permanent virtual circuit* (PVC) is a virtual circuit resembling a leased line in that invariant logical numbers identifying PVCs are dedicated to a single user. Thus, at a particular interface point a network service provider assigns a fixed number of virtual circuits to a user, each of which connects specific network/user interface points. Alternatively, a *switched virtual circuit* (SVC) permits a user to establish virtual circuits between arbitrary network interface points, much like direct-distance dialing in circuit-switched voice networks.

Although most packet-switched networks used for wide area or long-distance data communications offer users virtual circuit service, networks can be designed to offer users *connectionless* service where economics dictate simple switches and control procedures. It eliminates connection set-up, lowers overhead, and results in faster transmission times. Packets are routed independently over the network from source to destination and are delivered in whatever order they arrive at the destination. Connectionless modes are widely used in local area networks to reduce complexity and cost.

Figure 7.3 shows two methods for physically accessing packet switched networks. On the right-hand side of the figure is a host computer (i.e., any computer running a full protocol stack up to the Application layer) attached to a communications *front-end processor* (FEP) with integral PAD functional capabilities. The FEP is connected directly to a packet switch, which is either located on a customer's premises, or connected via digital access facilities.

FEPs, also called stored program communications controllers, are dedicated computers or systems of computers that control data communications between host processors and various types of data communications networks. FEP functions include route selection, multihost access, data switching, network management, message sequencing, and flow control. FEPs support both private and public data network operations. For example, the IBM FEPs support private IBM System Network Architecture (SNA) networks, but with network packet-switching interface programs, they can connect with public packet switched networks.

**Figure 7.3**
Access to packet-switched networks.

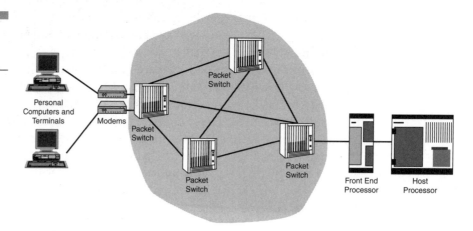

The left side of the figure shows how modems are used to connect terminals or other DTEs to remote packet switches and PADs. In this case either dedicated (leased) or public-switched dial-up voice network services can be used to access packet-switched network services using modems. With dial-up service, log-on to the packet network is required each time a user wishes to obtain service.

For the special case of the Internet, the packet switches and the FEPs are implemented as routers. *Routers* are described in the Layer 3 Network part of the "ISO Reference Model for OSI" subsection below. Router functions and operations are described later in this section and throughout the remainder of the book.

# Protocol Fundamentals

From the foregoing, it is evident that data communications networks require a high degree of compatibility and interoperability among DTEs and network elements, particularly with respect to physical and logical interfaces and controls. A challenge is presented by different vendor equipment and/or even different models from the same vendor, all of which must be interconnected.

In 1977, the International Organization for Standardization (ISO) established a subcommittee to develop a standards architecture to achieve the long-term goal of open systems interconnection (OSI). ISO is a voluntary international body concerned with developing standards for a variety of subjects. Data communications standards are developed through the workings of its Technical Committee 97. ISO membership is mainly composed of national standards-making organizations, for example, the American National Standards Institute (ANSI) in the United States, as discussed in Appendix A.

The term open systems interconnection denotes standards for the exchange of information among systems that are "*open*" to one another by virtue of incorporating ISO or other industry accepted standards. The fact that a system is open does not imply any particular system's implementation, technology, or means of interconnection but refers to compliance with applicable standards.

ISO has specified an *OSI Reference Model* that segments communications functions into seven layers. Each layer is assigned related subsets of communications functions implemented in a DTE required to communicate with another DTE. Each layer relies on the next lower layer to perform more primitive functions, and in turn provides services to support the next higher layer. Layers are defined so that changes in one layer do not affect other layers.

Information exchange occurs when corresponding (peer) layers in two systems communicate by means of a set of rules known as protocols. Protocols define the *syntax* (arrangements, formats, and patterns of bits and bytes) and the *semantics* (system control, information context or meaning of patterns of bits or bytes) of exchanged data, as well as numerous other characteristics such as data rates, timing, etc.

Defining the details of seven layers of protocols for data communications is an enormously complex task. Before delving into a technical discussion of the ISO layers, consider an example taken from a more time-honored form of communications. Figure 7.4 illustrates multiple layers of communications between two diplomats from different countries. The exchange of ideas between the two diplomats represents ISO Layer 7, user-to-user communications.

Since the diplomats have no common language, they each engage the services of a translator. The translator converts the message into a common language (e.g., French), writes it down on paper, places the letter in an addressed envelope, and mails the letter. The envelope is carried in sacks by trains between post offices in the originator's country. At each intermediate post office, the envelope is retrieved from the sack, the destination address is read, and the envelope is placed in a new sack on a different train to continue on its way to the destination. The process of receiving the envelope at each post office, verifying that it is in good condition, and passing it on to be routed to its destination represents ISO Layer 2, Link-Layer communications (the trains themselves are Layer 1, Physical-Layer communications). If Layer 2 is connection oriented, a message is sent to the post office at the origin end of the link noting the condition of the envelope. If the envelope was damaged, a new copy is sent on the next train.

**Figure 7.4**
Example of multi-layer peer-to-peer communications.

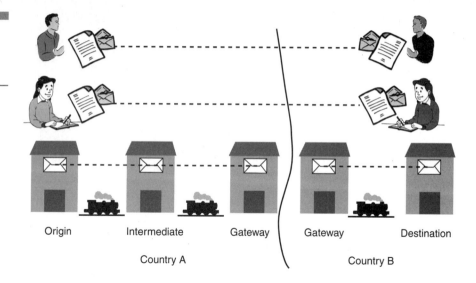

| Origin | Intermediate | Gateway | Gateway | Destination |

Country A          Country B

The reading of the destination address and the routing of the letter is one of the functions of ISO Layer 3, the Network Layer. The other function of Layer 3 occurs when the letter reaches its final post office (in this case, the gateway post office for the national postal system). In connection-oriented networks, the letter is verified to be in good condition and a message is sent back to the originating post office telling it that the letter arrived at the network boundary successfully. If the letter is damaged, this information is passed back to the originating post office, which will then resend a new copy of the letter. In connectionless networks, letters are passed along between networks without regard to their condition; other means are employed to notify the originator of problems with the message.

At the gateway boundary between the two countries, the letter is passed to the national postal network of the destination country using an agreed-upon process known as an internetwork protocol, a special case of Layer 3. The process of passing the letter from the gateway post office to the post office serving the second diplomat uses the Layer 3 and Layer 2 protocols of the destination country. Once the letter is delivered to the translator, he checks it for integrity. This layer of communications between translators is representative of ISO Layer 4, the Transport Layer. This layer is especially important since it deals with the end-to-end quality and reliability of the communications path between the users. Like the Network Layer, the Transport Layer may be connection-oriented or connectionless. A connection-oriented Transport Layer notifies the originating translator that the letter was received intact. If it was not, the originating translator (not the originating post office) resends a new copy of the letter. Once the letter's integrity is verified by the translator, it is passed to his diplomat for reading.

Note that the translators could change to English to write to each other without affecting either the Layer 1, 2, or 3 processes. Similarly, neither message integrity at Layer 3 nor the translation process of Layer 4 is affected should the physical transportation media change from trains to trucks.

## Tradeoffs in Protocol Design

At this point, it is worth stepping back and considering the implications of what we have just learned from the above example. We have seen that connection-oriented transmission ensures the successful receipt of the message between any two end-points where it is used. One might assume, then, that it would be used at every protocol layer in the system. This would be the case if there were not a considerable penalty to be paid in complexity, cost, and performance for using connection-oriented techniques.

In the above example, connection-oriented transmission is available at Layers 2, 3, and 4. At Layer 2, each post office must store copies of each letter sent on the outgoing trains and keep these copies until a message (known as an *acknowledgment*) arrives on a returning train that the letters were received correctly at the next station. At Layer 3, additional copies of each letter are stored at originating post offices waiting for return messages from destination post offices serving the recipient of the letter or serving a network boundary. Again, the letters are stored until the acknowledgments are received.

The process happens again at Layer 4, with the translator storing copies until he hears from the destination translator. In most cases, the sender will not send more letters to the same destination until he has positive acknowledgment of those already sent. The delays encountered in waiting for acknowledgment messages and the cost of storage are significant considerations for the use of connection-oriented protocols. When and where, then, are these techniques employed? If reliable transmission is required at Layer 4, and it usually is required, then it makes sense always to employ connection-oriented transport protocols. Given that, are these techniques required at the lower layers?

In the early days of digital transmission, error rates on links were relatively high. Since many links were required to complete a path across a network, connection-oriented protocols at Layer 2 were a necessity. Without them, the Layer 3 and above protocols would constantly be detecting errors and asking for retransmissions. The network would be clogged with the acknowledgment messages and the retransmissions, which would also contain errors.

Today, modern digital transmission systems are virtually free of errors and connection-oriented protocols are rarely used at Layer 2. Some systems still retain the connection-oriented paradigm at Layer 3 (e.g., the ITU-T X.25 protocol), but many rely solely on the Layer 4 transport protocol to detect the few errors that occur end-to-end at this layer. Following a discussion of the ISO layers, we present examples of today's most popular protocol suites.

# ISO Reference Model for OSI

Figure 7.5 illustrates the ISO Reference Model for OSI, the objective of which is to solve the problem of heterogeneous DTE and data network communications. However, the OSI model is not a product blueprint. Two companies can therefore build computers consistent with the model, but unable to exchange information. The model is only a framework—meant to be implemented with standards developed for each layer.

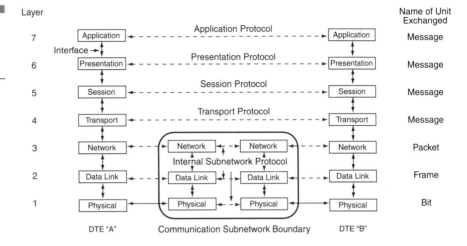

**Figure 7.5**
ISO reference model for open systems interconnection.

Standards must define services provided by each layer, as well as the protocols between layers. Standards do not dictate how the functions and services are implemented in either hardware or software, so these may differ from product to product.

The International Telecommunication Union (ITU) is a UN treaty organization that considers all technical, operational, and tariff matters for telecommunications worldwide. Its telecommunication standardization committee (ITU-T, formerly the International Consultative Committee for Telegraph and Telephone, or CCITT) functions as the international standards body for the industry.

The results of the ITU-T work are published every four years (following a plenary meeting) as "recommendations" in a series of books commonly referred to by the color of their covers (such as "orange book"). ITU-T recommendations are denoted by A.n, where A is a letter representing a series of recommendations (e.g., V for analog networks, X for digital networks), and n is an identifying number. (See Appendix A for more standards-setting information.)

In Figure 7.5 the protocol stacks to the left and right represent two DTEs connected by a communications subnetwork, shown in the middle. The names and numerical assignments for the seven layers are shown on the left. A summary description of the services specified for each layer follows.

# Layer 1: Physical

This layer provides mechanical, electrical, functional, and procedural characteristics to activate, maintain, and deactivate connections for the transmission of unstructured bitstreams over a physical link. The physical link can be connectors and wiring between the DTE and a DCE at a network access point, and fiber optic cable within a network. The ITU X.25 Recommendation defines the interface between *data terminal equipment* (DTE) and *data circuit terminating equipment* (DCE) for terminals operating in the packet mode over public data networks. DCE is a generic term for network-embedded devices that provide attachment points for user devices. Layer 1 involves such parameters as signal levels and bit duration. In the U.S., the RS-232 C standard is commonly used at Layer 1, and bits are the data units exchanged.

# Layer 2: Data Link

The Data-Link Layer provides for reliable transfer of data across the physical link. It provides for mapping data units from the next higher (network) layer to frames of data for transmission. Figure 7.1 presents the format for a typical data link frame. The addresses used at Layer 2 are known as *media access control* (MAC) addresses. The data link provides necessary synchronization, error control, and flow control functions. *Link Access Protocol-B* (LAP-B) is an option for Layer 2 in the ITU-T X series recommendations. It is a subset of the ISO-developed *high-level data link control* (HDLC) protocols. In many modern systems, the error-detection function normally associated with connection-oriented operation is performed at Layer 2, but frames with errors are merely discarded. The connection-oriented protocols at higher layers recognize frames not received and request retransmissions at the higher layer.

# Layer 3: Network

Layer 3 provides higher-level layers with independence from routing and switching associated with establishing a network connection. Functions include addressing, end-point identification, and service selection when different services are available. Examples of Level 3 protocols are the ITU-T X.25 recommendation and the Internet's IP protocol.

While Layers 1 and 2 can be described as local DTE (station) to DCE (network node) protocols, most of the Layer 3 dialogue is between stations and between nodes. For example, stations address packets to nodes for delivery through the network. There is also, however, a station-to-station aspect of Layer 3 protocols. Stations must provide networks with addressing and other information to route data to other stations.

The network devices that process Layer 3 protocols are referred to as routers. Routers perform the following steps on packets:

- Remove Layer 2 headers
- Check incoming packets for corruption
- Examine packet age and discard packets kept in the network too long
- Filter packets, as required, based on information in the packet
- Determine routes to destinations
- Build new Layer 2 headers
- Forward packets on appropriate output links.

Figure 7.6 illustrates how network DCEs can present a common Layer 3 protocol (in this case, X.25) to attached DTEs and still support different link protocols. This figure also shows how DCE Layer 1 local media connections on the network side (such as copper wire) can be interfaced with long-distance media (such as fiber optic cable). These conversions, together with the entire internal subnetwork operation are accomplished transparently to user DTEs, which are presented with an X.25 interface.

## Layer 4: Transport

In conjunction with the underlying Network, Data-Link, and Physical Layers, the Transport Layer provides end-to-end (station-to-station) control of transmitted data and optimizes use of network resources. This layer exists to provide transparent data transfer between Layer 5 session entities. In ISO terminology, an entity is the network processing capability (hardware, software, or both) that implements functions in a particular layer. Thus, entities are identified for each layer, e.g., the Layer 5 session.

Transport Layer services are provided to upper layers in order to establish, maintain, and release transparent data connections over two-way, simultaneous data transmission paths between pairs of transport addresses. The transport protocol capabilities needed depend upon the quality of the underlying layer services.

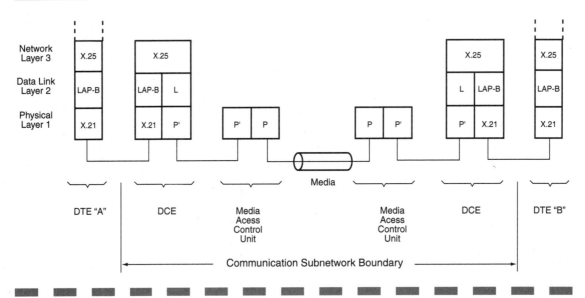

**Figure 7.6**   *Station-to-node and internal subnetwork protocol relationships.*

With reliable, error-free virtual circuit network service, a minimal Transport Layer is required. If the lower layers provide connectionless service, then the transport protocol must implement error detection and recovery, and other functions. ISO has defined five transport protocol classes, each consistent with a different underlying service.

## Layer 5: Session

A *session* is a connection between stations that allows them to communicate. For example, a host processor may need to establish multiple sessions simultaneously with remote terminals to accomplish file transfers with each.

The purpose of the Session Layer is to enable two presentation entities at remote stations to establish and use transport connections by organizing and synchronizing their dialogue and managing the data exchange. Very few protocol implementations today include a separate Session Layer; the functions of the Session Layer are combined in Layer 7, the Application Layer protocol.

## Layer 6: Presentation

The Presentation Layer delivers information to communicating application entities in a way that preserves meaning while resolving syntax differences. Toward this objective, Layer 6 can provide data transformation (e.g., data compression or encryption), formatting, and syntax selection.

*Virtual terminal protocol,* a Layer 6 protocol, hides differences in remote terminals from application entities by making the terminals all appear as generic or virtual terminals. When two remote host processors use virtual-terminal protocols, terminals appear as locally attached to either host. Like the Session Layer, the Presentation Layer is combined in the Application Layer protocol in most modern implementations.

## Layer 7: Application

The Application Layer enables a computer's application process to access the OSI environment. It serves as the passageway between application processes using open systems interconnection to exchange information. All services directly usable by the application process are provided by this layer. Services include identification of intended communications partners, determination of the current availability of the intended partners, establishment of the authority to communicate, agreement on responsibility for error recovery, and agreement on procedures to maintain data integrity.

## TCP/IP Protocols

In Chapter 2, the creation of the packet-switched network concept by the Advanced Research Projects Agency was chronicled. At the heart of the rapid growth of ARPA's original network, culminating in today's ubiquitous Internet, was a set of simple, yet powerful protocols that have stood the test of time to become the most popular solutions for internetworking ever devised. Although these protocols predated OSI reference model protocols discussed above, they implement equivalent procedures but in four layers as opposed to seven.

Figure 7.7 compares OSI and ARPA protocol models. The heart of ARPA's model is the Internet Layer, which manages message flow between host computers and intermediate packet switches. The protocol developed for this layer is known as the *Internet Protocol,* the "IP" in TCP/IP. IP is designed with the assumption that underlying communication subnetworks provide perfect communication channels (i.e., it is a connectionless protocol). ARPA's mission to support military networks dictated a reliable connection for the *Host-to-Host* Layer; ARPA responded by specifying *Transmission Control Protocol* (TCP) for applications requiring reliable end-to-end service.

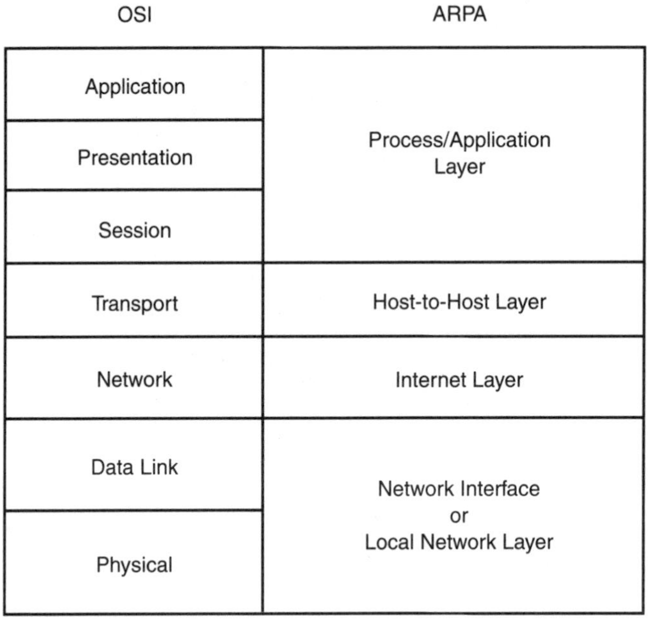

**Figure 7.7**
OSI and ARPA
protocol models.

| OSI | ARPA |
|---|---|
| Application | Process/Application Layer |
| Presentation | Process/Application Layer |
| Session | |
| Transport | Host-to-Host Layer |
| Network | Internet Layer |
| Data Link | Network Interface or Local Network Layer |
| Physical | Network Interface or Local Network Layer |

With TCP and IP providing underpinnings, a series of Process/Application protocols were developed to perform the real work for users. These include *Telecommunications Network* (TELNET) protocol to allow remote host access and terminal emulation, *File Transfer Protocol* (FTP) to transfer files between two host systems, *Simple Mail Transfer Protocol* (SMTP) to send electronic mail (e-mail) messages from one host to another, and the *Simple Network Management Protocol* (SNMP) to enable central management of network resources. The key to Internet growth is the widespread adoption of simple, efficient protocols that can be used across many computer platforms. This feat was and is accomplished not by any controlling government authority, but by a largely volunteer community operating under a self-governing structure designed to promote maximum user community participation and unbiased consideration of ideas.

At the center of this structure is the Internet Society and one of its components, the 13-member Internet Architecture Board (IAB). One of its task forces, the Internet Engineering Task Force (IETF), coordinates the technical aspects of the Internet and its protocols. The IETF produces numerous protocol standards, known as *Request for Comments* (RFC) documents. To become an Internet standard, a proposal undergoes several levels of testing and revision and is finally adopted by the IETF through a democratic voting process. Only after significant implementation and operational experience can a Draft Standard be elevated to an Internet Standard. Figure 7.8 depicts relationships among TCP/IP protocols and names the RFC reference for each standard.

**Figure 7.8**
Internet protocol summary.

| OSI | Protocol Implementation | | | | | | ARPA |
|---|---|---|---|---|---|---|---|
| Application | File Transfer | Electronic Mail | Terminal Emulation | File Transfer | Client/ Server | Network Management | Process/ Application Layer |
| Presentation | File Transfer Protocol (FTP) | Simple Mail Transfer Protocol (SMTP) | TELNET Protocol | Trivial File Transfer Protocol (TFTP) | Network File System Protocol (NFS) RFCs 1014, 1057, 1094 | Simple Network Management Protocol (SNMP) | |
| Session | RFC 959 | RFC 821 | RFC 821 | RFC 821 | | RFC 821 | |
| Transport | Transmission Control Protocol (TCP) RFC 793 | | | User Datagram Protocol (UDP) RFC 768 | | | Host-to-Host Layer |
| Network | Address Resolution Protocol (ARP) RFC 826 | | Internet Protocol (IP) RFC 791 | Internet Control Message Protocol (ICMP) RFC 792 | | | Internet Layer |
| Data Link | Network Interface Cards: Ethernet, Token Ring, ARCNET, WAN RFCs 894, 1042, 1201 | | | | | | Network Interface or Local Network Layer |
| Physical | Transmission Media: Twisted Pair, Coax, Fiber Optics, Wireless, etc. | | | | | | |

The following sections describe in detail the information contained in the headers for the IP Layer (as an example of a connectionless protocol) and the TCP Layer (a connection-oriented protocol). These examples provide insight into the workings of protocols in general as well as details of specific protocols themselves.

## Internet Protocol

IP was developed as a connectionless protocol at what we now refer to as Layer 3. As such, it is primarily concerned with delivering packages of bits from sources to destinations over interconnected systems of networks. As in all protocol layers, the "package of bits" includes original user information and any header and trailer bits added by higher-layer protocols. Header bits are control bits added to the beginning of a package for use by receiving protocol processors at the corresponding layer. Trailer bits are added at the end of the package for the same purpose and may or may not be used in any given protocol. IP uses only header bits to perform its functions. Figure 7.9 illustrates the structure of an IP packet.

**Figure 7.9**
Structure of an IP packet.

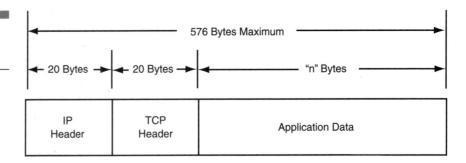

The major functions that must be dealt with by IP headers are addressing and fragmentation. *Addressing* is obviously needed to route packets to destinations, but what is *fragmentation* and why is it necessary? The local and long distance networks that IP packets must traverse may have different Layer 2 frame sizes. The complete IP packet, including the Layer 4 package plus the IP header, must exactly fit into these frames. If Layer 4 packages are shorter than the required length, they can be padded with null bytes. However, if too long, they must be broken into several pieces (i.e., fragments) that will fit. The IP header uses *fields* to help receivers reassemble fragments into original package formats.

Figure 7.10 identifies the specific fields within an IP header, using a standardized format for displaying protocol fields. Each horizontal group of bits (called a *word*) is 32 bits wide. The order in which bits are actually transmitted is from left to right and top to bottom. Note that the minimum header length is five words, or 20 bytes. The first word contains fields for IP version, header length, type of service, and total packet length. The second word comprises three fields supporting fragmentation and reassembly: a fragment identifier, a set of flags indicating whether a packet is the last fragment, and an offset to indicate where a fragment belongs in the complete message. The next word contains time to live which is decreased each time a packet passes through a router. When the TTL value reaches 0, the packet is destroyed. This prevents misaddressed packets from being routed forever. A protocol field identifies the higher-level protocol in use (e.g., TCP).

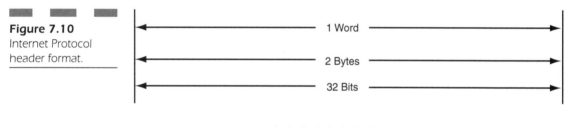

**Figure 7.10**
Internet Protocol
header format.

| Ver | IHL | Type of Service | Total Length | | |
|-----|-----|-----------------|--------------|--|--|
| Identifier | | | Flags | Fragment Offset | |
| Time to Live | | Protocol | Header Checksum | | |
| Source Address | | | | | |
| Destination Address | | | | | |
| Options + Padding | | | | | |

The fourth and fifth words contain 32-bit source and destination addresses, respectively. The destination address is used for routing; the source address can be used for security screening and filtering or other processing at destinations. Addresses are normally represented in dotted decimal notation, in which each byte is assigned a decimal number from 0 to 255 (e.g., 150.200.100.5). Each IP address is divided into network ID and host ID parts. A central authority assigns network IDs and local network administrators assign host IDs. Routers send packets to a network based on its network ID and that network completes the delivery to the host. The number of bits in the address assigned to a network ID depends on the size of that network.

It should be noted that protocol headers add *overhead* to information bytes carried by a network (overhead not present in circuit switched data networks). The packet size of Internet IP packets is 576 bytes, of which at least 20 bytes (3.5 percent) is IP overhead. The addition of Layer 4 and Layer 2 overhead bytes typically raises the packet overhead penalty to 8 percent.

## Transmission Control Protocol

As stated previously, IP does not guarantee reliable packet delivery. This function falls to the Layer 4 Transmission Control Protocol (TCP). In fact, TCP handles six functions: basic data transfer, reliability, flow control, multiplexing, connections, and precedence/security. Fields in TCP headers are shown in Figure 7.11. Headers include a sequence number used to ensure that data packets arrive in sequence, one requirement of reliable transmission. An acknowledgment number field verifies data receipt. TCP is a *Positive Acknowledgment with Retransmission* (PAR) protocol. When data are received correctly with expected sequence numbers, acknowledgment numbers are sent back to senders. If transmitting stations do not receive proper acknowledgements within specified times, they retransmit. No negative acknowledgments are sent.

Flow control is implemented using a header window field. Along with acknowledgment numbers, TCP segment receivers send *window size* data back to transmitters. *Window size* is the number of bytes of data receivers can accept and store in their buffers before sending acknowledgments. Small window size necessitates large numbers of acknowledgment transmissions, consuming bandwidth that could otherwise be allocated to user data. Large window sizes necessitate correspondingly large host buffers, a hardware penalty. Window size is determined during TCP connection setup procedures and can be changed by hosts during sessions as conditions change. Referred to as "sliding window operation" this capability enables flow rate control among hosts.

Two other fields in the TCP header specify source and destination port numbers. These port numbers correspond to specific end user processes (i.e., applications implemented by upper-layer protocols). The combination of a port number and an Internet address is called a *socket*. Since a given host can be a multi-function system (i.e., support several applications simultaneously), more than one socket can be active in a host. A TCP connection is the association of a pair of sockets in two machines. TCP provides true multiplexing of data connections through this mechanism.

**Figure 7.11**
Transmission Control
Protocol header
format.

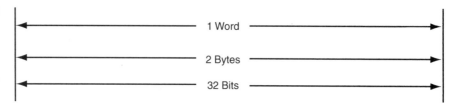

| 1 Word |
| 2 Bytes |
| 32 Bits |

```
                  1 1 1 1 1 1 1 1 1 1 2 2 2 2 2 2 2 2 2 2 3 3
0 1 2 3 4 5 6 7 8 9 0 1 2 3 4 5 6 7 8 9 0 1 2 3 4 5 6 7 8 9 0 1
```

| Source Port | Destination Port |
|---|---|
| Sequence Number | |
| Acknowledgment Number | |

| Offset | Reserved | Flags | Window |
|---|---|---|---|

| Checksum | Urgent Pointer |
|---|---|
| Options + Padding | |

## Other Protocols

Besides the application protocols mentioned above, other TCP/IP protocols exist at Layers 3 and 4. The *User Datagram Protocol* (UDP) provides
connectionless transport services for applications not requiring TCP reliability. Its shorter header and lack of connection setup overhead make it
more efficient when the amount of data to be transmitted is small.
Layer 3 protocols exist to perform address translations necessary to deliver packets to specific hardware addresses (e.g., the Address Resolution
Protocol and the Dynamic Host Configuration Protocol) and to communicate "network health" status among hosts (the Internet Control
Message Protocol). A detailed discussion of all TCP/IP mechanisms is
beyond this book's scope. Readers needing more details can consult the
library of RFCs available on the Internet at **http://www.faqs.org/rfcs**.

# New Directions

TCP/IP's greatest strength lies in its planner's (the IETF) flexible, distributed, and highly dynamic methods for implementing changes and adapting to market forces. The planners, vendors and manufacturers and Internet users have resisted numerous attempts to replace TCP/IP with legislated "standard" protocol suites (most notably OSI protocols). The current driving force for TCP/IP evolution is the growth of the commercial Internet with its virtually unlimited potential and its users' near insatiable appetite for innovative applications.

Two major factors are causing stress for the Internet today: its sheer size and the emergence of real-time streaming applications. First, the number of connected computers on the Internet is rapidly exhausting the currently defined IP address space. The four-byte address limitation and the way addresses are assigned to networks cannot support the current growth rate. For several years there has been pressure on the IETF to expand the address format. Second, streaming applications such as video, music, and voice are significantly increasing Internet traffic and the demand for low latency (i.e., low delay) that cannot be guaranteed by the current connectionless IP routing paradigm.

Latency across a network is just one of several performance parameters used to define Quality of Service (QoS) for a network. Different applications may require different QoS specifications. For example, users tolerate high error rates in voice traffic, but not excessive delay or varying delays. In contrast, electronic mail recipients won't tolerate message content errors but are relatively insensitive to reasonable fixed or variable delays. The Internet's current inability to support multiple QoS levels, in accordance with different application requirements, is a major shortcoming.

Solutions to the above problems come in two arenas. First, changes in the use of fields in the IP header to implement the means for supporting multiple QoS levels in the Internet backbone (some suitable for video, music, and voice applications) are being adopted by the IETF. Second, a new version of IP, known for historical reasons as IP Version 6 (IPv6), has been adopted to address long-term resolution of both address and QoS issues.

### DIFFSERV AND MPLS
The IETF is now examining two standards to help solve the IP QoS problem: Differentiated Services (Diffserv) and Multiprotocol Label Switching (MPLS). These techniques address the problem in very differ-

ent ways and have different implications on the Internet's architecture. While each can exist without the other, they can be used together.

Diffserv is a Layer 3 solution and uses IP's type-of-service field to carry information about IP packet service requirements. It relies on traffic conditioners at the *edge* (boundary) of networks to indicate each packet's requirements based on the needs of the application. For example, packets marked with an *expedited forwarding* (EF) indication receive better processing during the forwarding process than normal packets. This may include assignment to special priority queues. Of course, as the standard evolves, Diffserv-capable routers will have to be installed in the Internet infrastructure. One advantage of Diffserv is that router processing decisions are made on a per-packet basis, not on a per-session basis, allowing more flexibility for ISPs to configure routing algorithms.

Diffserv will be the first of these QoS mechanisms to be ratified as a standard. Since it specifies QoS at Layer 3, it will be implemented at the edge of the network in user devices, and be transported over any Layer 2 infrastructure. As an example, Microsoft is going to include Diffserv capabilities in its upcoming release of the Windows 2000 operating system.

MPLS, in contrast to Diffserv, maps Layer 3 traffic to connection-oriented Layer 2 transports. It adds a label containing specific routing information to each IP packet and allows routers to assign explicit paths to various classes of traffic. MPLS requires investing in sophisticated label-switching routers capable of reading new header information and assigning packets to specific paths. As such, it will likely be implemented at the core of carrier networks and may receive QoS packet requirement information from Diffserv fields.

Routing efficiency is obtained in networks by relieving each router in the path of the burden of running its own network-layer routing algorithm. In this alternative the routing path is calculated only once and encapsulated in a label, an extra 32 bits added to the front of current IP headers. Subsequent routers read the label and follow the path instructions. The path calculation done initially may depend on packet QoS requirements. Finally, since MPLS specifies complete paths for streams of packets, it can easily map such streams onto connection-oriented Layer 2 paths.

### IPv6

The traditional IP protocol described above carries version number four (the designation IPv4 is used when it is necessary to distinguish it from the new version six). The IETF began investigating options to address its shortcomings in 1990 and published its recommendations in January 1995. The revision was assigned version number six (an experimental protocol had been assigned version number five, but was never deployed) and is commonly known as IPv6. Of course, changes to IP cannot be done in isolation and at least 60 current TCP/IP standards must be revised to accommodate IPv6. Although many changes were and are being made, the most important for users are the expansion of the addresses and the inclusion of a flow label in headers.

*Flows* are defined as streams of packets associated with a particular application. As discussed above, the identification of flows and flow characteristics is an important part of implementing different QoS levels in the Internet. While IPv6 does not specify how flow labels in headers are to be used, it provides capabilities for source devices and routers to identify and process specific flows. As IPv6 is adopted, the Internet community will use these capabilities to implement what is necessary to support ever-evolving applications.

The major change in IPv6 is implementation of 128-bit addresses to replace current 32-bit addresses. Obviously, the four-part dotted decimal notation used for current IP addresses is no longer applicable. The preferred representation is:

```
x:x:x:x:x:x:x:x
```

where each "x" represents 16 bits. The 16 bits in each address part are represented using four hexadecimal digits (i.e., 0—9, A, B, C, D, E, F representing values from 0 to 15). For example, an IPv6 address could be:

```
FEDC:BA98:4387:3298:EFDA:AB65:4523:853A
```

Leading zeros are not required in representations for any address part. In addition, if long strings of zeros appear in the address (i.e., 0:0:0), a double colon "::" may be used to indicate multiple groups of 16 bits of zeros. The use of the double colon is restricted to one application in an address. Two examples of this address simplification are:

```
1080:0:0:0:8:800:200C:417A -> 1080::8:800:200C:4147A
0:0:0:0:0:0:0:1 -> ::1
```

Different options are still being considered for implementing hierarchies within address spaces, similar to the network-host hierarchy in IPv4. Many of these options include using a 48-bit network interface

ID, unique to each hardware interface card, as the lower 48 bits of the new IP address. In addition, special types of address formats have been defined that deal with address problems encountered while making the transition from IPv4 to IPv6. These formats are used at boundaries between IPv4 and IPv6 networks and use existing IPv4 node 32-bit addresses as the lower 32 bits of an IPv6 address for reference within the IPv6 network.

Note that expanding address bytes adds more overhead in TCP/IP systems. Retiring some fields and using optional extension headers for other functions has held the minimum size of the new IPv6 header to 40 bytes (vs. 20 bytes for the IPv4 header). Overall overhead penalties discussed previously are now 12 percent.

### TRANSITION TO IPV6

Clearly, IPv6 developers did not envision upgrading the Internet to IPv6 all at once. With millions of connected devices and exponential growth, the transition of the Internet to IPv6 represents the most ambitious undertaking of its kind in history. Since the Internet is made up of diverse systems from many manufacturers, it is expected that many systems may not be upgraded for years, if at all. Therefore, strategies have been defined to allow IPv4 and IPv6 networks and devices to coexist.

Two mechanisms have been proposed to accomplish this function: a dual IP layer, and IPv6-over-IPv4 tunneling. The dual-layer approach is the simplest and calls for both protocols to be implemented on new or upgraded devices. Such devices can then communicate to IPv6 devices using IPv6 and IPv4 devices using IPv4. Conversion capabilities make this kind of device applicable to gateway functions between network types.

*Tunneling* is an approach in which entire IPv6 packets are encapsulated inside IPv4 packets (i.e., an IPv4 header is put on top of an entire IPv6 packet, including the IPv6 header). This allows resulting packets to be routed through existing IPv4 networks. At the end of the "tunnel," dual-mode devices remove IPv4 headers and processes IPv6 packets.

For either scenario to operate, networks must provide information about both types of addresses, the configuration and addresses of gateways, and tunnel endpoints. The development of this infrastructure in a system as large as the Internet is an extremely difficult task. Hence, the first applications of IPv6 will probably be within isolated subnetworks with IPv4 used for transport over a wide area. When IPv6 becomes more widely implemented in commercial software, the transition will start to take place. As with all attempts to change TCP/IP and the Internet, the pace of change will be driven by a best-value criterion; there must be compelling IPv6 business advantages before commercial Internet users will bear conversion expenses.

# Network Technologies

A major IP strength and the greatest contributor to its success is its ready adaptability to a wide variety of network technologies. By network technologies, we mean the underlying Physical and Data-Link Layer systems and protocols that carry IP packets from source devices to destination devices. This process begins at network interface devices in user equipment and may involve changes of physical and data-link network technologies at boundaries between user premises equipment and wide area networks and even within wide area networks. The remaining sections in this chapter examine, from a protocol viewpoint, the most prevalent network technologies. Chapters 8, 13, 14, and 15 add physical-facility and system-level implementation as well as service-provision factors in evaluating these technologies.

## Local Area Network Technologies

Local area networks (LANs) are characterized by geography: they serve limited areas, usually confined to a single building, a single campus, or single organizations within a building or a campus. This limitation allows LANs to offer speed, transferring more information in the same amount of time with less delay. The major LAN requirement is the need to connect directly large numbers of user devices (including hundreds or even thousands of personal computers and servers).

It is clearly prohibitively expensive to connect every computer directly to every other computer (a conclusion arrived at in Chapter 1 regarding connecting all telephones directly to one another). However, for this application, circuit switch-based solutions, such as integrated voice/data PBXs or even yesterday's data PBXs are too expensive; and, as noted, circuit-switching technologies are not conducive to bursty computer-to-computer traffic applications.

Instead, LANs use a common transmission medium to which all served devices (DTEs) are attached. The earliest LANs employed coaxial cable media and a simple "Ethernet" protocol to share cable capacity efficiently.

The *Ethernet* protocol is based on principles and conventions people use in meeting situations. Consider people wishing to comment at meetings. The steps they might take are as follows:

- Listen to see if others are speaking
- When a pause occurs, begin talking
- Continue listening to make sure no one else has also begun speaking.

The Ethernet protocol goes by the name of *Carrier Sense Multiple Access with Collision Detection* (CSMA/CD). Carrier sense means a computer listens to the shared network medium to see if anyone is transmitting. If it detects nothing, it transmits its own data. However, due to delays in the medium, another computer may also have begun transmitting at the same time. In this case, both computers encounter interfering transmissions (a network collision). When this occurs, each system halts its transmission. Unlike humans, who handle such situations with lengthy discussions of who should talk first, under CSMA/CD protocols, computers take actions to retransmit while avoiding or minimizing recurring collisions. This is accomplished by delaying each computer's attempt to retransmit a randomly selected interval of time, up to some maximum interval. In the unlikely event that a second collision occurs, the delay before retransmitting is again randomized, but over a larger interval. This retransmission algorithm, although relatively easy to implement, is the most complex part of the Ethernet protocol, a fact that makes Ethernet network interface cards (NICs) the least expensive and most widely used to date.

Like the other protocols we have discussed, Ethernet's unit of transmitted data or frame imposes a specific group of fields and field structures. The original Ethernet frame structure is shown in Figure 7.12. The first 64 bits are a preamble that all attached DTEs use for synchronization. The next two fields contain destination and source addresses. These are 48-bit Media Access Control (MAC) addresses, not to be confused with Layer 3 IP addresses. LAN stations use these addresses to exchange frames. Obviously, no two DTEs can be assigned the same LAN MAC address. Unique MAC addresses are assigned to NICs during manufacture before shipping. To use Layer 2 Ethernet protocols to transport IP frames to LAN attached DTE destinations, IP network-to-LAN routers maintain maps between device MAC addresses and their corresponding IP addresses. Map-based address translation is one example of the type of background processing and interdevice communications needed to maintain networks, make different but connected networks interoperate, and transport data from end-user to end-user.

**Figure 7.12**
Ethernet frame structure.

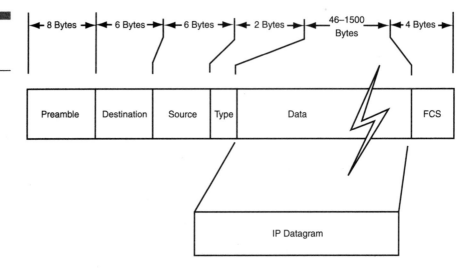

The next two bytes specify frame types, followed by 46 to 1,500 bytes of data (if the user data are less than 46 bytes, extra padding bytes are added to achieve 46 bytes). The last part of the frame is a four-byte Frame Check Sequence (FCS) that ensures frame integrity. When FCSs are not at expected values, receiving devices discard frames and let higher-level protocols request retransmission. Including all header and trailer bytes, the maximum length of an Ethernet frame is 1,544 bytes.

Ethernet was developed in 1973 by Digital Equipment Corp., Intel, and Xerox (known as DIX). In the early 1980s, DIX turned over the standard to the Institute for Electrical and Electronic Engineers (IEEE) whose project 802 was concerned with internetworking between LANs and wide area networks (WANs). The Ethernet standard became the basis for IEEE's 802.3 standard for CSMA/CD LANs. The frame format for 802.3 frames is similar, but not identical, to Ethernet frames. For example, the two type field bytes in Ethernet LANs are used as a frame length field in 802.3 LANs. Care must be taken when installing Layer 2 device drivers to select the proper frame type being used on the LAN. The 802.3 specification also uses eight bytes of the Ethernet data field for two network-control headers, leaving the data field to carry 38 to 1,492 bytes.

Despite simplicity and popularity, Ethernet LANs do not satisfy all LAN requirements. Because traffic is random and delay is caused by collisions, also random, Ethernet LANs cannot operate anywhere near maximum transmission bit rates supported by the media when the number of attached LAN stations is large and the aggregate traffic load high. As a consequence, the IEEE, based on a proposal submitted by IBM, approved a more deterministic access control scheme. The scheme uses a LAN ring-network topology and *token passing* to control when attached DTE devices can transmit. This standard, IEEE 802.5, is commonly known as Token Ring to distinguish it from yet other token bus alternatives. Figure 7.13 shows three workstations and a minicomputer on a Token Ring LAN. DTEs on this type of LAN take turns and transmit one frame only when their turn arrives. To determine when this occurs, DTEs pass "tokens" (a special kind of frame) to each other in sequence corresponding to the physical order in which DTEs are attached. When devices with frames ready to transmit receive tokens, they transmit a frame. Otherwise, DTEs pass tokens on to the next device. Even with large numbers of attached devices, Token Ring LANs operate at nearly 100 percent efficiency, as opposed to 40 percent for Ethernet LANs with large numbers of DTEs and large aggregate amounts of offered traffic.

**Figure 7.13**
Systems on a Token
Ring LAN.

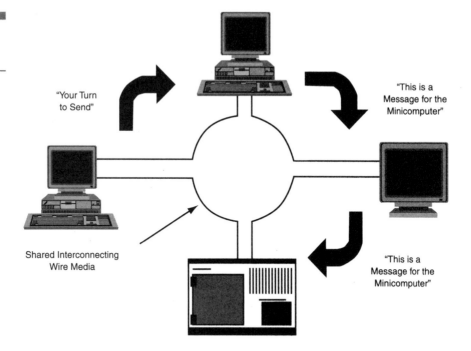

"Your Turn
to Send"

"This is a
Message for the
Minicomputer"

Shared Interconnecting
Wire Media

"This is a
Message for the
Minicomputer"

The performance advantage of Token Ring LANs under heavy traffic comes at a price: the NICs and cabling systems are considerably more expensive than those of Ethernet. In the end, the simpler, cheaper technology predominated and Token Ring systems are now mostly confined to organizations using IBM mainframes and IBM networking technologies.

Another protocol using a ring network topology, is the Fiber Distributed Data Interface (FDDI). FDDI is a high-speed LAN technology normally used as a local area backbone network to interconnect multiple, and usually lower-speed local area network segments. It also suffers from being relatively expensive and with the advent of high-speed, switched Ethernet solutions (see Chapter 13), FDDI popularity is waning.

## Frame Relay Networks

Wide area networks have characteristics diametrically opposite to local area counterparts. While geographical reach is far greater, they often operate at lower transmission speeds and/or cost more for the same speed. This makes it much more important to use wide area transmission resources efficiently. We have seen how the invention of packet switching and multi-layer protocol stacks constitute a revolutionary step forward in making wide area data communications viable and economical. Our explanations also disclose how different layers interact to process messages and other information for reliably delivery to users. With these basic discoveries well understood and developed, network engineers proceeded to enhance the efficient use of WAN resources by "tweaking" designs to achieve even higher performance levels under specified conditions.

The prevalent packet switching standard in corporate data networks in the late 1980s was the ITU-T X.25 standard, a Layer 3 connection-oriented protocol. The major problem with X.25 networks was one of performance. The amount of packet processing required to implement the standard at Layer 3, with then current switching and data processing technology, severely limits X.25 system throughput.

Two corporate network developments pointed the way to new and improved approaches. One development was the emergence of highly reliable transmission, obviating the need for sophisticated error recovery at Layers 2 and 3. Another was recognition that the magnitude of corporate point-to-point traffic justifies dedicated resources. X.25's virtual circuits address point-to-point traffic flow, but complex intermediate processing, even using today's high-speed technology, precludes achieving high throughputs attainable using other protocols and network designs.

X.25 planners responded by designing "express trains" through networks. Bags of mail with simple tags could be transferred from one train to the next at stations with only minimal processing delays. In protocol terms, this entails switching at Layer 2. The "bags" are called frames (analogous to the LAN frames, which perform the same function) and the switching technique is known as frame relay.

Frame relay is defined in ITU-T Recommendation I.122 Framework for additional packet mode bearer services, as a packet mode service. In effect, it combines the statistical multiplexing and port sharing of X.25 packet switching with the high-speed and low-delay characteristics of time division multiplexing and circuit switching. Because it implements no Layer 3 protocols and only the core Layer 2 functions, frame relay makes practical use of interswitch transmission rates of 1 to 2 Mb/s (e.g., DS1 rate services), considerably greater than the 56/64-kbps channels ordinarily used in X.25 networks.

Frame relay headers are simple, consisting of a destination address field and a frame check sequence. In contrast to Layer 3 addresses, which identify ultimate packet destinations, frame relay addresses merely identify the number of the outgoing facility at the next switch. This address, comprising only 10 bits, is known as the Data Link Connection Identifier (DLCI, see Figure 7.14).

A specific path through a frame relay network is known as a Permanent Virtual Circuit (PVC) and is implemented in frame relay switches by logic which recognizes an incoming DLCI value, substitutes the next DLCI value in the path for the next switch, and transmits frames on proper output facilities. The logic implementing PVCs is loaded into switches when PVCs are set up.

An added frame relay protocol feature is the ability to react to network congestion by controlling traffic flow. Three header bits interact to implement this feature: the Discard Eligibility bit (DE), the Forward Explicit Congestion Notification bit (FECN), and the Backward Explicit Congestion Notification bit (BECN). If a frame relay switch needs to discard frames (perhaps due to traffic congestion), it first discards frames with the DE bit set. With this bit, users can set priorities for frames. More commonly, this bit is used by operators to provide different levels of network service. Customers contract for a agreed-upon average throughput over a PVC (called the *Committed Information Rate*), and if they exceed this rate in a specified interval, the network "sets" the DE bit for frames using that PVC. Networks then carry those frames on a "best-effort only" basis. In addition, networks may use FECN and BECN bits to advise end-user systems about congestion, allowing them

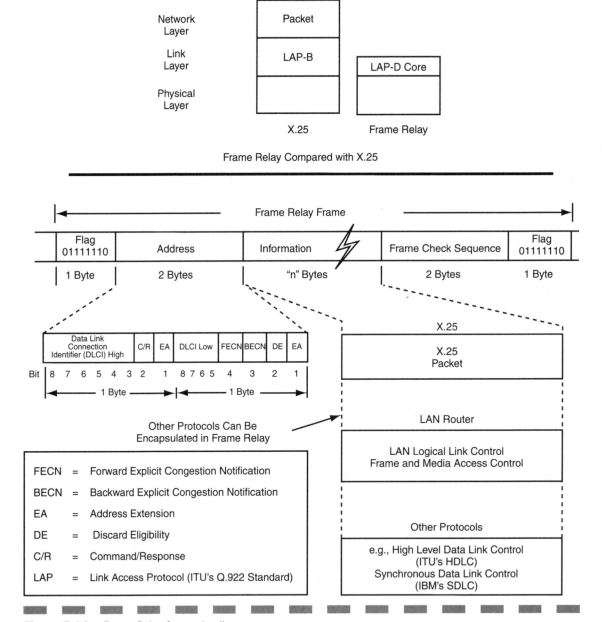

**Figure 7.14** Frame Relay frame details.

to adjust transmission rates accordingly. Chapter 14 discusses public frame relay service attributes, services that have become so popular that they are now corporate data-transport mechanisms of choice.

# Asynchronous Transfer Mode Networks

While frame relay was evolving, driven by the need for even faster data-transport switching technologies, a far more revolutionary concept was placed on drawing boards by the ITU-T and at leading research and development centers. This effort had and still has as its goal, eventual replacement of major portions of worldwide telecommunications infrastructures with facilities supporting unique voice, data, video, and other multimedia traffic requirements within a single network architecture. A set of standards, collectively known as Broadband Integrated Services Digital Network (BISDN), focuses on core transmission, switching, and multiplexing technologies required to achieve this goal. SONET transmission standards, discussed in Chapter 5, coupled with switching and multiplexing aspects of BISDN combine to create modern integrated switching, multiplexing, and transmission equipment paradigms called transfer modes. Two standards were developed: Synchronous Transfer Mode (STM), based on circuit switching principles, and Asynchronous Transfer Mode (ATM), based on packet switching concepts. Over time, ATM has become the preferred alternative in implementing integrated service architectures.

ATM uses fixed-length data units called *cells* to carry all types of traffic. The fixed cell size ensures that time-critical information such as voice or video is not adversely affected by long data frames or packets. The relatively short cell size (53 bytes, including header) enables realization of extremely low latency and very high-speed switching/multiplexing fabrics. As a packet-oriented technology, ATM implements dynamic bandwidth allocation for bursty traffic. Most important, ATM supports different classes of service designed to satisfy differing voice, video, and data traffic requirements. As a connection-oriented technology, it assigns virtual circuits with QoS levels appropriate to each traffic type. These basic notions are expanded upon below.

## ATM TECHNOLOGY

The ATM cell format consists of a five-byte cell header and 48 bytes of user information (the information bytes being called a *payload* in ATM terminology). Like other connection-oriented protocols, ATM supports both permanent and switched virtual circuits (SVCs). SVCs are connections established through signaling functions, much like circuit-switched connections. In addition to virtual circuits, which carry a single stream of cells from end-user to end-user, ATM also defines virtual paths, that is, bundles of virtual circuits routed in the same way. Virtual paths can be established in networks linking ATM switches and can be configured by virtual path cross-connect devices. As such, they provide a multiplexing function within ATM networks. The *Virtual Path Identifier/Virtual Circuit Identifier* fields of ATM calls combine to perform the same function for ATM switching as the DLCI does for frame relay switching. Figure 7.15 depicts the ATM cell format at user interfaces. Note that in this cell format diagram, horizontal rows are only 8 bits wide, not 32 bits wide as in previous packet-format diagrams. By convention, bits are numbered from 8 to 1 to reflect the time order in which they are transmitted.

**Figure 7.15**
ATM cell format at the user-network interface.

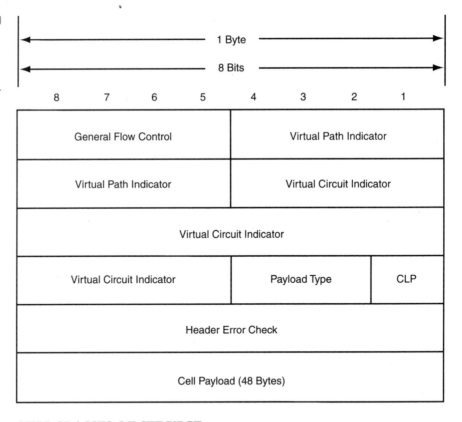

**ATM CLASSES OF SERVICE**

ATM's defining characteristic is its ability to establish and simultaneously transport several classes of user traffic. Standards bodies define five classes of service for ATM, each with different QoS parameters. Figure 7.16 lists these classes and includes a summary description of the QoS offered in each class. In essence, invoking a particular class of service ensures a type of cell behavior and treatment across networks which, in turn satisfies requirements associated with various types of traffic, as exemplified below.

**Figure 7.16**
ATM classes of service.

| Service Class | Quality of Service Description |
|---|---|
| Constant Bit Rate (CBR) | Emulates circuit switching. Cell rate is constant with time. Applications sensitive to delay variation. Used for voice. |
| Variable Bit Rate– Real Time (VBR-RT) | Cell rate varies with time. Statistical multiplexing used to conserve network resources. Low cell delay variation. Used for compressed voice and video. |
| Variable Bit Rate– Non-Real Time (VBR-NRT) | Similar to VBR-RT, but allows cell delay variation and does not synchronize transmitter and receiver. Used for reliable data transfer. |
| Available Bit Rate (ABR) | Only a minimum cell transfer rate guaranteed by the network. Rate-based flow control provided (i.e., user equipment must respond to rate control messages from the network). Used for general data transfer. |
| Unspecified Bit Rate (UBR) | No guarantees from the network (i.e., best effort service). Used for low priority data and control information. |

*Constant Bit Rate* (CBR) service is used to emulate dedicated circuits. The cell rate is constant over time and timing synchronization is maintained between transmitting and receiving devices. This class is used for applications sensitive to cell-delay variations, such as voice.

*Variable Bit Rate—Non-Real Time* (VBR-NRT) service allows users to send data at rates that vary with time. Networks can then invoke statistical multiplexing to conserve resources. Each virtual circuit using this service is characterized by a *peak cell rate* (PCR) and a *sustained cell rate* (SCR). The SCR is guaranteed and controlled by the same type of mechanism used to control CIR service in frame relay networks. Since applications using VBR-NRT are insensitive to cell-delay variation, synchronism between user devices is not maintained across the network.

*Variable Bit Rate—Real Time* (VBR-RT) service also allows users to send data at a rate that varies with time. This service is similar to VBR-NRT except that synchronism is maintained across networks and the SCR is maintained with less cell-delay variation than in VBR-NRT service. This service is often used with compressed video and compressed voice applications.

*Available Bit Rate* (ABR) service provides only a minimum guaranteed cell rate to the user. The user may transmit above this rate, but must respond to congestion messages from the network and control its rate. This mechanism is known as rate-based flow control and is distinct from flow-based control used for other services. If user equipment responds to network rate control messages, cell loss rates will be low.

*Unspecified Bit Rate* (UBR) service is a best-effort service, with no guarantees of either throughput or cell loss. Its low complexity (and, hence, low cost from ATM service provider's viewpoint) make it attractive for low-priority management traffic. Today, it is also being used for TCP/IP traffic over ATM, but its data-loss characteristics can cause problems at high congestion levels.

## ATM ADAPTATION LAYERS

We have seen that different types of traffic require different network treatment to achieve appropriate blends of QoS parameters. This is achieved by adapting application bit streams to ATM cells in different manners, based on ATM service class requirements of the application. This function is performed by the ATM *Adaptation Layer* (AAL). There are five AALs currently defined, each using bits allocated to the 48-byte ATM payload for additional, AAL-specific header information. The header contents are customized for the QoS requirements of the underlying service class.

*AAL1* is intended to map voice traffic onto CBR virtual circuits. The header contains clock-recovery bits and sequence numbers, but no error-control bits. Only one byte is required for AAL1 headers.

*AAL2* maps video traffic onto VBR-RT virtual circuits. It also contains sequence numbers, but not clock recovery since bit rates required by compressed video are variable. Video screens are sensitive to pixel errors, so AAL2 cells use an error-checking code for the whole cell. Also, since video screens are composed of many cells, cells are marked as the first, intermediate, or last cell in a screen. AAL2 headers are three bytes in length.

*AAL3/4* is used for mapping data traffic onto VBR-NRT virtual circuits. AAL3 is used for connection-oriented service and AAL4 for connectionless service. Like video, these services require error checking, sequencing, and identification of cells as parts of messages. The difference between this AAL and AAL2 is that synchronism is not maintained across networks (i.e., cell delay is not controlled on an end-to-end basis).

*AAL5* is also intended for use with connection-oriented data, but does not insert header information into every cell. Instead, information such as message length and an error-checking code for the whole message (not each cell) is appended to messages before fragmentation into cells. Most cells are then transmitted with a full 48 bytes of user data. AAL5 expects the underlying transmission to be very reliable. Its major advantage is simplicity and low implementation cost. In North America, AAL5 has supplanted AAL3/4 as the adaptation layer of choice for data transmission.

Adaptation layers are implemented at the edge of ATM networks by equipment that forms the basis of ATM's "multimedia magic." Once formatted into cells, various media-related data streams can be rapidly switched and/or multiplexed through networks, to be reconstructed by AAL equipment at the distant end.

# Summary

This chapter presents readers with basic conventional and fast-packet switching concepts and some salient implementation details. Chapters 8, 13, and 14 follow up by describing premises and metropolitan and wide area network packet-switching services (the mechanism by which most users reap the benefits of these advanced technologies), and additional hardware, software and protocol characteristics associated with today's most popular products and designs.

# CHAPTER 8

# Advanced Network Concepts

# Building Blocks for Integrated Digital Networks

In Chapters 1—7, we have covered a great deal of the telecommunications' landscape. Chapters 1 and 2 presented terminology, basic definitions and background material. Chapters 3 through 7 furnished descriptions of underlying concepts, techniques, and devices first, followed by explanations of what might best be described as "pillars of the telecommunications structure," namely its transmission, multiplexing, circuit switching, packet switching, frame relay, cell switching, and related management and signaling and supervision control mechanisms.

For each individual technology, these discussions systematically defined and explained salient technical and operational characteristics, important business applications, relevant standards, and major vendors and products. But it is pedagogically difficult, if not impossible in initial explanations, to impart (1) how such a large array of technologies interrelate or (2) how the technologies can be intelligently integrated to form responsive, reliable, and cost-effective composite telecommunications systems.

However, armed with an understanding of the telecommunications industry's unique vocabulary and knowledge of the operation of its major constituent elements, Figure 8.1's graphical representation now affords a means of gaining broader insight in these two important areas. First, with the amplifying discussion below, the figure not only depicts how and why individual technologies constitute alternative and complementary components in today's telecommunications systems, it also promotes a better appreciation of factors that drive the evolution of individual technology developments.

Chapter 1 defined telecommunications as any process that enables one or more users (persons or machines) to pass to one or more other users information of any nature delivered in any usable form—by wire, radio, visual, or other electrical, electromagnetic, optical, acoustic, or mechanical means. From earlier chapters, we know that to deliver these services, practical telecommunications systems require a number of essential subsystems. Shown in Figure 8.1 these elements are: customer premises equipment (information sources and sinks); transmission facilities; switching and multiplexing to make efficient use of transmission facilities; and network management and control mechanisms (including signaling and supervision) to respond to users' dialing instructions and enable efficient operation of all other resources.

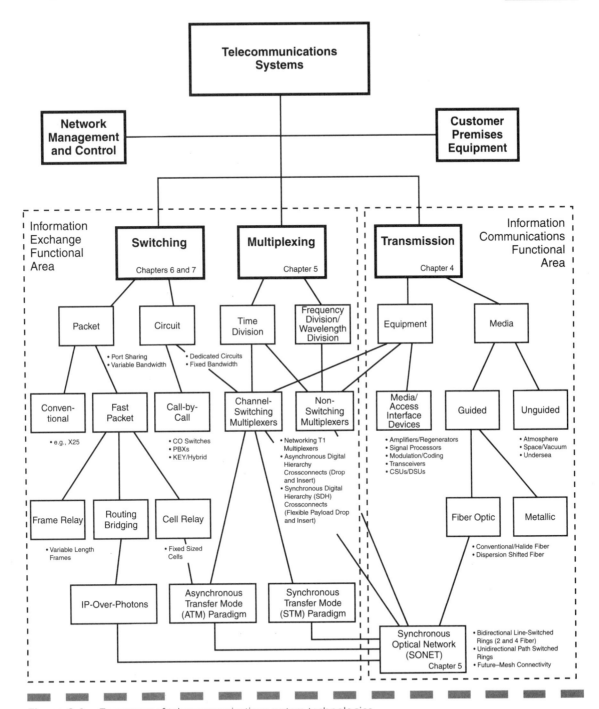

**Figure 8.1** Taxonomy of telecommunications system technologies.

Figure 8.1 focuses on two telecommunications technology functional areas. Dubbed information exchange (IX) and information communications (IC), elements of these constituent functional areas are enclosed in dashed boxes. The use of functional area terminology grew out of the need to specify technical and other performance requirements, independent of the particular technology used to satisfy those requirements.

Take circuit switching as an example. In the past, circuit switching functions have been satisfied with manual plug boards, manual mechanical switch matrices, electro-mechanical matrices, solid-state circuit-based analog switches, and finally modern stored-program control—digital matrix switches. As it turns out, from a performance standpoint, all important circuit switching characteristics can be specified independent of the technology or design approach pursued by any manufacturer.

For instance, what is important with circuit switching is capacity (the number of lines, trunks, or combinations that can be switched); blocking probability; signal bandwidth handling capability; reliability; maintainability; cost, and the like. Chapter 6 noted that most contemporary circuit switch designs implement about 80 percent of their functionality in software. Traditional manufacturers tend to view circuit switches as hardware devices with embedded information processors. A newer breed of competitors now views a circuit switch as a general purpose processor with special interface peripheral devices.

Using functional area (FA) specifications, circuit switch users/buyers can rise above the fray, state performance requirements, get the most innovative ideas and the best technology proposals from all bidders, and be confident that procured items meet business application needs. As it turns out, optimal use of FAs is achieved when classes or types of requirements assigned to each FA are as distinct, non-overlapping and as simply stated as possible. To this end the *information communications* (IC) functional area is defined as capabilities to move or transfer information from one location to another, and the *information exchange* (IX) functional area is defined as capabilities to switch, direct, route, multiplex, or inverse multiplex information. Overall, *functional areas* are subsets of information system capabilities that accomplish or support specified categories or subsets of information operations.

# Switching and Multiplexing Comparisons and Trends

Heading the IX-labeled dashed box in Figure 8.1 are "switching" and "multiplexing" with a taxonomy of technology alternatives for each beneath. Chapter 6 defined switching as equipment arranged to establish connections between lines, between lines and trunks, or between trunks; and, more generally, as capabilities to route transmitted signals in circuits between specific points in a network. Both the complementary nature of switching and multiplexing technologies and the rationale for their evolution to today's capabilities and future trends are revealed in the remaining parts of Figure 8.1's taxonomy and the following discussion.

In *circuit switching*—one of two switching subcategories—a user is assigned a dedicated circuit with a fixed bandwidth for some period of time. With *packet switching*, users share network ports, compete for shared transmission resources, and are provided variable bandwidth-on-demand capacity (i.e., they only use resources when they have information to transmit). Recalling earlier defined circuit, channel, and packet switching concepts gives meaning to these somewhat abstract terms, underscoring the definition's application to both voice circuit switching and data packet switching.

As noted previously, packet switching efficiently serves users with "bursty" traffic—long periods of inactivity punctuated by short transmissions. On the other hand, circuit switching is ideal for users with constant levels of traffic over extended periods, especially when the traffic demands minimal delay and nearly constant bandwidth occupancy, as with isochronous voice and video signals.

Circuit-switching subcategories include call-by-call and channel switching. In call-by-call circuit switching, a resource is selected and allocated upon each request for service. Channel switching provides the ability to modify multiplexers and network configurations remotely in response to outages or time-of-day traffic variations, or to accommodate growth and organizational changes. In yesterday's networks, channel switching was accomplished manually via patching or physical setup (known as provisioning or "technical control"). Today, programmable T1 multiplexers and digital cross-connect systems permit soft provisioning, and support the trend away from point-to-point private networks toward flexible virtual network topologies (see Chapter 10).

Packet switching subcategories include conventional and fast-packet techniques. In conventional packet switching, communications networks provide user interfaces which, as described in Chapter 7, furnish services associated with the bottom three ISO protocol layers. These networks have afforded years of reliable data communications, being designed to sustain error-free service using older, poorer-quality, lower-bandwidth analog circuits. Networks designed for X.25 typify this category, where attached DTE devices lack processing capabilities (dumb terminals) to support error control and other higher-level protocol services, so the networks themselves provide those functions.

*Fast packet* is a term referring to a number of broadband switching and networking paradigms. Implicit in the fast-packet technology is the assumption of an operating environment that includes reliable, digital, broadband, nearly error-free transmission, and intelligent end-user equipment. Because intelligent end-user systems have sufficient processing power to operate sophisticated protocol suites, it is now possible to perform error detection, correction, and requests for retransmission on an end-to-end basis. In combination with virtually error-free fiber optic transmission systems (bit-error rates of better than one error in a billion [$10^9$] bits without error control), network designs can be streamlined to emphasize speed of service and minimum message delay.

Again, as noted in Chapter 7, frame relay and cell relay are two fast packet technologies. Frame relay is defined in ITU-T Recommendation I.122 "Framework for Additional Packet Mode Bearer Services," as a packet mode service. In effect, it combines the *statistical multiplexing* and *port sharing* of X.25 packet switching with the high-speed and low-delay characteristics of time division multiplexing and circuit switching. Unlike X.25, however, frame relay implements no Layer 3 protocols and only the so-called "core" Layer 2 functions. With these simplifications, frame relay makes practical use of interswitch transmission rates of 1 to 2 Mb/s (e.g., DS1 rate services), considerably greater than the 56/64-kb/s channels ordinarily used in X.25 networks.

The basic units of information transferred are variable-length frames, using only two bytes for header information. Delay for frame relay is lower than for X.25, but it is variable and larger than that experienced in circuit-switched networks. This means that in the absence of special protocols, frame relay is generally not suitable for voice and video applications where appreciable and variable delays are unacceptable.

*Cell relay* is the process of transferring data in the form of fixed-length packets (cells). Cell relay is used in high-bandwidth, low-delay, packet-like switching and multiplexing techniques. It intrinsically addresses fixed and variable delay restrictions levied by isochronous traffic. The objective is to develop a single multiplexing/switching mechanism for dividing up usable capacity (bandwidth) in a manner that supports its allocation to both voice and video traffic and packet data communications services. Standards groups have debated the optimum cell size. Small cells favor low delay for applications needing it but involve a higher header-to-user information overhead penalty than would be needed for most data applications. The current specification is for a 53-byte cell which includes a 5-byte header and a 48-byte payload.

In addition to the difference in units of information transferred, frame and cell relay differ in terms of how and where the techniques are used in a network. Frame relay, like X.25, is a network interface specification. Cell relay, used at network cores, implements common switching/multiplexing for all types of information (including frame relay). Figure 8.2 summarizes these and other important technical characteristics that distinguish the switching technologies described above. The structure of the figure is such that entries should be read from left to right, and the spatial location of the entries roughly corresponds to the switching technologies with the same left-to-right justification.

For example, in the error-control row, conventional packet-switching networks (like X.25) provide comprehensive error detection and correction service; frame-relay networks provide error detection but no error correction; and networks using cell relay typically provide no error control, a characteristic shared by networks offering circuit-switching service. In terms of connection options, any of the packet-switching techniques can theoretically support *connectionless (datagram)* or *permanent* or *switched virtual circuit* connections. Circuit switching provides only fixed, physical circuit connections. Some frame-relay offerings are limited to permanent virtual circuit service.

For routing control, conventional packet switching includes explicit source and destination address information in each packet header. This permits networks to route each packet independently from source to destination. With frame relay, headers include only a *data link connection identifier* (DLCI), which permits networks to assign a fixed route for all frames between network end-points by simple "table look-up" procedures. Cell relay uses *virtual circuit identifiers* (VCIs) and *virtual path indicators* (VPIs) in much the same way that frame relay uses DLCIs. In circuit switching, networks select a fixed connection path prior to information exchange in response to user-provided called-party address (telephone number) information. Chapter 7 provides more details.

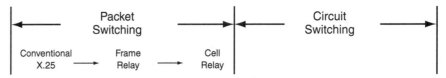

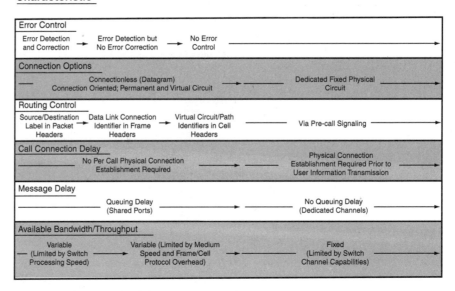

**Figure 8.2**
Switching characteristics and trends.

Because all packet-switching techniques employ port sharing, delays associated with establishing connections in circuit-switched networks (dialing and connection times) do not exist. On the other hand, message delays might occur in port-shared packet networks when many sources seek to use common switch and transmission resources simultaneously. In these cases, queues may develop subjecting individual messages to queuing delays, a phenomenon described in Chapters 7 and 11.

In circuit-switched networks, once a connection is established, the transmitting terminal has dedicated use of resources and hence experiences no queuing delay. Of course, circuit switching's no queuing delay comes at a cost. A grocery store that opened a separate checkout line for every customer that entered the store would have no queues, and the customers would love it, but such lavish service might not be affordable. Although admittedly somewhat extreme, this example illustrates why circuit switching and synchronous time division multiplexing exhibit analogous inefficiencies for certain types of traffic.

In terms of supplying bandwidth or throughput, packet switching provides users with variable bandwidth capabilities, ideal for bursty traffic. In contrast, circuit switching provides users with fixed bandwidth limited by the physical characteristics of connecting channels. Whereas useful bandwidth in conventional packet-switching networks is limited by switch-processing speed, in frame and cell relay, useful payload bandwidth is limited by transmission-medium speed and protocol overhead.

While concise, the above discussion reveals the major architectural factors affecting tomorrow's integrated digital network designs. As a result of the influence of high-quality fiber optic transmission and high-speed microprocessor technologies, it can be concluded that many of the circumstances that justified separate voice and data networks in the past are no longer valid. The rate of transition appears to be limited only by market economics and the deliberations of standards-setting bodies. Succeeding chapters discuss existing and emerging premises, metropolitan and wide area network services that incorporate the technologies described above. Accordingly, those chapters present additional telecommunications switching insights, viewed from the perspective of network carrier and equipment service applications.

Returning to Figure 8.1, the right-hand IC-labeled dashed box identifies pertinent transmission subsystem technology categories. A *transmission system* consists of a medium, any material substance or "free space" (i.e., a vacuum), that can be used for the propagation of suitable signals from one point to another, and termination equipment needed to generate and receive signals compatible with the medium selected. Termination equipment occurs in a wide variety of appearances to provide amplification in the case of analog signals; regeneration for digital signals; a myriad of signal processing, filtering, modulation, and coding functions; transceivers for *radio frequency* (RF) signals; and *channel and data service unit* CSU/DSU functions.

Consistent with trends toward miniaturization and functional integration, CSUs/DSUs may be embedded in multiplexer equipment, which, when appropriate, may in turn be embedded in digital switching equipment. As emphasized in Chapter 4, efficient and reliable transmission-medium use, is normally highly dependent upon sophisticated modulation and coding technologies, which likewise may be embedded in adapter cards or equipment made for other than transmission purposes.

The figure reminds us that modern telecommunications are often multimedia, meaning here that they employ more than a single guided or unguided transmission medium. Guided media, including paired metallic wire cable, coaxial cable, and fiber optic cable, constrain electromagnetic or acoustical waves within boundaries established by their physical construction. Unguided media are those in which boundary effects between free space and material substances are absent. The free space medium may include a gas or vapor. Unguided media, including the atmosphere and outer space, support terrestrial and satellite radio and optical transmission.

From previous discussions and in the remainder of this book, it is becoming increasingly clear that, for applications not requiring mobility (i.e., where cable-tethered equipment is acceptable), fiber optic cable is rapidly becoming the medium of choice. No major carrier is without aggressive fiber-development and deployment programs and all either sponsor intense R/D activities to advance the state of the art, or are in pursuit of promising start-up companies (e.g., the Ciena and Corvus Corporations) with existing advanced technologies.

The pioneering Broadband Integrated Services Digital Network (BISDN) and SONET standards work in the early 1980s has been and remains key to vigorous industry-wide developments and geometric growth in related equipment and tariff revenues. As such, SONET-conforming transmission networks are logical vehicles for broadband ATM and *synchronous digital hierarchy* (SDH) compatible signals.

When we understand underlying developments related to each technology, Figure 8.1 delivers a clear message about what led up to today's progress towards integrated digital networks (IDNs) and the likely shape of things to come. Ultimately, IDNs will exhibit three principal characteristics. First, tomorrow's IDNs will offer end-to-end digital connections. This will end the need for old analog voice-network-compatible modems and eventually extend broadband service to user premises.

Second, IDNs will accommodate and support stringent quality-of-service requirements associated with all information media (voice, bursty data, streaming data, video, imagery, etc.). Last and perhaps most important IDNs will capitalize on the ability to integrate switching, multiplexing, modulation, transmitting, and receiving functions, previously implemented in separate equipment, wherever reliability and cost-effectiveness benefits can accrue. The next section documents some of the first steps taken to achieve IDN goals.

# Integrated Services Digital Network

The concept of an ISDN was first formulated in the councils of the International Telecommunications Union—Telecommunications Standardization Sector (ITU-T), then known as the International Consultative Committee for Telephone and Telegraph (CCITT), before the 1976 plenary session. While the concept incorporated significant contributions from the U.S., it can fairly be said to have reflected a European view of telecommunications. In the early 1970s, European telecommunications administrations, and to some extent Japan, viewed evolving digitization of their telecommunication infrastructure—initiated by U.S.-based activities—as needing a standards framework to ensure future interoperability.

In those days of analog technology, switching and transmission systems were viewed as independent network segments. Digitization of the switching and transmission plants allowed the integration of many functions, particularly multiplexing, leading to the IDN concept. But this technological improvement could not reach customers until systematic provisions for access were established. Thus, the concept of user access to an existing IDN underlies the ISDN.

ISDN consists of a set of standards being developed by the ITU and various U.S. standards-setting organizations. The ITU formal recommendations, adopted in October, 1984, first defined ISDN as:

> "...a network, in general evolving from a telephony integrated digital network, that provides end-to-end digital connectivity to support a wide range of services, including voice and non-voice, to which users will have access by a limited set of standard multipurpose user-network interfaces."

Note that the ITU concept of ISDN was never meant to be a product or service specification, but rather a guideline for development of such offerings in a manner that would promote international connectivity and interoperability. ISDN characteristics include:

- **End-to-end digital connectivity**—All signals are transmitted in digital form from terminal to terminal.

- **Common channel signaling**—ISDN uses out-of-band signaling as described in Chapter 6. In this approach, messages containing addresses, network information, and protocol elements in standard formats, are transmitted over the network independently from user traffic. As CPE and network equipment adopt the same set of standards, function and feature transparency across both private and public networks will no longer be dependent on the use of proprietary products from a single vendor. (Note: The concept of customer-premises equipment has U.S. but no ITU relevance.)

- **Multipurpose user-network interfaces (UNI)**—ISDN permits the user to connect to voice, data, video, and other services by a single access mechanism, in contrast with the separate arrangements now required.

The U.S. version of ISDN as originally propounded by Bellcore was intended to support a wide variety of new products and services in evolutionary fashion, delivering them via two standardized user-network interfaces. It was also meant to establish a platform for future telecommunications utilities. ISDN was expected to reduce service costs through more efficient use of network resources, improve network management on the part of service providers as well as customers, and create a better-coordinated network environment.

# History

- 1968—The ITU Study Group D is formed to evaluate digital transmission systems.
- 1972—ITU Study Group D issues discussion of ITU functions and publications and issues recommendations and a proposal to study a worldwide Integrated Services Digital Network.
- 1976—ITU Study Group XVIII establishes questions for an ISDN ("Green Book." See Appendix A for "color book" definitions).
- 1980—ITU Study Group XVIII establishes a formal plan to define an ISDN conceptually at the next Plenary (1984).
- 1984—ITU issues the I series recommendations, defining the framework for ISDN ("Red Book").
- 1988—ITU issues the "Blue Book," further defining ISDN and outlining additional broadband ISDN studies.
- 1990—In the U.S., numerous field trials are under way, and working ISDN "islands" are in place; formation by the Corporation for Open Systems (COS) of the ISDN Executive Council; NIST (National Institute of Science and Technology) establishes the National ISDN Users Forum to broaden the utilization of ISDN and accelerate the development of applications.

■ 1991—Publication of Bellcore-developed report covering procedures referred to as National ISDN-1.

## Services

Figure 8.3 illustrates access and transport services provided by ISDN. Access service for smaller systems and equipment, such as voice/data workstations, is provided by the *Basic Rate Interface* (BRI). BRI consists of two 64 kb/s information bearer channels (B channels) and one 16-kb/s packet-switched data channel (D channel) which performs signal-

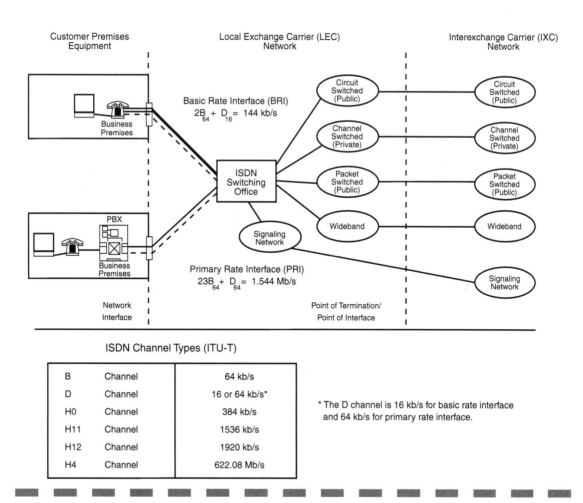

**Figure 8.3** Structures of Integrated Services Digital Networks (ISDNs).

ing for the B channels and furnishes a mechanism for packet switching user data. Note that ISDN defines new access arrangements, but that transport—in the intra-LATA and inter-LATA contexts for the U.S.—is provided by the existing IDN.

Access service for host processors and CPE switching systems (e.g., PBXs) is provided by the Primary Rate Interface (PRI). In the United States, PRI is based on the DS1 transmission rate of 1.544 Mb/s and consists of 23 B channels and one 64-kb/s D channel. In addition, six B channels may be bundled together to form a 384-kb/s H0 channel, or 24 B channels may be bundled to form a 1.536-Mb/s H11 channel. These bundles would be used for applications requiring high data rates, such as compressed video, or host-to-host bulk data transfers.

## Standards

The guiding principle for ISDN standards has been interoperability. While ITU is responsible for ISDN, ITU has adopted the International Standards Organization's (ISO) Open System Interconnection (OSI) reference model and incorporated it into ISDN data service recommendations. Three types of ISDN standards-based offerings are envisioned by ITU:

- Bearer services, described above, consist of network services delivered over the B channels and encompass the first three layers of the OSI model—i.e., the physical, data link, and network layers.
- Teleservices, implemented in the ISDN terminal equipment, may encompass any or all of the upper four layers of the OSI model— i.e., the transport, session, presentation, and application layers.
- Supplementary services are provided by the network for use with bearer services and teleservices.

## Interfaces

ITU has defined V, U, R, S, and T reference points to facilitate access and provide interface specifications, as depicted in Figure 8.4. Underlying the ITU concept of ISDN as access to the IDN, a basic notion is the NT, the *network termination*. As originally conceived, NT is service provider-owned equipment on customer premises which establishes the point of telecommunications service delivery by a carrier (U.S.A.)

or an administration (most of the rest of the world). However, in the U.S., the UNI must be at the U reference point, since the FCC NCTE Order mandates the separation of any and all forms of customer-premise equipment from the service provided by a carrier.

**Figure 8.4**
ISDN interfaces.

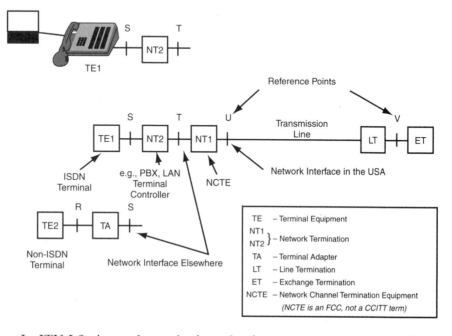

In ITU I Series parlance, basic and primary rate access are standardized at the customer side of the NT1, or as shown in Figure 8.4, the S or T reference points. The inscribed boxes, shown between the reference points on the figure, represent functional groupings defined by ITU. For example, NT2 and LT encompass all the functions needed to adapt the classic two-wire loop to a customer premises to support basic and primary rate signals. NCTE, a U.S.-only concept, implements NT1 functions. While the boxes in Figure 8.4 are not intended to define any particular equipment envelope, in the U.S. various manufacturers may incorporate any or all of the NT1, NT2, TE1, and TA functions in any particular equipment.

NT2 functions may be incorporated in single standalone terminals (telephones, data terminals, or other ISDN user devices) or digital PBX, LAN, or multiplexer equipment. ISDN-compatible terminals are designated Terminal Equipment 1 (TE1), conforming to the S and T reference points. Non-ISDN terminals are designated TE2, requiring a terminal adapter (TA) for physical connection to the network at the R reference point. The ANSI is developing ISDN standards for the U.S. ANSI T1.601-1992 is the BRI standard in the U.S. Since 1988, several manufacturers, notably Nortel and AT&T, have had BRI and PRI capability.

## Applications

While numerous potential ISDN applications have been identified, those that have actually made their way to market are listed below. Although generally unavailable when first postulated, most are now offered by LECs in other than ISDN services.

- **Incoming call identification**. Using Automatic Number Identification (ANI), the originating telephone number is passed over the D channel, allowing the terminating CPE to display the number, or use it to call up a database profile of the caller.

- **Call-by-call service selection**. Previously, digital facilities had to be preassigned to specific outbound or inbound services, e.g., T1 circuits for WATS and 800/900 services. Use of the ISDN D channel, however, allows dynamic B channel assignment on a real-time basis, potentially producing trunk cost savings of up to one third.

- **Outbound station identification.** ISDN can pass a PBX user's telephone number to the network, where previously only the PBX trunk group number was identified. This feature is valuable for resolution of billing issues.

- **Switched digital bandwidth.** ISDN offers dial-up speeds of up to 128 kb/s terminal-to-terminal using BRI and up to 1536 kb/s using PRI. Transmission rates through voice networks are currently limited to a maximum of 56 kb/s using V.90-type adaptive modems on switched voice-grade analog circuits.

- **Feature portability.** PBX features such as uniform numbering, call forwarding, and message waiting can be used in a wide area network of PBXs that perform as if they were a single system.

## Terminal Equipment

As a consequence of the NCTE Order making the U reference point the network interface (NI) in the U.S., the NT functions must be implemented in CPE. The NT functions performed for physical and electrical connection of an ISDN terminal to the BRI include:

- Full-duplex, 160 kb/s (the HDSL, 2B1Q modulation scheme described in Chapter 4) transmission, with echo cancellation, over two wires.
- Two-wire to four-wire conversion.
- Contention resolution for multiple terminal access (passive bus).
- Environmental protection (lightning and power cross).
- NT1 termination equipment manufactured by Lucent, Nortel, NEC, Ericsson, Siemens, and others. The adoption of ANSI T1.601-1992 enables this equipment to be compatible with all LEC service provision conforming to the standard.

## Carrier Activities

AT&T, MCI, and Sprint have implemented ISDN on a broad scale, both in virtual private networks for nationwide corporate networks, and to a lesser degree in government networks. These implementations consist mainly of PRI services supporting large corporate customers in the major metropolitan areas.

LEC offerings, however, are more limited. While 96 percent of all central offices in the U.S. are digital, stored-program control offices, only 42 percent are ISDN capable. Of all access lines, less than 1 percent are ISDN lines nationwide. It is estimated that online households use ISDN for about 6 percent of their access lines, but this does not constitute a high percentage since that is the target market for ISDN. Later sections discuss why ISDN penetration may not increase beyond present levels and may even begin to wane.

## Impediments to ISDN Growth

For years, widespread rollout of ISDN has been delayed by the "connectivity without compatibility" issue, together with the lack of portability of terminal equipment. Prior to the ANSI T1.601-1988, deployment of ISDN had been impeded by lack of uniform standards, which caused manufacturing delays and market confusion.

Interface issues have also caused problems, with IXCs providing only PRI service and LECs providing more BRI than PRI service. This is the result of PRIs being generally a long-haul network tool, and BRI being a CPE or local loop-related capability. Thus, ISDN as a concept is alien to the bifurcated intra-/inter-LATA carriage concept adopted in the U.S. as a consequence of divestiture and deregulation.

Use of ITU Signaling System No. 7 (SS7) in the underlying IDN is absolutely essential for the provision of ISDN and its service offerings. The AT&T and Bellcore versions of SS7 are commonly referred to as CCS 7 and SS7 respectively. Although these versions are functionally similar, gateways are required between LECs and IXCs to account for differences, a penalty of divestiture. Nevertheless, implementation of these signaling systems is gradually providing a foundation to link the ISDN islands that have grown up inside urban areas creating compatibility between inter-LATA and local-loop trunking facilities.

A major impediment to ISDN growth centers around the concept of interworking; the ability to maintain operational integrity when IDN networks providing ISDN access are connected to networks not supporting ISDN, or to other non-ISDN compatible systems or equipment. While possible, successful interworking is difficult to accomplish, consuming additional development and operational resources. A related issue is achieving compatibility and interoperability among PBX switches designed to be ISDN capable but produced by different manufacturers. This is a potentially significant cost item, constituting a barrier to market entry.

In many cases, LECs are unsure about the ability of their existing local-loop facilities—many of which are 40 or more years old—to support ISDN technically without large-scale, expensive rehabilitation. The T BRI reference point also demands an additional pair relative to the traditional analog one-pair-per-service arrangement.

Probably the biggest impediment to the spread of ISDN has been the phenomenal growth of premises data LANs. In 1984, when ISDN was being defined, there were a handful of proprietary, premises-based computer networks, led by Datapoint's Arcnet. Two years later, ISDN was still attempting to tackle 64-kb/s and 1.544-Mb/s transmission rates—already provided by non-ISDN services. LANs, meanwhile, had begun to proliferate around Ethernet and token ring technologies, offering high-speed premises data services for a relatively small investment.

Today, the promise of ISDN-capable PBX or LEC-provided premises services as ubiquitous, low-cost, high-speed local data pipelines has been overtaken by LAN marketplace events. As a consequence, PBX/Centrex systems for voice and LANs for data are the de facto standard for premises-based business communications needs. In light of these events, the often-touted advantages of integrated voice and data via basic rate access melts away, leaving only a yet-unproven residential demand to fuel future growth. Clearly the residential ISDN market is also shaky.

When ISDN was conceived, voice network-compatible modems lumbered along at 2,400 bps. In that era, the promise of 64- or 128-kb/s ISDN data communications looked extremely attractive. Today's V90 modems, selling for under $100, typically synchronize at between 32 and 56 kb/s, diminishing relative merits of ISDN rates. Moreover, numerous ASDL offerings are not only price competitive, but offer the promise of rates of 1.544 Mb/s or more. Worse yet, no ISDN service is appropriate for the burgeoning LAN-to-LAN and MAN/WAN data communications requirements outlined in Chapters 13 and 14. Finally, nearly all the advanced customer telephony features advocated in ISDN specifications are already available via non-ISDN LEC service offerings.

Of course in Japan and Europe, because government-controlled PTTs are not as affected by competition and market forces as service providers are in the United States, deployment of ISDN since the late 1980s has taken place on a much larger scale and the outlook for future use is more likely. The marginal utility that ISDN has been given in the United States doesn't negate or diminish the value or desirability of the goals and benefits that ISDN planners sought on behalf of both network users and operators. In fact, if some form of ASDL becomes the dominant mechanism for supplying residences and small businesses with broadband digital service, having removed subscriber-loop loading coils and branch taps for existing ISDN customers gives LECs a leg up in transitioning them to an ASDL-based service. Moreover, broadband ISDN standards, to be discussed next, still constitute the best-articulated framework for extending basic ISDN capabilities to integrated digital networks with the astonishingly large capacities needed to satisfy ever-increasing numbers of new "bandwidth-hungry" applications now appearing on an almost daily basis.

# Broadband Integrated Services Digital Network

There is an ever-growing need for high-speed MAN and WAN services driven by LAN-to-LAN interconnection and streaming multimedia Internet applications, with future digital video and imagery requirements promising to accelerate the trend. There also exist emerging requirements for technologies that economically support variable bit-rate traffic by providing bandwidth-on-demand services. Efforts to develop technologies and standards responding to these requirements, going beyond what some believe to be interim frame relay and SMDS solutions, are discussed later. ITU has and continues to develop Broadband Integrated Services Digital Network (BISDN) umbrella standards—incorporating underlying standards for integrated digital network switching, multiplexing, and transmission facilities—that will be able to meet expanding requirements well into the future.

In one of the first-draft ITU documents, BISDN is simply defined as "a service requiring transmission channels capable of supporting rates greater than the primary rate." In the U.S. the primary rate for narrowband ISDN (as the original standard is sometimes referred to) is 1.544 Mb/s. The intention behind ITU's Recommendation I.121 BISDN definition is to create the means for providing an all-purpose network to support a wide range of voice, data, and video services. To meet this objective, BISDN will use fiber optic transmission systems with bandwidths of 2,488.32 Mb/s or higher, with the potential of providing users with access to bandwidths hundreds of times greater than primary rate.

As with ISDN, the concept of BISDN is a network evolving from a telephony-integrated digital network that provides end-to-end digital connectivity to support a wide range of services, to which users will have access via a limited set of standard multipurpose network interfaces (UNIs). However, while existing IDNs possess capacities and operating modes for narrowband ISDN services, advanced switching and multiplexing technologies crucial to BISDN service are not generally available. As a consequence, much BISDN-related work is focused on standards and the development of core switching, multiplexing, and transmission technologies for the next-generation IDN. Two major initiatives introduced in Chapters 5 and 7 (SONET and ATM/STM) are treated below to further reveal how, in combination, BISDN planners envision their contribution to ultimate IDN structures.

# Asynchronous Transfer Mode (ATM)/ Synchronous Transfer Mode (STM)

ITU Study Group XVIII, the current focus for international BISDN, generically calls the switching and multiplexing aspects of BISDN transfer modes. The rationale behind selecting a terminology that merges switching and multiplexing aspects is the fact that in today's networks, switching and multiplexing functions are already often integrated in the same equipment. Two paradigms for blending switching and multiplexing techniques are ATM and STM. As depicted in part (a) of Figure 8.5, in the STM paradigm, circuit switching is combined with time division multiplexing. Except for higher bandwidth capacities needed to support BISDN services, STM is basically equivalent to current circuit switching and TDM techniques, as defined and described in Chapter 5.

Recall from Chapter 5 that modern circuit switches use time-slot interchange versions of time division multiplexing to implement their matrices. Moreover, today's digital PBXs and CO switches can eliminate the need for separate multiplexers by directly producing DS"N" compatible multiplexed signals. Both constitute existing product examples of STM, and both have amply demonstrated the efficiency and economic benefit of merging aspects of switching and multiplexing. With the emergence of synchronous digital hierarchy (SDH) conforming multiplexers, also described in Chapter 5, these arguments and benefits become even more compelling.

Using STM, when users request service, circuit switches provide routing through fixed-bandwidth multiplexers. If network loading prevents the allocation of needed resources on an end-to-end basis, the call is blocked. Also, for the duration of completed calls, or sessions, if the assigned user doesn't fully utilize the circuit bandwidth, the unused capacity cannot be made available to other users. For example, with variable bit-rate traffic, an STM derived circuit able to support peak bit rates must be assigned. Then during intervals when less-than-peak bit rates are generated, unused capacity goes wasted. Chapters 7 and 14 contain a broader discussion of variable bit-rate traffic and its impact on switching techniques.

This phenomenon is illustrated in a simplified way in part (b) of Figure 8.5. Here three information-generating devices are each fixed-assigned recurring time slots in a TDM bus. The situation compares to a long train in which every third boxcar is reserved for an assigned user whether or not the user has any material to put in the boxcars. So

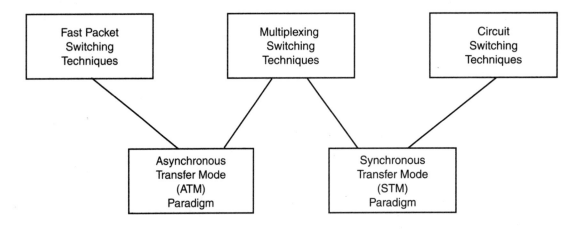

(a) Components of ATM and STM Paradigms

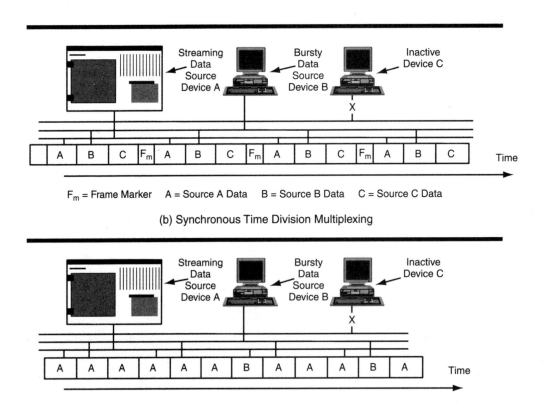

$F_m$ = Frame Marker    A = Source A Data    B = Source B Data    C = Source C Data

(b) Synchronous Time Division Multiplexing

(c) ATM Multiplexing

**Figure 8.5**    ATM and STM components and multiplexing characteristics.

the fact that one user device is inactive, essentially wastes the "time-slot" capacity since its capacity can not be reassigned to another user until the established connection is terminated.

Figure 8.5 part (c) shows how the asynchronous transfer mode paradigm combines fast packet-switching and multiplexing techniques. *ATM* is defined as a broad-bandwidth, low-delay, packet-like switching and multiplexing technique. The name selected by ITU is somewhat confusing since the transfer mode does not involve asynchronous (start-stop) transmission (defined in Chapter 3) associated with non-intelligent terminal traffic. Furthermore, ideally all ATM devices are perfectly synchronous with one another. The natural question is then, why did the ITU call this paradigm asynchronous?

The answer is revealed in Parts (b) and (c). In Part (b) and STMs in general, if the inputs are, for example, digital voice signal samples occurring every 125 microseconds, then they also occur every 125 microseconds in STM multiplexer output bit streams. Because TDM/STM output bit rates are typically higher than individual input signal bit rates, clocking rates of output signals are also higher. But the rate at which samples occur in multiplexer output bit streams is exactly the same rate at which they occur in input signal bit streams. That is, the input and output sample rates are both periodic and rate synchronous.

By contrast, to achieve ATM's advantage in conserving network resources when carrying bursty traffic like LAN-to-LAN and variable bit-rate video signals, no such synchronized ATM input-to-output relation can exist. Figure 8.5, Part (c) imparts some physical insight regarding how this happens. Note that in contrast with STM's fixed assignment of periodic time slots, ATM multiplexer outputs occur on the basis of traffic demand and priority, creating what can be termed asynchronous output signals, justifying the name asynchronous transfer mode.

ATM's advantage is that it statistically prevents wasted network capacity. When users request service, the network determines both the user's average and peak bandwidth and priority requirements. Knowing average and peak load requirements of other competing users, the network controller makes a determination, on a statistical basis, whether sufficient switching and multiplexing capacity is available to support the request.

With ATM, usable capacity is assigned dynamically, (on demand), during a session. This permits network engineering approaches that can accommodate bursty services, while guaranteeing acceptable performance for continuous bit-rate services such as voice and constant-rate video. In contrast with the fundamental notion of a fixed band-width channel, long associated with circuit-switched TDM/STM telecommunications, ATM defines new rules and procedures for dividing up bandwidth capacity and allocating it to users in support of a wide variety of services.

As noted in Chapter 7, ATM's advantage is achieved by adapting bit streams from various types of sources into fixed-sized information-bearing units called cells. Each cell contains header and information fields so that received cells can be associated with a particular infor-mation source, and user information—split among many cells—reassembled. Recall that in synchronous TDM, no headers have to be added to information segments contained in TDM's time slots. Once frame synchronized, then by convention, the first slot after a frame marker contains source "A's" data, the second time slot contains source "B's" data and so on, as illustrated in Part (b).

ATM headers contain a virtual channel identifier (VCI) that is used like a time-slot identifier in an STM approach. VCIs can be translated (much like time slots can be interchanged) at ATM interfaces, prior to being transported to another interface. Thus, ATM creates virtual channels, which can be manipulated like fixed STM channels, but offer variable rather than fixed bandwidth capacities.

It is interesting to observe that in ATM-based architectures, the sig-nificance of the word integrated, as in integrated digital networks, takes on the broadest possible connotation, encompassing both the ability transparently to accommodate voice, data, and video services, as well as the almost complete functional integration of switching and multiplexing within facilities.

Cell relay, unlike conventional packet relay, provides no error or flow control. As a consequence, cell relay is extremely efficient, mini-mizing processing requirements as a cell moves through a network. Figure 8.6 summarizes the comparison between packet switching/mul-tiplexing techniques. As noted, ATM service is connection oriented in that out-of-band call setup signaling is required prior to information transmission, and, once the call is established, all traffic is transmitted over the same logical connection between end-users. That logical con-nection however, can support either connection-oriented or connec-tionless user services.

Evolution to ATM involves significant investment and requires interworking strategies to permit gradual implementation within net-works, understanding that an entire network cannot be upgraded at once, and to permit interconnection with non-ATM networks. From a performance point of view, IDNs with ATM cores are now able to offer public data communications services that compete, on a cost basis, with large-user private data networks. Today, ATM and competing gigabit router-based carrier infrastructures form the basis for advanced business class data services described in Chapter 14.

| Characteristic | X.25 (Packet Relay) | Frame Relay | SMDS (Cell Relay) | ATM (Cell Relay) |
|---|---|---|---|---|
| Line Rates | ~64 kb/s | 56 kb/s/1.544 Mb/s | T1/T3 (1.544/44.736 Mb/s) | Up to 1.2 Gbps |
| Cell Packet/ Length | Variable | Variable (2 Byte Header) | Eventually the Same as ATM | Fixed 53 Bytes (5 Byte Header) |
| Error Control | Error Detection and Correction | Error Detection | Error Detection | None |
| Protocol | Layers 1–3 | Layer 1 Plus | Layers 1–3 | Layer 1 |
| Standards | ITU X.25 | ANSI T1.606/618/617 ITU I.122, Q.922/933 | IEE 802.6 and Bellcore TR-TSV-000772,3 | ITU I.121/2 |
| Delay | Large | Low | Low | Low |
| Traffic | Data | Voice/Data | Voice/Data/Video | Voice/Data/Video |

**Figure 8.6** *Comparison of switching/multiplexing techniques.*

## Synchronous Optical Network (SONET) Use within BISDNs

Having merged switching and multiplexing, ATM networks require wideband transmission. As an umbrella standard, ITU's BISDN Recommendation I.121 addresses switching, multiplexing, and transmission. The standards state that ATM can be supported by any suitable digital transmission system and cite SONET G.707/708/709 recommendations as an example. Figure 8.7 illustrates how ATM cells are mapped into SONET frames.

Within SONET's *synchronous transport signal*-n (STS-n) structure, a special "c" class, (STS-n"c") defines concatenated *synchronous payload envelope* (SPE) structures for ATM signals. Figure 8.7 for example, depicts how concatenated ATM cells are mapped to an STS-3c SPE . For *concatenated* signals, (like ATM) information in SPEs is treated as one continuous payload, and not as byte interleaved, lower rate signals described in Chapter 5's discussion of SONET/SDH (synchronous digital hierarchy) TDM operations. As a consequence, SONET equipment may not disassemble STS-nc SPEs into lower rate SONET signals, but must transport entire SPEs—unchanged. Contrast Figure 8.7 with Figures 5.17 through 5.19 to appreciate these differences.

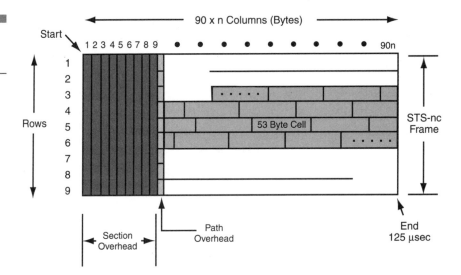

**Figure 8.7**
SONET frame with ATM cells.

Although flexibility in the choice of transmission within an IDN is an advantage, a standard BISDN access approach is fundamental to the provision of universal network interfaces for users. For users to capitalize on bandwidth-on-demand ATM network services, they will need wideband access to ATM cell relay switchesm which may involve extension of SONET-like capabilities, primarily developed for IDN core transport, directly to user premises. Amplifying details on SONET and ATM are in Chapters 5 and 7. Advanced and high-reliability SONET network topologies and designs are treated below.

# Ultra-high Reliability SONET Networks

Chapter 5 discusses the advantages and high capacity of SONET fiber optic transmission systems. Compared to the cable and microwave radio transmission systems of earlier years, the number of voice circuits that can be carried on a single transmission route has increased dramatically. A single fiber strand can carry 32,000 voice circuits, even without the latest wavelength division multiplexing techniques. Since multiple fiber strands are normally run in the same bundle, it is not unusual for a half a million voice trunks to be carried on the same transmission route.

In building transmission systems, a major constraint and expense is securing rights-of-way over which systems can be constructed. Also, the expense of trenching and burying wired transmission media does not vary significantly with the medium's bandwidth. These factors combine to reduce per-channel costs as medium capacity increases. In fact, advancements in fiber optic transmission have driven both local and long haul per-channel costs down by an order of magnitude in the last two decades.

However, advancements in single-medium capacity produced a *funnel factor* side effect that became a major reliability concern to carriers. In the past, when many different transmission systems were needed to satisfy internodal traffic requirements, a loss of any single circuit had a limited impact on overall operations. With the ability to "funnel" so much traffic into single fibers and under the competitive pressures that followed divestiture, carriers generally did not build more capacity than was necessary to satisfy projected traffic, jeopardizing the high reliability that had so long characterized telephone service.

While modern electronics provide innate transmission equipment reliability, equipment reliability alone does not prevent service outages in the face of anomalous events (acts of God and man). For example, microwave radio transmission fading and service disruption can be caused by extreme weather conditions. On the other hand, fiber optic systems fail most often when they are cut (a phenomenon affectionately known in the industry as "backhoe fade").

In the years following divestiture, several high-profile telephone system failures occurred, including the famous Hinsdale fire mentioned in Chapter 2. Since railbeds provide ideal rights of way for fiber systems a number of fiber optic system cuts have accompanied railroad accidents. In November 1988, one such cut along the Washington—New York corridor put 400,000 trunks out of service for an interexchange carrier; it required eight hours to reroute all the circuits and restore normal service.

Progress in automatic restoration of broadband systems since that time is dramatic. For instance, in August 1992 a similar fiber-cutting mishap occurred in Mississippi, disconnecting tens of thousands of channels. In this case, all disconnected channels were restored within five minutes. Today, multiple simultaneous cable cuts can often be restored in a small number of milliseconds, so quickly that users barely perceive the loss of a syllable. This striking performance is due principally to three factors. First, geographically diverse paths between nodes are required to protect against simultaneous cable cuts. Second, network designs must include sufficient reserve capacity and automatic protection switching to accommodate and reroute disconnected traffic. Finally, real-time and high-speed processor-based network management and control facilities are essential if faults are to be rapidly detected and corrective actions taken automatically (the only way millisecond restoration is possible).

To create highly reliable SONET-based transmission networks, most carriers have chosen ring architectures to interconnect major nodes. Figure 8.8 depicts the generic configuration used in all types of SONET rings, as these architectures have come to be known. Nodes on rings include SONET add-drop multiplexer (ADM) functions. As explained in Chapter 5, ADMs are ideally suited for ring operations. Input channels, called tributaries in SONET terminology, can be extracted and others inserted from rings with great efficiency. The

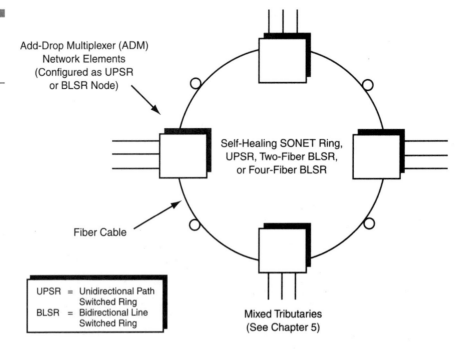

**Figure 8.8**
Generic SONET self-healing ring configuration.

Add-Drop Multiplexer (ADM) Network Elements (Configured as UPSR or BLSR Node)

Self-Healing SONET Ring, UPSR, Two-Fiber BLSR, or Four-Fiber BLSR

Fiber Cable

UPSR = Unidirectional Path Switched Ring
BLSR = Bidirectional Line Switched Ring

Mixed Tributaries
(See Chapter 5)

following sections describe the operation of several types of SONET rings, including their self-healing properties.[1]

## Self-healing SONET Rings

The concept behind self-healing SONET rings is the existence of two independent fiber paths passing through all the nodes. One path is used for the primary path of traffic (under normal circumstances) and the second path is used as a backup path to restore the ring upon failure of a segment. The transfer to the second path upon detection of a failure is accomplished within 50 msec, fast enough to ensure that circuit-switched connections on primary paths are not interrupted. Remember from Chapter 5 that in the absence of wavelength division multiplexing, a single fiber strand carries traffic in only one direction. Three SONET ring architectures differ in the number of fiber strands needed to establish the ring and the method of handling two directions of traffic during primary and backup operations. The three types of rings are:

- Unidirectional path switched ring (UPSR)
- Two-fiber bidirectional line switched ring (BLSR)
- Four-fiber bidirectional line switched ring

The traffic conditions and reliability factors used in selecting a particular approach are summarized in Figure 8.9. The rationale and operational characteristics are further amplified in the following discussion. In all the examples below, it can be assumed that ring signals conform to SONET OC-n specifications.

## Unidirectional Path Switched Rings

Adjacent nodes on a unidirectional path switched ring are connected with a single pair of optical fibers. Primary path traffic travels in only one direction on the ring. For full-duplex operation between two nodes, the forward traffic travels in one direction and the reverse traffic travels in the same direction (i.e., on the same fiber) around the other side of the ring.

---

1 For a more complete treatment of highly-reliable SONET rings configurations, see Nortel's *S/DMS Transport Node* (Document 56015.16/10-96 Issue 3), available from their Internet Web site.

**Figure 8.9**
Ring architecture
characteristics.

| Ring Architecture | Preferred Application |
|---|---|
| UPSR<br>(Unidirectional Path Switched Ring) | Access networks where most traffic terminates in a central office hub. |
| Two-Fiber BLSR<br>(Bidirectional Line Switched Ring) | Access and interoffice networks with a highly distributed traffic pattern. |
| Four-Fiber BLSR | Applications requiring ultra-high capacity and/or protection against multiple concurrent faults. |

A key characteristic of this configuration is that the capacity needed to sustain traffic between any two nodes must be reserved everywhere on the ring (between all the nodes on the ring). It is for this reason that UPSRs are suited to access arrangements where all nodes are communicating with a central node. When all or most of the traffic is pair-wise between a central node and each of the other nodes, the inability of two adjacent nodes to establish communications using only the segment of the ring connecting them involves little or no penalty. As seen below, this is not the case when traffic is more evenly distributed among all ring nodes.

The second UPSR fiber carries traffic in the opposite direction and mirrors the first fiber. The same traffic that flows clockwise on the primary fiber between two nodes flows counterclockwise on the backup fiber (see Figure 8.10). Hence, two copies of each node-to-node pair's traffic arrive at the end points, each via a separate path. The primary path's signal is "selected" until a failure or degradation is detected. At that time the backup path's signal is used. Figure 8.11 shows the traffic flow between the same two nodes as depicted in Figure 8.10, but after a fiber cut. The flow now resembles direct connection cases cited in Chapter 5, where each node receives and transmits traffic on different fibers. It is because fault detection and protection switching are administered for each path independently that the configurations depicted in Figures 8.10 and 8.11 are called *path switched rings.*

## Bidirectional Line Switched Rings

Bidirectional rings using two fibers carry primary-path traffic in the same manner as directly connected point-to-point fiber nodes, that is, separate fibers are used for each signal transmission direction. The important characteristic of BLSRs is that capacity required between

**Figure 8.10**
UPSR protection switching in fault-free state.

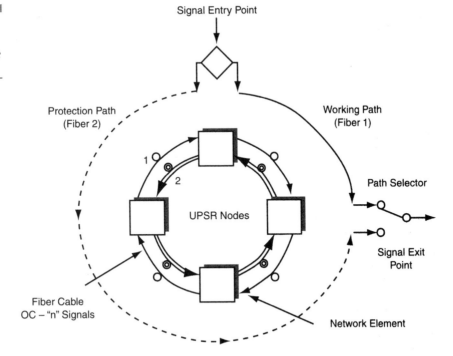

**Figure 8.11**
UPSR protection switching responding to a cable cut.

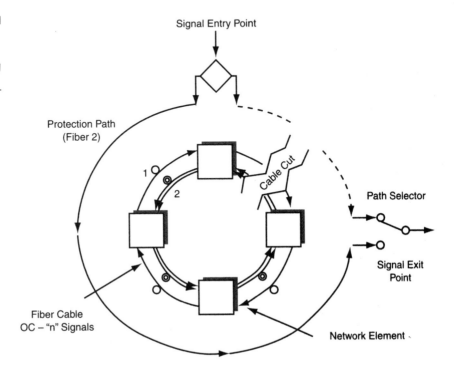

two nodes (i.e., the time slots dedicated to carrying traffic in both directions between those nodes) is available for reuse elsewhere on the ring. Figure 8.12 shows how three different inter-nodal service requirements can be assigned to one SONET time slot on a BLSR ring, whereas three different time slots are required using a UPSR ring.

**Figure 8.12**
Two-fiber BLSR ring illustrating time slot reuse.

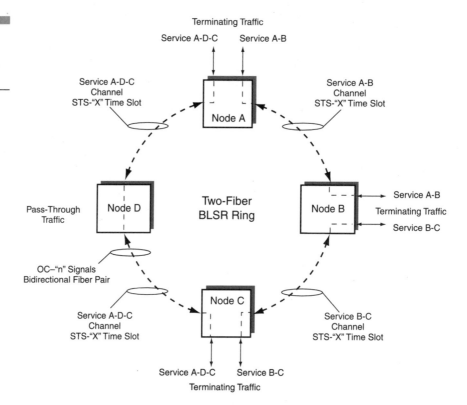

The magnitude of this BLSR versus UPSR advantage is dependent upon both ring size (the number of nodes) and traffic patterns. With BSLR, the aggregate of all internodal traffic rates is allowed to exceed a cable's maximum transmission rate. Thus, in BLSR rings, a cable's maximum throughput rate limits only the busiest node's aggregate capacity. In contrast, in UPSR rings the aggregate of all internodal traffic is the controlling factor. In a hub configuration, where all traffic passes through a central node, the two configurations have equal capacity. However, in a mesh traffic pattern, where nodes communicate randomly with other nodes, BLSR rings have the advantage. The BLSR best case is when traffic only occurs between adjacent nodes.

Figure 8.13 shows the BSLR-to-UPSR capacity advantage under various ring size and traffic pattern assumptions.

Since primary path traffic flows in both directions between nodes, one fiber cannot be reserved exclusively for backup traffic, as in the UPSR case. Instead, half the capacity in each fiber is reserved to restore service when failures occur. Once a fiber cut between two nodes is detected, time slots assigned to traffic between disconnected nodes must be mapped onto empty (reserved) time slots in signals carried on both fibers in the uncut portions of the ring. Traffic that may have been routed clockwise between two disconnected nodes, is routed among affected nodes in the reverse direction around the ring. Since all previously assigned time slots on fibers between affected nodes are rerouted as a group, BLSR's recovery operation is referred to as *line switching* as opposed to UPSR's path switching.

**Figure 8.13**
BLSR versus UPSR capacity comparison as a function of traffic type.

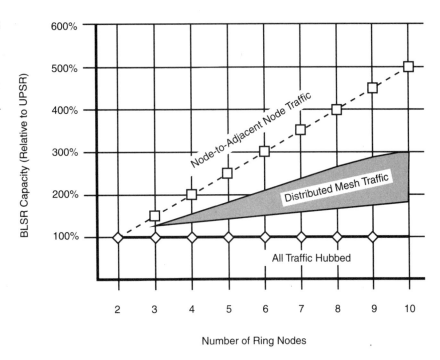

An additional BLSR design benefit is the ability to assign time slots reserved to restore "protected" traffic when failures occur to lower priority "unprotected" traffic during normal (no failures) conditions. This allows carriers to sell reserved protection capacity at discounted rates to customers who, in the unlikely event of massive failures, are willing to accept service outages. Unprotected traffic is automatically dropped when a protection switching event occurs. Not only does this add to a carrier's revenue stream, it also promotes rapid introduction of new services in advance of new fiber plant investment.

Four-fiber BLSRs are similar to two-fiber varieties except that two complete bidirectional fiber pairs are provided between adjacent nodes. In these configurations, fiber cuts result in both working-pair fibers being connected to a backup pair in the opposite direction. Since no working-pair time slots need to be reserved for protection, four-fiber BLSRs have twice the capacity of two-fiber BLSRs. Extra pairs of fibers between adjacent nodes also enhance the span and flexibility of restoration and recovery operations—for other than complete physical cable severing incidents. For example, when only working-pair performance degradation occurs (due perhaps to defective splices, faulty connectors, or equipment problems), protection switching actions affecting only adjacent nodes may be all that is needed. Since four fibers costs more than two, to date four-fiber BLSRs have found application principally in core interexchange networks where investment costs are more easily offset by the revenues generated from higher traffic volumes.

## Interconnected Rings

Practical networks cannot be implemented on single SONET rings. For one thing, SONET specifications limit rings to a maximum of sixteen nodes. In today's large networks, end-to-end connections are made via multiple, concatenated ring interconnections. In the early 90s, Sprint debuted a plan for 38 interlocking rings, with 16 nodes per ring, allowing hundreds of thousands of equivalent voice circuits to be restored almost instantaneously. In such networks, traffic passes from one ring to another through redundant gateway nodes, called matched nodes, providing high-reliability, high-survivability inter-ring service connections.

Figure 8.14 illustrates a matched node configuration. Since rings of different types and bandwidths must be interconnected, gateways operate at path (tributary) levels, maximizing interconnection flexibility. A path using a gateway is copied (extracted from one ring) onto a tributary to a connecting ring and is simultaneously transmitted to a redundant gateway on the original ring, where it is dropped and sent to the destination ring on a backup path. The primary gateway on the destination ring selects which path to use. In interconnected ring configurations, each ring continues to employ its native protection mode for traffic on the ring; the backup gateway path is used only in the event of a gateway failure.

While this treatment of SONET high-reliability networking has been necessarily brief, it relates all the essential elements of a well-articulated, standards-based strategy that has proven immensely successful in worldwide installations. For all its engineering precision, thoroughness, and relatively recent appearance, the pace of new developments is now so rapid that almost before the full impact of SONET's extraordinary capabilities have been fully realized, even more revolutionary competing initiatives are already being advanced. The next section outlines these developments.

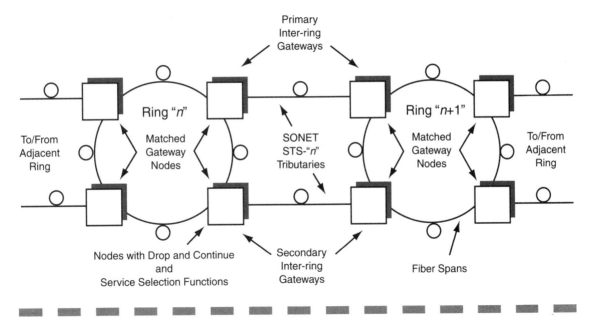

**Figure 8.14**   *Inter-ring service transport using survivable matched node gateways.*

# Advanced Network and Device Designs—Trends to All Optical Networks

Advances in telecommunications technologies that took place in the last few years of the twentieth century are almost certain to have enormous impact on twenty first-century developments. Beyond unprecedented technical performance and affordability gains, the rate of what has to be considered "breakthrough" discoveries and corollary product and service offerings has never been so rapid. In the not-too-distant past, the telecommunications industry was viewed as relatively understandable and stable, driven by a few principal monopolies and equipment/service providers, and overall, somewhat staid. So much for the past!

Today, telecommunications technologies and the variety of business ventures exploiting them are evolving so rapidly that many users (and even some industry participants and investors) are unable to keep up. Just ten years ago voice traffic dominated networks, with packet data traffic a minuscule part. Now, data constitutes over 50 percent of aggregate network traffic, and Internet traffic is growing at a fourfold annual rate. To accommodate increasing traffic volume, in the early 1980s DS3 equipment (45 Mb/s) began replacing slower DS1 (1.544 Mb/s) designs in carrier core networks. The arrival of SONET and fiber transmission plant in the mid-80s ushered in OC-12 (622 Mb/s) capabilities.

By 1991, OC-48 (2.5Gb/s) capabilities arrived. In 1994, the first two-wavelength, wavelength division multiplexers (WDMs using the 1,300- and 1,550-nanometer fiber optic cable windows) began to appear in carrier networks, doubling capacity to 5 Gb/s. Since then, most carriers have installed dense wavelength division multiplexers (DWDMs—the word "dense" is commonly applied when more than eight separate wavelength divided channels are packed into the 1,550-nanometer window—Chapter 5 has more details) as the preferred way to cope with the tide of rising demand for broadband service.

Why? Because today most fiber networks are still using fiber placed in the ground in the 80s and early 90s. Simply put, DWDM is a cost-effective "escape valve." By allowing carriers to meet rising demand using existing fiber, DWDM not only obviates the cost of burying new cable, it also shortens the time it takes carriers to respond to new service orders. The results have been astounding. In 1997, using Perelli amplifiers and WDM elements, MCI established a 4-wavelength OC-192 link (about 10 Gb/s) between St. Louis and Chicago. By 1999, with OC-48-based 16-wavelength equipment in some carrier systems, single strands of fiber supported 40-Gb/s data rates, nearly 1,000 times the throughput of mid-80s 45-Mb/s DS3 facilities. Manifesting its popularity, DWDM sales, which were $626 million in 1996, exceeded $1 billion in North America alone in 1997, and are projected to reach $4.6 billion by 2001.

So in beginning this section's treatment of advanced networks and devices, we start with installed baseline capabilities that just 20 years ago appeared to belong in the realm of science fiction. Beyond the bandwidth-expanding WDM/DWDM breakthroughs just described, we list just a few of the telecommunications technologies outlined in Figure 8.1:

- We have extensive fiber-based transmission facilities, most conforming to SONET standards and arranged in ultra-reliable bidirectional line switched rings (BLSRs);

- We have networks of very large public-circuit switches managed by powerful computers in centralized national control centers offering heretofore unachievable levels of technical and cost performance via virtual private networks (VPNs—see Chapters 10, 11, and 12);

- We have gigabit-per-second packet switching-based routers fueling the Internet's amazing ability to accommodate ever-increasing numbers of users and services; and

- We have an incredible repertoire of low-cost yet seemingly unlimitedly powerful digital signal-processing devices and techniques facilitating modulation and other signal processing operations—crucial to mobile, satellite, local access, and other modern network functions.

In conjunction with earlier chapter descriptions of baseline technologies, taking this opportunity to look ahead at the most important likely new developments establishes a framework for better understanding and assessing voice, data, video service options, and selection criteria contained in the remaining chapters. Among experts, reasonable people disagree—sometimes vehemently—about the merits of a number of possible advanced network approaches. While the details of technical arguments are often beyond what can be treated in this book, the major alternatives and related operational issues are of vital interest to all informed telecommunications business and individual users. To impart this knowledge we cite and analyze four currently prominent areas of research and development directed principally at core network applications.

## Electronic vs. Photonic Implementation

*Photonics* is a science and technology based on and concerned with the controlled flow of photons, or light particles. A photon is a quantity of electromagnetic energy equal to $h \times f$, where $h$ is a constant (Plank's) and $f$ is the frequency of the radiation. Photonics is an optical equivalent of electronics, the two technologies often coexisting in devices such optical disks, switches, modulators, electronic-to-photonic converters, and hardware used in fiber optic transmission systems.

Perhaps one of the most common catch phrases among carriers and photonic equipment manufacturers is the notion of *all-optical networks.* Proponents claim savings of up to 85 percent when compared to electronic-based long-distance networks. Lending credence to such claims, Corvis, a recent startup company and reportedly a take-over target of Lucent, Nortel, and Cisco that could cost $10 billion, claims that its erbium-doped fiber amplifier-based products (EDFA—see Chapter 4) can be spaced 2,000 miles apart versus 125-mile electronic amplifier separations. But within fiber networks, electronic versus photonic implementation tradeoffs pertain to all components— switches, DWDM multiplexers, wavelength converters, and so forth.

When we look across all network functions, there are currently no universal all-electronic or all-photonic "best" solutions. Figure 8.15 illuminates intermediate steps from hybrid electro-optical to all-optical possibilities as they apply to circuit or packet switching. Figure 8.15, part (a), represents photonic transmission and all-electronic switching. Thus, photonic signals entering switches from fiber transmission facilities must be converted to electronic format, and electronic switch output signals must be converted to photonic format for retransmission. Part (b) illustrates photonic transmission and switching with electronic switch control. Lastly, part (c) is an example of photonic transmission, switching, and control. Now, all other things being equal, the all-photonic approach eliminates photonic-to-electronic conversion equipment, and thus would be the best solution.

But life is not yet that simple. To date there are no practical all-photonic call-by-call circuit or packet switches that can compete with electronic counterparts. In the domain of optical cross connect (OXC) channel switching, however, the opposite conclusion appears to be justified as explained below.

## TDM vs. DWDM

Time division multiplexing (TDM) versus DWDM tradeoffs involve the following factors. Just above we referred to 16-wavelength DWDMs. Implied in that discussion was wavelength multiplexing of

**Figure 8.15**
Comparison of
photonic and
electronic
implementation
alternatives in
networks using fiber
transmission facilities.

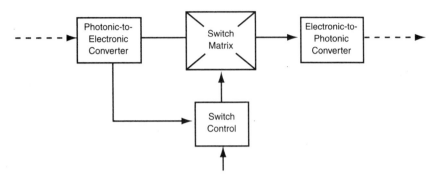

(a) Photonic Transmission—Electronic Switching and Switch Control

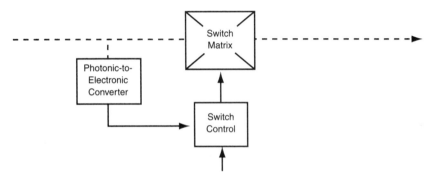

(b) Photonic Transmission and Switching—Electronic Switch Control

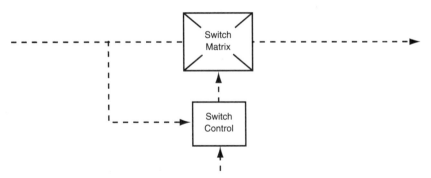

(c) Photonic Transmission, Switching and Switch Control

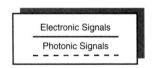

16 time division multiplexed OC-48 (2.48832-Gb/s) signals into a single
composite DWDM photonic signal, yielding a total throughput of

39.81312 Gb/s. (These precise figures are generally replaced with 2.5 and 40 Gb/s approximations.)

An alternative to achieving 40-Gb/s throughputs is to invoke higher-speed time division multiplexing and a reduced number of DWDM wavelength-based channels. For example, with potential OC-768 multiplexers, 39.81312-Gb/s signals could be placed in a single WDM channel, equaling the 40-Gb/s capacity of today's 16-wavelength DWDM systems. Figure 8.16 portrays the equipment differences in these two approaches with the understanding that the tradeoffs as presented below pertain mainly to isochronous or voice-network applications.

**Figure 8.16**
TDM versus DWDM transport alternatives.

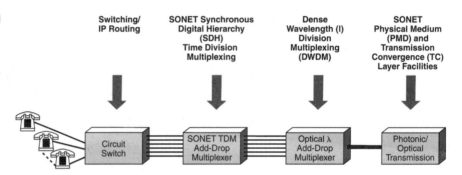

(a) Combined SONET/SDH time division and wavelength division multiplexing in an isochronous voice network traffic application.

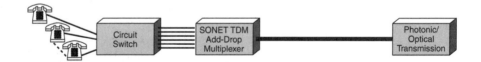

(b) SONET/SDH time division only multiplexing. Requires higher TDM bit rates to equal combined TDM/DWDM capacity.

So what are the tradeoffs between these two alternatives? First if cost-effective TDM multiplexers and/or packet switching-based routers are available that can directly generate and process 40-Gb/s bit streams, then the OC-768 approach enables the direct unpacking and processing of the 40-Gb/s payloads to the lowest statistically multiplexed payload or packet level in a single operation.

Although various techniques able to recognize packet headers at gigabit-per-second speed exist, either electronically or optically, it is still difficult to implement header processors (hundreds of lines of computer code per packet) operating at such high speeds to switch packets on the fly at every node.

In addition to processing hardware limitations, there is the consideration of a fiber transmission system's ability to handle very high-speed, single WDM channel signals. Because a significant portion of existing plant (much installed in the mid-80s and early 90s) uses standard single mode as opposed to *dispersion-shifted fibers* (see Chapter 4), the ability to accommodate OC-192 or 768 signals without offsetting penalties is not apparent. In the absence of some TDM/packet-switching breakthrough, today's proven OC-48/DWDM approach will continue to be the carrier's preferred core network approach.

Several factors support this conclusion. First, just around the corner are 40- and even 96-channel OC-48-based DWDM multiplexers. This will endow a single fiber strand with a 240-Gb/s capacity. There are added economies of scale. For example, a single EDFA (erbium-doped fiber amplifier) can amplify all DWDM channels, so that the progression from 16- to 96-channel operation leads to improved per-channel cost performance. Moreover, because OC-48 channelization does not stress fiber cable capabilities, the distance between EDFAs can be greater.

## Photonic Switching

As alluded to above, photonic replacements for all forms of electronic switching, that is call-by-call circuit switching, channel or cross-connect circuit switching, and cell, packet, or frame-relay switching, is at least theoretically possible. Today, however, except for *optical add/drop multiplexers* (OADMs) and *optical cross-connects* (OXCs) or wavelength routers, photonic switches are small and largely experimental. Nevertheless, OADM/OXC progress is impressive.

The December 1999 issue of *Business Communications Review* magazine observed that in what may be the last major telecommunications announcement of the twentieth century, Lucent introduced is Wavestar LambdaRouter. With 256 ports—each supporting 40 Gb/s—the device has an aggregate capacity of nearly 10 terabits per second (a terabit is a thousand gigabits or $10^{12}$ bits).

Just as important as its capacity is the product's use of *digital wrappers*, which are functionally analogous to packet headers. In lieu of manual patching associated with early cross connects, this new product embeds cross-connect switching instructions in fiber cable optical signals (along with ancillary data such as restoration and traffic-type information) so that cross connections can be established automatically. According to reports, this places the LambdaRouter in the class of pure optical switches, (Part (c) of Figure 8.15).

Recall from Chapter 7 that connectionless IP routing requires processing of each packet to recover and act upon its source and destination information. While this enables routing through networks on an individual packet-by-packet basis, such routing not only imposes significant processing requirements on every intervening switching node, but it is a switching regimen that cannot always deliver low and/or constant latency quality-of-service performance that meets QoS requirements of isochronous voice or video traffic.

Chapter 7 describes *multiprotocol label switching* (MPLS), an approach that solves the QoS problem by assigning "explicit" connection-oriented paths to various classes of traffic, obviating the need for packet-by-packet processing and switching. *Virtual path identifiers* (VPIs) in ATM and *data-link connection identifiers* (DLCIs) in frame relay exhibit the same packet aggregating and simplified connection-oriented routing qualities.

*Digital wrappers* achieve analogous results but use a form of *subcarrier multiplexed* (SCM) headers, wherein low bit-rate routing information is multiplexed on the same wavelength (optical carrier) that bears the payload data. SCM is essentially an optical mechanism for establishing virtual circuits with a granularity equal to the throughput rate supported by the channel-switched wavelength's channel bandwidth.

In a way, MPLS, *differentiated services* (Diffserv—also discussed in Chapter 7), digital wrappers and SCM, VPIs and DLCIs, are all compromises to implementing pure packet switching, particularly in carrier core networks that must transport all types of traffic. If one were to envision a spectrum of switching approaches, these approaches would fall between the two extremes of *pure packet* and *pure circuit* switching.

Most top-level tradeoff discussions of this issue emphasize the efficiency that packet switching's bandwidth-on-demand and statistical multiplexing operations achieve, while minimizing its 10—20 percent header-in-each packet throughput penalty and poor latency characteristics. With the advent of terabit network speeds made possible by fiber transmission, the limitations of pure packet switching, or of simply implementing current Internet Protocols in these high-speed networks, has led to a flurry of new technical papers and a great deal of research and experimentation seeking to discover practical alternatives.

Understanding that the need to execute hundreds of lines of code to process current IP headers on a hop-by-hop basis could produce bottlenecks, longer-term solutions for future terabit fiber network are being sought by designers and standards organizations. One method, dubbed optical tag switching, incorporates MPLS-like features. It assigns short fixed-length labels containing routing information, in "tags," to multiprotocol packets (e.g., IP, ATM, frame relay, etc.) for network transport. A tag-switched network consists of:

- Tag edge routers, located at network boundaries which generate tags.
- Tag switches to switch packets based on tag information.
- Tag protocols for distributing tag information among nodes.

All the above developments and arguments are germane to broader efforts to determine which protocol and multiplexing structures are most appropriate for tomorrow's advanced photonic networks, a topic treated next.

## SONET vs. IP-over-Photons

Perhaps the most heated debate among industry experts is now whether synchronous digital hierarchy (SDH), time division multiplexed SONET-based transmission networks will be appropriate when IP traffic dominates fiber networks. Internet service providers are faced with exponential growth in demand; adding capacity to their networks in a timely manner to meet this demand requires all the ingenuity they can muster. New technologies such as DWDM and photonic cross-connects and routers provide a new set of tools to attack the capacity problem head on. However, these new tools have called into question traditional architectures for providing reliable transport, most notably the ATM/SONET ring transmission architecture in core backbone networks.

The advantages of SONET rings are discussed earlier in this chapter. However, interconnection of high speed IP routers via SONET/SDH equipment raises important issues:

- Because protection time slots, in essence, reserve half of a ring's bandwidth for failure recovery, SONET rings are not considered to be as bandwidth efficient as other alternatives (identified below), at least in packet data traffic applications. Moreover, even in bidirectional line switched rings, ring speeds must be provisioned to accommodate peak traffic requirements between any two nodes on the ring. Unfortunately, that uppermost and fixed-allocated capacity is only used during peak traffic volume conditions since SONET's time division multiplex technology assigns time slots (or in the vernacular, consumes bandwidth) regardless of actual usage.

- With the limitation of 16 nodes on a ring, many interlocking rings with matched node gateways are needed to implement large networks. This means that intermediate rings must not only carry traffic originating and terminating on those rings, but transiting inter-ring traffic as well. So peak ring-throughput capabilities are driven higher than they might be in some mesh-connected network topologies. Furthermore, it must be admitted that fielding large, inter-connected ring networks poses highly complex design and management problems.

- SONET/SDH equipment, especially electronic add/drop multiplexers, is very expensive compared to IP routers operating at comparable speeds.

Beneath these high-level SONET-related considerations are more basic and yet overarching questions of how to provide switching, statistical multiplexing (including packet/cell switching), protection and management core-network functions in the most cost and technically effective manner. In today's highly competitive environment, correct answers are crucial to network operator survival. The current debate really centers on the phenomena first revealed in Chapter 7, namely, that many of the functions implemented at IP/ATM/SONET/DWDM levels are not only redundant, but involve significant technical and cost penalties. While Figure 8.17 doesn't retrench all the way back to the basic issues just elucidated, it does present near-term options for handling packet data (read IP traffic), some of which may be more effective in leading us to solutions that will.

Part (a) of Figure 8.17 is an approach in which IP router packets are segmented into ATM cells, which are then mapped into some level of SONET's STS synchronous TDM hierarchy, using add-drop multiplexing equipment. ADM outputs are in turn wavelength-multiplexed in DWDM equipment that generates signals conforming to existing physical medium and transmission convergence layer facilities (today, probably those conforming to SONET specifications).

Part (b) of Figure 8.17 eliminates SONET's usually electronically implemented ADM equipment and connects multiple ATM outputs directly to DWDM equipment. This option could correspond to Figure 8.7's example in which basic SONET framing is used to accommodate concatenated signals, but does not impose the complexities of SONET's add-drop TDM hierarchy.

Figure 8.17's Part (c) alternative maps IP router packets to STS TDM lower-level signals, preserving the option to use SONET's flexible add-

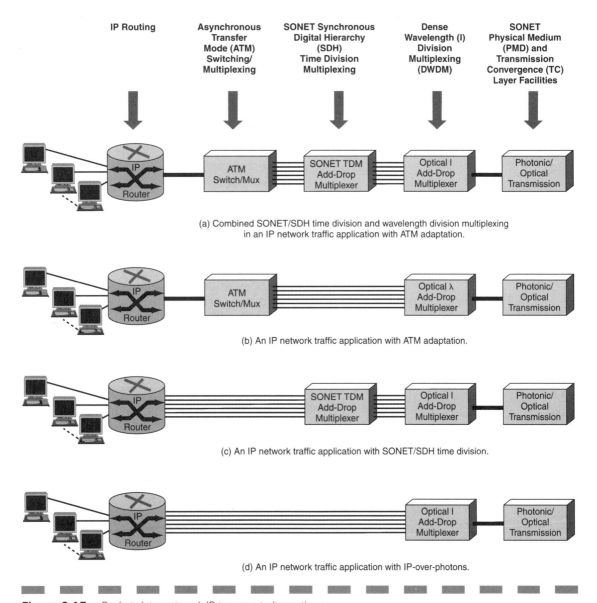

**Figure 8.17** Packet data network IP transport alternatives.

drop capabilities. Lastly, Part (d) represents the IP-over-photons option under which advocates envision the mapping of potentially terabit router packets directly to optical transmission structures. In the most likely near-term scenario, optical add-drop multiplexers precede protocol layer 1 facilities.

Regarding the four Figure 8.17 alternatives, the appearance of MPLS will allow IP routers to implement ATM-like quality-of-service levels without cell overhead. In addition, all optical cross-connects can provide high-speed transmission paths directly to IP routers, enabling them to implement at IP layers multiplexing and protection functions, equivalent to or better than SONET's functions. For example, IP-over-photon designs permit mesh connectivity among routers. Proponents now believe that mesh connectivity avoids bandwidth-wasting constraints of ring architectures while retaining multiple path/alternate routing reliability.

As Internet service providers rebuild core networks to handle ever-increasing IP traffic, these IP-over-photon architectures will take hold. However, the world is not all about IP (yet). Hundreds of millions of telephone instruments and the LEC central-office circuit switches and SS7 common-channel signaling systems supporting them remain. As carriers gravitate to IP-based core networks, some interworking provisions must be developed. Accomplishing this in a manner that preserves the integrity, grade-of-service, reliability, low cost, and easy-to-use qualities of today's ubiquitous voice networks is no simple task. Also, even if more and more services ride over IP-based core networks in the future, the ATM/SONET architecture is likely to endure at the edges of the network where bandwidth will be more scarce and applications more varied. Again, interworking with non-voice legacy systems is a necessity.

The Optical Internetworking Forum, a group chartered to define how gigabit routers and switches should interface with high-speed optical transport equipment and how such networks can be rapidly restored when fiber cuts or other failures occur, makes it clear that "although SONET multiplexers may be replaced at some level by high-speed packet switches or routers, IP-over-photons designs will be compatible with SONET physical (level 1) interfaces—long into the future."

Finally, since fiber transport performance is determined by the quality of the optical components that transmitted signals traverse on their way to destinations, as noted above, fiber cable imperfections and other impairments (jitter, cross-talk, amplified spontaneous emission [ASE], etc.) may limit practical upper single-channel speeds to rates less than what is needed to carry all offered traffic in a single ultra-wide wavelength band. In this case, multiple wavelength channels will continue to be needed, which in turn will mandate OADM and OXC equipment and channel-circuit switching. The conclusion is that SONET is not passé, as some assert, and that both SONET and some modification to current IP protocols that satisfies isochronous QoS traffic will be part of any new IP-over-photons network designs.

# Voice Services

This part describes voice services, underlying customer premises equipment (CPE), and network facilities that support them. In the not too distant past, separate facilities were furnished for voice, data, video, and other service types. Today, advanced technologies make multimedia switching and transmission possible. However, as indicated in Chapter 8, the main reason carriers vigorously pursue integrated service provision reflects not just the possibility for doing so, but the fact that such designs ultimately produce significant economies—factors crucial to maintaining leadership (or sometimes surviving) in highly competitive global telecommunications markets.

Another development since the first edition of this book affecting our treatment of voice services here is the emergence of hundreds of new mobile carriers and competitive LECs and IXCs. Beginning with ATT's divestiture and the Modification of Final Judgement (MFJ) in 1984, and accelerating after the 1996 Telecommunications Act, the telecommunications landscape has markedly changed. In the first edition, for example, it was correct to (1) ascribe to LECs the major voice service role within local access and transport areas (LATAs), and (2) state that BOCs were prohibited, by law, from offering inter-LATA, interexchange services.

Now, as explained in Chapter 2, under 1996 Telecommunications Act rules, BOCs who *fairly* make their facilities available to entities establishing competitive LEC (CLEC) enterprises may pursue long-distance, inter-LATA, or IXC business. Also, it is important to note that nearly all existing IXCs have strategic plans to bypass incumbent LEC facilities and provide "last-mile" and other local exchange services. Furthermore, as already noted, most mobile carriers seek to provide customers with both local and long distance service, many now offering the same *fixed* per-minute rate for local or long-distance calls.

It may happen that LATA and inter-LATA-based terminology, rules, and tariff-setting bases (so carefully enunciated just a few years ago), may gradually be abandoned. As a consequence, organizing a description of voice services in intra-LATA and inter-LATA terms, or in terms of carriers operating exclusively within those bounds, is no longer appropriate. For all the above reasons, organizing our explication of rapidly advancing telecommunications technologies, ever-changing regulatory and statute environments, and an almost unpredictable mix of carrier/service provider business structures—from the relatively stable viewpoint of "telecommunications services" (proffered in the first edition)—is more valid than ever.

Therefore, we identify and describe high-interest services first and enabling technologies second—as opposed to identifying and describing underlying technologies and then services. Why? Because in almost all cases, residential and business owners understand, relate to, are most interested in, and buy, *services* that satisfy their needs. Moreover, services, at least those of time-tested high utility, transcend changing technologies, carrier organizations and structures, and regulatory environments. Requirements for high-value *voice services* remain immutable even to integrating voice, data, and other multimedia capable facility trends.

In what follows we begin by presenting a taxonomy of voice services, shown in Figure P3.1. In Chapters 9 and 10, we describe specific current, popular premises and external network services, and the facilities and technologies on which they are based. In Chapter 11 we demonstrate how current Public Utility Commission and FCC regulations determine voice-service tariffed costs and explain basic traffic engineering principles that, used properly, lead to optimized network designs. Finally, Chapter 12 advises users how to procure voice service and how to optimize network designs, and offers an example request for proposal (RFP) outline for premises services.

Recall from Chapter 1 that a telecommunications service is a specified set of information transfer, and information transfer-supporting capabilities, delivered to a group of users by a telecommunications system. Remember also that applications are unique aggregations of telecommunications services satisfying particular business and residential needs. Using these definitions, it is possible to classify voice services independent of technology, service-provider, or regulatory considerations in the manner depicted in Figure P3.1.

At the highest level, voice services are segmented into *premises, local exchange,* and *interexchange* network services, and *management and control* services. Note that regardless of what current or future technologies are employed, or what type of mix of service-providing entities supply these services, the categories are always valid.

Management and control (M&C) applies not just to basic access and transport services, defined below, but to any of the extensive and growing list of advanced intelligent-network functions, (e.g., interactive voice response and mail, automatic call distribution, caller ID, custom call handling, and many others). Beyond basic service delivery, M&C supports technical, administrative, accounting, and security operations of entire networks as well as individual switches, multiplexers, transmission, CPE, software, and other telecommunications systems configuration items.

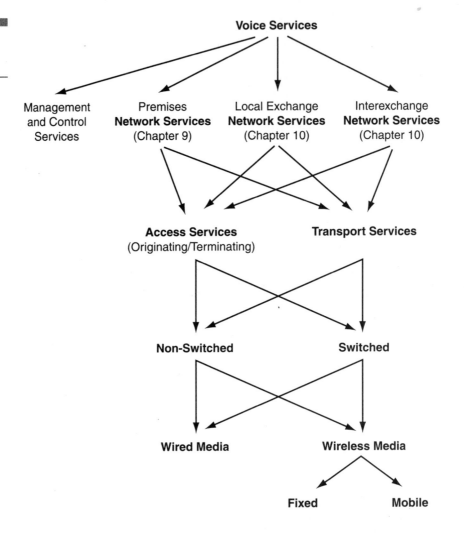

Formally, M&C is defined as capabilities to plan, organize, design, optimize, engineer, implement, operate, monitor, provision, maintain, synchronize, supervise, manage, control, and administer entities, systems, elements, processes, organizations, and events. Demonstrating the breadth of M&C functionality, as lengthy as this list is, each item implies additional or subsidiary capabilities. For example, in telecommunications systems, the ability to "monitor" normally implies comprehensive performance assessment facilities to detect, isolate, report, and record network faults; to measure offered and refused (busy condition) traffic; and to measure call-completion times, call duration, and, numerous other parameters critical to efficient operations.

In complex telecommunications systems with multiple-carrier, service-provider, and customer-owned subsystems, ideally *integrated* capabilities enable system-wide fault, performance, and configuration management and control. In this edition, specific M&C capabilities are discussed as parts of other service and network component and technology topics. The crucial role that common-channel signaling system No. 7 (SS7) plays in distributed, system-wide management and control is highlighted in Chapter 6.

Referring again to Figure P3.1, note that network services consist of *access* and *transport* services. *Transport services* are network switching, transmission, and related services that support information transfer between originating and terminating access facilities.

*Access services* are specified sets of information-transfer capabilities furnished to users at telecommunications network points-of-termination (POTs) to provide access to network transport services. Two examples are as follows. LEC *access services* includes loop transmission facilities from the LEC network POT (in this case more commonly known as a *network interface*—NI) to central office line-termination equipment (e.g., distribution frame cross-connects) as illustrated in Figure P3.2 below. Trunking between interexchange carrier points-of-presence (POPs) and local exchange carrier switching systems (the POT at the POP is identified as the *point of interface*—POI), is a second example of access service, illustrated in Figure P3.3 and explained in more detail in Chapter 10. End-to-end connections require *originating* and *terminating* access services.

The third level of categorization depicted in Figure P3.1 classifies voice services as either *non-switched* or *switched*. Incidences of switched transport services include *local calling*, message telecommunications service (MTS), wide area telecommunications service (WATS), and 800 service. Each of these, and other examples of switched and non-switched services, is defined and characterized in Chapter 10.

In the past, nearly all incidences of *wired* (metallic wire or optical fiber) *access service* fell into the "non-switched" category. In UTP subscriber loops, for example, access service is rendered via dedicated twisted-wire pairs. However, in wireless cellular systems, access service to subscribers must be "switched" as they roam from one cell to another. Note that in *switched transport voice service*, "switching" generally responds to subscriber dialing instructions to enable end-to-end connections with called parties. Cellular switched access service rendered to keep roaming subscribers connected has nothing to do with, and should be completely transparent to, subscribers and their end-to-end call establishment dialing instructions.

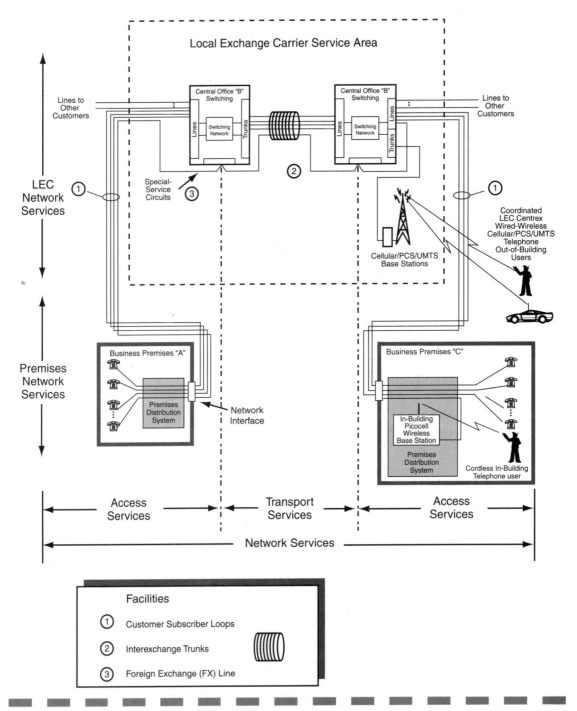

**Figure P3.2** Centrex local calling PSTN and foreign-exchange examples.

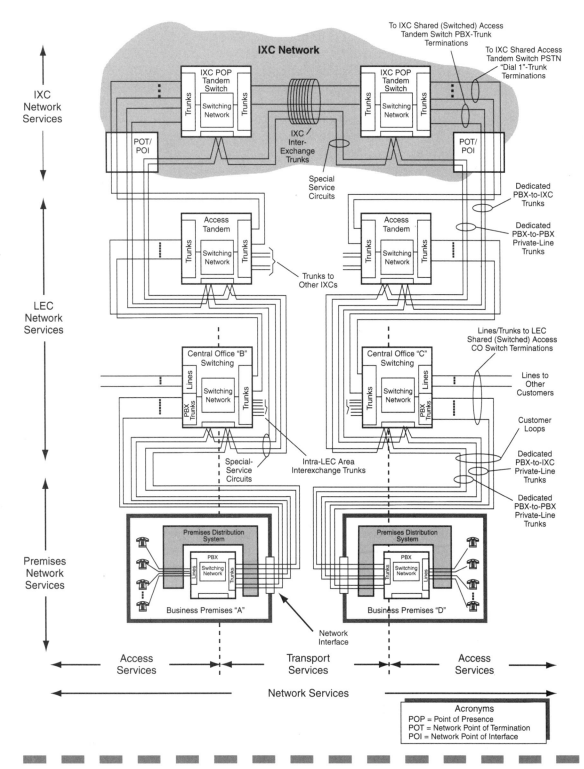

**Figure P3.3** PBX local and long-distance PSTN and dedicated private-line examples.

Similarly, after establishing end-to-end call connections, LECs and IXCs may invoke switching to reroute calls automatically in response to circuit failures or for traffic-handling convenience. But again, like cellular roaming-induced switching, that switching is ideally transparent to subscribers and has nothing to do with switched-service characteristics or per-call charges carriers offer to subscribers.

The next voice service subdivision shown in Figure P3.1 is wired versus wireless media. Historically, use of the term wire-line carrier gives rise to at least a technical misnomer. In the era of the MFJ, the term wire-line carrier was understood to be synonymous with existing operating telephone companies—as opposed to the swelling ranks of new cellular carriers, then referred to as radio common carriers. The reason why "wire-line carrier" may be considered something of a misnomer is that most carriers dubbed wire-line use both wired and radio-based or wireless transmission media. Today, due to FCC decisions to approve two cellular carrier franchises in most urban areas, we have the situation in which wire-line carriers operate about 50 percent of the country's wireless cellular telephone systems.

The terms wired and wireless occasion two additional potential technical misnomers. Fiber optic cable-based transmission systems could legitimately be called wireless, if by wire we mean metallic conductors only. If not literally correct, tradition describes either metallic or fiber optic transmission systems as wired. And just as clearly wireless does not refer to the use of fiber optic media, but rather is a category of transmission unencumbered by cable tethers of any kind. Specifically, by convention wireless means using electromagnetic (radio or lightwave), sound, or some other type of energy propagation and unguided media. Since it is too late to rename wireless, cableless and too cumbersome to substitute fiber or metallic wired for simply wired, we trust that with this explication our use of the terms wired, wire-lin" and wireless herein is unambiguous.

The final subdivision of voice service shown in Figure P3.1 acknowledges *fixed* and *mobile* manifestations of wireless service. Chapters 1 and 4, and particularly the discussion surrounding Figure 4.17, assign to mobile wireless any service that allows subscribers and subscriber terminals to move about freely. The cited sections draw attention to the fact that the rate (speed) and spatial extent (cell size) of movement give rise to subcategories and note the trend toward using common facilities for some fixed and mobile wireless services.

Figure P3.2 graphically portrays access and transport services associated with customer premises and LEC facilities. The figure exemplifies

Centrex, local calling, public switched telephone network (PSTN), and *foreign exchange* (FX) services within a single LEC serving area.

*Customer premises equipment* (CPE) facilities that underlie *access services* from business premises "A", (the left side of the figure), comprise telephone instruments and other voice network-compatible information-origination/termination equipment (such as telephones, answering machines, voice response, and facsimile devices) and a *premises distribution system* (PDS). A premises distribution system is a transmission network inside a building or among a group of buildings, for example an office park or a campus. A PDS connects desktop and other equipment with common host equipment (e.g., switches, computers and building automation systems), and to external telecommunications networks at NIs. PDS systems are described in detail in Chapter 9, directly following this Part 3 introduction.

LEC facilities supporting access services to premises "A" include loop transmission facilities from the NI to central office line-termination equipment (e.g., distribution frame cross-connects that tie local loops to LEC switching equipment) and for subscribers needing switched transport service, central office switch line-termination cards. Note that FX services are rendered via special service circuits and are terminated, but not provided with call-by-call switched services at the originating central office. FX, a service that provides a circuit between a user telephone and a central office other than the one that normally serves the caller, and other non-switched services are explained further in Chapter 10.

Facilities underlying access services to premises "D" on the right side of the figure are similar to those of premises "A" with the addition of wireless in-building PDS capabilities and coordinated wireless Centrex cellular access for out-of-building mobile subscribers. In-building wireless facilities include cordless telephones and in-building base stations that are wire connected to LEC NIs. Coordinated Centrex cellular mobile service is similar to ordinary cellular service except that subscribers can be reached through a single listed directory number—whether via wired in-office or out-of-building mobile telephones. With UMTS-capable portable telephones, the same instrument can establish radio connections to base stations inside premises, or any compatible outside mobile telephone service base station. Cordless telephone, cellular, PCS, and UMTS operations are addressed in Chapters 1, 4, and 16.

The top portion of Figure P3.2 illustrates both switched and non-switched transport services. Although for simplicity only two central offices are shown in the figure, in general switched transport service supports local calling among the many COs in large metropolitan areas. The *non-switched* transport service illustrated essentially extends *private or dedicated line* service originating at premises "A", to a CO switch "C" line-termination card, providing foreign-exchange service to Premises "A."

Figure P3.3 expands access and transport service notions to include both IXC service and PBX-based premises services. The right- and left-hand portions of the figure are mirror images that together enable readers to trace circuits and facilities from one business premises to another by which both LECs and IXCs deliver shared and dedicated access and transport services. In the bottom part of the figure, user telephones, PDS wiring, and PBX line termination cards comprise *access service* to on-premises PBXs. For office-to-office calls, PBXs provide in-house *switched transport* service.

For office-to-outside building/campus calls, PBXs provide connections to LEC NIs and LEC access services. In these cases LEC-shared (switched) access service is similar to Centrex-based service except that special LEC-PBX DID, DOD, and two-way trunk-termination cards are provided at the CO switch. Such connections afford PBX-equipped customers access to LEC service area transport and other services. (Direct Inward Dialing [DID], Direct Outward Dialing [DOD], and two-way services are explained further in Chapter 10).

The figure also shows how LECs provision dedicated access PBX-to-IXC and PBX-to-remote PBX private-line trunks. Dedicated access PBX-to-IXC trunks pass through LEC CO and tandem switch offices' cross-connect distribution frames (on a non-switched basis) directly to IXC shared access tandem switch PBX trunk-termination cards, from which IXCs render long-distance switched transport and other services. Special service-dedicated access PBX-to-remote PBX trunks similarly pass through LEC CO and tandem switch offices' cross-connect distribution frames (on a non-switched basis) directly to IXC cross-connect frames, and then through IXC dedicated (non-switched) long distance transport facilities.

When justified by high traffic levels, PBXs with tandem switching capabilities and interconnected by dedicated lines can be configured as *electronic switched networks* (ESNs) supporting *private switched transport* service among remote business locations. Additional facility-based private ESN details are found in Chapter 10. Again, although special service circuits are typically terminated at carrier office distribution frame cross-connect facilities and not afforded call-by-call switched service, carriers are free to use *nailed-up* switch connections in provisioning dedicated private line service.

The upper portions of the figure demonstrate how IXC networks obtain access to LEC networks. As shown in the figure and explained in Chapter 2, to comply with MFJ rules, LECs use traffic concentration and distribution *access tandem* switches to provide all IXCs *equal access* service. A *point-of-presence* (POP) is a physical location within a LATA at which an IXC establishes itself for the purpose of obtaining LATA access and from which a LEC provides access services to its intra-LATA networks. An IXC may have more than one POP within a LATA, and, as noted, POPs may support public and private, switched and non-switched services. In the future, even if literal LATA-based definitions and rules are abandoned, IXC and LEC facility arrangements similar to those depicted in the figure will continue to exist within LEC service areas.

Note that ordinary PSTN shared access LEC subscriber "dial 1" long distance traffic also terminates on IXC tandem switch trunk termination cards from which full SS7 call processing and other features are available.

As alluded to above, beyond basic voice network services is an almost endless list of advanced intelligent network (AIN) special and programmable services and features that add value to basic services. The grand scale of today's service and feature options is made possible by powerful network-embedded and standalone computers, the ingenious and expansive capabilities of SS7, and modern network management and control systems. AIN operational concepts and many Centrex, PBX, KTS, and SS7-based features are covered in Chapters 6 and 11.

# Premises Network Services

Chapters 9 and 10 furnish service, facility, and technology details of the voice services outlined in the taxonomy just presented. Using that taxonomy's organization, Figure 9.1 serves as a roadmap of *premises network services* treated in this chapter. The two major Chapter 9 voice network services subdivisions are premises access and premises transport services.

**Figure 9.1**

Taxonomy of voice premises network services.

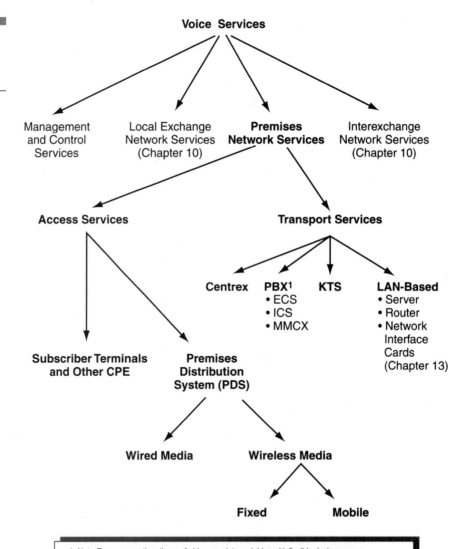

1  Note: To overcome the stigma of older proprietary, rigid, and inflexible designs, some manufacturers are abandoning "PBX" descriptors. Lucent, for example, has named two of their products with advanced PBX functionality ECS, Enterprise Communications System, and MMCX for Multimedia Communications Exchange. Nortel has dubbed one of their products ICS, which stands for Integrated Communications Server.

# Premises Access Services

Facilities underlying business and home *premises access services* comprise telephones and other voice-network-compatible origination/termination equipment (such as answering machines, voice-response, and facsimile devices), and a premises distribution system (PDS). Telephones and other terminal equipment are described in Chapters 3, 6, and 16. Premises distribution systems are discussed next.

## Premises Distribution Systems

*Premises distribution systems* range in complexity from simple residential telephone wiring to multimedia skyscraper installations serving thousands of offices. As noted in Chapter 4, premises distribution systems may be wired, wireless, or hybrid transmission networks, inside buildings or among buildings on a campus. These networks connect desktop and other terminal equipment with common host equipment (e.g., switches, computers, and building automation systems), and to external telecommunications networks at network interfaces. The top-level diagram in Figure 9.2 depicts major PDS interfaces. The reason why both voice and data equipment are included in this figure (in a section focusing on voice services) is explained below. PDS is used here as a generic term, even though AT&T at one time used it to describe a specific product offering.

**Figure 9.2**
Premises distribution system relation to other facilities.

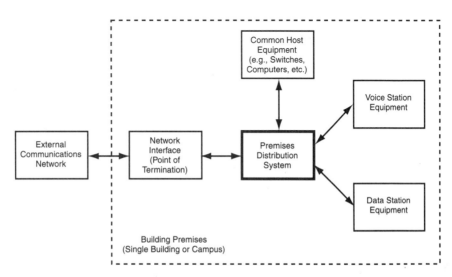

### EIA/TIA 568-A STRUCTURED CABLING STANDARDS AND DEFINITIONS

Because a modern PDS represents between 15 and 50 percent of total premises telecommunication system costs, it is commanding a growing share of management attention. For new buildings, costs are minimized with detailed design and planning and PDS installation during building construction. PDS upgrades in existing buildings are normally more expensive, and in some cases (such as in historical buildings), prohibitively so.

As a consequence, a prime PDS objective is to provide uniform wiring capable of supporting voice and data in multiple product/vendor environments, thus minimizing the need to modify wiring as user/tenant requirements change or when technology upgrades are justified. In the past, this has been nearly impossible. At one time, for example, IBM alone used over 50 different types of cabling for its product lines. Recent innovative developments now permit the use of a single type of unshielded twisted pair (UTP) copper wire (versions of which have been used in telephone service for years), for premises voice and data transmission. Objectives are to achieve useful PDS life cycles of 10 to 20 years.

To meet these objectives, in July 1991 the Electronic Industries Association[1] (EIA) and the Telecommunications Industry Association (TIA) published the Commercial Building Telecommunications Wiring Standards (EIA/TIA 568 and 569) to enable planning and installation of building wiring with little knowledge of the telecommunication products that will be installed. These standards define the following PDS components illustrated in Figure 9.3 (which, for purposes of example, includes a PBX):

■ **Horizontal cabling**—The connection between telecommunications outlets in work areas and telecommunications closets. It includes the wall-mounted outlets, horizontal cabling itself, and mechanical terminations and patch cords (or jumpers) that comprise horizontal cross-connects mounted in telecommunications closets. EIA/TIA 568A specifies a "star" cabling topology, that is, dedicated connections from a horizontal cross-connect radiating out to each individual work area outlet.

---

1 In March 1998 EIA, the Electronic Industries Association changed its name to Electronic Industries Alliance.

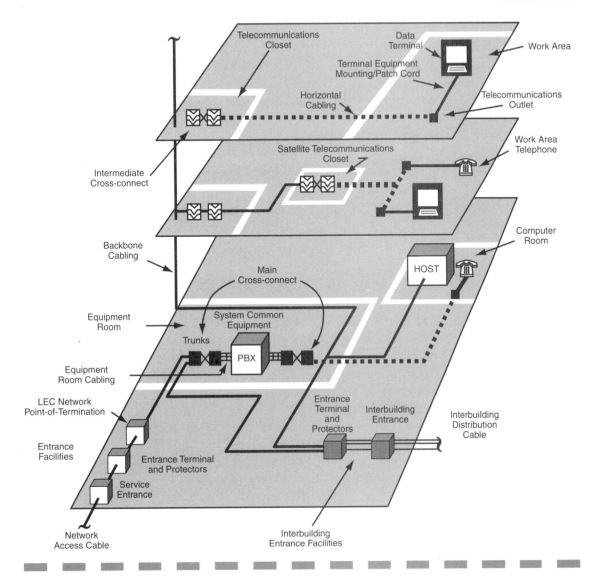

**Figure 9.3** Premises distribution system components.

- **Backbone cabling**—The connection between telecommunications closets and equipment rooms and entrance facilities within a building, as well as between buildings in a campus. In the past, because backbone wiring includes cabling between floors, it was also referred to as "vertical wiring" or the "riser subsystem." EIA/TIA-568A specifications apply to backbone cables, intermediate and main cross-connects, and backbone-to-backbone cross-connection patch cords or jumpers.

- **Work area**—An area containing telephones, workstations, or other terminal equipment. The WA includes mounting or patch cords connecting terminal equipment and telecommunications (information) outlets.

- **Telecommunications closet**—An area for connecting horizontal and backbone cabling and containing cross-connects and other active or passive equipment. Typically, at least one TC location is allocated per building floor.

- **Equipment room**—A special purpose room(s) with access to the backbone wiring for housing telecommunications (PBXs, routers, servers, hubs, etc.), data processing, security, and safety alarm equipment.

- **Entrance facilities**—The point of interconnection between the building wiring system and external telecommunications facilities (LEC networks, other buildings, etc.). Telecordia defines interfaces with LEC networks as points of termination (POTs)/network interfaces (NIs). EIA/TIA-607, Commercial Building Grounding and Bonding Requirements for Telecommunications, specifies telecommunications main grounding busbar (TMGB), telecommunications grounding busbar (TGB), and telecommunications bonding backbone (TBB) sizing and bonding. Bonding ensures low-resistance metal part connections for direct current and low-frequency electric currents. There are two types of grounding, earth and equipment grounding. Because experts maintain that improper grounding is a major cause of telecommunications problems, EIA/TIA has seen fit to devote an entire specification to this important design and installation activity.

- **Administration**—The requirements for specifying labeling, identification, documentation, etc., for the entire telecommunications cabling system. Documentation includes creating system-level documents needed to manage and maintain end-to-end connectivity. EIA/TIA-606, an Administration Standard for Telecommunications Infrastructures of Commercial Buildings, specifies uniform labeling, color coding, and recording data for the administration of pathways/spaces, cabling (media, terminations, splices), and grounding/bonding. A number of PC-based PDS management systems are available that include highly automated building-cabling drawing packages using graphical-object-oriented relational databases.

Figure 9.3 places main cross-connects in the equipment room, and intermediate cross-connects in telecommunications closets. A *cross-connect* is used to connect and administer communications circuits. In a cross-connect, jumper wires or patch cords are used to make circuit connections between horizontal and backbone cable segments.

Older punch-down "66-type" models use only jumper wires and require special tools and training. These are still the most popular cross-connect systems. More recent, 110-type termination systems, as well as modular plug-and-jack types, use patch cords with attached plugs. These provide flexible circuit labeling arrangements, and can be easily changed and administered by non-technical personnel.

Typically, for a building with 1,000 work areas, the main cross-connect involves tens of thousands of wire connections. Many organizations require *moves, adds, and changes* (MACs) of station equipment at 50 percent of their work areas per year. Such high churn rates and the need to accommodate growth and equipment upgrades makes automated PDS management systems essential for most medium and large installations.

To convey an appreciation for the manpower-intensive nature of wire management, consider this example. In 1984 AT&T employed over 20,000 technicians just to operate distribution frames (cross-connects in telephone buildings).

Cross-connects are arranged *hierarchically*; that is, main cross-connects feed several intermediate cross-connects, as shown in Figure 9.3. Special cross-connects are available for coaxial and fiber optic cable. Both these types of cable offer greater bandwidths over longer distances than twisted-pair wire and find application for high-speed data, wide-bandwidth video (e.g., cable TV), and multichannel (multiplexed) voice applications.

Fiber optic cables (lightguides) are direct replacements for coaxial and twisted-pair wiring. At present they are more expensive per connection but do find PDS application in multichannel backbone wiring, where cost can be spread over large numbers of circuits. Fiber has significant advantages for backbone wiring, including inter-building connections between central and remote switch modules, as well as for long runs in large, multi-floor single buildings. With underground duct bank construction costs in the range of $100 per foot, a major advantage accrues when two strands of fiber displace copper pairs serving 500 or more voice channels.

## CABLE, CONNECTOR, AND OUTLET SPECIFICATIONS

To comprehend the magnitude of the benefits of using standards-based cabling one need only imagine the chaotic situation that would exist if every electrical appliance had unique plug, outlet, and power wiring requirements. The order that EIA/TIA structured cabling now makes possible is in no small part due to specification of a relatively limited number of affordable, standard cable-media type connectors and outlets.

Figure 9.4 depicts EIA/TIA-recognized cabling media and Figure 9.5 summarizes approved cable and plug-and-jack (outlet) connector arrangement options and their relation to defined building spaces. The EIA/TIA 568A standard recommends two telecommunications outlets for each work area, as shown in the top, right-hand part of Figure 9.5. Telecommunications outlets are defined by the following:

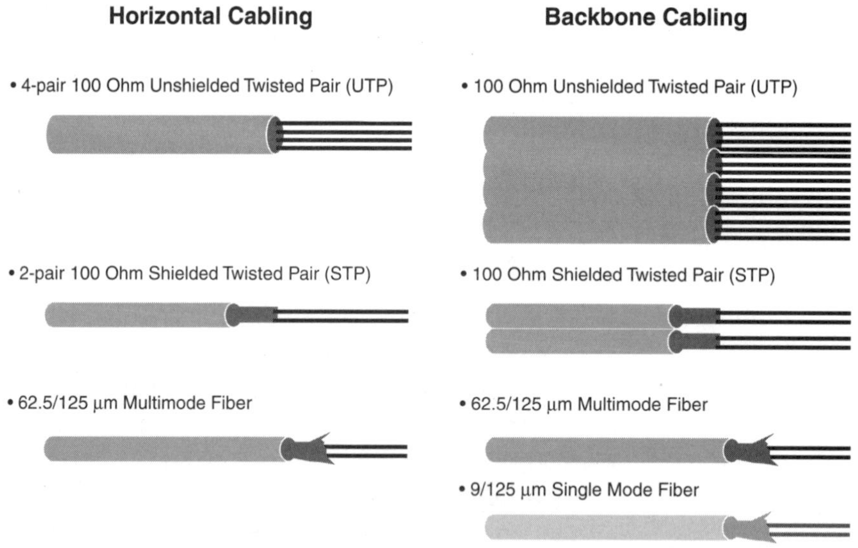

**Figure 9.4**
*EIA/TIA recognized cabling media.*

**Horizontal Cabling**

• 4-pair 100 Ohm Unshielded Twisted Pair (UTP)

• 2-pair 100 Ohm Shielded Twisted Pair (STP)

• 62.5/125 µm Multimode Fiber

**Backbone Cabling**

• 100 Ohm Unshielded Twisted Pair (UTP)

• 100 Ohm Shielded Twisted Pair (STP)

• 62.5/125 µm Multimode Fiber

• 9/125 µm Single Mode Fiber

- One outlet shall be supported by four-pair unshielded twisted-pair (UTP) cable. This medium meets Institute of Electrical and Electronic Engineers IEEE 802.3 10BaseT and other related technical specifications. IEEE 802.3 10BaseT specifies a class of standalone data local area networks (LANs) supporting 10 Mb/s data rates. Additional details and other 100/1000BaseT (100/1000 Mb/s) variants are discussed later in this section and in Chapter 13. (Adapting UTP for high bit rates is also treated in the digital subscriber loop sections of Chapter 4). Incorporating these specifications into EIA/TIA stan-

**Figure 9.5**
PDS backbone, horizontal, and work area cabling interconnection.

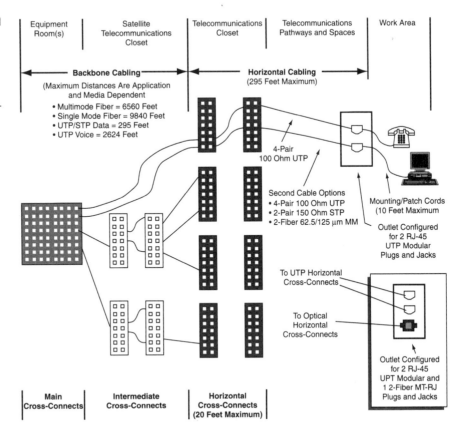

| Equipment Room(s) | Satellite Telecommunications Closet | Telecommunications Closet | Telecommunications Pathways and Spaces | Work Area |

**Backbone Cabling**
(Maximum Distances Are Application and Media Dependent)
• Multimode Fiber = 6560 Feet
• Single Mode Fiber = 9840 Feet
• UTP/STP Data = 295 Feet
• UTP Voice = 2624 Feet

**Horizontal Cabling**
(295 Feet Maximum)

4-Pair
100 Ohm UTP

Second Cable Options
• 4-Pair 100 Ohm UTP
• 2-Pair 150 Ohm STP
• 2-Fiber 62.5/125 μm MM

Mounting/Patch Cords
(10 Feet Maximum)

Outlet Configured
for 2 RJ-45
UTP Modular
Plugs and Jacks

To UTP Horizontal
Cross-Connects

To Optical
Horizontal
Cross-Connects

Outlet Configured
for 2 RJ-45
UPT Modular and
1 2-Fiber MT-RJ
Plugs and Jacks

| Main Cross-Connects | Intermediate Cross-Connects | Horizontal Cross-Connects (20 Feet Maximum) |

dards makes possible the use of a single telecommunications outlet for either voice or data, a significant development.

- The second outlet can be supported by either the four-pair UTP cable, by two-pair shielded twisted-pair (STP) cable, or by coaxial cable. The latter cables are specified by IEEE 802.5 and 802.3 10Base2 specifications and correspond to other classes of data LANs. Although 50-ohm coaxial cable is recognized in current specifications, it is not recommended for new installations and is expected to be removed in EIA/TIA 568A revisions.

Recommended plugs and jacks for terminating UTP cable are of the *modular RJ-XX* variety where XX are numbers identifying particular designs. RJ-series modular plugs and jacks are really an outgrowth from the 1968 Carterfone decision in which the FCC determined that Carterfone and other telephone devices, not manufactured or supplied by telephone companies, could be connected to the public network. Following this, by the mid-1970s, it became clear that what is needed is a simple means by which technically untrained customers can connect telephone devices—purchased at phone-center stores—to telephone networks (without a qualified telephone company technician residence visit).

The result is the RJ-series, a series of *registered jacks*—registered, that is, under Part 68 of FCC Rules. RJ-11, a 6-pin (or connection point) plug and jack is now the most popular in the world being used on virtually all telephone instruments, modems, and other devices designed to be connected to ubiquitous worldwide voice-telephone networks. EIA/TIA 568A specifies 8-pin RJ-45s with pair-to-pin assignments as shown in Figure 9.6. Cleverly, RJ-11s and RJ-45s are designed so that RJ-11 plugs attached to most telephones can be directly connected to RJ-45 outlet jacks or sockets. Thus wiring an office for EIA/TIA recommended dual RJ-45 wall outlets does not mandate changing all RJ-11-equipped device plugs, or even the use of conversion cord devices.

**Figure 9.6**
EIA/TIA 568A RJ-45
connector
configuration.

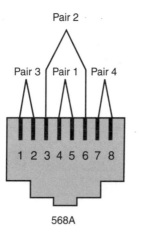

568A

**Pair-to-Pin Assignments**

| | | | |
|---|---|---|---|
| Pair 1 | Pins 4 and 5 | White Blue/Blue | Voice |
| Pair 2 | Pins 3 and 6 | White Orange/Orange | Data |
| Pair 3 | Pins 1 and 2 | White Green/Green | Data |
| Pair 4 | Pins 7 and 8 | White Brown/Brown | Voice |

### UTP "N"BaseT CABLE CLASSIFICATION CATEGORIES

In late 1991, the EIA/TIA issued Technical Systems Bulletin (TSB) 36, which defined transmission performance standards for UTP cable up to 100 MHz. This major milestone has galvanized the UTP cable industry, providing a potent weapon in industry attempts to delay the inroads of fiber to the desktop. Overall, EIA/TIA has defined five levels or categories of cable performance, and is developing new Category 6 and 7 specifications. Category 3, a cable quality met by most existing premises UTP installations in the United States and originally intended for voice service only, is capable of supporting 10BaseT operation within cable-run distance limitations shown on Figure 9.5.

EIA/TIA's Category 5 standard, issued in 1995, imposes stricter performance-degrading parameter requirements and supports 100BaseT and higher speed operation. Figure 9.7 illustrates the run-length versus data-rate capability associated with Category 3, 4, and 5 cables. It is important to note that EIA/TIA, ISO, or other UTP performance standards have nothing to do with cable manufacturing processes or materials. It is incumbent upon manufacturers who make those design decisions to ensure product "category compliance," normally accomplished via certification testing. Also, other than free-market selection or rejection forces, there are no means to take punitive action against manufacturers' misrepresenting their products. Thus, it is always advisable to include post-installation verification testing as a formal contract requirement in any PDS procurement.

**Figure 9.7**
Unshielded twisted pair (UTP) data rate capabilities.

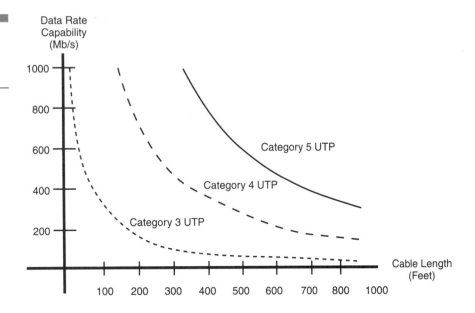

Both business managers and technical staff should possess at least a rudimentary understanding of cable performance-limiting factors when purchasing a PDS. One such factor is attenuation or the power loss an electrical signal incurs as it traverses a cable from transmitter to receiver. The decrease happens as a result of absorption, reflection, diffusion, scattering, deflection, or dispersion processes discussed in Chapters 3 and 4. EIA/TIA 568A limits attenuation to 24 dB for a 100

MHz signal. A 20-dB loss means that received signals are just 1/100th of transmitted signal power levels, illustrating how weak received signals can be relative to transmitted levels. Note that in UTP installations, transmitted signals are introduced on one pair and intended for reception and detection only on that pair. Unfortunately, it often happens that traces of signals introduced on one pair can be detected on other pairs in the same sheath or bundle.

Chapter 4 introduced the notion of *crosstalk*, in this case, the undesired signal energy radiated from one UTP pair, coupling to and interfering with desired signals carried in adjacent pairs in the same cable bundle. In particular, Chapter 4 identified near-end crosstalk (NEXT) as the key crosstalk specification. When both strong locally transmitted signals and weak received signals from remote transmitters appear at cable-bundle ends, the degrading impact of crosstalk is greatest.

While NEXT is a dominant factor, when an attempt is made to squeeze the highest data rates from UTP, both *far-end crosstalk* (FEXT) and *return loss* become critical. FEXT is defined as crosstalk interference at the far end from a transmitter (intended receiver locations) generated by other remote or distant transmitter signals. Return loss is the amount of transmitter signal power reflected back to transmitters caused by mismatches between transmitter and cable-network impedance. A thorough explanation of the impact of *impedance mismatching* is beyond the scope of this book, but it is sufficient for our purposes to simply understand that power reflected back to a transmitter cannot be used by receivers. Thus, to the extent it exists, it degrades link performance.

Other cable-performance parameter terms that appear in the literature are *Equal Level Far-End Crosstalk* (ELFEXT) and *Power Sum* ELFEXT. ELFEXT measures the interference resulting from a single remote transmitter, transmitting at a power level equal to that of the desired signal's transmitter power. PSELFEXT sums the total interference power resulting from all adjacent cable pairs.

Return loss and FEXT have minimal effect when Category 5 cable is used for 10BaseT signals but have a significant impact on 100BaseT (Fast Ethernet—see Chapter 13) and 1000BaseT signals. As mentioned, field testing for return loss, ELFEXT, and PSELFEXT before accepting delivery of or attempting to use a PDS installation for 100—1000 Mb/s signals, is highly recommended. EIA/TIA TSB-67 specifies basic PDS testing procedures and requirements. Specifications for return loss, ELFEXT, and PSELFEXT is slated for publication in TSB-95.

What makes possible 1,000 Mb/s data rates over even high quality UTP is the application of efficient modulation/coding techniques.

Moreover, practical, low-power, small-size, efficient modulation/coding devices in turn depend upon the availability and simultaneous application of both sophisticated digital signal processing and very high-speed and high-density solid-state integrated-circuit technologies. Just a decade ago, the price-performance levels of today's complex modulation/encoding devices (found in DSL, set-top cable modem, PDS, mobile telephone, and other products)—now considered "commodities"—would have been beyond reach.

### INSTALLATION PRACTICES AND FIRE SAFETY

Beyond selecting UTP cable with appropriate electrical characteristics as described above, ensuring installed performance demands proper installation. Returning to Figure 9.5, based on the use of EIA/TIA compliant cabling, the figure indicates maximum cable run distances, including work-area mounting/patch cords and cross-connect jumpers. But perhaps the most sensitive performance-determining UTP cable attribute is *pair-twist* uniformity and tightness. In one of the first installation tasks, cables are pulled from telecommunications closets to work-area outlets. EIA/TIA 568-A specifies a maximum pulling tension of 25 pound-feet to prevent stretched cable segments that tend to untwist pairs.

In running cable through walls, concealed overhead and under-floor air-handling spaces (plenums), conduits, raceways, etc., corners, and bends are bound to be necessary. Again EIA/TIA 568-A specifies that for four-pair Category 5 cable, the bend radius for each turn must be greater than one inch. Category 5 cables with more than four pairs must follow a bend radius that exceeds ten times the outside diameter of the cable. Like excessive pulling tension, tight bends tend to flatten or untwist pairs. Aberrations in twist uniformity cause impedance discontinuities that increase return loss—and other degrading effects. Flattened twists among adjacent pairs also tend to increase NEXT and FEXT crosstalk coupling.

One of the last installation steps involves stripping away jacketing material and making metallic cable-to-plug/jack connections. Here EIA/TIA 568-A specifies that Category 5 cable should never be untwisted more than 1/2 inch from points of termination. Excessive untwisting once more increases crosstalk and susceptibility to EMI/RFI (electromagnetic and radio frequency interference commonly encountered when telecommunications cabling is in close proximity to medical, air conditioning, or other high power equipment).

A final important cable installation issue is fire safety. Cable manufacturers typically offer cables with at least four pairs in a common jacket sheath of either PVC (polyvinyl chloride, a tough water- and flame-resistant thermoplastic insulation material), or perfluoropolymer plenum-rated construction. National Fire Protection Association (NFPA) standards require that only low-smoke, *noncombustible* (e.g., steel, glass, and concrete) or *low-combustible* materials be used in plenums. However, frequent recabling—driven by the rapid growth in intra-building LAN data communications requirements—can result in an accumulation of higher fuel-load, combustible material in plenums.

For over 20 years, most plenum-rated cables used 100 percent perfluoropolymer insulation for cable cores (such as fluorinated ethylene propylene [FEP]) that meet fuel-load requirements for limited combustible materials. Now, changes in cable construction and the use of combustible materials is dramatically increasing fuel-load contribution from plenum cables. Fuel load is extremely important since it contributes to the growth rate, size, and the ultimate destructive power of a fire.

According to a Dupont white paper[2], "Fast growing fires are the most dangerous kind overall. Also, fires with high fuel-load combustibles in concealed spaces are especially difficult for firefighters to control as evidenced by several catastrophic fires during 1996, including one at the Dusseldorf Airport involving multiple fatalities. Current flame-retardant polyolefin insulation materials sometimes used as a substitute for 100 percent FEP in CMP-listed (common plenum) cables can have 400 to 600 percent higher fuel-load than FEP. Non-flame retardant polyolefins can have 800 percent higher fuel-loads." An added advantage of FEP (trade name Teflon®) is that it is smooth and slippery, a quality that makes it ideal for cable pulling.

Strong industry support for UTP-based structured wiring is indicated by the number of standards organizations contributing to its development. Besides EIA, TIA, and NFPA, participants include ANSI, CSA (Canadian Standards Association); IEC, (International Electrotechnic Commission); ISO (International Organization for Standardization); and UL (Underwriters Laboratories, Inc.) EIA/TIA-compliant cables are manufactured by Berk-Tek, Belden, Lucent and others. "N"BaseT-compatible adapter card, hub, cross-connect, and other hardware are offered by 3COM, Cisco, Compaq, Hewlett-Packard, Intel, Standard

---

2 Dupont white paper number H-74278, published in March 1997.

MicroSystems, Ascenté, CNet Technology, Netopia, Linksys, Matrox Networks, and many others.

### FIBER OPTIC AND OTHER CABLES

Neither shielded nor coaxial cables are appropriate for voice applications, and their use is not consistent with product and vendor independence objectives of uniform wiring. However, solutions in the form of adapters called *baluns*[3] have been developed which permit carrying coaxial cable-based IEEE 802.5 and 802.3 10Base2 LAN traffic over standard four-pair UTP wiring. What this means is that older installations designed for coaxial cable can be adapted to UTP by using baluns. Since EIA/TIA is phasing out the inclusion of coaxial cable in its specifications, no additional discussion of its properties is provided here.

Optical fiber cable is an optional EIA/TIA horizontal cabling medium. This is putting fiber optic capabilities into work areas for high-speed Computer Aided Design (CAD)/Computer Aided Manufacturing (CAM), high resolution graphics, video, and other wideband PC terminal applications. For this reason, AT&T introduced a *composite cable* constructed with two four-pair, 24-gauge, UTP cables, combined with two multimode fiber cables and matching work-area outlets. The inset in Figure 9.5 portrays an outlet accommodating two RJ-45 UTP jacks and a single MT-RJ fiber optic jack that connects two fiber strands.

Costs of installing fiber cable are declining. Replacing complex and field labor-intensive epoxy-based and strand-end polishing splicing requirements, new small form factor connector (SFFC) approaches reduce single fiber connector assembly times to less than one minute and accomplish jacketed cable assembly in just three minutes. Vertical cavity surface emitting lasers (VCSELs) are now furnishing laser power and speed at yesterday's lower performance LED prices. Other enhancements in the realm of higher fiber-cable glass quality, lower-cost electronics, and techniques to offset signal degradation due to

---

3 A balun (BALanced/UNbalanced) is a small passive device that connects a balanced line, such as a twisted pair, to an unbalanced line, such as coaxial cable. A balun permits UTP building cabling to be used to interconnect equipment originally designed for coaxial interconnection. A balanced line is one in which voltages on its two conductors are equal with respect to ground. With UTP, ideally both conductors exhibit equal ground-to-conductor voltages. By contrast, coaxial cable outer metallic sheaths are generally connected to equipment or earth grounds, so that traffic signals on the center conductor are measured relative to ground.

differential mode delay (DMD) are accelerating the pace toward cost equivalency among metallic-wire and fiber-cable installation options.

Fiber-cable manufacturers include Alcatel, Siecor, Lucent, AMP, 3M, and General Photonics. Manufacturers of fiber-cable network-related electronics include 3Com, Cabletron, Cisco, Extreme Networks, Fore Systems, Foundry Networks, Hewlett-Packard, IBM, Lanart, Lucent DNS, Lucent Microelectronics, Methode, Nortel Networks, Polycor, Transition Networks, and XLNT.

### WIRELESS PREMISES DISTRIBUTION SYSTEMS

Wireless (radio or lightwave-based) approaches exist to replace or supplement cabled premises distribution systems. Introduced in Chapter 4 as home or pico-cell examples of fixed wireless transmission systems, *wireless premises distribution systems* (WPDSs) have arisen principally from two initiatives: cordless telephone and *wireless local area network* (WLAN) developments.

In home and office buildings, cordless telephones communicate with one or more base stations connected directly to telco network interfaces or indirectly via PBX/KTS equipment. Cordless telephones have progressed from early analog models—operating in the FCC allocated 50 MHz portion of the radio frequency spectrum—to digital models operating in the 900 MHz and the Industrial Medical Scientific (ISM) 2.4 GHz bands. None of these allocations require FCC licensing and the 2.4–2.5 GHz bands are license exempt in Europe and Japan as well.

Chapter 4 identifies example applications and vendor products that combine both cable-based and wireless PDS benefits. Chapter 6 describes a Siemens feature-laden, completely wireless KTS product. Recall also Chapter 4's description of *universal mobile telephone system* (UMTS) initiatives to integrate indoor and outdoor mobile service in single instruments. Most home cordless telephone products handle voice traffic only, although the European Digital Enhanced Cordless Telecommunications (DECT) standard includes a 1.152-Mb/s data-rate specification.

For data and multimedia traffic, proprietary WLAN products—operating in the 900-MHz, 2, 4, and 5-GHz bands—have been available since the early 1990s. In June 1997, IEEE established 802.11, a standard specifying medium-access control procedures (MAC is a subset of a WLAN Layer 2 data-link protocol—see Chapter 7 for definitions) and two radio-based and one infrared PHYs (physical Layer 1 protocols) supporting either 1- or 2-Mb/s data rates. In July 1998, IEEE approved a wireless PHY standard for 11-Mb/s operations, with a fallback rate of 5.5 Mb/s.

Another IEEE working group is developing a different PHY to handle higher data rates. This development follows an amendment to Part 15 FCC rules that makes available 300 MHz of bandwidth in the 5-GHz part of the spectrum, an allocation selected for use by a new *unlicensed national infrastructure* (UNII) category of equipment. This new standard targets data rates in the 6- to 54-Mb/s range. With that upper boundary, conforming WLANs would be able to carry the SONET 51.84-Mb/s synchronous digital hierarchy STS-1/OCn-1 (Synchronous Transport Signal/Optical Carrier Level) signals described in Chapter 5 for use in broadband fiber optic transmissions systems. Figure 9.8 summarizes IEEE WLAN standards.

**Figure 9.8**
IEEE wireless LAN standards.

| Area | Spectrum | Standard | Data Rate (Mb/s) | Modulation |
|------|----------|----------|------------------|------------|
| North America | 2.4–2.4835 GHz | IEEE 802.11 | 1, 2 | BPSK/QPSK |
| North America | 2.4–2.4835 GHz | IEEE 802.11a | 1, 2, 5.5, 11 | BPSK/QPSK |
| | | | | |
| United States (UNII Lower Band) | 5.150–5.250 GHz | IEEE 802.11b | 1, 2, 5.5, 11 | OFDM |
| United States (UNII Middle Band) | 5.250–5.350 GHz | IEEE 802.11b | 1, 2, 5.5, 11 | OFDM |
| United States (UNII Upper Band) | 5.725–5.825 GHz | IEEE 802.11b | 1, 2, 5.5, 11 | OFDM |
| OFDM = Orthogonal Frequency Division Multiplexing<br>UNI = Unlicensed National Information Infrastructure | | | | |

IEEE specifies *orthogonal frequency division multiplexing* (OFDM) as the basis for its new standard. A form of discrete multitone modulation (DMT) explained in Chapter 4, OFDM splits high-rate data streams into a number of lower-rate streams transmitted simultaneously over a number of subcarriers (52 in this case). Unlike DMT, in this case each carrier frequency differs from all others by an integer number. This orthogonality property is particularly useful in mitigating multipath interference, which occurs when signals, traveling over multiple paths of differing loss and length, arrive at receivers and either reinforce or cancel one another.

Just as received broadcast TV signals often consist of direct and reflected signal components (reflected perhaps by buildings, airplanes, etc.) causing "ghosts" or reducing received power levels, multiple WLAN signals reflecting off walls, floors, metal framing, and other building appurtenances, traveling different paths toward receivers often result in similar signal degradation. OFDM, previously used primarily in digital video and other non-packet-switched applications, has been adopted by the European Telecommunications Standards Institute (ESTI) and the Japanese Ministry of Post and Telecommunications for application in their WLAN standards.

Related ETSI standards include HIPERLAN Type 1 and HIPERLAN Type 2, a standard the newly formed ETSI working group, Broadband Radio Access Networks (BRAN) is developing. HIPERLAN Type 2 is an indoor wireless LAN specification incorporating QoS provisions. Recall from Chapters 1, 5, and 7 that when applied to packet-switched networks and Voice-Over-Internet Protocols (VoIP), QoS normally refers to special design features ensuring compliance with strict latency and delay requirements demanded by voice and other isochronous traffic. So, in developing HIPERLAN 2, BRAN anticipates users who may want to combine voice and other multimedia traffic with the data traffic that most LANs are designed to handle.

Wireless LAN product manufacturers include Lucent, Apple, Netgear (a subsidiary of Nortel), Harris (integrated circuit chips), Home Wireless Networks, ShareWave Inc., and Kyushi Matsushita Electric Co.

The article "New High-Rate Wireless LAN Standards" by Richard van Nee, et al, published in the December 1999 issue of the *IEEE Communications Magazine*, presents a comprehensive compilation of worldwide WLAN developments.

## Typical PDS Costs

Figure 9.9 illustrates how PDS costs compare to per-line PBX costs. The costing information corresponds to a feature-rich, multi-building, high-technology industrial center campus. In this program, buildings are to be added over a period of time, so vendor responses were requested to show the decrease in per-line costs as the system grew from 100 lines to over 5,000 lines.

Unlike charges for switching, voice mail, administrative, and other services, and features that benefit from economies of scale, the per-line cost for PDS, being largely manpower driven, remains relatively constant as more lines are added.

The PDS costs shown in the figure correspond to the following horizontal-wiring and work-area options:

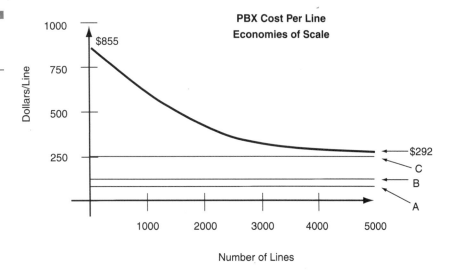

**Figure 9.9**
Comparison of PDS and other costs.

| PDS Configurations and Costs* | | | |
| --- | --- | --- | --- |
| | Fiber Outlets | Copper Wire RJ-45 Outlets | PDS $ Per Workstation |
| A | 0 | 1 | $ 93 |
| B | 0 | 2 | $150 |
| C | 2 | 2 | $268 |

*Average telecommunications closet-to-work area run length of 295 feet.

**A** One RJ-45 jack and no fiber outlets

**B** Two RJ-45 jacks and no fiber outlets

**C** Two RJ-45 jacks and two terminated fiber outlets

The fiber option relates to factory or distributor pre-cut and polished fiber cable and the use of low-work (quick mount) factor connectors. PDS cost calculations are based on an average telecommunications closet-to-work-area run length of 295 feet.

For large systems and EIA/TIA-recommended dual modular, eight-conductor jack configurations, PDS costs are about 50 percent of other costs. For the fiber option, PDS costs are nearly equal to other costs. These results alone dispute the belief that wiring is the "low-tech" part of telecommunications business systems, and therefore of lesser importance. Analyses treating operations, administration, and maintenance life-cycle costs lead to similar conclusions.

Reflecting the magnitude of life-cycle PDS costs, a number of cable management system products have reached the market. These products generally incorporate computer-aided design (CAD) capabilities and support all PDS activities from design and installation through day-to-day MAC and configuration management. The more comprehensive products involve interactive graphics and relational database tools that provide automated inventory tracking for all voice, data, video, and PDS assets, initial and on-going documentation and report generation, troubleshooting, trouble ticket and work order generation, other maintenance support capabilities, and help desk functions.

Some products include high-performance graphical drawing capabilities that electronically import building architectural engineering drawings. In these cases one can visually display floorplans and locations of equipment, user terminal outlets, equipment room layouts, cross-connects, conduit systems, power systems, etc. Capabilities also may include storing textual data in databases in a way that directly relates to user-selected graphical objects in engineering drawings. For example, with these products, when one double clicks on a cross-connect icon appearing on a floor plan drawing in a computer-driven display, a textual list of all cables, work areas and work-area equipment connected to that cross-connect might appear automatically.

Industry studies and surveys indicate that cable management systems can reduce costs associated with MACs (which typically cost around $200 each) by as much as 30 percent. For locations with 1,000 MACs per year, cable management system savings for MACs alone can amount to $50,000 per year. Vendors include IMAP Corporation, Exan Technologies Ltd., ISI Inc., and Lucent. Lucent, for example, recommends its Systimax SCS (structured connectivity solution) as an integrated graphics and data-based management system for structured copper or fiber cabled and wireless installations, of up to 20,000 outlets.

## Premises Transport Services

Recalling that network services consist of *access* and *transport* services and that *transport services* are network switching, transmission, and related services that support information transfer between originating and terminating access facilities, note that in Figure 9.3 transport services are furnished by a PBX. For intra-building voice telephone communications, originating and terminating access services are provided by calling and called-party telephones and all the PDS components just described.

Smaller enterprises may use key telephone systems instead of PBXs, and large enterprises may elect to use telephone company-provided Centrex services in lieu of PBX facilities—as explained in Chapter 6. PBX, KTS and Centrex premises voice access and transport service capabilities and features are presented in detail in Chapter 6. The relationship between premises and LEC provided network services is covered in Chapter 12, an overall guide to voice services.

As noted in Chapter 6, because most circuit switches implement 80 percent or more of their functionality in software (as opposed to hardware), and because new Voice Over Internet Protocols (VoIP) are proving viable, packet switched-based alternatives to PBX, KTS, and Centrex service provision are emerging. Note in the PDS discussion above that an assumption implicit in the EIA/TIA structured wiring plan is that separate UTP pairs are allocated to voice and data services. In a VoIP world, PBX, KTS, and Centrex facilities could be replaced by voice-capable, packet switched-based LAN-attached server/routers. In this scenario, voice and data traffic would be mixed and traverse the same UTP pairs or fiber optic strands.

Thus, the historical practice of using embedded computer hardware and proprietary hardware and software designs to implement circuit-switched functions is, at a minimum, being challenged. In their place, some manufacturers are already offering high-performance general-purpose computer hardware (so called "un-PBX-type" products), along with applications written to be executed by Microsoft Windows, NT, or other popular operating system-compatible software. Even if the trend to packet-switched VoIP is slowed by offsetting improvements in high-speed circuit switching, ultimately a single switching-multiplexing technique, capable of handling all traffic types, is likely to emerge. When this happens, the need for separate traffic or service-related access and transport facilities will tend to vanish. Chapters 13 and 14 revisit service integration trends from the data service development point of view.

# Home Premises Distribution Systems

So far, this chapter has focused on PDS systems and technologies with business applications in commercial buildings, multiple-dwelling apartments, condominiums, or hotels. *Home PDS* installations can employ any of the technologies described but actual home PDS designs are often driven by emphasis on different requirements, by more-limited budgets, and by much older legacy wiring—some dating back nearly a century.

With respect to legacy wiring, in the early parts of this century—and perhaps as late as the 1970s—home wiring consisted of simple "loop-wired" telephone cabling, electrical power wiring for lighting and appliances, and some low-voltage wiring for door-bell chimes and heating, ventilating, and air conditioning (HVAC) control.

By *loop wiring* we mean not subscriber loops extending from LECs to residences, but rather the practice of starting intra-home telephone wiring at network interfaces, running it from one jack to another and finally back to the interface. While this satisfies most simple home telephone or voice requirements, it complicates and limits data-service connection performance relative to dedicated pair(s), star-wiring of PCs-to-telecommunications closet cross-connects, or *home-run* (direct PC-to-host computer/server) wiring, now specified in EIA/TIA structured wiring standards.

Although in recent years the installation of two or more residential telco lines has grown dramatically, most dwellings do not have a single unused twisted pair in locations where they are needed for new applications, and retrofit is expensive. Fortunately, ADSL-like modulation schemes are being adapted for relatively high-rate data traffic using legacy "loop wiring" and Category 3-rated, or less, cable and installation procedures.

## Home PDS Requirements and Applications

The left side of Figure 9.10 lists popular applications driving today's home PDS requirements. Connection to external LEC/IXC-provided POTS tops the list, but especially in homes with more than one line or cordless telephones, local intercom, conferencing, paging, and remote door-entry communications are increasingly sought after.

The next class of emerging requirements is for some form of residential computer networking—among multiple intra-building located computers and peripherals—and to the Internet or other external data-service networks. Stimulating vendor interest are the already 17 million United States homes with two or more computers. Chapter 4 describes how advanced digital subscriber line (DSL) techniques are

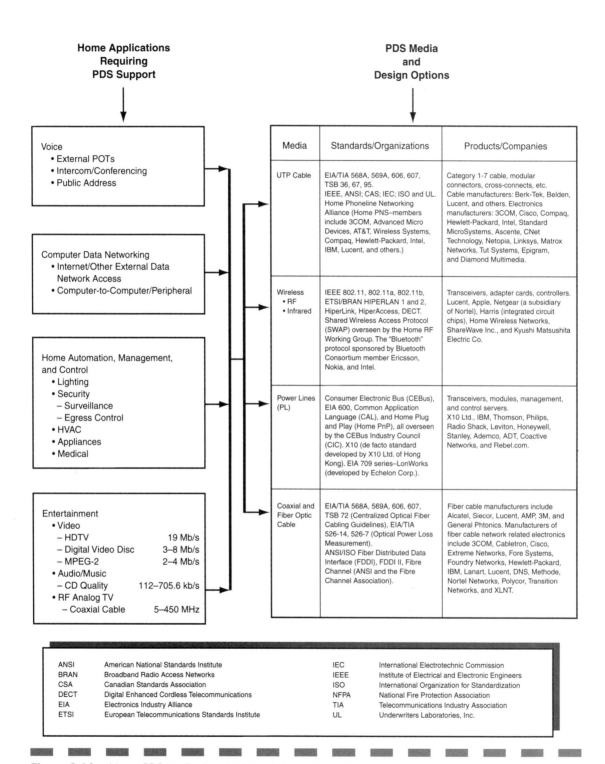

**Home Applications Requiring PDS Support**

Voice
- External POTs
- Intercom/Conferencing
- Public Address

Computer Data Networking
- Internet/Other External Data Network Access
- Computer-to-Computer/Peripheral

Home Automation, Management, and Control
- Lighting
- Security
  - Surveillance
  - Egress Control
- HVAC
- Appliances
- Medical

Entertainment
- Video
  - HDTV      19 Mb/s
  - Digital Video Disc    3–8 Mb/s
  - MPEG-2     2–4 Mb/s
- Audio/Music
  - CD Quality    112–705.6 kb/s
- RF Analog TV
  - Coaxial Cable    5–450 MHz

**PDS Media and Design Options**

| Media | Standards/Organizations | Products/Companies |
|---|---|---|
| UTP Cable | EIA/TIA 568A, 569A, 606, 607, TSB 36, 67, 95. IEEE, ANSI; CAS; IEC; ISO and UL. Home Phoneline Networking Alliance (Home PNS–members include 3COM, Advanced Micro Devices, AT&T, Wireless Systems, Compaq, Hewlett-Packard, Intel, IBM, Lucent, and others.) | Category 1-7 cable, modular connectors, cross-connects, etc. Cable manufacturers: Berk-Tek, Belden, Lucent, and others. Electronics manufacturers: 3COM, Cisco, Compaq, Hewlett-Packard, Intel, Standard MicroSystems, Ascente, CNet Technology, Netopia, Linksys, Matrox Networks, Tut Systems, Epigram, and Diamond Multimedia. |
| Wireless • RF • Infrared | IEEE 802.11, 802.11a, 802.11b, ETSI/BRAN HIPERLAN 1 and 2, HiperLink, HiperAccess, DECT. Shared Wireless Access Protocol (SWAP) overseen by the Home RF Working Group. The "Bluetooth" protocol sponsored by Bluetooth Consortium member Ericsson, Nokia, and Intel. | Transceivers, adapter cards, controllers. Lucent, Apple, Netgear (a subsidiary of Nortel), Harris (integrated circuit chips), Home Wireless Networks, ShareWave Inc., and Kyushi Matsushita Electric Co. |
| Power Lines (PL) | Consumer Electronic Bus (CEBus), EIA 600, Common Application Language (CAL), and Home Plug and Play (Home PnP), all overseen by the CEBus Industry Council (CIC). X10 (de facto standard developed by X10 Ltd. of Hong Kong). EIA 709 series–LonWorks (developed by Echelon Corp.). | Transceivers, modules, management, and control servers. X10 Ltd., IBM, Thomson, Philips, Radio Shack, Leviton, Honeywell, Stanley, Ademco, ADT, Coactive Networks, and Rebel.com. |
| Coaxial and Fiber Optic Cable | EIA/TIA 568A, 569A, 606, 607, TSB 72 (Centralized Optical Fiber Cabling Guidelines), EIA/TIA 526-14, 526-7 (Optical Power Loss Measurement). ANSI/ISO Fiber Distributed Data Interface (FDDI), FDDI II, Fibre Channel (ANSI and the Fibre Channel Association). | Fiber cable manufacturers include Alcatel, Siecor, Lucent, AMP, 3M, and General Phtonics. Manufacturers of fiber cable network related electronics include 3COM, Cabletron, Cisco, Extreme Networks, Fore Systems, Foundry Networks, Hewlett-Packard, IBM, Lanart, Lucent, DNS, Methode, Nortel Networks, Polycor, Transition Networks, and XLNT. |

| | | | |
|---|---|---|---|
| ANSI | American National Standards Institute | IEC | International Electrotechnic Commission |
| BRAN | Broadband Radio Access Networks | IEEE | Institute of Electrical and Electronic Engineers |
| CSA | Canadian Standards Association | ISO | International Organization for Standardization |
| DECT | Digital Enhanced Cordless Telecommunications | NFPA | National Fire Protection Association |
| EIA | Electronics Industry Alliance | TIA | Telecommunications Industry Association |
| ETSI | European Telecommunications Standards Institute | UL | Underwriters Laboratories, Inc. |

**Figure 9.10**   Home PDS applications and media/design options.

addressing the "last-mile" problem of extending high-speed network capabilities to users. Within residences, this challenge takes on the dimension of the "last 100 feet." Assuming that DSL service is available at residential NIs, the task of extending that service to one or more desktops is a key home-PDS objective.

Although computer networking is a requirement in both home and commercial PDS installations, business applications swamp most home requirements in terms of numbers of attached devices, peak, and aggregate traffic levels.

Currently, however, the largest market for home networking centers on home automation. Today, this includes innovations that manage and control lighting, HVAC apparatus, appliances, infant and elderly monitoring and medical notification systems, pumps and sprinklers, remote utility system control and meter reading, and a growing list of video/motion detection surveillance, egress control, and other security mechanisms. Because there are so many devices and appliances to monitor and control, neither existing telephone wiring, new telephone wiring, nor dedicated low-voltage wiring constitutes practical "device-to-controller" home automation communications-channel alternatives.

Fortunately, since the mid-1970s, techniques using electrical power-line wiring as communications channels have evolved to such an advanced state that they easily satisfy numerous sensor/device-to-remote actuator or computer monitoring and control equipment connection requirements. Some experts believe this market will exceed $2.75 billion in 2000.

Entertainment-signal distributions impose additional, wide-bandwidth and high data-rate home networking requirements. In the digital video category, high definition TV (HDTV), digital video disc (DVD), and MPEG-2 coding demand, respectively, 19-Mb/s, 3—8-Mb/s and 2—4-Mb/s capabilities. Coaxial cable analog TV-signal distribution typically uses the 5—450-MHz spectrum. Depending on compression technique, CD-quality music requires 112—705.6-kb/s digital signaling rates. Analog audio distribution for music, intercom, or public address imposes minimal home networking requirements other than the physical need for one pair per channel (two pairs for stereo, five pairs or more for home theater setups). A Yankee Group marketing specialist projects that by 2001 the home computer networking and entertainment market will reach $725 million annually.

# Home PDS Components and Design Options

Figure 9.11, a high-level functional-block diagram illustrates modern home PDS components, and internal and external interconnections among them. Device-access service connections are symbolically portrayed by single "thick lines" with arrows at each end. These access-line symbols may represent one or more physical connections and possibly more than one type of connecting medium.

**Figure 9.11**
Home PDS
components.

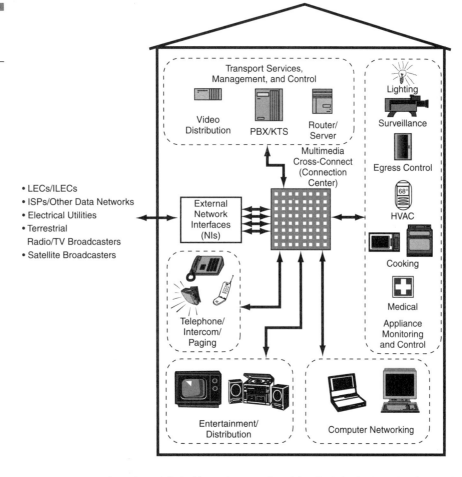

For example, the thick line from the block labeled "telephone/ intercom/paging" may represent a dozen UTP cables connecting as many modular wall-jacks together—and directly to a centrally located telco network interface or to an NI via a cross-connect. The same thick line may also represent radio frequency (RF) connections from cordless telephones to base stations that are at some point wire-connected to the UTP network.

Similarly, the thick line from the block labeled "appliance monitoring and control" may represent connections using *power line communications*. PL communications normally involve a module designed to be inserted into electrical sockets; into this a device, be it a lamp or coffee pot, is itself plugged. Most of these modules conform to X10 Ltd.'s X10 design specifications. Since X10 devices were introduced in the 1970s, more than 100 million have been sold. Device modules generate and receive signals (are transceivers), designed to overcome noise and other PL transmission problems, and use them to communicate with other remote, single, or centralized multi-device actuator/controllers, equipped with matching modules—also connected at some point in a dwelling to its electrical power lines.

Except for the challenges of establishing error-free communications using relatively noisy and hostile PL media, power lines are ideally suited to home automation appliance monitoring and control since nearly all such devices require connection to power lines, whether or not they are remotely controlled.

If we refer now to the table on the right-hand part of Figure 9.10, the first column identifies four home networking media options: UTP cable, RF or infrared wireless electromagnetic transmission, power line, and coaxial and fiber optic cable. The second and third columns list relevant standards, standards-setting or industry organizations, products, and product manufacturers.

To complete the PL discussion, Figure 9.10 lists the Consumer Electronic Bus (CEBus), the Home Plug and Play (Home PnP), and the Common Application Language (CAL) specifications, all overseen by the CEBus Industry Council (CIC), to ensure development of interoperable home-automation hardware and software products. CEBus is now incorporated in EIA 600, and LonWorks, a home automation operating system developed by Echelon Corporation, is now an open standard under both EIA 600 and 709.

In the *UTP Cable* row, Figure 9.10 lists all the standards, standards-setting organizations, products, and manufacturers that pertain to this chapter's earlier description of commercial building PDS systems, with some additional home PDS-relevant citations. For instance, the Home Phoneline Networking Alliance focuses on UTP cable-based home networking in general, and innovative methods of using legacy wiring for high-speed digital data networking in particular. Formed in

1998, Home PNA now has more than 100 members including founders 3Com, Advanced Micro Devices, AT&T, Compaq, Hewlett-Packard, Intel, IBM, Lucent, and Wireless Systems.

In Figure 9.10's *Wireless* row, standards introduced for commercial building PDS systems are all viable candidates in home applications. Two standards not yet mentioned are the Shared Wireless Access Protocol (SWAP—overseen by the Home RF Working Group) and the "Bluetooth" protocol sponsored by Bluetooth Consortium members Ericsson, Nokia, and Intel. See Chapter 16 for more information on the Bluetooth initiative.

The last *Coaxial and Fiber Optic Cable* row relates mainly to entertainment distribution home-networking requirements. Today, virtually all RF and video TV signal distribution is by means of coaxial cable. The reason for this is that in the absence of high-quality (Category 5 or better) UTP, affordable fiber optic cable and connectors, or broadband local wireless networks, there is no other practical medium for wide-bandwidth analog or high-speed digital signals. For home applications this includes single-channel analog NTSC composite video signals (see Chapter 3), 100—200-channel analog FDM multiplexed RF TV signals, or high-quality full-motion digital video signals.

## Fiber—Is It the Ultimate PDS Medium?

With pre-cut and polished fiber cable, quick-mount modular connectors and overall declining costs, fiber alternatives are looking very attractive for wide-bandwidth analog and high-speed digital signals. For example, as cable and connector costs converge, since each RJ-45 connector has eight pins and terminates eight strands, and fiber replacements two, fiber may involve less labor-intensive and lower installed cost than UTP alternatives, and far superior transmission, interference immunity, and security characteristics than UTP, coaxial, or wireless media.

Resisting the trend to fiber is the reality that in nearly all current applications, selecting fiber necessitates the use of electrical-to-optical signal converters. Countering this argument, however, is the need for complex, signal processing-intensive modulation-coding adapters to achieve multi-megabit-per-second data rates with UTP. Also, even wide-bandwidth analog coaxial cable-based systems require high-quality and therefore relatively expensive amplifiers and other termination facilities. Thus, with mass-produced and low-cost electrical-to-optical signal converters, it is conceivable that a single universal fiber connector/cable-based approach could become the PDS mechanism of choice, at least in new installations.

The increasingly apparent advantages of integrating voice, data, and video service supports trends toward universal cabling. Since even single fibers have the capacity to carry all aggregated voice, data, and video home-networking traffic, a single universal fiber connector/cable-based PDS approach may very well emerge, satisfying virtually all *cable-tethered* home networking PDS requirements.

Certainly, over the long term, some form of structured cabling will prevail. Although in 1999 only 50,000 homes had structured cabling, that number is expected to mushroom 16-fold by 2003. An essential structured cabling feature is the use of multimedia (if more than one medium is necessary) cross-connects or connection centers, as shown in Figure 9.11. The figure is consistent with the new Home Director product line upon which IBM is attempting to persuade customers and consulting and construction companies to pursue structured cabling. While the advantages may not be apparent for low-end home networks (for example, those involving only a small number of telephones and one or two cable TV outlets), for high-end Figure 9.11-type installations, structured and star-based cabled connection centers are crucial.

## Automated/Intelligent-Home Transport, Management, and Control Services

From an "intelligent home" point of view, access to or connections with a myriad of devices is only half the battle. To make high-end integrated systems function, centralized management and control and telecommunications transport services (depicted in Figure 9.11's top-left functional block) are needed. The "router/server" symbol in the figure is a generic representation of computer-based processing facilities for managing and controlling home automation, computer networking, telephone, and entertainment functions and devices. For example, a dedicated processor, an HVAC controller, and power line and infrared interface modules are included as IBM Home Network Controller components.

For intra-building computer networking and access to Internet and other external data networks, router and communications server and gateway (see Chapters 7 and 13) services are needed. Coactive Networks,

Rebel.com and other companies offer standalone products for these purposes, but they can be implemented in general-purpose PCs using LonWorks and other network operating system software.

As in many commercial buildings, for voice and other circuit-switched traffic, high-end intelligent-home installation may also require telecommunications transport services. As stated, in those cases small PBXs, KTSs, or Centrex services may be employed. Once installed, these facilities can readily provide conferencing, intercom, paging, and the usual extensive list of telecommunications and adjunct functions and features described in Chapter 6. Finally, some high-end installations may require video distribution facilities in support of both entertainment and closed-circuit TV signals.

To probe home networking technologies further, see "Networks for Homes", an article in December 1999's *IEEE Spectrum* magazine, that contains more details and a list of references.

# CHAPTER 10

# Voice Network Services

The introduction to Part 3 defines major voice-service categories and provides high-level technical and operational characteristics peculiar to each type of service. The important concepts of access and transport, local exchange and interexchange, and switched and nonswitched services are explained. Chapter 9 discussed voice services relating to devices and telecommunications operations within customer premises.

In this chapter, we extend the discussion to the realm of telecommunications carrier voice services, which traditionally are subdivided into local exchange and interexchange network service categories. This subdivision roughly corresponds to commonly understood notions of "local" and "long-distance" service, but current FCC deregulatory activities described in Chapter 2 are blurring even these boundaries. In what follows, *local exchange services* refer to services provided by the carrier serving the premises of the user. *Interexchange services* are services provided by another carrier, usually to satisfy user requirements outside the first carrier's serving area.

The presentation centers around a graphic depicting each network service type. The graphic identifies telecommunications facilities associated with each service and the point of termination (network interface, NI) between customer-premises equipment and carrier facilities. Principal applications supported by the service, pricing elements (how the service is paid for), and technical characteristics are summarized on the graphic and discussed in the narrative.

# Local Exchange Network Services

The taxonomy of local exchange network services is shown in Figure 10.1 to consist of two subgroups: access and transport services. The services are provided using local exchange carrier facilities. Access services provide communications between user premises and transport services rendered at local exchange carrier central offices (COs). Transport services provide communications among COs that connect originating and terminating access services.

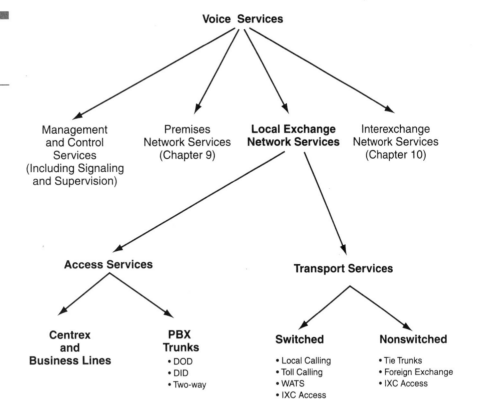

**Figure 10.1**
Taxonomy of local exchange network services.

## LEC Access Services

*LEC access services* are available either as dedicated "lines" to a single user device, or as "trunks" to a customer premises-located switch (e.g., a PBX). The latter case typically enables many user devices to share trunk facilities. Options under these two categories are described in the following sections.

### BUSINESS LINE AND CENTREX SERVICES

Figure 10.2 depicts *LEC business line and Centrex services.* The underlying facilities include LEC voice frequency (VF) loops that connect central-office switches to business-premises network interfaces (NIs), customer premises equipment (CPE), and premises distribution systems that connect CPE to network interfaces.

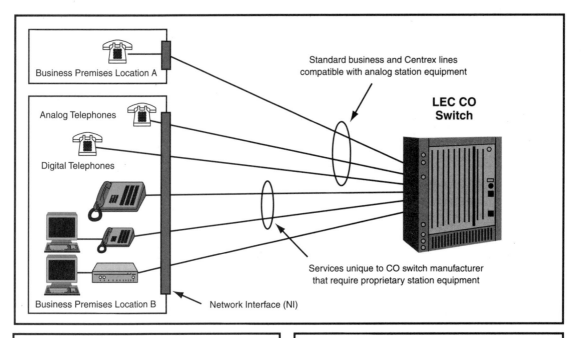

Standard business and Centrex lines
compatible with analog station equipment

**LEC CO
Switch**

Business Premises Location A

Analog Telephones

Digital Telephones

Business Premises Location B

Network Interface (NI)

Services unique to CO switch manufacturer
that require proprietary station equipment

---

## Principal Applications

**Business Line:**
- Access to the PSTN
- Access to custom calling and CLASS* features

**Centrex:**
- Access to the PSTN
- Station-to-station calling
- Attendant services
- Access to PBX-like features
- Access to CLASS* and other enhanced features
- Access to OA&M and adjunct applications and features

*CLASS = Customer Local Area Signaling Service

## Technical Characteristics

**Access Facility Description:**
- Single-channel VF analog transmission facility
- Two-wire
- Metallic twisted-pair interface at the NI

**Call Direction Modes:**
- Two-way

**Signaling Interfaces:**
- Loop signaling

**Signaling Techniques:**
- Facilility dependent, direct current
- Supervisiory
  - Loop-start origination
- Addressing
  - Dial pulse
  - Dual-tone multifrequency (DTMF)

## Price Elements

- Installation charges
- Monthly line charges
- Feature charges
- In most service areas, discounts via special contracts are available to Centrex customers willing to make volume and term commitments.

---

**Figure 10.2**   *Business line and Centrex service.*

The principal business applications supported by these services are:

- Access to LEC-provided local and toll calling transport services
- Access to LEC central-office features, e.g., custom calling and CLASS
- For the Centrex case, access to intra-premises station-to-station calling, feature packages, and attendant services.

Custom calling features include call forwarding, call waiting, and others, depending upon the serving office. *Custom local area signaling service* (CLASS) is a set of local calling enhancements to basic exchange telephone service. CLASS features use digital switching and signaling to provide automatic callback, automatic recall, selective call forwarding and call rejection, customer-originated call trace, calling number identification, bulk calling number delivery, and others. Note that while these features augment both access and transport services, they are a subset of a management and control category of voice services as indicated in Figure 10.1. Chapter 6 provides details.

Costs for business and Centrex line access services include one-time installation, monthly line, and feature charges. (Message unit charges are allocated to transport services and discussed below.) *Subscriber Line Charges* (SLC), fees mandated by the FCC to assure that costs for local loop service are adequately provided for, are charged monthly on a per-line basis. Through special contracts, large Centrex customers can acquire discounts in most serving areas in exchange for long-term and minimum-line-size commitments.

Technical characteristics associated with these access services are as follows. The loop facility is a VF analog carrier transmission system, consisting of bundled two-wire, copper twisted-pair wiring terminated at an NI. For large customers LECs may elect to use either digital loop carrier or remote switch modules located on or adjacent to customer premises. These alternatives are transparent to customers; appearances at NIs are identical.

Individual voice circuits are two way and full duplex enabling both call origination and termination. Loop-start supervision and either dial pulse or DTMF tone address signaling are supported by business and Centrex line access service. The "Signaling System Fundamentals" part of Chapter 6 explains "address" and other signaling terms and operations.

Analog standard and proprietary phones can be used with these services. Although Figure 10.2 indicates Centrex support for digital telephone and data interface units, this is accomplished using analog loop facilities (as described in Chapters 4 and 6), special CO line interface cards, and station equipment designed for the switch. Where available, integrated services digital network (ISDN) central office capabilities permit the use of digital station equipment from any vendor complying with ITU and ANSI ISDN standards. Recall that whereas standard analog central office switch interface cards permit either voice or data modem traffic, proprietary digital or ISDN models do not support modem traffic.

### PBX-TO-CENTRAL OFFICE TRUNK ACCESS SERVICE

Figure 10.3 depicts *LEC PBX trunk access service.* In addition to the analog voice frequency (VF) loops used for business line/Centrex access services, DS1 digital facilities are also common. Since the trunks are connected to switches at the customer premises end, there might be additional requirements for facilities that pass addressing information to premises switches.

The principal business applications supported by this service are access to local and toll calling. Three different types of trunks are normally furnished by LECs: *direct inward dial* (DID), *direct outward dial* (DOD), and *two-way CO* trunks. DID allows incoming calls to ring specific stations without attendant assistance. DOD allows outgoing calls to be placed directly from PBX stations. Two-way trunks support attendant console operation so on-premises operators can place and receive calls.

Access to custom calling and CLASS features is not possible with PBX-to-central office trunk service.

Costs for PBX-to-central office trunk access services include one-time installation, monthly line, and DID station number charges. (Message-unit and other usage charges are transport services charges, discussed later.) Since PBX trunks carry more traffic per circuit than business lines, they are normally more expensive than business lines. However, many fewer of them are required, as explained in Chapter 11. LECs normally don't grant discounts for large customers because they are often in competition with vendors offering PBXs as alternatives to LEC Centrex.

For this type of access, the SLC is charged monthly on a per-trunk basis. Here, as with other charges, PBXs have an advantage over Centrex since the number of PBX trunks is normally less than the number of Centrex lines by 70 to 90 percent. However, many Centrex contracts now credit customers with the difference between PBX and Centrex SLC charges, a result of deregulating Centrex service and a desire to make it more competitive with PBX service.

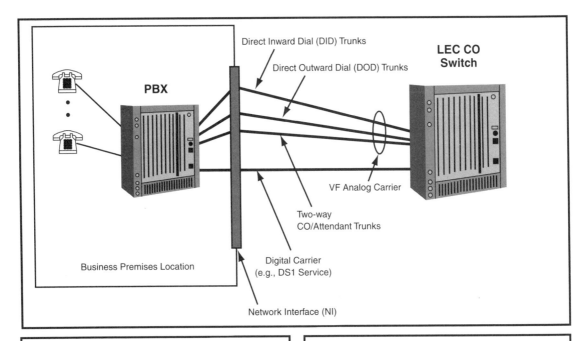

**Business Premises Location**

Direct Inward Dial (DID) Trunks

Direct Outward Dial (DOD) Trunks

**PBX**

**LEC CO Switch**

VF Analog Carrier

Two-way
CO/Attendant Trunks

Digital Carrier
(e.g., DS1 Service)

Network Interface (NI)

**Figure 10.3** PBX-to-central office trunk service.

---

### Principal Applications

**Direct Inward Dial Trunks:**
• Call termination access to the PSTN

**Direct Outward Dial Trunks:**
• Call origination access to the PSTN

**Central Office/Attendant Trunks:**
• Call origination/termination access to the PSTN

**Note:**
• Access to custom calling and CLASS features is not available with PBX-to-CO trunk service.

### Price Elements

• Installation charges
• Monthly line charges
• Station number charges (DID only)

### Technical Characteristics

**Access Facility Description:**
• Single-channel VF analog transmission facility
• Two-wire
• Metallic twisted-pair interface at the NI
• Optional DS-1 carrier service

**Call Direction Modes:**
• Two-way for CO/attendant trunks
• One-way for DID and DOD trunks

**Signaling Interfaces:**
• Loop signaling

**Signaling Techniques:**
• Facility dependent, direct current
• Supervisiory
    – Ground or loop-start origination for DOD trunks
    – Ground start for CO/attendant trunks
    – Wink-start for DID trunks
• Addressing
    – Dial pulse
    – Dual-tone multifrequency (DTMF)
• Optional robbed-bit signaling for DS1 connections

Technical characteristics of PBX trunks are similar to those of business lines. For sites with heavy traffic, customers normally elect to use LEC digital carrier transmission services. Typically the price of 24-channel DS1 service is less than that for 24 separate analog channels. Also, with digital PBXs, a single T1-type carrier connection supports 24 voice channels, reducing interface complexity and OA&M costs. Some LECs offer fractional T1 (that is, digital service for less than 24 channels) at lower than full T1 prices, making the move to digital carrier even more attractive.

For analog PBXs, D4 channel banks can be used with digital carrier service. In cases where all the channels are not needed for voice service, spare DS 1 capacity can be used for data or other services. Digital channel banks provide analog interfaces for the PBX identical to those associated with LEC VF carrier service described above.

Call direction, signaling interface, and signaling techniques used with PBX-to-CO trunks are listed in Figure 10.3. Figure 10.4 summarizes the popular analog and digital LEC access service options discussed above. It is certain that ongoing digital subscriber line (DSL) development activities will produce additional LEC access service options. Chapter 4 describes issues related to DSL design approaches that satisfy new high-speed *data access service* requirements, while maintaining existing *voice access service* capabilities and NI technical interface characteristics. Considering the enormous investment in hundreds of millions of existing telephone instruments, there is no question that LECs will continue to support the analog voice access services listed in Figure 10.4 for the foreseeable future.

# LEC Transport Services

*LEC transport services* among COs are characterized by being either switched (i.e., shared) or non-switched (i.e., dedicated) services. Again, these services are provided by the local exchange carrier serving the user's premises. The two categories of services are described next.

### LEC SWITCHED TRANSPORT SERVICES

Figure 10.5 illustrates *LEC public switched transport services*. Access to LEC CO switches is provided using one of the methods described in the previous section. The underlying transport facilities include LEC central office and tandem switching, and interoffice transmission. From a services viewpoint, the details regarding these intermediate switches and transmission facilities are transparent to users. For this reason, switched transport facilities are typically represented as "clouds" in network diagrams. For incumbent LECs still subject to the restrictions of the Modified Final Judgment, transport services can only be provided within a LATA.

**Figure 10.4**
Summary of LEC access voice services.

| LEC Access Service | Call Direction | VF Facility | Signaling Technique | Digital Facility |
|---|---|---|---|---|
| Business Line | Two-way | Two-wire | Loop Start | N/A[1] |
| CENTREX | Two-way | Two-wire | Loop Start | N/A[1] |
| Coin Operation | Two-way | Two-wire | Ground Start | N/A[1] |
| PBX Trunks | | | | |
| Direct Inward Dialing (DID) | One-way | Two-wire | Wink Start | Four-wire[2] (24 Channel DS1 Signals) |
| Direct OutwardDialing (DOD) | One-way | Two-wire | Ground or Loop Start | Four-wire[2] (24 Channel DS1 Signals) |
| Central Office | Two-way | Two-wire | Ground Start | Four-wire[2] (24 Channel DS1 Signals) |

1  Digital voice access service is not normally available from LECs except with proprietary station equipment from the CO switch manufacturer. Digital access is also possible where ISDN basic rate interface (BRI) service is available.

2  Robbed-bit signaling is normally used for direct digital carrier connections to PBXs. Available digital channel units for digital carrier, provide a full set of signaling features, including loop-start, ground-start, and E and M leads options. COs with SS7 capabilities may offer *digital subscriber signaling system* (DSS) signaling in accordance with Q.920/921 and Q.930//931 recommendations.

The principal end-user applications supported by these services are local and toll calling. Local services include local calling, 411 directory and operator assistance, and 911 emergency service.

Toll service includes message telecommunications service (MTS), 555-1212 directory and operator assistance, and wide area telephone service (WATS). In all cases, redundancy and sizing of LEC facilities results in transport grade-of-service and reliability performance significantly better than affordable private network facilities.

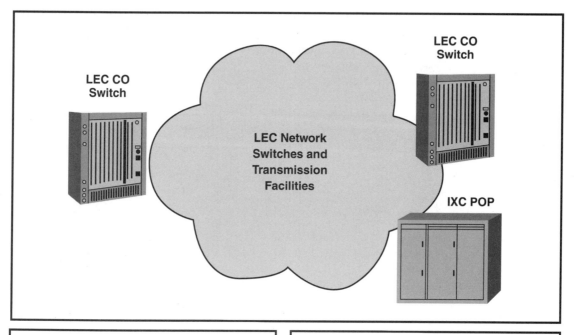

**Principal Applications**

**Local Service:**
- Local calling
- Directory assistance (411), operator assistance
- Emergency services (911)

**Toll Service:**
- Message telecommunications service (MTS)
- Directory assistance (555-1212), operator assistance
- Wide-area telephone service (WATS)

**Switched Acccess to IXC Services:**
- Economical access for low volume users

**Technical Characteristics**

**Facility Description:**
- Central office and tandem switches
- Trend to digital switching and transmission systems
- Increasing use of fiber optic transmission
- Redundant paths with alternate routing

**Signaling Techniques:**
- Multiple frequency (MF) inband signaling used by older systems
  - Addressing
  - Supervisory
- Common channel signaling (CCS) now prevalent
  - Addressing
  - Supervisory
  - Feature messaging

**Price Elements**

- Local message unit charges (banded)
- Toll usage charges
- WATS (discounted toll)
- Switched access usage charges

**Figure 10.5**   Local exchange switched transport service.

Pricing elements include distance-banded, message-unit charges for local calling, toll charges, and WATS charges. Toll charges are usage charges based on duration and distance. WATS charges are essentially discounted toll charges—that is, the service billing is on a bulk basis rather than for individual calls.

A special kind of local exchange switched transport service exists when one end of the service is an interexchange carrier's POP instead of a connection to a user premises. Since this type of service ultimately connects a user to another type of transport service (i.e., interexchange transport service), it is more consistent to refer to the combination of user-to-LEC access service, and LEC-to-IXC transport service as *interexchange switched access service.* We discuss interexchange switched access service characteristics later in this chapter.

## LEC NON-SWITCHED TRANSPORT SERVICES

Figure 10.6 depicts *LEC non-switched transport services.* As with switched services, access to LEC COs is provided using one of the methods described in previous sections. At the CO, however, access facilities are not connected to a circuit switch, but rather are connected directly to network transport transmission facilities using multiplexer and/or cross-connect equipment. The distinguishing characteristic of these services versus switched services is the fact that dedicated transport circuits are reserved for and available only to customers leasing the facilities, even during intervals when customers have no traffic to transmit.

The principal applications supported by non-switched services are:

- Private lines for voice, teletypewriter, alarms, data, etc.
- Tie trunks for connecting PBXs
- Foreign-exchange (FX) circuits
- Trunks for dedicated access to IXC services.

Foreign-exchange circuits terminate in a distant LEC central office and provide switched services as if the user premises at the originating end of the FX line were located in the vicinity of the distant office. For further elaboration see the Introduction to Part 3.

The basic technical characteristics of facilities associated with non-switched services are the same as for other LEC services. In some cases, enhanced capabilities, such as sophisticated network management and control capabilities may be provided.

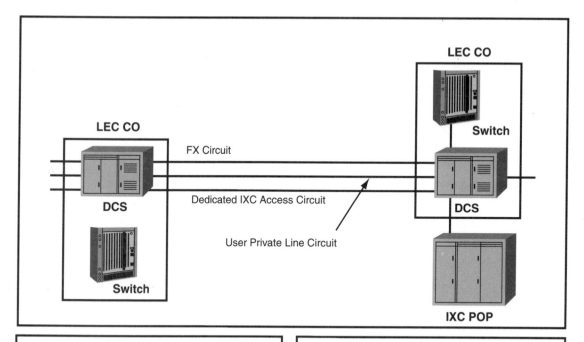

Figure 10.6 shows Local exchange non-switched transport service with the following labels: LEC CO, Switch, DCS (top right), FX Circuit, Dedicated IXC Access Circuit, User Private Line Circuit, LEC CO, DCS, Switch (left), IXC POP.

### Principal Applications

**Private Lines:**
• Voice, teletypewriter, alarm, and other applications

**Tie Trunks:**
• Direct connection to PBX on customer premises
• Requires signaling services

**Foreign Exchange (FX) Circuits:**
• Access to local calling in another CO area

**Non-switched Access to IXC Services:**
• PBX-to-POP connections for access to:
  – IXC switched services
  – IXC non-switched services

### Price Elements

• Installation charge
• Monthly termination charges
• Mileage dependent monthly transmission charges
• Signaling charges (if required)

### Technical Characteristics

**Facility Description:**
• Digital transmission facilities (e.g., DS1, DS3) used in the transport network
• Digital cross-connect system (DCS) used to "groom" connection arrangements on a per-customer basis. DCS connections:
  – Are not changeable on a call-by-call basis
  – Can change on a preset time basis
  – Offer rerouting protection in case of disasters

**Figure 10.6** Local exchange non-switched transport service.

Most LECs use programmable digital cross-connect systems (DCS) to "groom" connection arrangements on a per-customer basis. *Grooming* refers to establishing central office cross-connections that correspond to individual customers' needs. For example, a customer may require multiple channels of dedicated transmission facilities to be used for combinations of voice, data, and other applications. He may also define routing to establish connections between specified business premises, which may be different for each application, and vary on a site-by-site basis.

The LEC can use a DCS to combine traffic from multiple customers, over common, high-speed digital network facilities. Such "static" connection arrangements could be accomplished with manual cross-connects, but using programmable DCSs reduces cost and enhances performance and flexibility. As discussed in Chapter 5, DCSs are processor-controlled devices that permit individual DS0 channels to be cross-connected among multiple DS1 or higher (i.e., DS2, 3 or 4) level terminating signals. The operation is illustrated in Figure 10.6 where, for example, the DCS could establish tie trunk connections between PBXs using private lines, foreign-exchange (FX) connections to a remote central office, and dedicated PBX-to-IXC POP connections for access to IXC services.

DCS connection arrangements can be programmed to change on a time-of-day or day-of-week basis. They do not provide call-by-call switching. DCSs offer flexible rerouting capabilities in case of disasters or failures. With them carriers can offer excellent dedicated, point-to-point network availability and reliability.

Non-switched LEC transport service pricing elements include installation charges, central-office termination charges, mileage-dependent transmission charges, and signaling charges where applicable. All recurring charges are not dependent on usage. FX service incurs normal exchange usage charges at the remote CO, but these charges are not considered part of non-switched service.

As with LEC switched transport, a special kind of local exchange non-switched transport service exists when one end of the service is an interexchange carrier's POP instead of a connection to a user premises. Again, since this type of service ultimately connects a user to another type of transport service (i.e., interexchange transport service), it is more consistent to classify this service as an interexchange non-switched access service which is discussed below.

# ■■ ■■ Interexchange Network Services

The taxonomy of interexchange network services is shown in Figure 10.7. The services are provided using both local exchange and interexchange carrier facilities. As with local exchange network services, the two major subcategories of interexchange network services are access and transport services. Access services provide communications between user premises and interexchange carrier POPs. Interexchange transport services provide communications among POPs and, hence, connect originating IXC access to terminating IXC access services. Differences between LEC and IXC access services are not so much technical, but rather derive from the fact that LEC facilities are used to deliver IXC access service.

**Figure 10.7**
Taxonomy of
interexchange
network services.

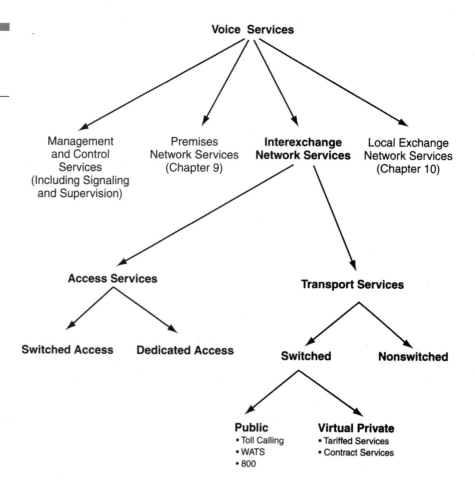

As explained in Chapter 2, the primary reason for this is the historical division of telecommunications providers into local and interexchange carrier business and territorial domains—and restrictions imposed by divestiture accords. Under Telecommunications Act of 1996 provisions, in the future a single carrier or business entity may provide both local and long-distance and access and transport services. Until the Act's provisions are fully implemented, however, the structure of interexchange services described herein will continue to exist.

Transport services are either switched or non-switched. In the interexchange arena, switched services are further divided into public and virtual private network varieties. Public services are outbound and inbound switched services to or from user locations. Virtual private network (VPN) services are replacements for the private networks of the 1970s and 1980s. In that era, some corporations or government agencies with large amounts of traffic among geographically dispersed locations procured or leased their own tandem switching and transmission facilities, forming "closed" or private networks.

Users and locations with authorized access to private or VPN networks are referred to as "onnet" users/locations. Access service is required to reach tandem switches in private networks or IXC service delivery points in VPNs. In some private networks' dual function PBXs, that is PBXs with tandem switching capabilities, both access and tandem switching transport services are provided in the same machine. More VPN details are provided later in this chapter.

## IXC Access Services

Access services from user premises to POPs may be either switched or non-switched. In actuality, these services are provided by special cases of local exchange transport services where one end of the service is an interexchange carrier POP. Because these services carry predominately interstate traffic, they are regulated by the FCC, instead of by state public utility commissions that regulate other local exchange transport services.

### IXC SWITCHED-ACCESS SERVICES
In Figure 10.5, facilities supporting *IXC switched-access service* include LEC switches and interconnecting transmission resources. From a LEC customer point of view, for those with limited long-distance traffic volume (most residential and small business users), using LEC switch-

ing and transmission facilities to obtain IXC access service is the most economical alternative. From an IXC's viewpoint, high-capacity POP-to-nearest LEC tandem switch circuits represent the most economical mechanism for IXCs to gain access to each and every LEC customer telephone within that LEC's serving area. LECs only charge IXCs for actual user connection times, and IXCs are allowed to recover those charges by including them in IXC-to-calling party long-distance (transport service) telephone bills. See Chapter 11 for details.

With the onset of divestiture and competition in the long-distance market, the technical problems of how to direct outbound calls to IXCs chosen by calling parties arose. Initially, customers of IXCs other than AT&T were required to dial a separate IXC access number, wait for a second dial tone, and then dial a destination telephone number. Since this access method was decidedly unequal to AT&T's, LEC access service charges to other IXCs were less than those to AT&T. Those reduced costs led to a temporary flurry of cut-rate IXC ventures. Eventually, however, the present method of associating a *primary interexchange carrier* (PIC) with each user's telephone and automatically routing long-distance calls accordingly, has accomplished what is known as equal access, now almost universally available.

Switched access can also be used with VPN services by designating a VPN service as a PIC. In this case, subscriber lines authorized to use a VPN are considered onnet and have access to all VPN features. However, since most PBXs or Centrex service-based locations generate traffic volumes large enough to cost-justify dedicated user-to-POP access facilities, switched IXC access service is normally selected only by single or key telephone systems users.

### DEDICATED IXC ACCESS SERVICES

Figure 10.6 illustrates three different categories of *LEC non-switched transport service*. As mentioned earlier, these services are often used to provision *dedicated IXC access services,* as depicted in the figure. Although "non-switched" and "dedicated" are both terms used in the industry, the service described herein is most commonly referred to as "dedicated access." Circuits from user premises to POPs are reserved in the LEC network for that user's IXC access traffic, whether or not at any particular time the user is actually generating IXC access traffic. (Note: LECs provide no call-by-call switching service for these circuits.)

At a POP, dedicated access-service circuits can be connected to either switched or non-switched IXC transport service. In accessing switched transport services, the tradeoff between using dedicated versus switched access service is an economic one. With switched access, usage charges are assessed for each in- or outbound call minute per month. With dedicated access, enough circuits must be ordered to satisfy monthly busy-hour required grades-of-service (see Chapter 11 for a discussion of how this quantity is determined). For large-volume users, fixed-rate per-circuit-per-month dedicated access costs are generally less than cumulative switched access usage charges. Chapter 12 provides real-world examples of these economic tradeoffs.

Dedicated access economies and the availability of computer-controlled switches to route calls over dedicated access circuits are two primary drivers justifying the use of IXC virtual private network service. Large sites with PBX or Centrex service often generate sufficient levels of traffic to warrant dedicated-access network connections. Even in companies with a few large and mostly small sites (large sites generate a high percentage of total corporate traffic), using dedicated access at those locations and VPN service is frequently the most cost-effective approach.

A final reason for selecting dedicated access is to provide integrated access to several IXC services over a common high-capacity access circuit. Multiplexers on customer premises can be used to support multiple low-speed access circuit channels (possibly used to access different IXC services) over a single high-speed access facility. Chapter 12 presents the economic rationale for such arrangements.

## IXC Transport Services

*IXC transport services* among POPs are again either switched or non-switched, and rendered by subscriber-selected interexchange carriers. Within the switched-service category, important examples include public switched services, toll-free services, and virtual private network services.

### IXC PUBLIC SWITCHED TRANSPORT SERVICES

Figure 10.8 illustrates *IXC public switched transport services*. Access to POP switches and these services is via either switched or dedicated access service, as described above. IXC public switched services are used primarily by residential and single-location small business customers. The principal applications supported are MTS, dial-1 WATS, and dedicated-access WATS.

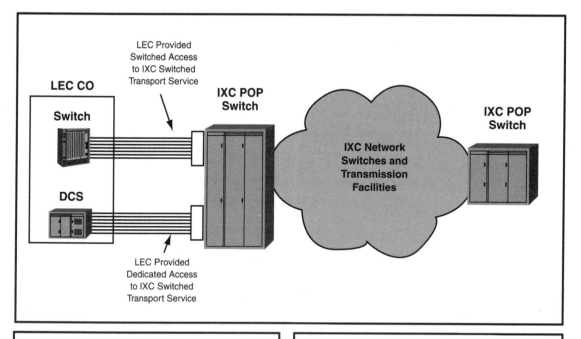

LEC Provided
Switched Access
to IXC Switched
Transport Service

**LEC CO**

**Switch**

**DCS**

**IXC POP
Switch**

**IXC Network
Switches and
Transmission
Facilities**

**IXC POP
Switch**

LEC Provided
Dedicated Access
to IXC Switched
Transport Service

### Principal Applications

**Message Telecommunications Service (MTS):**
- Reliable transport between POPs
- IXC buys access service from LECs

**"Dial 1" WATS Service:**
- Volume discounted billing of MTS-like service

**Dedicated Access WATS:**
- Dedicated customer premises-to-POP access provided
  by IXC or the customer

### Price Elements

- Installation charges
- Call duration usage charges, often mileage-based
- Off-peak time period discounts
- Volume discounts (WATS only)

### Technical Characteristics

**Facility Description:**
- Digital stored program control tandem switches
- Mostly digital IXC interswitch transmission
- Increasing use of fiber optic transmission
- Redundant paths with alternate routing

**Signaling Techniques:**
- Common channel signaling (CCS) now prevalent
  - Predominantly ANSI SS7
  - No CCS to PBXs in the United States

**Figure 10.8** Interexchange public switched transport services.

*MTS* is standard (direct-distance dial) or operator-assisted long-distance service, employing switched access for call origination and termination. For originating calls, the IXC carrier of choice is normally preselected by customers. Callers can override IXC preselection by dialing additional IXC carrier identification code digits. As noted above, IXC portions of customer bills include LEC charges to IXCs for access service to their networks.

With *dial-1 WATS*, all calls are completed as MTS calls but customer billing is made on a bulk rather than a call-by-call basis, with discounts proportional to traffic levels. Dial-1 WATS calls are provisioned using LEC switched access.

As the name implies, *dedicated-access WATS* employs dedicated access circuits (trunks) between business premises and IXC POPs, bypassing LEC switched-access facilities. Since companies opting for dedicated-access WATS usually generate more long-distance traffic than companies using dial-1 WATS, discounts are also typically greater.

These applications are described here in terms of generic IXC services. However, each IXC offers a menu of similar services with variations under private label names, e.g., AT&T's MEGACOM is a dedicated-access WATS-type service.

For these services, IXCs use high-capacity, digital, stored program-control tandem switches and digital network transmission facilities. Although microwave radio and satellite carrier systems are also employed, the trend is to fiber cable on all major network routes. The signaling trend is toward common-channel signaling versions of ITU SS7.

Pricing elements for MTS and dial-1 WATS include connection or installation charges, call duration and mileage-based usage charges, time-of-day and day-of-week discounts, and for WATS, volume discounts. Dedicated-access WATS involves installation charges, special-access dedicated-line charges, call duration and mileage-based usage charges, time-of-day, and day-of-week and volume discounts. As stated previously, rates for LEC services used to access IXC services are governed by tariffs established by the FCC. Rates for IXC services are also governed by FCC tariffs.

## IXC TOLL-FREE SWITCHED-TRANSPORT SERVICES

The public switched services described above are called *outbound* services because call originators pay for the call (e.g., public switched transport service subscribers). In 1967, AT&T introduced a business service wherein call recipients (the business service subscribers) pay for inbound calls. Companies find this attractive since it provides incentives for potential customers to procure merchandise or services from them. A special area code, 800, was assigned for phone numbers associated with this service. Since subscribers receive rather than originate calls, the service is called an *inbound* service. The general public uses the term "800 calls" to refer to these calls. Since calls made are at no charge to the callers, various services in this class are referred to as "toll-free" services.

By 1981, common channel signaling systems made it possible to access centralized databases where toll-free call processing is provided in accordance with specific subscriber instructions. This processing is not limited to translating 800 numbers to subscriber public numbers for routing (a basic process of any toll-free service), but includes translation to various numbers based on time of day or originating-call location. The centralized database was maintained by AT&T, the only provider of 800 service at that time.

After divestiture in 1984, multiple interexchange carriers began offering toll-free services, each with its own subscriber information databases. A major problem appeared, however, in routing 800 calls. Previously, any call dialed as 1-800-NNX-XXXX was routed to AT&T for toll-free processing. After divestiture, local exchange carriers had no way of telling which 800 numbers belonged to which IXC. A system was devised to assign specific NNXs (the 4th, 5th, and 6th digits of an 800 number) to specific IXCs. The local exchange carriers could then route 800 calls based on NNX portions of dialed numbers.

While this approach allowed calls to be completed, it suffered from two major flaws. First, subscribers had to change 800 numbers when they changed toll-free service providers. Since businesses spend large amounts of money advertising their 800 numbers, changing numbers is a major inconvenience and expense. To make matters worse, companies traditionally requested vanity 800 numbers (i.e., numbers spelling business-related catch-phrase words when keypad letters are used for dialing). At a minimum, changing 800 numbers when changing toll-free service providers prevents the continued use of well-known, business-associated numbers and necessitates significant stationery, business card, and other media updates.

Both these side-effects combined to create great inequities between the service provided by AT&T (with all the incumbent customers) and the competitive services offered by new IXCs. In 1991, the FCC ordered the industry to make 800 numbers "portable" among IXCs by mid-1993. This resulted in the creation of a national database system known as

Service Management System/800 (SMS/800). LECs now access this database system to determine IXCs associated with particular 800 numbers. As a consequence, 800 service-providing carriers can now be changed without 800-number changes, and toll-free service markets are much more competitive. Demand for toll free numbers is increasing so dramatically that a new toll-free prefix (888) was assigned in 1996 and another (877) in 1998. Today, almost half the long-distance calls made in the United States during daytime are calls to toll-free numbers.

Figure 10.9 illustrates *IXC toll-free switched service* facilities. Calls to toll-free numbers are made from public switched networks. The signaling system recognizes calls as toll free by their special area codes and interrupts call processing to query toll-free databases at *Service Control Points*. Identification of the toll-free carrier servicing the dialed number, and any other routing instructions, are returned to originating switches. Calls are then handed off to appropriate carriers for further processing. At the terminating end of a call, access from the IXC POP to its subscriber premises can be via dedicated or switched access, as with other IXC services.

For small businesses, or even individual users, switched access is a more economical choice. For reasons previously discussed, high-traffic volume users normally employ dedicated access lines to reduce per-call costs.

While toll-free services have always been an important business tool, they've never been as prominent as they are now. Today, toll-free calls are the lifelines for an ever-increasing number of large business-call centers implementing a wide variety of highly automated applications. Modern call centers employ hundreds of agents who take orders, provide customer service, and render myriad of other crucial business functions.

Due to the efficiency of telecommunications, these call centers can often be located away from the main business headquarters—closer to customers or in areas where agent labor costs are lower. In very large organizations, with multiple, geographically dispersed call centers, modern toll-free services provide a variety of advanced features that seamlessly integrate multi-center operations—giving customers the impression they are dealing with a single center.

Toll-free features are classified as either pre-connection or post-connection. Pre-connection features efficiently route calls to the "best" agent in the network. This routing is controlled by data and software residing at carrier *Network Control Points* (NCPs). Routing plans are specified by customers and may be static or dynamic. Static routing

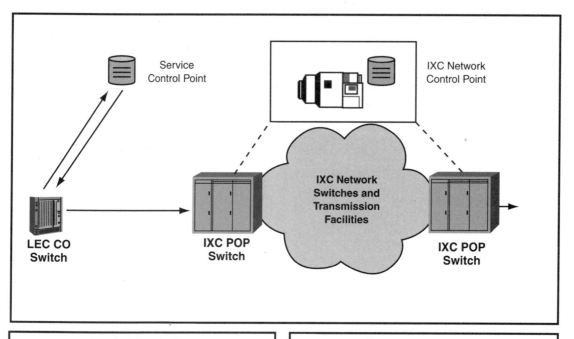

Service
Control Point

IXC Network
Control Point

**IXC Network
Switches and
Transmission
Facilities**

**LEC CO
Switch**

**IXC POP
Switch**

**IXC POP
Switch**

### Principal Applications

**Toll Free Service:**
• Feature rich environment provided by database-driven
  call processing

**Dedicated Access Toll Free Service:**
• Dedicated customer premises-to-POP access provided
  at the destination

### Technical Characteristics

**Facility Description:**
• Digital stored program control tandem switches
• Mostly digital IXC interswitch transmission
• Increasing use of fiber optic transmission
• Redundant paths with alternate routing

**Signaling Techniques:**
• Common channel signaling (SS7)
• Access to Service Management System/800 database
  at signal control points
• IXC provides advanced features at Network Control Points

### Price Elements

• Installation charges
• Call duration usage charges, often mileage-based
• Off-peak time period discounts
• Volume discounts
• Feature charges

**Figure 10.9**    *Interexchange toll-free switched transport services.*

plans are stored at NCPs and executed in accordance with criteria established by subscribers. Criteria may include time of day, caller location or other parameters.

Dynamic routing plans automatically adjust to changing conditions. An example of a sophisticated system employing dynamic routing is shown in Figure 10.10. The "Intelligent Call Manager," a product marketed by Cisco Systems since its acquisition of GeoTel, (the original developer), uses real-time call center equipment-derived information to make routing decisions. This information may include estimates of agent loading, customer queuing delays, or the number of agents available to serve callers at any instant.

**Figure 10.10**
Advanced toll-free routing application example.

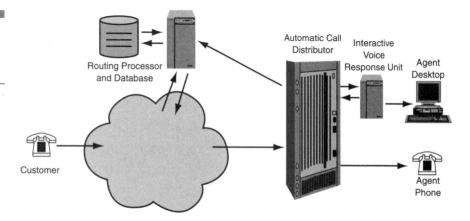

The routing processor gathers real-time information on agent loading and other factors from automatic call distributors and stores the data in a database along with customer profiles and routing plans.

A toll free call is first routed to the routing processor which uses the stored information (and possibly information about the caller from the Caller ID) to determine the most efficient routing for the call.

Routing information is passed back to the network, which routes the call to the proper destination ACD.

The system can also request caller identification numbers, access customer account and profile database information, and complete other call-facilitating actions prior to making routing decisions. Information in user profile databases may also be forwarded to agent monitors as a call is delivered. These and numerous other features are designed to help businesses provide the best possible customer call response service.

Post-connection features take effect once initial connections to agents or *interactive voice response* units (IVRs) are established. With new post-connection caller information, a more appropriate destination for some calls may be warranted. When this occurs, rather than just forwarding calls to new numbers, and tying up unnecessary network resources in the process, a message is sent to a network call-monitoring processor. The original call path is disconnected, freeing current network resources, and rerouted to a new destination. Reconnection actions requiring explanations to callers can often be delivered using customized, prerecorded announcements.

Given the high visibility with their customers that toll-free services provide for businesses, it is not surprising that they are the fastest growing of all carrier voice services. In the future, businesses are likely to combine voice-network and Internet telecommunications-based customer service centers. When this happens, a whole new generation of call-processing capabilities may emerge in which agents and customers will be able to communicate using voice, data, graphics, and video media.

Initial inquires made using the Internet, for example, may at the click of a mouse permit customers to establish a voice-network connection to an agent who could subsequently assist in understanding a Web page from which the voice call was placed, or direct customers to more helpful information. Conversely, from initial voice-network calls, agents might direct Internet access-equipped customers to Web pages that both customers and agents can view and discuss simultaneously. These and other innovations will greatly expand a businesses' ability to help customers make purchasing decisions or otherwise service their needs.

### VIRTUAL PRIVATE NETWORK SERVICES
As noted above, the advent of centralized computer-controlled call processing (that is call processing not residing in switches) and centralized database-driven call routing has given rise to another revolution in the provision of large customer switched services.

The access service examples discussed above show that dedicated access services can be more cost effective than switched services for large volume users, due principally to excessive per-call usage charges incurred with high calling volume.

Prior to the mid-1980s, this situation also existed for transport services. In response, some large users purchased or leased backbone switches and dedicated interswitch circuits (trunks) to interconnect business locations. Access to the backbone switches was also dedicated. Since only one company's traffic is carried on a private network, features and routing plans can be customized for each organization.

However, there are some negative attributes of this approach. First, due to the statistical nature of traffic demand, during off-peak hours, some "paid-for" network capacity is often wasted, partially offsetting anticipated private network savings. Second, a large set of skilled personnel is needed to design and manage private networks (several hundred staff are often required).

Although many organizations implementing private networks were very large (IBM's internal network served over 300,000 users), even they could not approach the economy of scale realizable in public telephone companies. In that timeframe, the problem with public carrier facilities was that technologies did not exist to sense calls originating from individual organizations and subsequently provide the special, customer-defined call processing and other custom features of interest to large corporations. By contrast, all these capabilities were possible in facilities-based private networks.

Centralized, database-controlled public network architectures changed the rules. It is now possible to perform many advanced call-processing operations, all customized in accordance with rules defined by customers. Dialed numbers can be translated and calls routed to specific places or restricted based on originating locations. Capacity between customer locations can be engineered for specific customer traffic flows, but can also be shared with other customers. The net effect is that virtual private networks can be created "in software" that benefit from the economies of scale of the entire public network. All that is needed is that carriers pass the savings on to large users in the form of lower usage rates.

Divestiture provided the killing blow to facilities-based private networks. The creation of LATAs made the long access lines from many user locations to the nearest private backbone switch prohibitively expensive. In addition, access lines had to pass through IXC POPs on the way to backbone switches, a requirement that makes IXC switched service less expensive than private alternatives.

In 1984, AT&T introduced its Software Defined Network (SDN) service. MCI and Sprint soon followed with vNet and VPN services, respectively. Competition quickly drove prices down and private networks for voice disappeared from the corporate landscape by the end of the decade. Since these services provided private-network functionality on a shared (or "virtual") basis, and since their services were virtually indistinguishable from those of facilities-based private networks, they became known—generically—as virtual private network services.

Figure 10.11 illustrates virtual private network services. The defining characteristics are the existence of subscriber locations using dedicated access on both ends of calls and the involvement of a database-controlled central processor in the routing of every call.

Today, virtual private networks are provided on both a tariffed and a contract basis. Beginning in the early 1990s, the FCC allowed rates proposed in competitive bids to be filed as special tariffs not subject to the same regulatory restrictions as normal tariffs. This applied in particular to AT&T, which in that timeframe was still being regulated as a dominant carrier. These "contract tariffs" became the principal vehicle by which competing IXCs marketed VPN services to large government and commercial enterprises. Chapter 12 discusses tradeoffs and processes for selecting voice services in more detail.

### NON-SWITCHED TRANSPORT SERVICES

Figure 10.12 depicts *IXC non-switched services.* These are similar to ones offered by local exchange carriers except that they provide dedicated transmission between POPs. As such, they cover much longer distances than their LEC counterparts. DCS equipment is used extensively to groom and aggregate lower-speed, non-switched circuits by successive multiplexing onto higher-speed, long-haul transmission facilities in a manner analogous to county roads feeding onto state highways that feed onto large interstate superhighways.

Non-switched transport is rarely used for voice interswitch trunks because of the existence of the much more economical shared alternatives described above. Imagine leasing a whole lane from coast to coast on the interstate highway system. Instead, most non-switched transport today is supplied as private lines connecting computers or private packet switches and used for data rather than voice. Historically, the industry practice of procuring or leasing facilities-based private data networks has outlasted that practice for voice networks. The primary reason is that common-user, public data network facilities of a size that dwarf private data networks and produce VPN-like economies of scale, are yet to be developed. Chapter 13 sheds light on current private-versus-public switched data network service tradeoffs.

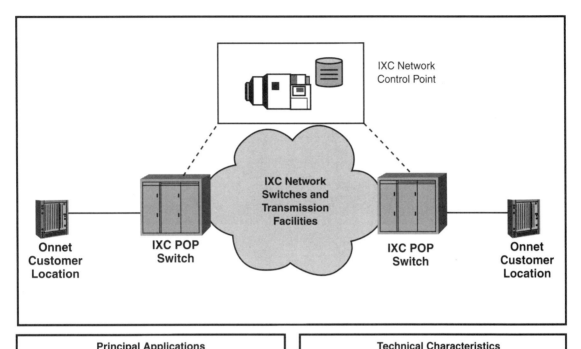

**Figure 10.11** Virtual private network services.

## Principal Applications

**Virtual Private Network Service:**
- Feature rich environment provided by database-driven call processing
- Calling among onnet customer locations supported
- Onnet locations may use dedicated or switched access

## Price Elements

- Installation charges
- Call duration usage charges, often mileage-based
- Off-peak time period discounts
- Volume discounts
- Feature charges

## Technical Characteristics

**Facility Description:**
- Digital stored program control tandem switches
- Mostly digital IXC interswitch transmission
- Increasing use of fiber optic transmission
- Redundant paths with alternate routing

**Signaling Techniques:**
- Common channel signaling (SS7)
- IXC provides advanced features at network control points

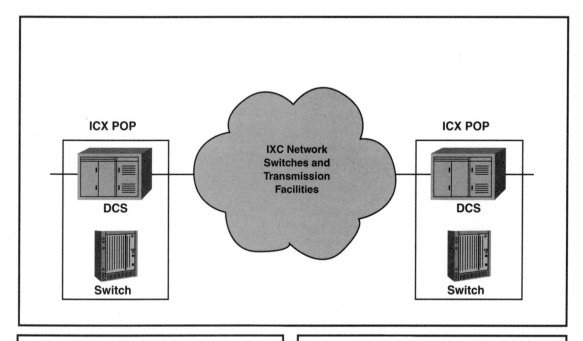

**Principal Applications**

**Private Lines:**
• Voice, data, and other applications

**Price Elements**

• Installation charges
• Mileage dependent monthly transmission charges

**Technical Characteristics**

**Facility Description:**
• Digital transmission facilities (e.g., DS1, DS3) used in the transport network
• Digital cross-connect system (DCS) used to "groom" connection arrangements on a per-customer basis. DCS connections:
  – Are not changeable on a call-by-call basis
  – Can change on a preset time basis
  – Offer rerouting protection in case of disasters

**Figure 10.12**  Interexchange non-switched transport service.

# CHAPTER 11

# Cost Structures
# and Network
# Design

The network design challenge is to connect combinations of switches, transmission media, and terminal equipment to allow users to exchange information. As we have seen in the previous ten chapters, economics motivates the development and use of switches and multiplexers to share expensive transmission resources. In telecommunications, the nature of cost functions and statistically efficient use of resources governing performance favors large aggregations of traffic. That is, at equivalent performance levels, a single network resource handling a large quantity of traffic is less expensive than several resources with smaller amounts of traffic. Properly designed, networks permit traffic aggregation while efficiently routing traffic to destinations in response to user demand.

To be successful, network design must determine the type, quantity, and location of resources needed to meet system-performance requirements at minimum cost. Costs include initial equipment and installation investment and recurring costs for use of telecommunications resources. Often, tradeoffs are involved between these cost elements. Telecommunications carriers provide resources primarily through tariffed services. Consequently, knowledge of both tariffs and network design is crucial in implementing networks that meet all business requirements while minimizing costs. An introduction to principles underlying tariffs is presented in the next section. Following this, principles of *traffic engineering,* a branch of network design science which predicts and measures performance and can be used to evaluate how well tariffed offerings satisfy business requirements, are introduced. Chapters 12 and 15 deal with voice and data network design in more detail.

# Tariffs

A *tariff* is the published rate for a specific telecommunications service, equipment, or facility that constitutes a public contract between the user and the telecommunications supplier or carrier. Tariffed rates are established by and for telecommunications common carriers in a formal process in which carriers submit filings for government regulatory review, possible amendment, and approval. Historically, tariff filings provide the most complete and precise descriptions of carrier offerings. Further, tariffed offerings cannot be dropped or changed without regulatory approval.

All IXCs were originally regulated by the FCC. In the Competitive Carrier proceeding (Docket 70-252), the FCC decided to forbear from regulating *other common carriers* (OCCs). As the dominant IXC, AT&T was still regulated by the FCC and required to file tariffs for its IXC services. This regulation required AT&T to justify not only that its rates were not excessive (i.e., protecting consumers), but also that it was recovering its costs (i.e., it was not using its monopoly power to drive its nascent competitors out of business). When the IXC market became sufficiently competitive, AT&T was also relieved of much of its regulatory burden. Although this freedom from regulation no longer requires IXCs to file tariffs, most of them do file rate structures for basic services as tariffs. If and when the FCC no longer accepts tariffs, IXCs will be required publicly to post their rates and other tariff-like information for customer review. Hence, the ensuing discussions of tariff rates in this chapter will continue to apply to all IXCs.

LECs are also regulated and required to file tariffs for most of their LEC services. The regulating entity is normally the state Public Utility Commission, with one exception. As noted in the last chapter, LEC services providing users access to IXC inter-LATA services and carrying inter-LATA traffic are regulated by the FCC under a separate set of tariffs. These services provide access from a user's premises to the IXC POP on a shared or dedicated basis. Dedicated access is provided under *special access* tariffs, and shared access is provided under *switched access* tariffs.

Figure 11.1 summarizes different types of tariffs and regulatory bodies responsible for each. The jurisdictional partitioning shown in the figure assumes that inter-LATA networks carry interstate user traffic. For networks entirely within the boundary of a single state, intrastate versions of the inter-LATA tariffs apply and are under PUC jurisdiction. The segmentation of tariffs into "access" and "transport" parts, shown in the figure, is consistent with and traceable to the technical performance-based subdivision of telecommunications *network services* into *access* and *transport* service segments.

Recall from Chapter 1 (reinforced in Part 3) that network services needed to complete calls include originating access service, transport service, and terminating access service. *Originating access service* includes all facilities and related services needed to support information transfer from an information source to some network's transport service. *Transport service* includes switching, multiplexing, transmission, and any other facilities and related services needed to support information transfer between originating and terminating access services. *Terminating access service* includes all the facilities and related services needed to support information transfer from some network transport service to an information sink. Information *sources* and *sinks* are, respectively, those parts of telecommunications systems that transmit and receive information.

**Figure 11.1**
Types of tariffs.

| Intra-LATA | | Inter-LATA (Interstate) | |
|---|---|---|---|
| LEC | | LEC | IXC |
| Access | Transport | Access | Transport |
| Dedicated Services — Private Line (PUC) | | Special Access (FCC) | Private Line (FCC) |
| Shared Services — Switched Services (PUC) | | Switched Access (FCC) | Switched Services (FCC) |

In cases where multiple or concatenated networks are used to connect end-user source/sink equipment, the facilities underlying access service to higher-level networks may comprise customer-owned equipment and both access and transport services from connecting networks. This is precisely the case where business or residential premises use intervening LEC local serving area networks to gain access to IXC long-distance networks. It is the access portion of IXC inter-LATA tariffs regulated by the FCC in Figure 11.1 that compensate LECs for LEC-switch-to-IXC-switch access service. Note that here LECs use LEC dedicated (special) or shared (switched) transport service to render IXC-access service to its customers. Figure P3.3 allows the reader to trace and analyze the above LEC and IXC interconnections on a complete end-user-to-end-user basis.

During the transition period between the MFJ-controlled world of LATAs and restrictions on ILECs, and a fully competitive world—where boundaries between local and long-distance service are blurred-LATAs will continue to exist and will define boundaries between what we commonly call *IXC access* (user-to-POP communication) and *IXC transport* (POP-to-POP communication). When today's LECs begin to offer both local and long-distance service, the role of local access and transport areas (LATAs) in establishing tariffs or regulatory boundaries may become superfluous. However, as long as long-distance carriers exist that must rely on another carriers' local access and transport area facilities to offer service to nationwide subscribers, such IXCs will have to establish POPs within those LEC serving areas and compensate them for rendering subscriber-to-IXC access service.

In the remainder of this book, we account for current LATA-driven tariffs and regulatory boundaries but present material in the context of rapidly evolving regulation-free, competition-driven developments.

## Cost Recovery Structures

Whether determined by regulation or by competitive pressures, the rates charged for telecommunications services are designed to recover the costs of providing the services. Historically, correlation between rates for a particular service and the cost of providing that service has not been high. In a monopoly environment, one service (like long-distance calling) could subsidize another (like local phone service) with little overall impact on the monopoly service provider. In today's highly competitive, post-divestiture environment, subsidy is economically unwise. Hence, current rate structures reflect underlying costs and attempt a fair distribution of costs among service users.

From a pricing viewpoint, the primary distinction is between dedicated services and shared services. A *dedicated service* assigns resources to users for exclusive use on a full-time basis. The rates for such a service are on a monthly basis and independent of actual quantities of user traffic carried by the service. A *shared service* provides resources shared by many customers. Customers may request resources at any time, and pay in accordance with actual service usage.

Pricing for IXC-provided services is the most complex today, since IXCs typically rely on LECs to provide access services to the end-users. There are three distinct pricing environments for each IXC service: originating LEC access, IXC transport, and terminating LEC access. Each environment can be priced on a dedicated or shared basis, and each combination defines a specific IXC service offering. Generally, high-volume users can justify dedicated resources while low-volume users are more economically served by shared resources.

When dedicated LEC services are used to access IXC services, LEC charges are normally paid directly by the user. When shared LEC services are used to access IXC services, LEC charges are usually paid by the IXC and recovered in IXC usage rates. Figure 11.2 illustrates three examples of IXC price structures and the underlying facilities used to support them. These examples are discussed further in Chapter 12. Figure P3.3 presents additional facilities details involved in IXC access and transport service provision, including typical PBX-based premises to LEC interconnections.

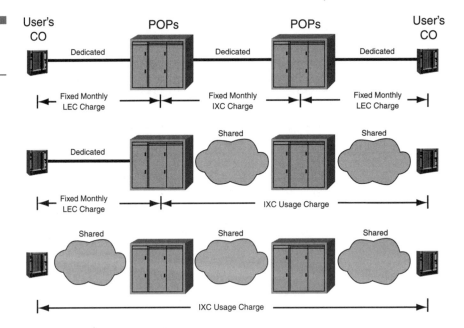

**Figure 11.2**
LEC/IXC tariff examples.

## Elements of Tariffs

In the remaining parts of this chapter, some mathematical formulae are introduced. Although not necessary to an understanding of tariff concepts and applications, these formulae are included to furnish additional clarification for use by telecommunications professionals.

Tariffs are based on common elements used to recover costs of specific telecommunications resources. Understanding the terms and concepts behind these elements facilitates tariff analysis.

Dedicated services require carrier interface resources and assignment of transmission facility bandwidth between user premises. Accordingly, the pricing structure for a dedicated service contains a *fixed-charge* element and a *distance-sensitive* element. The unit cost per mile may be constant or may decrease with increasing mileage (reflecting use of more efficient, high-volume long distance transmission media). Moreover, carriers may give volume discounts in exchange for long-term monthly minimum purchase commitments.

Shared services also require interface resources and thus a fixed charge element, but the bulk of shared service charges reflect usage. *Usage charges* are derived from a unit price (or a set of unit prices based on distance), multiplied by the number of units actually used. These charges may be discounted by time of day or day of the week. Carriers charge less during low-use time periods since additional traffic can be carried during those periods without adding shared network facilities. Again, certain services may include volume discounts.

Since many tariffs are distance sensitive, there must be an established, simple mechanism for determining distances as a basis for price determination. Fortunately, it's not necessary to know the latitude and longitude of every building in the country to price telecommunication services. Rather, in modern tariffs, distances are computed only between key physical network elements employed to provide services. These elements, introduced earlier, include LEC central offices serving users, and IXC POPs. Figure 2.3 illustrates the physical structure, but it should be noted that several COs may reside in the same LEC building.

For tariff purposes, *wire centers,* locations of one or more local switching systems and a point at which customer loops converge, are assigned specific coordinates (*wire-center coordinates*) on a grid system. Each IXC POP is also assigned a coordinate on the same grid. Known as the V&H Coordinate System, the grid was developed to permit accurate mileage determination in North America. Its origin is in Greenland and vertical (V) and horizontal (H) numerical coordinates increase toward the west and south in a somewhat skewed manner as shown in Figure 11.3. Distance is determined using a modified version of the Pythagorean formula:

## EQUATION 11.1

$$D = \sqrt{\frac{((V1 - V2)^2 + (H1 - H2)^2)}{10}}$$

Note: The division by 10 is performed before the square root is taken.

Distances for LEC tariffs are computed between end-point-serving wire centers, while distances for IXC tariffs are computed between IXC POPs. V&H coordinates for the IXC POPs are either published in IXC Tariffs (e.g., AT&T Tariff #10) or are available directly from the IXC. Wire-center coordinates are uniquely defined by the first six digits of public telephone numbers.

**Figure 11.3**
V&H (vertical and horizontal) tariff coordinate map.

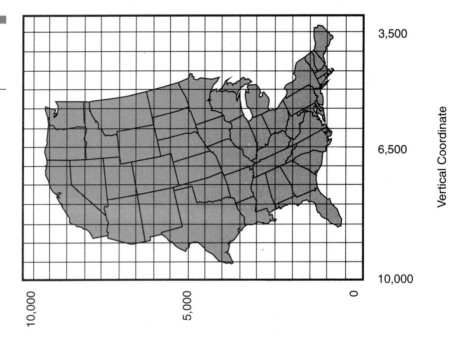

The North American numbering plan uses 10 digits to identify each telephone. The first three digits represent the *numbering plan area* (NPA) portion (commonly known as the *area code*). The next three digits represent the *exchange* portion of the number. In historical Bell System parlance, N refers to a digit which can assume the values 2—9 and X is a digit which can assume values 0—9. Using this convention, NXX has become a commonly used term referring to the exchange portion of a telephone number. NPA-NXX is sufficient user location information to determine distance for the purposes of computing tariffs. Electronic databases associating NPA-NXX with V&H coordinates, LATAs, serving telephone companies, cities, and states are available from Telecordia and several commercial companies. Such databases are essential for tariffed service analyses.

## Examples of Tariff Elements

Figure 11.4 illustrates two examples of services and the tariff elements used to price those services. The first example corresponds to end-to-end, dedicated private-line service, while the second example portrays dedicated private-line access to shared, switched service. The multiple fixed and mileage-dependent elements indicate the complexity involved with large network analysis. Ordinarily, large networks must support a plurality of different end-to-end services. They involve many LECs, each with different PUCs and tariffs, and potentially a mix of alternative carriers, again with different tariffs.

While carriers render computer-based network modeling and design optimization assistance, normally they treat only their own service offerings. Neutral organizations such as Mitretek Systems of McLean, Virginia offer end-users comprehensive network modeling and design optimization services, and since they offer no telecommunications carriage services, they provide unbiased assessments of all carrier offerings. Chapters 12 and 15 further detail tariff elements for a wide range of LEC and IXC services.

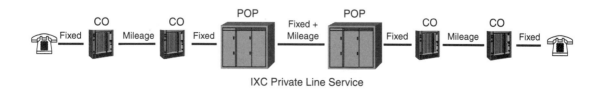

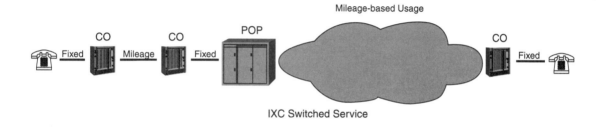

**Figure 11.4**    Examples of tariff elements.

# Traffic Engineering Basics

The concept of sharing telecommunications resources through switching immediately raises the engineering question of how many shared resources must be provided to ensure adequate user performance (however one chooses to define adequate). If too many resources are provided, many are never used, extra costs are expended by carriers, and rates charged for service are higher than necessary. On the other hand, if too few resources are provided, service might not be available when requested. In such cases, users are forced to try later or join queues, waiting for a resource. Traffic engineering experience teaches that in network performance determination and analysis, the following factors are key:

- User demand for resources
- User behavior upon finding all resources busy

It is interesting to note, and helpful to understanding underlying principles, that these factors and the mathematics used to quantify performance apply to other than telecommunications resource sharing. Although our focus is on telephone callers and shared trunks, interactive behavior between grocery shoppers and checkout clerks or customers and bank tellers is analogous. Visualizations based on everyday events help both managers and engineers to see beyond the apparent complexity imposed by mathematical rigor and to gain a practical insight into the business consequences of traffic engineering factors and characteristics.

Characterizing demand for resources is the first important step in traffic engineering. Demand has two components: first, how often users "arrive" or request resources (termed *arrival rate*); and second, how long they use resources before releasing them (termed *holding time*). (Note: It is assumed that resources in use by one user are unavailable to others until released.) The product of these quantities is a measure of traffic demand on resources. As shown below, both components can be statistically modeled, and experience shows that model predictions correlate well with actual measured telecommunications system performance.

As users randomly arrive to request service, two outcomes are possible. If a resource is free, it is seized by the user for a period of time. If no resource is free (all resources are busy), the user may or may not leave without receiving service. This "busy" phenomenon is known as *blocking* in telephone networks and users who leave without receiving service following a busy indication are said to employ a *lost-calls-cleared* discipline.

Most voice networks operate in this manner. The measure of performance in such a system is the probability of blocking that users experience when entering the system. As discussed in Chapter 6, this probability is commonly called *grade of service* (GOS)—although, as previously noted, GOS can also relate to indicators of customer satisfaction with other aspects of service, such as noise or echo. In the traffic context, a grade of service of 1 percent, or 0.01 means that for every 100 users entering the system, on the average, one will find all resources busy.

While voice networks operate in the above manner, data networks (and grocery stores) use a different discipline when users encounter busy resources. Here users join queues (lines) and wait for service. In this discipline, called *lost-calls-delayed*, the measure of performance is *average delay*, that is the elapsed time between when a user enters a system and when a service is completed.

## Beginning Theory

The discipline of computing performance for a given traffic demand, number of resources, and probable distribution of arrivals and service times is called *queuing theory*. A comprehensive discussion of this topic is beyond the scope of this book, but basic assumptions and procedures yield useful answers to most traffic engineering problems, and a rudimentary understanding of the theory provides valuable insight into the relationship between network design and business operations which the networks support.

In the previous section, we characterized arrival of users and the time spent in the system as statistical quantities. Figure 11.5 helps illustrate the importance of this characterization. In the figure, the first example shows three users presenting a combined load of one hour of traffic during an hour of time. Each user makes one call with a holding time of 20 minutes. The calls are scheduled so that they occur on the hour, at 20 minutes past the hour, and at 40 minutes past the hour. With such arrival scheduling and controlled call duration, a single circuit is sufficient to carry the traffic load with no blocking.

The second example shows the same average load over the hour, but all arrivals occur during the first 20 minutes. One circuit results in two of three calls being blocked over the hour even though the average offered load is the same as in the first case. The situation would be even more complicated with varying call duration. The statistical distribution of arrival times and call lengths plays a major role in calculating expected performance. Fortunately, a large body of analysis has been accomplished and a workable characterization developed in terms of both call *interarrival* and *call-duration* distribution.

**Figure 11.5**
Call traffic examples.

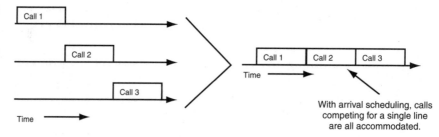

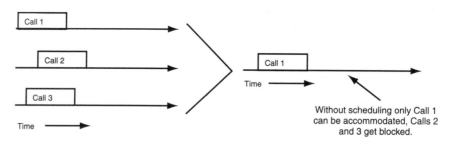

(a) Scheduled Call Arrivals

(b) Random Call Arrivals

As it turns out, both quantities have been shown to follow a negative exponential distribution shown in Figure 11.6. For example, in the figure, the average call duration is three minutes. As shown, call durations with values much larger than average values can occur, but only with ever-decreasing probability. Calling patterns with negative exponential distribution interarrival times are said to follow a *Poisson (random) distribution*. All the performance formulas presented herein are based on assumptions that traffic obeys these statistical distributions. As a consequence, users need only calculate *average arrival rate* and *average holding time* to characterize traffic demand. Traffic demand or intensity is commonly called *offered load*.

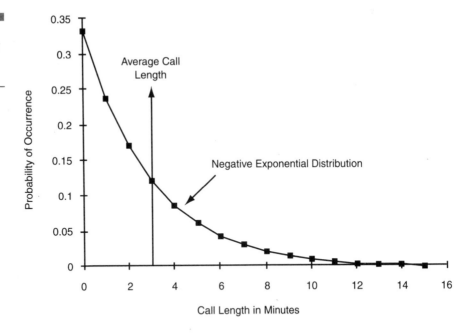

**Figure 11.6**
Statistical variation of random call lengths in traffic networks.

The international dimensionless unit of traffic demand, obtained by multiplying arrival rate by holding time, is called an *erlang*, named after A. K. Erlang, a Danish mathematician acknowledged to be the father of queuing theory. Erlang's formulae, derived in 1917, remain the backbone of queuing performance evaluation today. Quantitatively, one erlang of traffic is equivalent to a single user who uses a single resource 100 percent of the time. Alternatively, one erlang of traffic is generated by 10 users who each occupy a resource 10 percent of the time. An example traffic intensity (or offered load) calculation for a group of 100 users each making an average of three voice calls per hour, with an average length of three minutes per call is as follows:

## EQUATION 11.2

- Erlangs = (call arrival rate) × (holding time)
- Erlangs = (100 users) × (3 calls per user per hour) × (3 minutes per call)
- Erlangs = (300 calls per hour) × (0.05 hours per call)
- Erlangs = 15

Note that in the equation, all dimensional units cancel out (the 3 minutes per call had to be converted to 0.05 hours for this to occur). It is essential for all quantities to be measured in the same units when using traffic and queuing formulae. One erlang is equal to 36 CCS (*centi call seconds*), another dimensionless traffic measure defined in the example cited in the PBX section of Chapter 6. Either measure can be used in characterizing network performance, providing care is taken to ensure that consistent dimensional units of measure are used in calculations.

Besides traffic, the other quantity needed to compute performance is the number of circuits (or more generally, *resources*) available to handle traffic. Circuits act like tellers in a bank; they are "servers" satisfying shared service demand. In traffic engineering, there are two categories of circuits: those dedicated to a single user or piece of equipment (i.e., dedicated circuits or lines) and those shared among users using switches or concentrators (i.e., shared circuits or trunks). A *trunk group* is traffic engineered as a unit for the establishment of connections within or between telephone company switching systems in which all paths are interchangeable. A trunk group can be searched in sequence until a free trunk is found.

The basic voice traffic engineering problem is to determine the probability of blocking of a trunk group given an erlang load and the number of circuits in the group. Since each call utilizes a trunk in both directions, the erlang loads offered to both ends of a trunk group are added together to compute the total load. Erlang derived a formula to compute blocking using the above statistical assumptions. For *lost-calls-cleared*, the formula is called the *Erlang-B formula* and is as follows:

## EQUATION 11.3

$$B(a, n) = \frac{(a^n / n!)}{\sum_{i=0}^{n} (a^i / i!)}$$

Here B(*a,n*) is the probability of blocking (grade of service) which n circuits provides under an offered traffic load of "*a*" erlangs.

Unfortunately, this closed-form equation is very difficult to compute for large values of *a* and *n* (even a computer will encounter overflow and underflow errors working with numbers like 100 to the 100th power divided by 100 factorial). This difficulty is overcome using a recursive form of the equation, which can be implemented using a simple loop in a computer program:

## EQUATION 11.4

$$B(a, k) = \frac{a \times B(a, k - 1)}{k + a \times B(a, k - 1)}$$

$a$ = erlangs
$k$ = circuits
$B(a,k)$ = Probability of blocking
(lost-call-cleared)

Example computer code:

```
b = 1.0
do i = 1 to n
    b = (a x b)/(i + (a x b))
enddo
```

Here $B(a,k)$ is the probability of blocking which $k$ circuits provide under an offered traffic load of $a$ erlangs. In statistical analysis, a recursive estimation technique is one in which successive (recursive) estimates are based on previous estimates until a sufficiently accurate value of some desired numerical quantity (in this case blocking) is determined. It should be noted that in lieu of computer evaluation of Erlang's formulae, tables relating number of servers and blocking probability to traffic load, in terms of traffic load measured in either erlangs or CCS, are published in many electronic reference data books, such as the *Reference Manual for Telecommunications Engineering*, by Roger L. Freeman, published by Wiley-Interscience.

The recursive algorithm is also convenient for determining the number of circuits required to achieve a given grade of service by simply executing the recursive formula until blocking is less than a desired objective value. The formula is very nonlinear, i.e., identical changes in input parameter values do not result in proportional, or linearly scaled, changes in output parameter values. Such nonlinear performance reveals the reason why larger trunk groups are considerably more efficient than smaller ones in terms of carrying traffic at any given call-blocking performance level. At the heart of network design, this nonlinear property is discussed in more detail below in the section on behavior of traffic systems and illustrated in Figure 11.15. As a final Erlang-B observation, note that performance does not depend on arrival rate or holding time individually, but only on the product. For lost-calls-cleared numerous short calls have the same effect as fewer long calls.

*Lost-calls-delayed* performance, used for packet-switched systems, is characterized by the *Erlang-C* equation and associated equations for delay. It should be noted that the term "packet" in this context refers to the entity that is being switched as a unit (and held in queues). This may vary with the particular system under consideration. Early message-switching systems switched an entire message as a unit. As explained in Chapter 7, modern systems fragment messages into frames, packets, and cells as switched entities. In the processes presented below, the packet statistics should correlate with the entity appropriate for the system under consideration.

The relevant quantities here are *packet arrival rate, average packet length,* and a third quantity, circuit transmission speed or *line speed.* In this case, average packet length and transmission speed are treated separately since packet-length distribution is a demand parameter (not controllable by the system designer), while transmission speed is a parameter within a system designer's control.

As before, a usual assumption is that interarrival time and the packet length follow a negative exponential statistical distribution, and again the erlang load is the arrival rate multiplied by holding time. In this instance, however, holding time is packet length divided by line speed. Normally, packet length is specified in bytes while line speeds are specified in bits per second. The number of bits per byte depends on the mode of transmission. For asynchronous systems, such as analog modems, 10 bits are used to transmit each byte (8 bits of information plus a start and a stop bit). For synchronous digital systems, only the 8 bits of information are required. An example traffic-intensity (or offered load) calculation for a group of data terminals generating an average of 10 packets per second, with an average packet length of 360 bytes and using a 24,000 bps analog modem speed, is as follows:

## EQUATION 11.5

- Erlangs = (packet arrival rate) × (holding time)
- Erlangs = (packet arrival rate) × (packet length) ÷ (line speed)
- Erlangs = (10 packets per second) × (3,600 bits per packet) ÷ (24,000 bits per second)
- Erlangs = 1.5

Note again, that when consistent units are used all dimensional units cancel out.

For lost-calls-delayed, the probability of finding all circuits busy upon arriving and thus joining the queue is given by the Erlang-C equation:

## EQUATION 11.6

$$C(a, n) = \frac{B(a, n)}{1 - \frac{a}{n} \times (1 - B(a, n))}$$

a = erlangs
k = circuits
B(a,n) = Erlang-B equation
C(a,n) = Probability of blocking
(lost-calls-delayed)

Note that Erlang-C is given in terms of Erlang-B, which can be computed recursively as before.

The average delay time that a packet spends in a network (queuing time plus transmission time for the packet) is given by:

## EQUATION 11.7

$$T = \frac{L}{C} + \frac{C(a, n) \times L}{(1 - \frac{a}{n}) \times n \times C}$$

a = erlangs
L = packet length in bits
n = circuits
C = line speed in bits per second
C(a,n) = Probability of blocking
T = average packet delay time

It should be noted that the performance of multiple low-speed circuits is not equivalent to a single high-speed circuit. The basis for this is revealed in the two components of the delay equation presented above. The first component (L/C) takes into account the time required to transmit a single packet once a circuit is available. This delay component is referred to as *transmission time* and is dependent only on line speed, not the number of circuits in a trunk group. It represents a minimum trunk group delay. The second component corresponds to the time a packet spends in queues waiting for circuits to become available. This delay, queue time, is dependent on both line speed and the number of circuits in a trunk group. Note also that delay is not independent of packet length as in the lost-calls-cleared (circuit-switched) case where many short calls have the same effect as fewer long calls. In this case, for a given line speed, short packets result in less delay than long packets.

Queue lengths at opposite ends of a circuit are totally independent. Each direction acts like an independent circuit with its own traffic and performance. The only relation between two directions of transmission occurs when a circuit is purchased, when usually both directions of transmission are procured at the same bit rate.

Lastly, the average queue length, in packets, is given by:

### EQUATION 11.8

$$q = \frac{C(a, n) \times \frac{a}{n}}{(1 - \frac{a}{n})}$$

$a$ = erlangs
$n$ = circuits
$C(a,n)$ = Erlang c equation
$q$ = average number of packets in queue

The average time a packet spends in a network and queue length, in packets, are the performance measures of greatest interest to users.

## Traffic Engineering in Practice

The preceding section outlined basic traffic engineering theory. While this (and much more) is required for engineers designing and administering large national carrier networks, end-users typically need a more practical approach to estimating facility demands from more recognizable business need parameters. Ask telecommunications managers how many erlangs their sites generate in the busy hour and you will probably receive blank stares. However, they can usually tell you the number of call minutes on their monthly phone bills and whether their operations work nights and weekends. This section outlines a step-by-step approach to single site/trunk group traffic engineering based on typical business operation utilization characteristics.

Before analyzing examples, it is important to discuss how traffic demand varies as a function of time of the day and day of the week, and how these variations affect traffic engineering results. Since the relationship between demand and performance is nonlinear, performance under peak load conditions may degrade if the network design is based on average load. To understand the tradeoffs involved, let's first examine daily and hourly voice-traffic profiles for a typical corporate location.

Figure 11.7 shows the number of measured call minutes per day for a typical month. A complete month of traffic is the minimum that should be used for sizing studies. Weekends and holidays are apparent from this chart. It would also be obvious when measurement instrumentation does not capture all traffic for a day, as sometimes happens near the end of a month.

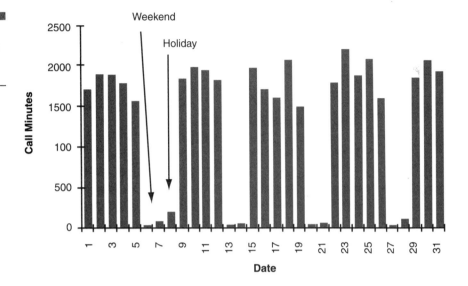

**Figure 11.7**
Example distribution of call minutes per day.

It is not necessary to examine every hour of every day explicitly. Nor, as indicated, is it advisable to base designs on average day traffic, i.e., total monthly traffic divided by the number of days. When this approach is pursued, the fact that traffic is low and performance is great on weekends (when no one is calling) is of no consolation to users routinely denied service on workdays during busy periods. To circumvent such deficiencies, it is standard practice to introduce the notion of *busy days* and *busy hours.*

Ordinarily, the next step would be to determine average *business day traffic* (i.e., the total business day traffic divided by the number of business days). Generally this is acceptable if the level of traffic for all business days is nearly equal. However, for some businesses, specific days of the week are considerably busier than others (such as having many customer phone-in orders on Mondays). To account for this phenomenon, yet not react to isolated single-day peaks with no enduring statistical significance, it is preferable to use *busy-day loads,* defined to be the average traffic of the five busiest days of the month. The effective number of busy days in a month is the total traffic in a month divided by the busy-day load.

After calculation of the busy-day load, the next level of analysis is to examine hourly *traffic profiles*. Again, good nighttime performance is not what users desire. An hourly profile for a typical corporate user is shown in Figure 11.8. Business hours and lunchtime behavior are readily determined from this chart. On the chart, calling that extends beyond the normal business day is normally due to multiple-time-zone traffic. Ordinarily, there are four busy hours, two in the morning and two in the afternoon, with approximately the same traffic level. Each busy hour normally accounts for between 12 percent and 16 percent of a day's total traffic. If no other information is available, the busy-hour traffic level can be assumed to be 14 percent of a busy day's traffic. By convention, performance is specified over the busy hour. Hence, the erlang load of a trunk group is the number of hours of traffic offered during the busy hour of a busy day.

**Figure 11.8**

Example distribution of call minutes per hour of the day.

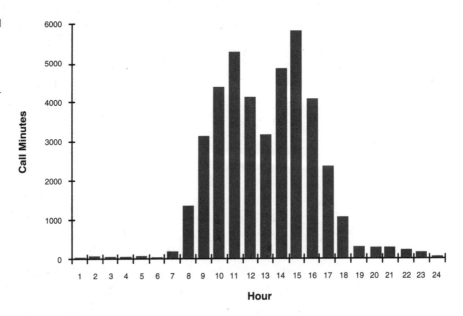

Different approaches may be taken to determine the traffic requirements of a system. For a completely new system with no historical traffic records, monthly minutes can be estimated and standard assumptions used to derive busy-day and busy-hour traffic levels. If a current system exists, monthly bills or even a month's *call detail records* in electronic form can normally be obtained. Actual daily and hourly distributions can be plotted from call-detail records, available from PBXs and other switching machines, and the actual busy-day and busy-hour loads calculated. Especially in large networks with many locations, this is by far the preferred method for establishing quantitative requirements.

Estimating requirements for data communications networks involves additional considerations. In gathering records, the basic units to be measured are packet arrival rates and packet lengths. Note that line speeds of current circuits do not constitute "requirements" for future or upgraded networks, but are merely part of the current solution. Too often with data networks, managers mistakenly state requirements in terms of circuits and line speeds as opposed to actual traffic units.

Also, it is important in data networks to identify requirements for both directions of traffic at each location so that each direction can be engineered separately. A large percentage of the total volume of data traffic may be via *batch transfers* at night. Since this traffic is not time sensitive, network capacity is normally engineered for *interactive traffic*—or query-and-response traffic usually involving human operators—with the objective to achieve specified delay-performance during busy hours. Once engineered, this capacity is typically sufficient for batch traffic, the transmission of which can span multiple hours after business closing times. Delay is the time interval between the instant at which a network station seeks access to a transmission channel to transmit a packet, and the instant that the network completes delivery of the packet.

The next two sections present voice and data examples separately, using illustrative worksheets which can serve as guides for other, more general traffic engineering exercises.

## Circuit-switched Example

The first example involves a 2,000-telephone line PBX to be installed in a new building. A group of central office trunks, known as a *two-way trunk group,* is used for outgoing and incoming local and long distance calls. If incoming calls are on a separate trunk group from the CO, then each group must be engineered separately. The configuration is shown in Figure 11.9. The worksheet shown in Figure 11.10 documents the computation (described in the following paragraphs) of the number of trunks required in the CO trunk group to achieve a grade of service of 0.02.

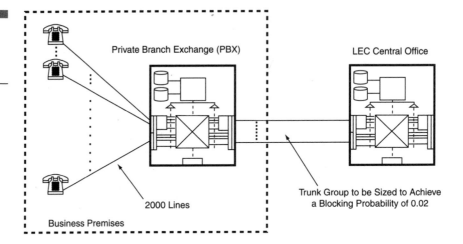

**Figure 11.9**
Example PBX trunking configuration.

| | Total | Incoming | Outgoing |
|---|---|---|---|
| Number of Stations | 2,000 | 2,000 | 2,000 |
| Minutes per Month per Stations | 400 | 200 | 200 |
| Minutes per Month | 800,000 | 400,000 | 400,000 |
| | | | |
| Busy Days per Month | 22 | 22 | 22 |
| Minutes per Day | 36,364 | 18,182 | 18,182 |
| | | | |
| Busy Hour Percent | 14% | 14% | 14% |
| Busy Hour Minutes | 5,091 | 2,545 | 2,545 |
| Busy Hour Erlangs | 84.85 | 42.42 | 42.42 |
| Overhead Factor | 1.1 | 1.1 | 1.1 |
| Adjusted Erlangs | 93.33 | 46.67 | 46.67 |
| | | | |
| Grade-of-Service | 0.02 | 0.02 | 0.02 |
| Number of Circuits | 106 | 57 | 57 |

**Figure 11.10**
Circuit-switched traffic example worksheet.

For this example no measured traffic statistics are available, but experience has shown that 400 minutes per month of originating and terminating "outside" traffic per telephone line is typical for business traffic. This results in a total of 800,000 minutes per month. Note that this estimate does not include station-to-station traffic supported by the PBX, since this traffic does not use trunks to the CO.

In the absence of measured or inferred business application data to the contrary, an assumption is made that half the traffic is derived from calls originating from the site. Also in the absence of measured traffic, it is assumed that there are 22 busy days in the month. Thus, busy-day load is computed by dividing 800,000 minutes per month by 22 busy days per month yielding 36,364 minutes per busy day. Similarly, using 14 percent as the busy-hour percentage of daily traffic yields a busy-hour load of 84.85 erlangs in both directions (42.425 in each direction). That is, 14 percent multiplied by 36,364 minutes per busy day equals 5,091 minutes per busy hour; and, 5,091 minutes per hour divided by 60 minutes per hour equals 84.85 erlangs (recalling that an erlang is equivalent to one circuit occupied full time).

One final adjustment that needs to be made is the conversion from *conversation time* (also called *billable time*) to *circuit usage time*. Traffic derived from billing sources or estimated from projected usage accounts only for the time users actually engage in conversation. In telephone parlance, this is the elapsed time between *distant-end answer* (or called party going off hook at the start of a call) and either party going on hook at the end of a call. However, circuits are actually tied up during call setup before the distant party answers. They are also occupied for uncompleted calls (busy signals or ring-no-answer calls), for which no billing records are generated. Statistics for these cases indicate an average overhead factor of 10 percent should be added to billing time to estimate total circuit usage. This factor is applicable for circuit sizing only and should not be applied to traffic used for estimating tariff costs, since carriers charge only for conversation time.

Using the Erlang-B formula, it is determined that 106 circuits are required in the two-way trunk group. Had one-way trunks been required, 57 circuits would be required for each direction, indicating the efficiency of larger groups, even for groups of this size. Differences are more marked at smaller sizes, as demonstrated below. When traffic begins flowing, actual loads and blocking performance should be monitored daily to determine if design adjustments are required. After the initial month, a monthly review of performance is adequate Monthly load profiles should also be tracked for possible seasonal variations. Using measured per-station statistics, the impact of increasing the facility size (e.g., the number of served stations) can be predicted and additional trunks provisioned ahead of time, minimizing the chance for level-of-service disruptions.

## Packet-switched Example

The second example considers an interactive switching system based on Internet technology. Fifty terminals are connected to a router. The terminals are requesting Web pages that are downloaded through the router to terminals. The average incoming packet length is assumed to be 530 bytes which, after overhead is added, is equivalent to the maximum Internet packet size. The outgoing packet length is much shorter, a 20-byte query in this example. While these assumptions do not correlate to normal negative exponential distributions for packet lengths (the fixed maximum length precludes very long packets that would otherwise occur, albeit infrequently, if the packet duration statistics followed the negative exponential distribution), standard equations nevertheless yield useful upper bounds on the delay performance.

Unfortunately, no standard assumptions are available regarding terminal activity, since the traffic is strictly dependent on the particular information system applications being used. If network service already exists, the best source for traffic information is actual characters per month (or packets per month). See Chapter 7 for data communications definitions and operational descriptions which can be derived from billing data. When historical data are not available, traffic is best derived on a per-terminal basis. This involves estimating peak bytes transferred per minute, based on specific knowledge of each major application's characteristics and organizational workloads.

The example Web application corresponds to the downloading of 50,000 bytes per minute (equivalent to one typical Web page per minute of incoming traffic). The terminal uploads one query packet for each page, resulting in 1,887 bytes per minute of outgoing traffic. Under this method, we characterize the busy hour and then use assumptions to derive circuit requirements. The 50 terminals receive 2,500,000 bytes per minute during the busy hour. This equates to an arrival rate of 78.62 packets per second, with an average length of 530 bytes. As in the voice example, it is necessary to account for packet overhead. In this case, overhead corresponds to extra bytes needed to implement data communications protocol functions. Again, these functions are discussed in detail in Chapter 7. For now, it can be assumed that 46 bytes per packet are required for overhead functions (typical for a TCP/IP implementation), yielding an adjusted average packet length of 576 bytes.

Prior to computing data-packet traffic statistics, the number of circuits and line speed must be chosen. As previously described, the average time a packet spends in a network and queue length, as measured in packets, are the performance parameters of interest. Line speed can be a crucial factor in providing acceptable response times or user packet delays (a half-second round-trip delay is usually required to avoid interactive-user annoyance). While the number of circuits in a trunk group multiplied by the line speed equals the total bandwidth available for packet traffic, in multiple-line arrangements the composite bandwidth is not available to any given packet. Consequently, to minimize transmission time and keep packet delays within bounds, long packets are better served by fewer high-speed lines than by larger numbers of low-speed lines.

For a packet length of 576 bytes, Figure 11.11 illustrates transmission times for typical line speeds in North America. For our example 576-byte packet lengths, line speeds at or above 28,800 bits per second achieve acceptable transmission times. We next size the trunk group at candidate line speeds as a prelude to selecting one that yields acceptable queue performance at minimum cost. Rules of thumb under maximum expected traffic loads are to maintain:

- Average queue times nearly equal to transmission times.

- Average queue lengths less than a single packet.

▬▬  ▬▬  ▬▬

**Figure 11.11**
Transmission times
for a 576-byte
packet.

| Technology | Line Speed (Bits per Sec) | Transmission Time (Milliseconds) |
|---|---|---|
| Modem | 14,400 | 400.0 |
| Modem | 28,000 | 200.0 |
| Modem | 56,000 | 102.9 |
| ISDN | 128,000 | 36.0 |
| DSL | 640,000 | 7.2 |
| T1 | 1,536,000 | 3.0 |

*1 millisecond = 0.001 seconds*

To continue the calculation, erlang load, which depends on line speed, is computed (see Equation 11.5 and Figure 11.12 worksheets for values). Since all traffic is treated as lost-calls-delayed, the number of circuits provided must be equal to or greater than offered traffic erlangs. If this is not the case, infinite queues for circuits will form (analogous to grocery store checkout lines before a snowstorm). Starting with the number of circuits determined by the erlang value (rounded up to the next integer), formulae for delay and average queue length are computed for additional circuits until both queue-time and queue-length criteria are met.

Economic factors relating to available transmission services in the United States dictate the following progression of trunk group choices:

- One analog circuit
- One 56-kb/s digital circuit
- One 128-kb/s ISDN circuit
- One 640-kbps ADSL circuit (where available)
- Multiple 56-kbps circuits
- One DS1 (1.544 Mb/s) circuit

An analog circuit can carry data up to 56 kbps with the proper modems. We contrast this with ADSL service providing 640 kbps downstream and 64 kbps upstream. The worksheets in Figure 11.12 illustrate the required number of circuits for the two configurations, imposing the queue length criteria. Based on progression of trunk group costs and the transmission and queue time performance, one ADSL circuit is the best choice by far for this packet switched example.

| Line Speed = 56,000 Bits per Second | Incoming | Outgoing | Line Speed = 640,000/64,000 Bits per Second | Incoming | Outgoing |
|---|---|---|---|---|---|
| Number of Terminals | 50 | 50 | Number of Terminals | 50 | 50 |
| Bytes per Minute per Terminal | 50,000 | 1,887 | Bytes per Minute per Terminal | 50,000 | 1,887 |
| Bytes per minute | 2,500,000 | 94,344 | Bytes per minute | 2,500,000 | 94,340 |
| Average Packet Length (Bytes) | 530 | 20 | Average Packet Length (Bytes) | 530 | 20 |
| Overhead Bytes | 46 | 46 | Overhead Bytes | 46 | 46 |
| Packet per Second | 78.62 | 78.62 | Packet per Second | 78.62 | 78.62 |
| Adjusted Packet Length | 576 | 66 | Adjusted Packet Length | 576 | 66 |
| Line Speed (Bits per Sec) | 56,000 | 56,000 | Line Speed (Bits per Sec) | 640,000 | 64,000 |
| Busy Hour Erlangs | 8.09 | 0.93 | Busy Hour Erlangs | 0.57 | 0.81 |
| Transmission Time (msec) | 103 | 12 | Transmission Time (msec) | 7 | 8 |
| Number of Circuits | 11 | 11 | Number of Circuits | 1 | 1 |
| Actual Queue Time (msec) | 9 | 0 | Actual Queue Time (msec) | 10 | 15 |
| Actual Time in System (msec) | 112 | 12 | Actual Time in System (msec) | 17 | 23 |
| Actual Number in Queue | 0.72 | 0 | Actual Number in Queue | 0.74 | 1.2 |

**Figure 11.12**  Packet-switched traffic example worksheets.

## Behavior of Traffic Systems

The previous sections focused on traffic engineering theory and the mechanics of calculating key network performance attributes. The purpose of this section is to illuminate tradeoffs between performance and resource utilization efficiency, that is, *cost effectiveness*. If cost were not a factor, it is clear that for any given level of traffic, very low blocking probability and packet delay can always be obtained by simply providing more—and higher speed facilities. The most valuable lessons learned with respect to practical network design and implementation reflect the nonlinearity of traffic systems, described above in the context of equations used to calculate performance, and how that nonlinearity impacts cost. To illustrate the consequences of nonlinearity and its implications, both circuit- and packet-switched cases are treated.

Circuit-switched examples are considered first. The probability of blocking, a function of offered load and the number of circuits, is depicted in Figure 11.13 for a 10-circuit trunk group. Below a certain traffic level (approximately 4 erlangs), essentially no blocking occurs. Blocking then rises exponentially over a range of traffic levels before approaching the limit of 100 percent (every call blocked). The interesting portion of this performance curve takes place between blocking probabilities of 0.01 and 0.1 (1 percent and 10 percent).

**Figure 11.13**
Blocking probability versus traffic load.

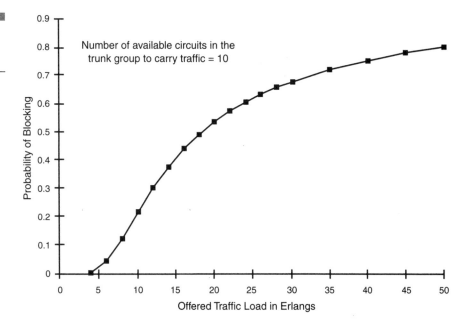

Another way of presenting this nonlinearity is to keep the load constant and calculate and plot the number of circuits required as a function of blocking. Figure 11.14 shows the results for an offered load of 10 erlangs. From this presentation some of the important nuances of nonlinear behavior begin to emerge. Surprisingly, accepting poor performance (10 versus 1 percent blocking) results in only a 30 percent reduction in the number of circuits (and therefore circuit cost). The relevance of this observation is that designing for 10 percent blocking to reduce expenses is not a viable solution. At 10 percent blocking, performance degradation is so noticeable and annoying to users that typically it prompts even higher traffic load from repetitive blocked-call retries. Good practice requires a grade of service of 5 percent or less, and preferably a level in the 1—2 percent range.

**Figure 11.14**
Blocking probability versus number of available circuits.

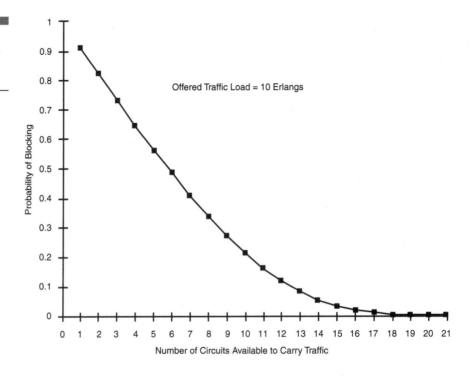

The most revealing way to present traffic network performance from a cost-efficiency standpoint is illustrated in Figure 11.15. The array of numbers in the boxed area represents the amount of *carried traffic,* measured in hours per circuit per month, under varying offered load levels (the vertical variations) and varying values of network design blocking probability (the horizontal variations). Corresponding offered-load levels are indicated on the vertical axis and blocking probabilities on the horizontal axis. In the array, values of carried load correspond to traffic resulting from hourly traffic distributions with characteristics shown in Figure 11.8.

Examination of Figure 11.15 reveals that there are two ways to improve trunk-group utilization efficiency. First, for any given offered load, accepting higher blocking probabilities yields higher per-circuit utilization efficiency. Second, for any given blocking probability, aggregating more traffic yields higher per-circuit utilization efficiency.

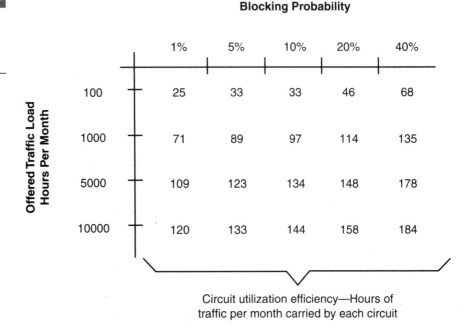

**Figure 11.15**
Carried traffic in hours per circuit per month.

In understanding the nature of traffic networks, the two most important lessons are:

- For a given level of traffic, designing for a better grade of service translates directly to poorer resource utilization efficiency and higher cost per unit of delivered service.

- There are economies of scale; at any specified level of performance, greater amounts of traffic (i.e., traffic aggregation) translate to reduced cost per unit of delivered service.

Assuming that poor GOS performance is not a viable solution, and that the offered load described for the PBX in the circuit-switched example above is determined by its stations, at first it appears that the important parameters are outside the control of designers and network users. In fact, however, use of a PBX can be viewed as a means of aggregating local traffic from 2,000 station lines and concentrating it into a 93-circuit trunk group. In this example, recall that an alternative to using a PBX (i.e., Centrex) would necessitate 2,000 separate lines between a customer's premises and a LEC central office, each with a very poor utilization of only about 6 hours per month. While it is true that PBXs afford significant facility cost advantages, a user must pay for the PBX equipment, and overall the Centrex/PBX cost trade-off involves numerous other factors.

The lessons learned also apply to private-switched network design. If private networks are designed to route overflow traffic automatically (traffic that would be blocked by the private network alone) to public or other secondary paths, then the private network can be designed for a level of blocking that is high enough to result in efficient use of its trunks. The operation of such designs is transparent to users who never encounter blocking at levels below those offered by the PSTN, normally a level significantly better than any private network can affordably support. In these hybrid designs, some of the private-network savings are offset by higher per-call costs encountered using the public network. Network engineering design tradeoffs are discussed further in Chapter 12.

Packet-switched systems exhibit the same kind of nonlinear behavior as circuit-switched systems. Figure 11.16 depicts average packet delay as a function of load for a typical single-circuit trunk group. The nonlinear or *threshold effect* is readily observed as delay approaches infinity when the erlang load approaches 1.0. The implications of this lost-calls-delayed performance are similar to those for lost-calls-cleared voice traffic in that designing a network for acceptable performance under "peak loads" usually means that resources are poorly used during "average load" traffic conditions.

For most situations this means that networks are inefficiently used most of the time, and only become "efficiently loaded" during busy hours. As we will see in future chapters, this effect is best mitigated by employing switches and routing techniques in a manner that aggregates traffic onto the largest (and most efficient) trunk groups possible, consistent with costs. The field of traffic engineering that deals with determining the optimum network topology and circuit sizes is called network design and optimization. Chapters 12 and 15 present examples of classic problems in network design for voice and data networks and some of the techniques used to solve these problems.

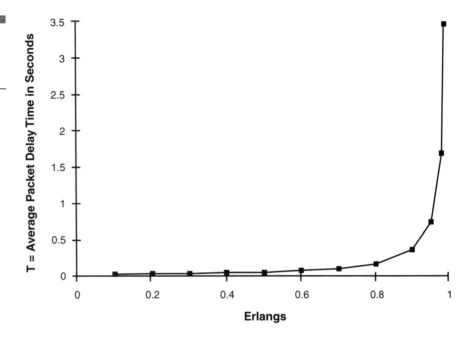

**Figure 11.16**
Packet delay as a
function of offered
traffic load.

# CHAPTER 12

# Selecting
# Voice Services

This chapter brings together the themes of the previous three chapters and applies them to the principal task of the voice networking professional: the proper selection of network and premises services. Network services are considered first, using Chapter 11 concepts of traffic engineering, tariff analysis, and cost optimization. Premises services are discussed afterward with emphasis on competitive procurement practices.

# Selecting Voice Network Services

Network design has two principal objectives. First, networks must provide specified services at performance levels (grades of service) that satisfy business requirements. Second, these performance levels must be satisfied in the most economical manner. This is especially true as telecommunications and networking services become more and more integrated into strategic business systems. The savings realized from efficient telecommunications services are a competitive business tool and a mainstream goal for large and small businesses. As explained in Chapter 11, aggregation of traffic (higher traffic volumes) always leads to more efficient use of telecommunications resources. It can be argued, therefore, that users with large traffic volumes deserve lower unit prices for public network services since they contribute disproportionately to the total traffic volume, making those facilities more efficient for all users.

Indeed, current tariffs are structured to encourage businesses to select service alternatives that minimize costs at higher levels of traffic. However, it is left to users to design their networks to make the most cost-effective use of the wide range of public (shared) and private (dedicated) carrier service offerings. This chapter examines tariffs used to price LEC and IXC network services, building on the basic concepts presented in Chapters 10 and 11. Here we examine tariff structures, how tariffs are applied, and, most important, the relationships between tariffs and network designs that minimize cost while satisfying business needs.

As noted in Chapter 11, a primary distinction among tariffs is jurisdictional. In the post-divestiture era, there are two classes of tariffs, one pertaining to LEC intra-LATA services and the other pertaining to IXC inter-LATA services. Predivestiture intrastate and interstate service distinctions remain significant in terms of actual rate determination. Recall that, under terms of the Modification of Final Judgment (MFJ), only IXCs may carry traffic between LATAs, even when those LATAs are in the same state. The provisions of the Telecommunications Act of 1996 will eventually remove these restrictions; hence, as in Chapter 10, we discuss separately selecting services from local exchange carriers and services from interexchange carriers.

Each jurisdictional tariff category contains two subcategories—one pertaining to *switched services,* and the other to *dedicated services.* As noted, when opting for public-network switched services, low-traffic volume users share the cost of expensive transmission and switching facilities. However, when on a nearly constant basis, users generate sufficiently high traffic levels among two (or more) points in a network, it may be to their advantage to obtain services on a non-shared or dedicated basis.

The two network points might be either two user premises, a user premises and a carrier POP, or two carrier POPs. In the following examples, tradeoffs between *dedicated-service fixed charges* and *switched-service usage-based charges,* are the common theme—with traffic volume usually the determining factor. For dedicated-service options, Chapter 11's traffic engineering principles are applied to optimize technical and cost performance in terms of "fixed charges" published in tariffs. For switched-service options, usage charges—taking volume discounts into account—are likewise obtained from published tariffs. Here, as in Chapter 15, quantitative results and derived figures are based on nationwide tariff data obtained from CCMI's Telview[1] service, now available to subscribers via the Internet.

## Local Exchange Network Services

Currently, unlike the highly competitive interexchange business arena, in most localities there are few, if any, competing local exchange service providers. Consequently, service selection involves examination of alternative offerings from a small number of, or possibly only one, providers. Reflecting the underlying access and transport facility distinctions defined earlier, tariffs applying to either switched or dedicated services are similarly segmented in terms of *access* and *transport* components and treated separately in what follows.

---

1 Center for Communications Management, Inc. (CCMI), 11300 Rockville Pike, Rockville Maryland 20852.

## LEC ACCESS SERVICES

Local exchange access services are local loops bundled with local calling privileges over defined geographic areas. Different types of access services and charges are typically offered to residential, single-line business, and multiline business customers. In most jurisdictions, customers may choose to pay for local calls individually or under terms of partial or fully unlimited calling plans. Unlimited calling plans may not be available to businesses. Other access options include PBX trunk-versus-line and digital-versus-analog service offerings.

As an example LEC access-service charge analysis, we consider premises-located PBXs to LEC central office trunk connections. Because shared trunks exchange signaling and supervisory information, specific signaling requirements determine the type of trunk ordered. As discussed in Chapter 11, trunks may be DID, DOD or two way. On Direct Inward Dialing (DID) trunks, along with incoming voice traffic CO switches send *dialed station number digits* to the PBX so that calls can be forwarded directly to individual stations without operator assistance. By contrast, non-DID PBX trunks route incoming calls to operators who must subsequently establish station connections. Although DID trunks are more expensive, automatic connections are preferred by most calling and called parties, and they generally save money by reducing the need for, or the number of, PBX operators.

Figure 12.1 illustrates charges for multi-line business PBX trunks in the state of Georgia as filed in BellSouth's General Subscriber Service Tariff (BellSouth's Georgia tariffs are comparable to those found in other jurisdictions and are used here as an example). The rate groups are based on the number of subscriber lines existing in the local calling area where the PBX is located. The upper portion of the figure presents charges for *unlimited* service (i.e., service that allows unlimited calling with no usage or time-dependent charges), the preferred alternative for high-traffic locations. Inward-only trunks can be equipped for direct inward dialing by installing a trunk termination and purchasing DID station *number blocks* large enough to serve PBX populations. DID is the major reason that users obtain separate *incoming* and *outgoing* PBX trunk groups.

The lower portion of Figure 12.1 presents an alternate charging plan for outgoing or combination trunks, providing what are known as measured trunks. The fixed charge per trunk is lower than in the unlimited case, but usage charges apply after an allowance of 75 messages per trunk. The usage charges, sometimes referred to as *message units*, depend on whether calls are within basic service areas or in expanded service areas (but do not depend on rate groups). In the example shown, for rate group 2, the break-even point between measured trunks and unlimited trunks is 183 calls per trunk per month for calls within the basic service area. Given normal traffic engineering assumptions, this corresponds to less than one external call per day per PBX station.

Residential and single-line business rates follow similar structures with the exception that DID charges are not applicable to these services.

**Figure 12.1**
LEC access tariff example.

| Service | Installation | Monthly | | | |
|---|---|---|---|---|---|
| | | Rate Group 2 | Rate Group 5 | Rate Group 7 | Rate Group 12 |
| **Unlimited Business PBX Trunk:** | | | | | |
| Outbound only | $58.25 | $39.66 | $50.70 | $59.50 | $77.10 |
| Inbound only | $58.25 | $36.16 | $47.20 | $56.00 | $73.60 |
| Combination | $58.25 | $39.66 | $50.70 | $59.50 | $77.10 |
| | | | | | |
| **Measured Business PBX Trunk:** | | | | | |
| Outbound only | $58.25 | $26.46 | $33.47 | $39.06 | $50.24 |
| Combination | $58.25 | $26.46 | $33.47 | $39.06 | $50.24 |
| | | | | | |
| **DID Service:** | | | | | |
| Establish trunk group and first 20 numbers | $915.00 | $4.00 | $4.00 | $4.00 | $4.00 |
| Additional block of 20 numbers | $15.00 | $4.00 | $4.00 | $4.00 | $4.00 |
| Trunk termination | $90.00 | $40.00 | $40.00 | $40.00 | $40.00 |
| | | Per Call | Per Minute | | |
| **Usage Charges:** | | | | | |
| Within basic service area | | $0.12 | $0.00 | | |
| In expanded service area 0–10 miles | | $0.04 | $0.03 | | |
| In expanded service area 11–22 miles | | $0.04 | $0.05 | | |
| In expanded service area 23–40 miles | | $0.04 | $0.06 | | |
| In expanded service area 41–55 miles | | $0.04 | $0.08 | | |

## LEC SWITCHED-TRANSPORT SERVICES

This section focuses on nonlocal services, or services beyond the local subscriber calling area as defined by the LEC. The primary nonlocal LEC switched services are standard MTS and LEC versions of WATS. With MTS, calls use the same premises-to-LEC CO exchange access lines (loops) used for local calls. Tariff rates are *mileage*-sensitive and discounted during evening and nighttime hours. *Evening* hours are defined as 5:00 P.M. to 11:00 P.M., and *nighttime* hours are from 11:00 P.M. to 8:00 A.M. Nighttime rates apply on Saturday and until 5:00 P.M. on Sunday (Note, tariffs in different jurisdictions may vary in both interval definition and discount rates). Many tariffs, including the example below, combine evening and nighttime periods into a single "offpeak" period. The daytime period is then referred to as the "peak" period.

For MTS, there is a charge for an initial time interval (usually 30 seconds) that is higher than the charge for additional periods (usually 6 seconds). Figure 12.2's rates are typical for LEC-provided MTS, but the actual numbers are from the BellSouth Georgia tariff used above. Note that peak usage charges average 24 cents per minute for toll calls over 16 miles (which would comprise the majority of non-local calls for most users). Some jurisdictions offer special calling plans with lower per-minute costs, in exchange for customer volume commitments. *Mileage* for MTS calls is ordinarily calculated using serving wire center V&H coordinates, as explained in Chapter 11.

**Figure 12.2**
LEC MTS rate
example.

| Mileage | Initial 30 Seconds | Additional 6 Seconds |
|---------|--------------------|-----------------------|
| 0–10 | $0.060 | $0.004 |
| 11–16 | $0.070 | $0.006 |
| Over 16 | $0.120 | $0.024 |
| Offpeak Discount = 40% | | |

WATS rates are based on volume rather than distance. Figure 12.3 is a representative outward-calling WATS rate table. The *monthly hours of usage* column refers to total call-hours per month for a group of WATS circuits at a particular location. Call hours are totaled over all rate periods (peak and offpeak); the total hour value is then used to select an hour band (row). The applicable rates for each time period (column) in this row are then multiplied by the corresponding traffic in each time period to calculate the total usage charges. Total cost per month is the sum of usage charges plus fixed charges for access lines—which can only be used for WATS traffic. Fixed access rates are independent of customer location and priced at $25 per month per circuit.

The following calculation illustrates how Figure 12.3's rate table is used.

**Usage:** 120 hours per month; 90 hours peak, 30 hours offpeak
**Hour band:** 80–120 hours
**Peak cost calculation:**
   90 hours (5,400 minutes)   @ $0.125/min = $675.00
**Offpeak cost calculation:**
   30 hours (1,800 minutes)   @ $0.115/min = $207.00

**Total Usage:**

120 hours at a total cost of $882.00

**Figure 12.3**
LEC WATS rate
example.

| WATS Line Fixed Charges: | $25/line/month | |
|---|---|---|
| Usage Charges: | | |
| | **Per Minute** | |
| Hours Per Account | Peak | Offpeak |
| 0–15 | $0.150 | $0.140 |
| 15–40 | $0.145 | $0.135 |
| 40–80 | $0.140 | $0.130 |
| 80–120 | $0.125 | $0.115 |
| 120–170 | $0.100 | $0.090 |
| 170–320 | $0.097 | $0.085 |
| 320–500 | $0.092 | $0.082 |
| 500–1200 | $0.089 | $0.079 |
| 1200–2500 | $0.083 | $0.075 |
| Over 2500 | $0.077 | $0.070 |

The overall usage cost in this example is 12.25¢/minute. If three access lines are ordered to carry this traffic, then additional access costs are 3 × $25 = $75, or 1.0¢/minute. For customers with the above traffic volume, the resulting unit cost of 13.3¢/minute is clearly preferable to MTS alternatives that may approach 24¢/minute. Note that savings occur only when customers have enough traffic to achieve volume discounts and offset fixed access charges, with break-even points occurring at about 10 hours of traffic per month. Note also that a poor busy-hour grade of service over a small number of WATS access lines can be improved by switching equipment that will automatically route WATS calls away from busy access lines, back to MTS service. While this forfeits some cost savings, it minimizes user annoyance and high call-retry rates, which often further degrade GOS.

## LEC PRIVATE-LINE SERVICES

Private-line services can be supported by LECs using either voice-grade (analog) or digital facilities. Technical characteristics (for example, 2-wire/4-wire, frequency response, signaling arrangements, etc.) distinguish analog facility and service offerings. Figure 12.4 relates voice-grade service designators to applications in terms of technical characteristics.

**Figure 12.4**
Characteristics of voice-grade private lines.

| Service | Application |
| --- | --- |
| Type 2230 | 2-wire interface with 2-wire facilities—voice transmission, private line, mobile radio, supervisory use |
| Type 2231 | 2-wire interface with 2- or 4-wire facilities—PBX OPX, signaling required |
| Type 2432 | 2- or 4-wire interface with 4-wire facilities—tie line, PBX-PBX signaling required |
| Type 2434 | 2- or 4- wire interface with 4-wire facilities—tie lines, Centrex-Centrex (with E & M signaling) |
| Type 2435 | 4-wire interface with 4-wire facilities—voice transmission, multipoint service |
| Type 2260 | 2-wire interface, 2-wire facilities, half duplex data services |
| Type 2261 | 2-wire interface with 2-wire facilities—Dataphone Select-a-Station or Telemetry Alarm Bridging Service |
| Type 2462 | 4-wire interface with 4-wire facilities—Dataphone Select-a-Station or Telemetry Alarm Bridging Service |
| Type 2463 | 4-wire interface with 4-wire facilities—analog data services, multipoint service |
| Type 2464 | 2-wire interface with 4-wire facilities—analog data services, multipoint provided |

Tariff service rate elements vary from state to state, and equivalent rate elements may bear different service designator names. The trend, however, is to common cost-based rate elements and designators. For example, in most jurisdictions, one of the private-line service pricing elements is the *local-channel rate* element. This specifies charges for local loops from the customer's premises to serving wire centers. Local-channel charges are normally not mileage dependent. Where local-channel elements are mileage dependent, users rarely have access to the premises-to-wire center distance information (1 mile on the average) needed to calculate costs independently and must therefore rely on estimates.

A local-channel charge is assessed at both originating and terminating COs. An *interoffice channel* connects two serving COs (more accurately, serving wire centers). The interoffice-channel private-line service rate element is mileage dependent, and wire-center V&H coordinates are used to calculate mileage charges. If interoffice channels are used to support switched-network services, rate-element charges for signaling equipment are assessed at each end. Other charges are assessed for special options—for example, line conditioning to improve circuit performance for data communications applications.

Unfortunately, the realm of tariffs has developed (over many decades) a unique vocabulary that neither conforms to terminology emanating from the MFJ, nor is consistent among the separate LEC and IXC tariff domains. For instance, in contrast with the definition in LEC tariffs cited above, in IXC tariffs, interoffice channels connect IXC serving POPs, and local channels include all LEC or bypass network components used to connect customer premises to serving IXC POPs. Other unique and often contradictory tariff terminology will be defined as encountered in the information that follows.

Returning to the LEC private line service discussion, Figure 12.5 shows representative local channel, interoffice channel, and signaling rate elements for LEC voice-grade analog circuits. Attention should be paid to the mileage table for interoffice channels. In most states, and in all IXC tariffs, total interoffice mileage is used to compute mileage-dependent components, which are then added to the fixed monthly charge. As an example, an interoffice channel charge for a 50-mile circuit would be:

$32 + 50 × $1.95/mile = $129.50

Some states still employ an older approach in which each mileage band is used. In this case, charges for a 50-mile circuit would be calculated as:

$32 + 8 × $2.05/mile + 17 × $2.00/mile + 25 × $1.95/mile = $131.15

Not all states use serving wire centers as the basis for pricing. Some still use exchange, or *rate-center* pricing in which connections are considered intra-exchange if they are in the same rate center. Interexchange circuits have mileage-based pricing components calculated on the basis of rate-center coordinates. Careful tariff reading is required to ascertain how to apply mileage-based rate tables.

**Figure 12.5**
LEC voice-grade
private line rate
example.

| Local Channels | Nonrecurring Charge | | Monthly Rate |
|---|---|---|---|
| Per Point of Termination | First ($) | Additional ($) | ($) |
| Type 2230 | 345.00 | 115.00 | 25.00 |
| Type 2231 | 345.00 | 115.00 | 25.00 |
| Type 2432 | 400.00 | 145.00 | 45.00 |
| Type 2434 | 160.00 | 83.00 | 10.00 |
| Type 2435 | 370.00 | 130.00 | 45.00 |
| Type 2260 | 415.00 | 160.00 | 30.00 |
| Type 2261 | 575.00 | 245.00 | 24.00 |
| Type 2462 | 565.00 | 235.00 | 38.00 |
| Type 2463 | 415.00 | 160.00 | 50.00 |
| Type 2464 | 415.00 | 155.00 | 50.00 |

| Local Channels | Nonrecurring Charge | Monthly Rate | |
|---|---|---|---|
| Fixed and Mileage Charges Applicable | Per Channel ($) | Fixed ($) | Per Mile ($) |
| 1–8 Miles | 105.00 | 32.00 | 2.05 |
| 9–25 Miles | 105.00 | 32.00 | 2.00 |
| Over 25 Miles | 105.00 | 32.00 | 1.95 |

| Signaling | Nonrecurring Charge | | Monthly Rate |
|---|---|---|---|
| Per Local Channel | Initial ($) | Subsequent ($) | ($) |
| Manual Ringdown | 40.00 | 215.00 | 11.00 |
| Automatic Ringdown | 15.00 | 72.00 | 10.00 |
| E & M Signaling | 45.00 | 185.00 | 10.00 |
| Type A: (0–199 Ohms) | 43.00 | 135.00 | 6.00 |
| Type B: (200–899 Ohms) | 42.00 | 135.00 | 6.00 |
| Type C: (900 or more Ohms) | 12.00 | 135.00 | 3.00 |

It is also important to distinguish between the use of "intra-" and "interexchange" as used in intra-LATA and inter-LATA tariff contexts. Whereas today the term interexchange, as in IntereXchange Carrier, applies strictly to inter-LATA services, the central office connotation of exchange, as reflected in local and state tariffs and many telephony reference books, dates back almost a century.

Digital circuits have tariff structures similar to analog circuits. Differences involve physical facilities used to provide circuits. Today, most facilities are derived from 1.544-Mb/s DS1 facilities, described in Chapter 5. In LEC networks, these circuits are provided directly to users at the 1.544-Mb/s rate under generic names such as, "high-capacity," or trademarked names like "Megalink Service." Digital local channels and interoffice channels can often be substituted for analog counterparts. Figure 12.6 displays rate tables for digital facilities from the same jurisdiction as the voice-grade example above. The interoffice DS1 channel cost for a 50-mile circuit is:

$$\$85 + 50 \times \$31.00/\text{mile} = \$1,635$$

equivalent to the cost of just 13 analog voice-grade circuits. Since DS1 service provides 24 voice-grade circuits, it is more cost-effective than analog service whenever more than 13 channels are required. This crossover point varies from state to state, but, in general, this comparison is typical.

**Figure 12.6**
LEC 1.544 Mb/s
circuit rate example.

| | Nonrecurring Charge ($) | Month to Month ($) | 36 Months ($) | 60 Months ($) | 84 Months ($) |
|---|---|---|---|---|---|
| **Digital Local Channel** | | | | | |
| Each | | | | | |
| First 1/2 Miles | 300.00 | 82.00 | 81.00 | 81.00 | 81.00 |
| Each Additional 1/2 Mile, | | | | | |
| or Fraction Thereof | — | 35.00 | 34.00 | 32.00 | 30.00 |
| | | | | | |
| **Interoffice Channels** | | | | | |
| (Furnished Between COs) | | | | | |
| | | | | | |
| Each Channel 0–8 Miles | | | | | |
| Fixed Monthly Rate | 100.00 | 65.00 | 65.00 | 65.00 | 65.00 |
| Each Airline Mile or | | | | | |
| Fraction Thereof | — | 35.00 | 34.00 | 32.00 | 30.00 |
| | | | | | |
| Each Channel 9–25 Miles | | | | | |
| Fixed Monthly Rate | 100.00 | 70.00 | 70.00 | 70.00 | 70.00 |
| Each Airline Mile or | | | | | |
| Fraction Thereof | — | 33.00 | 32.00 | 30.00 | 28.00 |
| | | | | | |
| Each Channel Over 25 Miles | | | | | |
| Fixed Monthly Rate | 100.00 | 85.00 | 85.00 | 85.00 | 85.00 |
| Each Airline Mile or | | | | | |
| Fraction Thereof | — | 31.00 | 30.00 | 28.00 | 26.00 |

In addition to point-to-point use between user premises, LECs now offer CO multiplexing and dedicated 1.544-Mb/s DS1 services as multichannel interoffice backbones for multiple lower-speed circuits from customer premises. This configuration is shown in Figure 12.7. LEC charges take the form of port charges for each low-speed circuit terminated on the input side of the multiplexer and the 1.544-Mb/s circuit connected to the output side of the multiplexer.

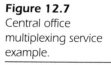

**Figure 12.7**
Central office
multiplexing service
example.

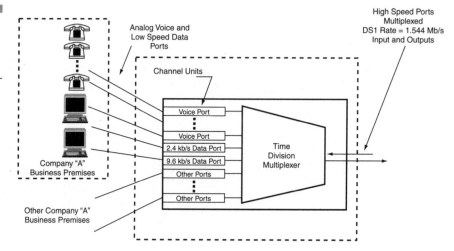

Historically, lower-speed digital data services (circuits with line speeds of 2.4, 4.8, 9.6, and 56 kb/s) have been provisioned with separate, synchronous facilities using special hubs and end-to-end timing. More recently, 56-kb/s service is offered on channels derived from T1 facilities. Both options may exist in the same service area under different service names.

Lower-speed *synchronous digital services* are engineered to superior bit-error-rate and error-free-seconds specifications when compared to analog circuits with modems or T1-derived circuits, but they are more expensive. Users generally find that performance improvements do not justify the more-expensive facilities, although careful examination of technical specifications for each application is mandatory. Figure 12.8 shows a rate structure for synchronous digital data service. Here, circuits must pass through at least one LEC node office for test and maintenance purposes, increasing mileage and termination charges.

Carrier circuits used for digital transmission require user-premises network channel terminating equipment (NCTE) to provide for digital signal processing and to protect the network from harmful signals. Following the FCC's NCTE Order in 1983 (see Appendix A), this equipment is no longer supplied by the LEC. Users must provide the equipment, which has to be designed in accordance with LEC interface specifications. For digital circuits, channel service units (CSU) and data service units (DSUs) are required, and, as described in Chapter 5, these functions are usually combined in a single piece of equipment. For analog facilities, users must provide modems, described in Chapter 3.

**Figure 12.8**
LEC synchronous
digital private line
rate example.

| | Nonrecurring Charge | | Monthly Rate |
|---|---|---|---|
| | First | Additional | |
| **Digital Local Channel** | ($) | ($) | ($) |
| Per Local Channel | | | |
| 2.4, 4.8, or 9.6 kb/s | 340.00 | 118.00 | 50.00 |
| 56 kb/s | 340.00 | 118.00 | 70.00 |
| | | | |
| **Node Channel Termination** | | | |
| Per Local Channel | | | |
| 2.4, 4.8, or 9.6 kb/s | 48.00 | 42.00 | 11.00 |
| 56 kb/s | 48.00 | 42.00 | 30.00 |

**Digital Interoffice Channel**
Furnished between serving wire center and Node CO or between Node COs.

| | Nonrecurring Charge | Monthly Rate | |
|---|---|---|---|
| | | Fixed | Per Mile |
| | ($) | ($) | ($) |
| 2.4, 4.8, or 9.6 kb/s: | | | |
| 0–8 Miles | 79.00 | 20.00 | 2.05 |
| 9–25 Miles | 79.00 | 20.00 | 2.00 |
| Over 25 Miles | 79.00 | 20.00 | 1.95 |
| | | | |
| 56 kb/s: | | | |
| 0–8 Miles | 79.00 | 40.00 | 4.10 |
| 9–25 Miles | 79.00 | 40.00 | 4.00 |
| Over 25 Miles | 79.00 | 40.00 | 3.90 |

The decision to include dedicated (private) facilities in business telecommunications networks is nearly always driven by the desire to reduce costs, thus, traffic volume and applicable tariffs ultimately dictate the design, apart from privacy and other reasons. To ensure high utilization of private facilities during busy hours, traffic that would be blocked on private facilities may first be routed to WATS and lastly to MTS. In this way, businesses can employ a hierarchy of services to minimize costs while maintaining grade of service. PBXs and other switching equipment offer *least-cost routing* features, which support the automatic selection of preferred service options in response to offered traffic.

## Interexchange Network Services

The current interexchange network environment is highly competitive both in the number of service providers and the range of services offered. The current lack of regulation has also resulted in less application of published tariff rates and more reliance on the competitive procurement of custom packages of services. This section presents the pricing structure of current carrier offerings and explores the new realm of *special contract* service packages. As before, access and transport services are treated separately.

### IXC ACCESS SERVICES

As explained in Chapter 10, access to IXC services (from user premises to the IXC point-of-presence) can be either dedicated or switched. When access to IXC POPs is gained through switched LEC network facilities, users employ the same local loops used for local and LEC toll-switched service. LEC charges for this switched access are paid by IXCs and are included in IXC service rates. Because of this, IXCs appear to offer service from originating to terminating CO, and mileage is measured accordingly (from and/or to the rate center of the CO using switched access).

In reality, of course, this is not the case, and the IXC provides end-to-end service using LEC originating and terminating transport, with IXCs fully compensating LECs for the use of their switched-network services. Since these LEC services provide access to interstate services, they are included in *Switched Access* tariffs that fall under FCC, not state public utility commission, jurisdiction.

When dedicated access to IXC switched or dedicated services is used, users pay for access circuits directly. While LEC facilities employed are the same as those used for LEC private-line service, pricing structures for dedicated access are different. Again, this is because the circuits are used to access interstate services and, hence, fall under FCC jurisdiction. Rates for these circuits are included in LEC *Special Access* tariffs. They are also mileage based, but use different terminology for their components (e.g., *channel termination* instead of local channel). The appropriate mileage in this case is the mileage from the user premises' wire center to the wire center serving the IXC POP.

Although users can order the dedicated-access circuits discussed above directly from LECs, there are disadvantages to doing so. Users would be responsible for coordinating service orders, installation dates, trouble isolation, and other joint activities involving both access and IXC services. In addition, there would be separate bills for the access and transport portions of the network, complicating *charge-back costs* to specific end-users. As a result, it is common practice today for IXCs to serve as customer agents and handle all aspects of access-circuit provisioning. In this case, access prices are filed by IXCs in FCC tariffs. The prices generally reflect underlying LEC charges plus small surcharges, called an *access coordination fees.* Current practice for IXC access pricing is to list a fixed price for each access-circuit type based on area code and exchange (NPA-NXX) numbers associated with user premises. For example, AT&T's access prices are found in AT&T's FCC Tariff #11.

### MTS, WATS, AND TOLL FREE SWITCHED SERVICES

IXC switched services, provided through IXC POPs, enable sharing of IXC facilities and involve rate structures based on usage. Interexchange MTS is the long-distance service of choice for residential and low-volume business users. Here, access to IXC facilities is gained through switched LEC network facilities at both the originating and terminating ends of calls. In the recent past, MTS was provided under tariff and consisted of a rate table with 11 distinct mileage bands and three time-of-day periods. The new IXCs (other than AT&T) used rate tables that were exact mirrors of AT&T rate tables, including the same mileage band limits, which allowed easy comparison of the tables and identification of the new IXC's lower rates.

Today, however, deregulation, equal access, and intense competition have created a completely different market environment. There are still "standard" tariff rates, but the mileage bands have disappeared. Interexchange carriers now use myriads special calling plans designed to attract a special class of customer (or to simply sound attractive) in all-out competition for residential customers. Indicative of the rate of new service offerings is the fact that AT&T has filed 28 special calling plans in the FCC Tariff #27, Domestic MTS. In some cases, carriers have been able to persuade customers to accept attractive-sounding offerings while reaping high profits from unsuspecting customers (more on this later).

Figure 12.9 illustrates the range of unit costs currently being offered for MTS service as well as the special cost structures employed that require special attention by potential customers. A typical marketing ploy is to advertise a low cost per minute for usage, but require a fixed monthly charge, a minimum monthly charge, or a fixed charge per call (often expressed as a fixed charge for the first several minutes of a call). Monthly fixed charges must be divided by the number of minutes the customer uses in a month to determine the effective unit cost. For low-usage customers, this can add a substantial amount to the advertised unit rate. For example, consider the first two services in Figure 12.9:

- The first service charges a flat rate of 8.5 cents/minute with no other charges or minimums.

- The second service charges 7 cents/minute plus a fixed charge of $3.00 per month. For a customer with more than 200 minutes of MTS calling a month, the $3.00 fixed charge adds 1.5 cents/per minute or less to the rate. The second service is preferable for these customers. However, for less than 200 minutes per month of calling, more than 1.5 cents/minute is added to the basic rate and the first service is preferable. At a usage rate of only 50 minutes per month, 6 cents/minute would be added to the basic rate for an effective rate of 13 cents/minute, almost double the advertised rate.

**Figure 12.9**
IXC MTS rate example.

| Plan | Unit Price Cents Per Minute | Fixed Charge Per Month | Notes |
|---|---|---|---|
| Plan 1 | 8.5 | $0.00 | |
| Plan 2 | 7 | $3.00 | |
| Plan 3 | 5 | $0.00 | $1.00 for First 20 Minutes Per Call |
| Standard Tariff | 26 Day 16 Evening 11.5 Night | $0.00 | |

Likewise, the service that advertises "one dollar for all calls up to twenty minutes" effectively charges 5 cents/minute for 20-minute calls, 10 cents/minute for 10-minute calls, and 20 cents/minute for 5-minute calls. Users must consider their individual calling patterns and volumes when evaluating competing offers from carriers. However, the overall news on long-distance calling charges is good. Steadily decreasing unit costs since divestiture have resulted in an overall unit cost of only 10 cents/minute for all calls made in the U.S. in 1998. The current benchmark for MTS prices for callers with moderate volume is between 8 and 9 cents/minute.

The above phenomenon has caused a paradigm shift in the evaluation of WATS, toll-free, and other high-volume services. While these services previously sported low per-minute rates at high volumes, at published tariff rates they are no longer competitive with discounted MTS rates. Figure 12.10 shows tariff rates for AT&T's MEGACOM service, a dedicated-access WATS service. Assuming the third mileage band and a business mix of 85 percent daytime traffic, the undiscounted rate averages 24 cents/minute. Even after the maximum volume and term discounts listed in the tariff are applied, the unit cost for this service is still 16 cents/minute, well above the benchmark rate established by the MTS calling plans. As discussed below, service integration and competitive acquisition techniques are ways for businesses to achieve the best prices in today's market.

**Figure 12.10**
IXC dedicated-access
WATS rate example
(AT&T MEGACOM).

| Mileage | AT&T MEGACOM Service | | | | | |
|---------|----------------------|---|---|---|---|---|
| | Initial 18 Seconds | | | Each Additional 6 Seconds | | |
| | Day | Evening | Night | Day | Evening | Night |
| 0–55 | $0.0585 | $0.0411 | $0.0372 | $0.0195 | $0.0137 | $0.0124 |
| 56–292 | $0.0690 | $0.0486 | $0.0432 | $0.0230 | $0.0162 | $0.0144 |
| 293–430 | $0.0753 | $0.0525 | $0.0468 | $0.0251 | $0.0175 | $0.0156 |
| 431–925 | $0.0828 | $0.0576 | $0.0510 | $0.0276 | $0.0192 | $0.0170 |
| 926–1910 | $0.0888 | $0.0615 | $0.0552 | $0.0296 | $0.0205 | $0.0184 |
| 1911–3000 | $0.0936 | $0.0654 | $0.0576 | $0.0312 | $0.0218 | $0.0192 |
| 3001–4250 | $0.0936 | $0.0654 | $0.0576 | $0.0312 | $0.0218 | $0.0192 |
| Over 4250 | $0.0936 | $0.0654 | $0.0576 | $0.0312 | $0.0218 | $0.0192 |

Toll-free services represent a special case in the public-switched services arena. As concluded in Chapter 10, a toll-free service offering is a great deal more than simply "reversing the charges" on calls from public networks. Advanced features of toll-free service can be used to provide innovative, high-quality business services while producing significant savings. The largest portion of call center operations expense is the labor cost of the "agents" talking with customers. The number of agents required to provide a given grade of service for specified customer calling rates and service times is a traffic engineering problem not unlike those described in Chapter 11.

Advanced routing operations that allow many call centers to appear as one *virtual call center* use agents more efficiently, thus reducing the total number of agents required. Because of this phenomenon, it is extremely important to estimate traffic demands and model and design call-center operations—before buying toll-free service. Detailed requirements form the basis for competitive procurements where competing carriers propose not only toll-free transport and access service, but also features that implement the most efficient system. Accordingly, since feature charges can be a large part of toll-free service costs, estimated usage of *features* must also be part of any RFP preparation and competitive proposal evaluation processes.

As with WATS service, published tariff rates for toll-free services are just a starting point for competitive negotiations. Figure 12.11 shows the tariff rates for AT&T's MEGACOM 800 service. Note that these are per-hour rates rather than the more common per-minute or per-6-second rates for other services. The average undiscounted cost for a business customer under this tariff would be 30 cents/minute. Again, discounts provided by tariff provisions do not approach the benchmark level. Competitive processes must be employed to obtain more attractive price levels.

**Figure 12.11**
IXC dedicated-access toll-free service rate example (AT&T MEGACOM 800).

| AT&T MEGACOM 800 Service | | | |
|---|---|---|---|
| **Bands** | **Charge Per Hour** | | |
| | Day | Evening | Night |
| 1 | $15.84 | $12.60 | $12.60 |
| 2–6 | $18.72 | $15.12 | $15.12 |

## VIRTUAL PRIVATE NETWORK SERVICES

Chapter 10 demonstrated how dedicated private-line access to IXC POPs leads to and supports the concept of virtual private network (VPN) service offerings. Instead of leasing private lines between carrier POPs, the user shares the IXC carrier's network, pays usage charges, and yet enjoys equivalent benefits of facilities-based private networks—with the only exception being that no physical facilities are dedicated exclusively to any user. VPN rate structures are similar to dedicated WATS services, except that there are more end-to-end call-connection alternatives. As in any private network, users at locations served by dedicated access to IXC POPs can call each other, but unlike dedicated WATS, the dedicated access lines carry both originating and terminating traffic. The dedicated access locations are called on-net locations and calls between such locations are referred to as *on-net* calls. Calls using switched access to IXC VPNs for either origination, termination, or both are termed *off-net.* This leads to four possible types of traffic:

1. On-net to on-net
2. On-net to off-net
3. Off-net to on-net
4. Off-net to off-net

Rate schedules are published for each traffic type. Figure 12.12 shows A, B, and C rate schedules for AT&T's Software Defined Network (SDN). Schedule A pertains to traffic type 4 and includes charges for both originating and terminating LEC switched-access service. Schedule B pertains to traffic types 2 or 3 and includes a single LEC switched-access service charge for either call origination or termination. Schedule C is used for traffic type 1 and includes no LEC access charges. In this case LEC private-line access services are separately arranged and paid for. Note that the difference between Schedules A and B and the difference between Schedules B and C should represent the average cost of switched access from LECs (that is, switched access at each end of the call). In the case of the rate schedules shown, this difference is about 8 cents/minute per end. The current price level (as of 12/31/99) of switched-access charges is 1.4 cents/minute and has been dropping continuously since divestiture (it was 2.0 cents/minute as of 6/30/98). Why, then, is the switched-access charge built into the rate tables so high?

The answer is another variant of the phenomenon encountered above with WATS tariffs. Consider the following outbound traffic distribution, typical of large business customers:

- 85% daytime
- 15% evening
- 25% on-net to on-net
- 70% on-net to off-net
- 5% off-net to off-net

Using the undiscounted rate schedules for SDN, the average VPN usage cost would be 24 cents/minute. As discussed above, discounts of at least 70 percent would have to be negotiated or competitively procured to make this service cost effective. Since discounts apply to all price schedules, the effective switched-access cost for one end in the tables is 2.4 cents/minute (more in line with true costs). This is another indication that no carrier intends to sell service at tariff rates in this competitive market.

Remember that VPN is a true network service with the objective to be indistinguishable, by users, from facilities-based private networks. As such there is a wide range of selectable features, each with defined charges. The customer can use these features to tailor his network to business needs and use it as a business tool. Hence, it is not completely fair to compare unit prices of services like VPN with the plain vanilla

███  ███  ███

**Figure 12.12**
Virtual private
network rate
example (AT&T
Software Defined
Network).

## AT&T Software Defined Network Service

| Mileage | Schedule A (Off-net to Off-net) | | | | | |
|---------|---------|---------|---------|---------|---------|---------|
| | Initial 18 Seconds | | | Each Additional 6 Seconds | | |
| | Day | Evening | Night | Day | Evening | Night |
| 0–55 | $0.0852 | $0.0840 | $0.0840 | $0.0284 | $0.0280 | $0.0280 |
| 56–292 | $0.0927 | $0.0840 | $0.0840 | $0.0309 | $0.0280 | $0.0280 |
| 293–430 | $0.0990 | $0.0840 | $0.0840 | $0.0330 | $0.0280 | $0.0280 |
| 431–925 | $0.1029 | $0.0840 | $0.0840 | $0.0343 | $0.0280 | $0.0280 |
| 926–1910 | $0.1080 | $0.0840 | $0.0840 | $0.0360 | $0.0280 | $0.0280 |
| 1911–3000 | $0.1080 | $0.0840 | $0.0840 | $0.0360 | $0.0280 | $0.0280 |
| 3001–4250 | $0.1080 | $0.0840 | $0.0840 | $0.0360 | $0.0280 | $0.0280 |
| Over 4250 | $0.1080 | $0.0840 | $0.0840 | $0.0360 | $0.0280 | $0.0280 |

| Mileage | Schedule B (On-net to/from Off-net) | | | | | |
|---------|---------|---------|---------|---------|---------|---------|
| | Initial 18 Seconds | | | Each Additional 6 Seconds | | |
| | Day | Evening | Night | Day | Evening | Night |
| 0–55 | $0.0555 | $0.0459 | $0.0459 | $0.0185 | $0.0153 | $0.0153 |
| 56–292 | $0.0654 | $0.0522 | $0.0522 | $0.0218 | $0.0174 | $0.0174 |
| 293–430 | $0.0711 | $0.0564 | $0.0564 | $0.0237 | $0.0188 | $0.0188 |
| 431–925 | $0.0780 | $0.0612 | $0.0612 | $0.0260 | $0.0204 | $0.0204 |
| 926–1910 | $0.0846 | $0.0669 | $0.0669 | $0.0282 | $0.0223 | $0.0223 |
| 1911–3000 | $0.0864 | $0.0681 | $0.0681 | $0.0288 | $0.0227 | $0.0227 |
| 3001–4250 | $0.0975 | $0.0750 | $0.0750 | $0.0325 | $0.0250 | $0.0250 |
| Over 4250 | $0.1002 | $0.0774 | $0.0774 | $0.0334 | $0.0258 | $0.0258 |

| Mileage | Schedule C (On-net to On-net) | | | | | |
|---------|---------|---------|---------|---------|---------|---------|
| | Initial 18 Seconds | | | Each Additional 6 Seconds | | |
| | Day | Evening | Night | Day | Evening | Night |
| 0–55 | $0.0318 | $0.0219 | $0.0219 | $0.0106 | $0.0073 | $0.0073 |
| 56–292 | $0.0411 | $0.0285 | $0.0285 | $0.0137 | $0.0095 | $0.0095 |
| 293–430 | $0.0468 | $0.0327 | $0.0327 | $0.0156 | $0.0109 | $0.0109 |
| 431–925 | $0.0549 | $0.0384 | $0.0384 | $0.0183 | $0.0128 | $0.0128 |
| 926–1910 | $0.0600 | $0.0432 | $0.0432 | $0.0200 | $0.0144 | $0.0144 |
| 1911–3000 | $0.0654 | $0.0459 | $0.0459 | $0.0218 | $0.0153 | $0.0153 |
| 3001–4250 | $0.0732 | $0.0513 | $0.0513 | $0.0244 | $0.0171 | $0.0171 |
| Over 4250 | $0.0765 | $0.0534 | $0.0534 | $0.0255 | $0.0178 | $0.0178 |

MTS services. Nevertheless, unit prices for flagship services like VPN can be obtained for less than even the most aggressive MTS calling plans. The message for our readers is that such desirable results are no longer automatic rewards that accompany high-volume and long-term customer commitment concessions. To optimize technical, business operation, and cost performance, end-users must be equipped to exploit today's competitive marketplace aggressively. This ultimately demands informed buyers which, after all, is the purpose of this book.

### IXC PRIVATE LINE SERVICES

Like LEC counterparts, tariffs for IXC private lines are categorized as either analog or digital. In dedicated private line service, IXCs only provide physical facilities for *interoffice channels* (IOCs) between POPs.

There are three rate elements for IXC private-line services. First, mileage-dependent charges are assessed for IOCs. Next are charges for *central office connections* (COCs), which are charges for connecting IOCs to other facilities such as LEC-provided local channels. Certain carriers now use the terminology *access connection* when referring to a central office connection. A COC is required at each end of an interoffice channel. Finally, charges are assessed for certain interoffice channel options such as line conditioning, signaling, diverse routing for higher reliability, etc. AT&T's Tariff #9 is used below to represent IOC, COC, and optional elements of IXC private line tariffs. AT&T's Tariff #9 basically addresses elements of IXC private-line services rendered using IXC-owned facilities.

A radical change has occurred in dedicated-circuit facilities provisioning as the underlying IXC physical plant has evolved from analog to digital transmission plant. In the past, low-cost analog circuits and considerably more expensive low-rate digital circuits provided over separate, synchronous networks were the norm. Today DS1 service facilities dominate, and most analog voice private-line service is actually delivered using 64-kb/s-channel DS0 channels derived from DS1 digital facilities. Consequently, IXCs now offer 64-kb/s digital private-line service to users at analog prices. While performance specifications are not as high as with the older synchronous networks, 64-kb/s DS1-derived circuits have proven adequate for data communications at a much lower cost. IXCs have expanded these offerings to include multiples of 64-kb/s DS0 service in increments up to 768 kb/s. The resulting service, commonly known as Fractional T1 service, is described in Chapter 5.

Heavy competition in the IXC arena has precipitated up to 60 percent discounts for high-volume, long-term commitments for private DS1 digital services. In contrast, discounts for high-volume, long-term commitments for lower-rate digital and analog services are limited to 20 percent, with fractional T1 services falling in between. At present, the situation is so dynamic that thorough readings of current tariffs and prudence in considering long-term commitments are mandatory.

The rate tables for IXC private-line services follow the same format as the newer LEC rate tables; that is, the applicable mileage band is determined and fixed and mileage charges are computed. Figure 12.13, a representative table for fractional T1 circuits, illustrates IOC and COC charges. The tariff provides for volume and term discounts up to 36 percent for fractional T1 circuits. Note that originating and terminating COC charges and any optional charges must be added to IOC charges to determine total inter-LATA costs. Some IXC discount plans apply to total inter-LATA charges, while others apply only to the IOC components.

**Figure 12.13**
IXC fractional T1 rate examples.

| Interoffice Channel Charges | | | |
|---|---|---|---|
| **Channel (b/s)** | **COC Monthly** | **Fixed** | **Per Mile** |
| 9.6K | $25.00 | $444.00 | $0.35 |
| 56/64K | $25.00 | $444.00 | $0.35 |
| 128K | $32.75 | $799.00 | $0.63 |
| 192K | $34.30 | $1,172.00 | $0.92 |
| 256K | $38.45 | $1,527.00 | $1.20 |
| 320K | $41.45 | $1,865.00 | $1.47 |
| 384K | $43.90 | $2,184.00 | $1.72 |
| 448K | $46.75 | $2,486.00 | $1.96 |
| 512K | $49.35 | $2,771.00 | $2.18 |
| 576K | $49.85 | $3,037.00 | $2.39 |
| 640K | $50.90 | $3,286.00 | $2.59 |
| 704K | $51.95 | $3,516.00 | $2.77 |
| 768K | $54.80 | $3,730.00 | $2.94 |

For high-speed circuits (T1 transmission speed and above), AT&T and other IXCs have instituted rate tables that list a specific price for all POP-POP pairs offered by the carrier. While this table is large for carriers with several hundred POPs, it does allow the carrier the flexibility to offer prices in line with the underlying cost of physical facilities and avoid the competitive risk penalties that occur when prices are averaged and offered everywhere.

### COMPETITIVE CONTRACT SERVICES

Each individual service discussion above points to a need to use competition to obtain special IXC pricing. This cannot be over-emphasized in any discussion of selecting services; no one paying the slightest attention to the interexchange marketplace would pay tariff prices for services today. In contrast, the way to obtain the best prices is to aggregate all telecommunications service requirements (including outbound voice, toll-free voice, data, and private lines) and offer them as the prize in a competitive procurement.

This multi-service approach was introduce by AT&T in the early 1990s as a way to compete with new IXCs for large corporate customers. AT&T was still under regulation and could not offer the same service to different customers at different rates. In response to the growing corporate market for virtual networks, AT&T introduced a new tariff called Virtual Telecommunications Network Service (VTNS), filed with the FCC as Tariff #12. It encompasses both outbound and inbound switched-voice services, circuit-switched data services, and private-line services. The rate components for these services used nomenclature completely unique to this tariff. A common section of the tariff contained rate tables for each individual service when obtained as a whole. However, these rates were rarely, if ever, applied. AT&T used this tariff as a repository for rates it offered to specific customers as part of a competitive process.

If AT&T won a competition, it filed the rates for that customer as an *option* in Tariff #12, along with any rate-affecting terms and conditions. Other IXCs quickly protested that this was a thinly veiled method of providing *like services* at anti-competitive rates. The FCC upheld AT&T's position, leading the way toward the notion that rates resulting from competitive processes were inherently valid and could be filed as tariffs without detailed cost justification, even by AT&T. Since these tariff rates resulted from corporate contracts, Tariff #12 and other similar tariffs have become known as *contract tariffs*.

Part of the FCC ruling on contract tariffs was that any other *similarly situated customer* (i.e., a customer meeting the terms and conditions of the contract/tariff) could also order service under that tariff option. As a result, contract tariff options evolved to contain many constraints tailored to the original customer and designed to keep most other entities from ordering under the option. These include restrictions on time-of-day distribution of traffic, mileage distribution of circuits, maximum quantities, geographic limitations and distributions, etc. This effectively reduced the incentive to use another's option and led the market to the current multi-service competitive procurement paradigm.

As in any highly competitive situation, vendors are willing to accept low profit margins if customers show commitment and if the risks of providing services at guaranteed prices over time are low. In the contract tariff arena, commitment comes in the form of a *Minimum Annual Commitment* (MAC), usually in the form of a revenue commitment. If the MAC is not met in a given year, the customer owes the carrier the difference between the actual revenue and the MAC. In commercial contracts, this MAC is usually 75—80 percent of the total expected revenue from the contract. Users should attempt to commit the minimum required to obtain good prices and keep the remaining traffic as "swing" traffic to use as leverage in obtaining good prices over time.

Some traffic may even be given to a second service vendor with periodic price reviews to reevaluate price levels and possibly move traffic from one vendor to another. A final "price management" technique is to use an index of market prices to benchmark the rates obtained in one's own contract. Fixed prices for five or more years are good protection against future price increases due to inflation, but the interexchange telecommunications market has been characterized by rapidly falling prices since divestiture and shows no signs of reversing itself. Hence, some form of annual rate review or most-favored customer clause should be included in negotiated contracts to move contract prices downward when the market forces make original contract price noncompetitive.

An illustrative example of price trends for a competitively procured and managed set of services comes from the experiences of the federal government. In 1988, the General Services Administration (GSA) procured services from industry under the Federal Telecommunications System 2000 (FTS2000) program. This ten-year program represented the combined non-military federal government requirements for voice, toll-free, data, video, and private-line services. Two contracts were awarded to AT&T and Sprint with an initial 60—40 percent split of traffic.

At two points in the contract, recompetitions (known as "price redeterminations") were held to solicit new prices and possibly readjust traffic splits based on the new price levels. There was also a "publicly available price cap" mechanism in force that compared the FTS2000 unit prices with the best commercial contract tariff prices over time. Figure 12.14 shows the unit price for switched voice service under FTS2000 over the life of the contract. The three competitive events (the initial competition and the two price redeterminations) each had significant contributions to the program as a whole. The combined government savings resulting from the three events (compared to the prices in force before each event) amounted to over $3 billion. In 1998, GSA recompeted the contracts under the follow-on FTS2001 program and obtained prices for the next eight years that continue the seemingly unstoppable downward trend.

**Figure 12.14**
Special contract service price-history example.

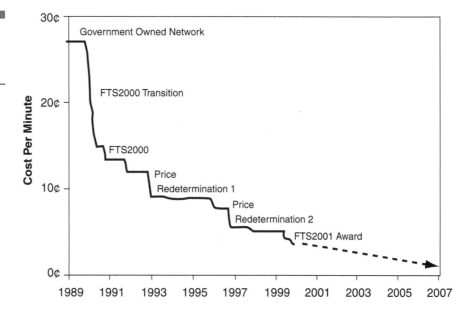

The message to telecommunications managers everywhere is clear: you get what you negotiate! Define requirements as precisely as possible to lower a bidder's perception of risk (GSA provided computer files of historical and projected traffic over the contract's lifetime). To the extent possible, keep competitive pressures on providers throughout contract periods of performance—and keep as many options open as possible.

Thus far, this chapter's focus is on selecting voice network services. The remaining sections provide details on how to use competition in selecting voice premises services. The lessons learned on the proper generation and use of Request for Proposals (RFPs) for premises services, to be presented next, apply as well to competitive network services procurements just described.

# Premises Services Procurement

This section discusses practical aspects of planning and implementing premises telecommunications products and services. Principal activities in this regard include determining user requirements, defining the project, preparing technical specifications and RFPs, evaluating proposals, and selecting approaches that best meet business needs. Following requirements determination and project definition, the next step is RFP preparation. Three key RFP ingredients are required to ensure the success of the balance of the selection and implementation process, namely:

- Accurate, unambiguous technical, management, and cost requirements articulated in industry-accepted terminology.
- Explicit instructions on specifying mandatory proposal information content.
- Clearly defined proposal format instructions.

For the specified system to conform with user requirements, RFP statements must be consistent with those needs and phrased in a manner that minimizes possible misinterpretation. Next, it is crucial to ask the right questions of prospective bidders. If information needed to conduct thorough and systematic bid evaluations is not explicitly requested, it probably will not be included in the proposal, particularly if it might expose the less-competitive aspects of a bidder's offering. Finally, lack of uniformity in proposal formats often creates insurmountable obstacles in comparing and evaluating competing bids.

Although an in-depth treatment of the telecommunications product/service selection process is beyond the scope of this book, this chapter outlines major RFP sections, with example exhibit and bidder formats for key areas. The outline and the formats have evolved over years of successful use in actual procurements. The chapter closes with important lessons learned from recent procurements.

## RFP Example Outline

RFP size and composition varies with the size and complexity of the telecommunications project. Especially in large projects, effective RFPs begin with explicit and clearly articulated project descriptions and objectives, and afterward present specific and unambiguous technical, operational, and contractual requirements. Often overlooked but highly important are precisely enunciated *mandatory proposal preparation requirements* and "up-front" statements that proposals not conforming to those requirements may be considered non-response and unacceptable. An RFP for large installations (in the 1,000-to-25,000 workstation locations range in single or multi-building campus applications) covering voice, data, premises distribution and ancillary services and systems, that exhibits these desirable qualities follows:

1.0 Introduction
    1.1 Purpose of the RFP
    1.2 Site description, referencing any attached drawings
    1.3 Points of contact, overall schedules and delivery dates

2.0 General Requirements
    2.1 Project scope.
        • Top-level System Functional, Operational and Capacity Requirements
    2.3 Project definitions
    2.4 Proposal format and content requirements
    2.5 Treatment of proprietary data

3.0 Evaluation Criteria
    3.1 How the proposal will be evaluated, including areas of focus, precedence, weighting, and necessary bidder qualifications

4.0 Specific Performance and Technical Specifications
    (This includes Configuration Item [hardware and software], mandatory feature and Reliability, Maintainability and
    Availability Requirements)
    4.1 Voice System
    4.2 Data System
    4.3 Premises Distribution System
    4.4 Fire/Life Safety/Security System (as required)
    4.5 Video System (as required)

5.0 Environmental
    5.1 Requests details on support needs of proposed system, imposed on buyer or seller
        • Floor space
        • Ceiling height
        • Floor loading
        • Wire and cable access
        • Normal, UPS and back-up A.C. power. Input Power facilities,
        • Grounding system
        • HVAC (BTU per hour demand), Temperature and Humidity requirements,
        • Dust Control
        • Lighting
        • Fire Protection
        • Convenience information system and power outlets
        • Desks and furnishings
        Bidders are requested to provide floor plan layouts to accommodate proposed telecommunications, supporting and ancillary equipment, and office and storage spaces.

6.0 System Implementation and Installation Requirements
    6.1 Implementation and Installation Schedule
- Equipment, communications, and support rooms ready
- Shipping and delivery of equipment
- Installation start and completion
- Equipment hardware testing
- Installation of site distribution and local area network cable
- Installation of distribution frames
- Carrier and local operator access arrangements complete system cross-connection and testing
- Overall system acceptance testing

    6.2 Bidder Project Plan Submittal Requirements
- Management Plan
- Staffing Plan
- Third Party List
- Contingency Plan
- Planning Continuity

    6.3 Contractor Responsibilities
- Location and Access to Existing utilities
- Surveying
- Preservation
- Clean Up

    6.4 Logistics
- Right of Access
- Operations and Storage Areas
- Temporary Buildings

    6.5 Compliance Issues
- Legal
- Permits, Licenses, and Easements
- Other Contracts, Subcontracts

    6.6 Testing and Acceptance
- System Tests
- Testing Adequacy
- Testing Certification
- System Acceptance

    6.7 Documentation
    6.8 Training

7.0 System Acquisition Requirements
    7.1 Payment Schedule
- Invoices

    7.2 Additions to Purchase Agreement
    7.3 Assumption of Risk
    7.4 Buyer Indemnification
    7.5 Confidential Information
    7.6 Effect of Work Stoppages
    7.7 Liquidated Damages
    7.8 Contractual material

8.0 System Operation, Maintenance, Administration, and Management Support Requirements
    8.1 Maintenance concept.
- Failure Categories and Response Times.

    8.2 Maintenance Service
- Trouble Response

- System Problems
- Agreements and rates
- Personnel
- Maintained Base
- On-Site Spare Parts
- Trouble Reporting
- Scheduled Maintenance
- Remote Maintenance
- Indemnification
- Catastrophe Plans
- Work Stoppage
- Pricing

8.3 Maintenance information.
8.4 Tools and test equipment.
8.5 Progress reporting
8.6 Contingency planning
8.7 Moves, adds, and changes (MAC) work plan
8.8 Service intervals
8.9 Post-warranty services
8.10 Additions to warranty and maintenance agreements

9.0 Pricing
9.1 Elements of bidder's price proposal
9.2 Pricing format
9.3 Pricing of mandatory versus optional items
9.4 Warranty service pricing
9.5 Post-warranty service pricing
9.6 Allowable price increases
9.7 Pre-cutover/post-cutover unit costs
9.8 Purchase (Including Warranty)
9.9 Lease with Purchase Option (Including Warranty)

10.0 Instructions to Bidders

Appendices and Attachments

This portion of the RFP contains additional information necessary for bidders to submit a proposal, e.g., site survey schedule, architectural, mechanical/electrical/interior layout (furniture) drawings, together with details of any unusual or critical customer requirements.

As outlined above, the RFP calls out technical specifications to be addressed in bidder proposals covering the engineering, installation, testing, and cutover (placing into operational service) of the particular telecommunications project. Each major portion of the required deliverable is described in the RFP. RFP instructions should direct bidders to demonstrate how their proposed telecommunications solutions address the following general characteristics:

- Incorporation of industry standards and "open system" designs
- Application of modular hardware/software designs
- Use of commercial off-the-shelf products and services
- Ease of management

■ Optimized life-cycle costs/benefits

Content of the sections in the above list is implicit in the titles. It has become common practice in the last several years to break Section 4, Specific Requirements and Technical Specifications, into at least three separate parts, one each for voice, data, and premises distribution requirements. Less than a decade ago, voice and data vendor products demanded unique wiring and cabling. Consequently, premises distribution system requirements were incorporated separately into voice and data system requirements sections.

As noted in Chapter 9, the current approach favors *universal wiring,* intended to support vendor voice and data systems as well as future upgrades to new technologies. Hence most contemporary RFPs include standalone PDS sections, which indirectly impose compliance requirements on voice and data systems. Recall from Figure 9.9 that per-line PDS costs can be a major portion of per-line PBX system costs, and that in cases where fiber cable to desktops is required, PDS costs nearly equal PBX costs—another factor justifying separate PDS section treatment.

Within the voice part of the *Specific Requirements and Technical Specifications* section 4 of the RFP, there should be one or more pages devoted to a detailed description of the required system size at cutover, and its expansion capacity throughout the operational life-cycle. A sizing specification for a typical digital telephone system is shown in Figure 12.15. Note that the specification can be met by either a PBX switching system or by Centrex service. It is important not to slant the description in order to retain maximum flexibility of choice.

Beyond sizing, where records are available, bidders should be provided with traffic data (such as busy-day/busy-hour data provided in Figures 11.7 and 11.8, and the overall division of voice traffic (and data traffic, where applicable), both for cutover and projected expansion capacity levels. The division of traffic is expressed in terms of *intrasystem* (station-to-station), *incoming* (calls originating outside the system), and *outgoing* (calls terminating outside the system).

Information on division of traffic makes it easier for the bidders to propose the most efficient, cost-effective trunking arrangements meeting individual business needs. If this information is not available, bidders will generally presume an equal division of traffic: one-third intrasystem, one-third incoming, and one-third outgoing.

System grade of service (GOS), discussed in Chapters 6 and 11, should also be specified in the RFP. In cases where records from a current system exist, GOS requirements can be expressed in terms of available traffic statistics. In the absence of traffic statistics, a rule of thumb is to specify a GOS of P=0.005 (1 call in 200 blocked) per station for intrasystem calls, and P=0.01 (1 call in 100 blocked) per station for incoming and outgoing calls.

**Figure 12.15**
Example RFP telephone system sizing exhibit.

| | System Capacities | | | |
|---|---|---|---|---|
| **Item** | **——At Cutover——** | | **System** | |
| | **Equipped** | **Wired** | **Capacity** | |
| Stations: | | | | |
| Analog | * 1000 | 2000 | ** | |
| Digital | * 200 | 2000 | ** | |
| Total | * 1200 | 4000 | ** | |
| | | | | |
| Off-Premises Stations | | | | |
| (All Analog) | 100 | 300 | ** | |
| | | | | |
| DID Trunks | 40 | 120 | ** | |
| | | | | |
| DOD Trunks: | | | | |
| Band 5 WATS | 20 | | | |
| Local, DDD, IDDD | 15 | | | |
| Total | 35 | 100 | ** | |
| | | | | |
| Two-way Trunks: | | | | |
| LDN | 20 | | | |
| DISA | 10 | | | |
| Total | 30 | 90 | ** | |
| | | | | |
| Radio Paging Trunks | 10 | 30 | ** | |
| | | | | |
| Tie Trunks: | | | | |
| Incoming | 20 | 60 | ** | |
| Outgoing | 20 | 60 | ** | |
| Total | 40 | 120 | ** | |
| | | | | |
| PFCT Trunks | | | | |
| (All Two-way Trunks) | 2 | 6 | | |
| | | | | |
| Attendant Console | 2 | 4 | ** | |

\* Bidder shall quote prices for the following options:
1. All analog stations
2. Analog/digital mix shown above.

\*\* Bidder shall state maximum size to which the system can be expanded through cabinet additions, etc.

Notes:
1. Equipped = Installed and working or spare

2. Wire = Can be equipped by adding PCBs only

3. PFCT = Power failure cut-through trunks

4. DISA = Direct inward system access

5. LDN = Listed directory number

6. Tie lines to be dial repeating 4-wire E&M for access to non-digital public and private network facilities

Recall from Chapters 6 and 11 the relationship between GOS and centi-call seconds (CCS). CCS expresses the average time during the busy hour that a station line is busy (in hundreds of seconds). A line used 100 percent of the time represents 36 CCS. As stated in Chapter 11, typical per-line voice and data offered traffic loads are 4—6 and 18—36 CCS, respectively. Line and trunk CCS traffic loading is the basis on which GOS is specified and measured.

Bidders should be requested to state the combined (based on division of traffic) per-voice (or data) station CCS capacity proposed in the *wired-for* system configuration. The wired-for configuration represents system traffic-handling resources available at cutover to support future expansion without adding shelves or cabinets. For large systems, bidders should also be asked to state the maximum number of busy-hour call attempts that the proposed system can handle.

The average telephone system is engineered to provide between four and seven CCS per line at P=0.01. For greater traffic-handling capacity, optional pricing may apply, even for Centrex service. Bidders should be requested to state any optional CCS levels available in the proposed system, together with associated price premiums.

Next, it is important to pay close attention to the structure of the voice-system dialing plan. Cutovers have been delayed because of clashes between telephone numbers and access codes for features and network facilities. Simple dialing plans are best, with single-digit access codes used wherever possible. Feature access codes generally include a * or # prefix, followed by one or two digits. So long as these prefixes are used, clashes with station numbers and network facilities will be avoided. Station numbers vary in digit count, based upon the size of the system. A three-digit dialing plan supports approximately 800 stations; a four-digit plan supports 8,000; a five-digit plan supports 80,000. Most PBXs employ three-digit or four-digit station-to-station dialing. A example PBX system dialing plan is shown in Figure 12.16.

**Figure 12.16**
Example PBX dialing plan.

| Digits | Service |
| --- | --- |
| 1 | Reserved |
| 2-XXX | Stations |
| 3-XXX | Stations |
| 4-XXX | Stations |
| 5-XXX | Stations |
| 6 | Reserved |
| 7 | Reserved |
| 8-XXX-XXXX | On-Net |
| 9-XXX-XXXX | Local or On-Net |
| *9-XXX-XXXXXXX | Long Distance or Off-Net |
| 0 | PBX Attendant |

*May require dialing "1" or "0" if not automatically inserted by switching equipment.*

The station-to-station dialing feature of a Centrex switch has potentially wider scope than that of a PBX. The CO switch's normal seven-digit dialing plan (single NXX) can serve around 800,000 separate telephone numbers. Thus, if a Centrex customer has several offices, all served by the same CO switch or multiple switches with digital city-wide Centrex capabilities, it is relatively easy to develop an abbreviated station-to-station dialing plan of five or fewer digits, if sufficient station (-XXXX) numbers are available within one of the exchange (NXX) codes provided by that CO switch.

Under such a uniform dialing plan, the customer would make local calls using seven (or 10 digits in some urban areas) digits, intracompany calls using three, four, or five digits, and intercom calls using one or two digits (generally using CPE telephone equipment).

Dialing-plan capacity can give Centrex an advantage over a multiple PBX configuration, which requires additional hardware in the form of tie lines and/or OPXs (both of which are expensive to operate and maintain) and additional tandem network software to integrate multiswitch operations.

The PBX system dialing plan shown in Figure 12.16 also applies to the Centrex environment. As previously noted, in a distributed (city-wide) Centrex system using multiple COs, planning for a universal dialing plan can be complicated by the lack of available CO exchange codes, station numbers within exchange codes, and differences in features and operations from one CO switch to another.

The *voice* part of the Specific Requirements and Technical Specifications section of contemporary RFPs usually includes subsections for requirements in the following areas, some of which were just discussed:

- System line and trunk sizing
- Cutover date
- System expansion
- Division of traffic
- Grade of service
- Dialing plan
- Survivability
- Station equipment
- Network connections
- System management
- Types and quantities of deliverables
- Computer common-control architecture
- Power supplies (including any requirements for uninterruptible power supplies)

- Network interface characteristics
- Operational parameters
- Peripheral equipment
- Reliability/maintainability/availability
- Regulatory compliance
- System security
- Features

Feature specifications normally include required features and bidder-suggested features. The RFP should provide a brief description of each required feature to clarify expected general characteristics. Where necessary, specific details on feature operation and capacities are requested from the bidders. Bidders should be advised to state their own nomenclature for each feature and to describe any differences in feature operation from the descriptions of mandatory features in the RFP.

At the beginning of a features section in the RFP, the following information should be requested from the bidders, where applicable, for each required feature:

- Method of activation and deactivation.
- Capacities, such as maximum number of intercom groups/codes per group, speed calling lists, etc.
- Features that preclude or limit the use of other features, together with a statement of the degree of impact.
- Features requiring a station line assignment.
- Features that cannot be provided to all station users at a particular system size (cutover configuration versus wired capacity).

It is important to request that bidders state whether a particular feature is standard or optional in the proposed system, and if optional, whether it requires hardware, software, labor, and/or other chargeable cost elements. Optional costs should be identified by the bidder, and itemized in the *pricing* section of the proposal, but not incorporated into the total system price.

Where a bidder suggests additional features which are not specified in the RFP, but which the bidder deems appropriate to the system, such features should be described and associated prices quoted as above, but not incorporated into the total system price.

# Lessons Learned from Recent Procurements

This subsection examines the telecommunications product/service selection process by analyzing proposals from a large PBX manufacturer and a midwest RBOC, who submitted bids for a fourth-generation PBX and a digital city-wide Centrex system, respectively. The telecommunications project involved a new high-technology industrial park. The project was scheduled to begin with a single building served by 120 lines and to grow to an eventual 5,000-line, 25-building campus. Vendors were instructed to depict per-line proposal costs as a function of project growth. To attract the broadest spectrum of tenants, particularly those engaged in the provision of high-technology products and services, the park operator wanted to evaluate a range of modern telecommunications capabilities, together with various levels of park management involvement in providing shared tenant services. The discussions that follow focus on the PDS and switching equipment portions of the project.

Overall, the objective of competitive procurements is to identify and select alternatives optimized to meet functional requirements and to minimize costs. Comparison and ranking approaches must therefore evaluate technical performance against specified criteria, employing discriminators that support a rational, repeatable selection making process. Because today's technology equips PBX vendors and RBOC/LECs alike with the ability to satisfy technical requirements (rarely do proposals take exception to any major performance stipulation), discrimination among bid alternatives comes down to quantitative cost differences and usually more qualitative technical management and risk factors.

The PDS bids for the above project are a case in point. The PBX vendor and the RBOC submitted nearly identical PDS price quotes, which are summarized in Figure 9.9. The RFP specified unshielded twisted-pair (UTP) wiring conforming to standards described in Chapter 9. In addition, the RFP included a mandatory optional requirement for fiber optic horizontal wiring direct to desktops. The request for fiber quotes was intended to assist the park manager in deciding whether to make an additional initial investment to increase the likelihood of attracting tenants with needs for 100-Mb/s desktop connections (EIA/TIA TSB-36 had not yet been published). The results of the PDS bid and evaluation process are discussed below.

The upper portion of Figure 9.9 reflects costs associated with three different PDS cabling approaches, labeled A, B, and C in perspective with those of switching equipment. Note that the switching cost per station decreases as system size increases, while the PDS cost per station remains constant.

Illustrated in the lower portion of Figure 9.9, PDS alternatives A and B provide unshielded twisted-pair (UTP) copper wire only, conforming to IEEE 10/100BaseT and EIA/TIA standards. Alternative B implements the EIA/TIA standard of two UTP wiring runs terminating in separate RJ-45 jacks at each workstation outlet. Each jack is capable of supporting voice or data/LAN operations at a 100-Mb/s rate, with developing standards potentially raising even that rate. Alternative C adds two strands of multimode fiber to each workstation.

Figure 9.9 furnishes some useful insights. First, PDS costs do not materially change with growth, since they are linear, and driven by per-foot labor and material costs not subject to economies of scale. Also, the economy of UTP horizontal wiring is underscored, in light of the significant-cost "step function" presented by fiber-to-the-desktop.

In the interbuilding PDS backbone cable application, however, fiber cable became extremely cost effective. Potentially high per-building station counts were involved in this campus wiring scenario. The costs of installing long runs of multiple UTP backbone cables consisting of up to 1,200 pairs each, together with the costs of constructing ductbank (typically around $100 per linear foot) to house the cables, made fiber very attractive. Used for linking remote switching modules (one per building) with a single two- or four-pair multimode fiber cable, the PDS cost-per-station using fiber backbone cable was a small fraction of the UTP copper alternative. For reasons explained in Chapter 6, PBXs in remote switch module configurations are better equipped than Centrex to capitalize on such benefits.

In the end, the PDS selection decision was not based on technical or cost differences in bid responses, but rather on options to:

- Not plan for greater than 100-Mb/s desktop service.
- Bank on the success of new UTP technologies under development for greater than 100-Mb/s desktop service.

As noted in Chapters 9 and 13, a number of reputable companies—united in a UTP Forum—have since developed UTP that can support 1,000 Mb/s-rates in horizontal PDS wiring. Yet developments had not progressed to the point where the park manager could rely on fixed price quotes to assist him in his selection decision. As indicated at the start of this section, selection could not be based on technical or quantitative cost differences among competing proposals, but rather had to be based on harder-to-quantify performance risks associated with an emerging technology.

Similar results surrounded the choice between PBX and Centrex for premises switched services. A 60-month, one-dollar buyout lease approach for the PBX alternative had been specified in the RFP to facilitate proposal comparison with the Centrex bid, and to normalize requirements for initial capital investment. On a technical, cost, and implementation schedule basis, the Centrex and PBX approaches proved nearly identical. Figures 12.17 and 12.18 summarize categories of PBX and Centrex bid costs as the project grows to 5000-line capacity. The RFP for this procurement requested bidders to provide proposal information in the formats illustrated in Figs. 12.17 and 12.18, in order to facilitate a thorough, service-by-service complete life-cycle (initial, recurring, operations, maintenance, administration, etc.) cost comparison. Note that the total cost per line is nearly equal over the expansion period.

| Date | Basic Costs | | | O/M&A Costs | | Total Costs | |
|------|-------------|------|------|-------------|------|-------------|------|
| | # Additional Stations/ Total Stations | Subtotal Cost/Month | Subtotal Cost/Station Month | Subtotal Cost/Month | Subtotal Cost/Station Month | Total Cost/Month | Total Cost/Station Month |
| Start Date | 120/120 | $3478 | $28.98 | $2675 | $22.29 | $6153 | $51.27 |
| 3 Months | 425/545 | $12670 | $23.25 | $10745 | $19.72 | $23415 | $42.96 |
| 6 Months | 800/1345 | $21488 | $15.98 | $18360 | $13.65 | $39848 | $29.63 |
| Year 1 | 445/1800 | $26463 | $14.70 | $19216 | $10.68 | $45679 | $25.38 |
| Year 2 | 3200/5000 | $61118 | $12.22 | $40096 | $8.02 | $101214 | $20.24 |
| Year 3 | 0/5000 | $61118 | $12.22 | $54871 | $10.97 | $115989 | $23.20 |
| Year 4 | 0/5000 | $61118 | $12.22 | $56160 | $11.23 | $117278 | $23.46 |
| Year 5 | 0/5000 | $61118 | $12.22 | $57834 | $11.57 | $118953 | $23.79 |
| Year 6 | 0/5000 | $11935 | $2.39 | $59503 | $11.90 | $71439 | $14.29 |

**Figure 12.17**   PBX life-cycle costs (60 month lease with $1 buyout).

Depending on the individual system, life-cycle costs are influenced by the following:

- **System management**—Traditionally less costly for Centrex, this is now changing with more CPE management provided by the LEC under tariff or contract.

| Date | Basic Costs | | | O/M&A Costs | | Total Costs | |
|------|-------------|---|---|-------------|---|-------------|---|
| | # Additional Stations/ Total Stations | Subtotal Cost/Month | Subtotal Cost/Station Month | Subtotal Cost/Month | Subtotal Cost/Station Month | Total Cost/Month | Total Cost/Station Month |
| Start Date | 120/120 | $4900 | $40.84 | $290 | $2.42 | $5190 | $43.25 |
| 3 Months | 425/545 | $16515 | $30.30 | $1323 | $2.43 | $17838 | $32.73 |
| 6 Months | 800/1345 | $36010 | $26.77 | $3269 | $2.43 | $39279 | $29.20 |
| Year 1 | 445/1800 | $46401 | $25.78 | $4378 | $2.43 | $50779 | $28.21 |
| Year 2 | 3200/5000 | $119777 | $23.96 | $12163 | $2.43 | $131939 | $26.39 |
| Year 3 | 0/5000 | $119777 | $23.96 | $12163 | $2.43 | $131939 | $26.39 |
| Year 4 | 0/5000 | $119777 | $23.96 | $12163 | $2.43 | $131939 | $26.39 |
| Year 5 | 0/5000 | $119777 | $23.96 | $12163 | $2.43 | $131939 | $26.39 |
| Year 6 | 0/5000 | $119777 | $23.96 | $12163 | $2.43 | $131939 | $26.39 |

**Figure 12.18**   Centrex life-cycle costs.

- **Switching equipment maintenance and repair**—These costs are included in the Centrex tariff but must be added for the PBX after the warranty period expires.
- **Moves, adds, and changes**—Incurred for both Centrex and PBX, but may be lessening for Centrex because of trend toward "total package" LEC support.

In the campus system described above, system management and maintenance support provided at no additional charge by the LEC had to be added to the PBX costs, resulting in nearly equal total costs. For this reason, system selection rationale shifted toward assessment of two risk factors. The first was the added client burden of managing the PBX and third-party support contractors. The second was effectiveness and supportability of the PBX after 60 months of use.

The key decision driver with respect to the first factor was that the LEC offered significant, second-year-and-out support services that were bundled into the Centrex line rate. The key decision driver with respect to the second factor was that significant cost savings could be achieved in the outyears, if the PBX's useful life could be extended past the 60-month buyout point.

As in the PDS example, the selection decision could not be based on technical or cost differences among competing bids. It became dependent on assessment of:

- Residual risks associated with operating and managing a large telephone system (even after competent third-party support contractors had been identified and related costs had been taken into account).

- Risks involved with the predictions of technological sufficiency and continued supportability of five-plus-year-old PBXs.

PBX vendors make strong arguments that large telephone system operations present business opportunities, not risks, and that their modular, software-driven PBX designs ensure seamless expansion far into the future. LECs counter with equally convincing arguments that economics of scale favor using shared public facilities, that customer operation of large telephone systems is prone to failure and that Centrex service provides the ultimate insurance against technical obsolescence.

Although differences in proposed technical and management approaches will continue to challenge the evaluator on a project-by-project basis, the above examples demonstrate the important role that RFP content and format instructions play in the telecommunications product/service evaluation and selection process.

# Data Services

This part describes *data services* and the underlying customer premises equipment (CPE) and network facilities that support them. As in the voice section, we identify and describe high-interest services first, and then present details of salient facility technologies used to implement them. At the highest levels, the taxonomy of data services, shown in Figure P4.1, closely resembles the categories comprising voice services. By industry convention, the treatment of data services introduces some new terms to the telecommunications lexicon, namely the formal distinction among *local, metropolitan*, and *wide area networks*—LANs, MANs, and WANs. IEEE definitions and the rationale concerning why they are generally applied in data communications but not in voice communications network discussions are explained in Chapter 14. Two categories of services, *application service provider* (ASP) services and *security/public key* services, although not unique to telecommunications, are so crucial to Internet and Internet-based virtual private networks, that they merit the special treatment afforded them below.

To form a more complete understanding of modern data communications services and technologies, it is helpful to understand predecessor limitations and to review the evolution of facility capabilities and user requirements that prompted their development

## Requirements and Facility Environmental Trends

To place data communications system design trends in perspective, it is instructive to contrast data service requirements/environments of the 1970s, 80s, and early 90s with today's situation, and make projections for the future. In the early 70s, communications between so called "dumb terminals" and central host computers dominated data network traffic. These terminals could support only simple interface protocols, so to achieve reliable transport service, networks had to supply error control, addressing, connection establishment, and other high-level protocol services.

As a result, existing wide area data communications networks employed SNA, X.25, or circuit-switched-based architectures. *SNA (Systems Network Architecture)*, IBM's proprietary description of the logical structure, formats, protocols, and operational sequences for transmitting information units (packets) and controlling network configuration and operations, dominated business data communications until the early 1990s. Also, since the 1970s, voice transmission facilities have been the primary mechanism for carrying residential and small-business user terminal data traffic. Whereas acceptable quality digital voice traffic is relatively insensitive to transmission errors (one error in 1,000 bits is tolerable), data communications demands near error-free transport.

**Figure P4.1**

Taxonomy of data services.

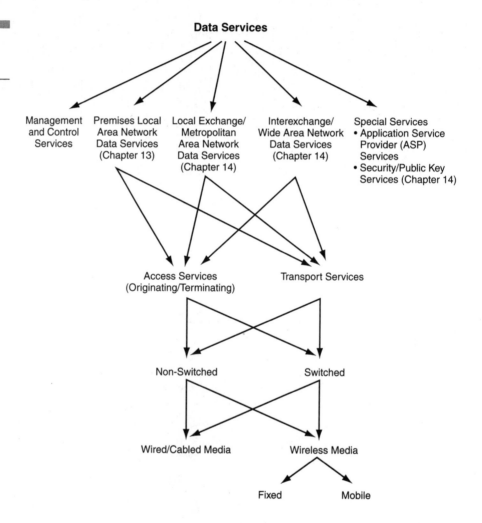

Powerful error-detection and correction techniques used in SNA and X.25 packet networks furnish end-users with reliable, error-free transport, masking poorer-quality transmission facilities. The price paid, however, is low throughput and high message delay due to extensive processing within networks. As noted in Chapter 7, the combination of virtually error-free fiber optic transmission (bit error rates of less than one error in a billion [$10^9$] bits without error control) and intelligent end-user CPE devices now permits streamlined network designs that significantly enhance speed and reduce message delay.

Other important factors distinguishing current and past situations are an increased amount of data traffic and the higher bandwidth requirements associated with that traffic. As noted before, just ten years ago voice traffic dominated networks, with packet data traffic a minuscule part. Today, data traffic accounts for nearly 50 percent of aggregate network traffic; while the demand for voice service is increasing at 3 to 5 percent annually, demand for data service is expanding at a greater than 100 percent annual rate. Furthermore, in contrast to interactive applications with modest bandwidth requirements (9.6 to 56 kb/s) of the past, some current and future data communications applications require orders-of-magnitude more bandwidth (1 Mb/s to 1 Gb/s). Examples include LAN-to-LAN, digital imagery, high-speed computer-aided design (CAD), and video applications.

Expanding bandwidth requirements need faster MAN and WAN response times. In the past, interactive service users had to be content with slow screen refresh times and transaction response characteristics. Today's users are accustomed to near-instantaneous response available on high-speed LANs. When 100-Mb/s LANs are interconnected by 1.5-Mb/s MAN or WAN lines, users become frustrated with response time degradation occurring when they are logged on to remote LANs.

Another development driving future network designs is the need for variable versus constant bit-rate services. Traditional voice and data transmission requirements are best satisfied by *constant bit-rate* (CBR) services. For example, without *digital speech interpolation* (DSI) techniques as described in Chapter 3, conventional PCM voice encoders produce output bitstreams at fixed 64-kb/s bit rates, independent of pauses in speech activity. CBR services are similarly suited for file transfers at constant rates over entire sessions.

Some new applications, however, can be more economically supported with *variable bit-rate* (VBR) services. For example, users may need to scan files or documents, briefly viewing pages until they find what they are seeking. While they are in the *scan mode*, average bandwidth must be high enough to build display screens rapidly. Once desired objects are located, the user pauses or enters an *editing mode*, during which time greatly reduced average bit rates are sufficient (actual rates being driven by how fast users think and type). Such applications are said to exhibit *VBR/Start-Stop* characteristics. That is, during a data communications session, periods of intense activity may be interspersed with periods of less and/or no activity.

Another VBR example involves forms of digital video encoding. Without compression, encoding standard National Television System Committee (NTSC) television analog signals produces CBR digital signals at bit rates of about 140 Mb/s. Encoders employing conventional compression algorithms produce digital output signals at constant bit rates less than TV's peak 140 Mb/s bandwidth requirement. Such encoders take advantage of the fact that during still periods (little image motion) as low as 1 Mb/s may be adequate to transmit all essential image information. Since these encoders generate constant bit rates too low to preserve full 140-Mb/s NTSC digital quality, the algorithms used are termed *lossy*. During rapid image motion intervals, lossy algorithms compromise image resolution, smooth image motion tracking, or both. The motivation behind the use of such encoders, and the lower than 140-Mb/s output rates, is to provide acceptable quality while conserving expensive transmission bandwidth capacity.

Newer classes of video encoding algorithms generate outputs at a variable bit rate. These new algorithms are *lossless;* they involve no loss in resolution quality or motion tracking capabilities. During periods of rapid image motion, the new encoders can produce peak bit rates enabling full image quality. During periods of little motion, these algorithms produce lossless encoded video outputs at as low as 1 Mb/s.

For networks to support VBR traffic with maximum economy (equivalent to handling the greatest number of VBR signals possible within a fixed maximum bandwidth and a specified grade of service), they must employ switching and multiplexing designs that provide users with VBR or *bandwidth-on-demand* services. Today's circuit switching networks cannot provide bandwidth-on-demand services. Rather, as noted earlier, following call establishment, *fixed bandwidth* channels are dedicated to single users for entire call durations. In circuit-switched networks, the fixed bandwidth would have to be at least as great as the peak VBR bit rate. This means that for those intervals where actual VBR rates are less than the peak rate, some or most of the fixed circuit-switched channel capacity is wasted. That is, the channel is not fully used by the connected parties, and the network has no way of allocating the unused capacity to other users.

Packet-switched networks can be designed to provide bandwidth on demand for users with VBR traffic. But for the network to determine how many VBR users can be simultaneously accommodated within a transmission facility of fixed bandwidth, it needs to know both the *peak* and the *average bit-rate* statistics of each user's traffic sharing the facility. The problem of designing VBR-capable networks to ensure grade-of-service levels is not unlike the circuit-switched process described in Chapter 11—except, of course, that it is a good deal more complicated. Blocking probability is not simply a function of the traffic intensity expressible in numbers of fixed-bandwidth call arrivals and numbers of available channels, but must take into account both the number of calls, and peak and average bit rates involved with each call.

Today, *asynchronous transfer mode* (ATM) networks are equipped to provide true bandwidth-on-demand services for users with VBR traffic. ATM networks will accept or reject connections based on traffic already being carried by the network, and the new user's average and peak bandwidth requirements. ATM networks also implement policing mechanisms to ensure that users don't violate peak or average bit-rate commitments. ATM networks provide flexible and efficient service for LAN-to-LAN, compressed video, and other VBR bandwidth-on-demand applications. Service providers benefit from ATM since their networks are able to carry more traffic than circuit or non-ATM packet-switched alternatives. Users benefit from ATM services since service providers pass on some of the savings.

ATM not withstanding, the rise of the Internet to the point where it touches the lives of everyone, directly or indirectly, is by far the single most influential factor in the evolution of data communications over the last five years and will remain so for the foreseeable future. It is only due to the Internet's popularity that we now have a public switched data network (PSDN) that rivals the size and breadth of the public switched telephone network (PSTN), providing almost universal connectivity and nearly unrestricted information flow across all borders.

As popular and as useful as the Internet is, as explained below, it currently is not equipped to efficiently meet the QoS requirements for isochronous (voice and video) or VBR service delivery, nor does it permit service providers to guarantee throughput rates to their customers. Conversely, frame relay and ATM-based services that have begun to replace dedicated private line service in private networks *can* support throughput guarantees. Moreover, ATM is capable of delivering a spectrum of QoS levels. Even so, it is important to keep in mind that at this point in time, only the Internet offers truly public packet-switched services on a national and a worldwide basis.

The reasons behind these statements are fully explained in Chapter 14 where all varieties of both LEC/metropolitan area network and IXC/wide area data network access and transport services are addressed. In the IXC/WAN category, Internet Protocol (IP)-based transport services, and in particular those rendered by local and national Internet service providers (ISPs) and backbone or core Internet network service providers (NSPs), are described. As a prelude to those elucidations which treat individual access and transport services separately, Figure P4.2 provides an end-user-to-end-user connectivity "road map" for overall context.

Note in the figure that various vertically cast data access network facilities are arrayed across the top of the figure. Examples range from analog modem and newer xDSL-based data access services for residential and small-business customers, to synchronous T1/T3 and HDSL2-based access services, more appropriate for higher traffic-volume business customers (see Chapter 3 for xDSL and HDSL2 explanations). All access service representations at the top of the figure relate to fixed customer premises and are furnished using wired or cabled (unshielded twisted pair [UTP] or fiber optic cable) facilities. Mobile data access services are represented in the lower left-hand corner and use wireless facilities introduced in Chapters 1 and 4, and treated in more detail in Chapters 16 and 17.

Transport services are depicted horizontally in the figure. The left side illustrates switching and other transport service delivery-point facilities of local exchange carriers, mobile carriers, and Internet service providers. Since most residential and small-business data access services are provided via UTP subscriber loops, analog modems and other voice network facilities, circuit-switched traffic (i.e., the LEC/MAN-to-IXC/WAN connection paths) are shown as thick dashed lines. Packet-switched traffic paths are thick dark black lines. LEC central office-to-access tandem office arrangements correspond to voice network designs described in Chapter 2. In many instances, SONET rings now interconnect central and access tandem switching offices—facilities that may also be used to offer SONET ring service to local ISPs and high traffic-volume businesses needing local broadband transport service. SONET ring service is described in Chapter 14.

The middle right-to-left portions of Figure P4.2 represent East Coast LEC/MAN/ISP facilities, connected to West Coast counterparts via long distance IXC/WAN transport services. IXCs offer a variety of

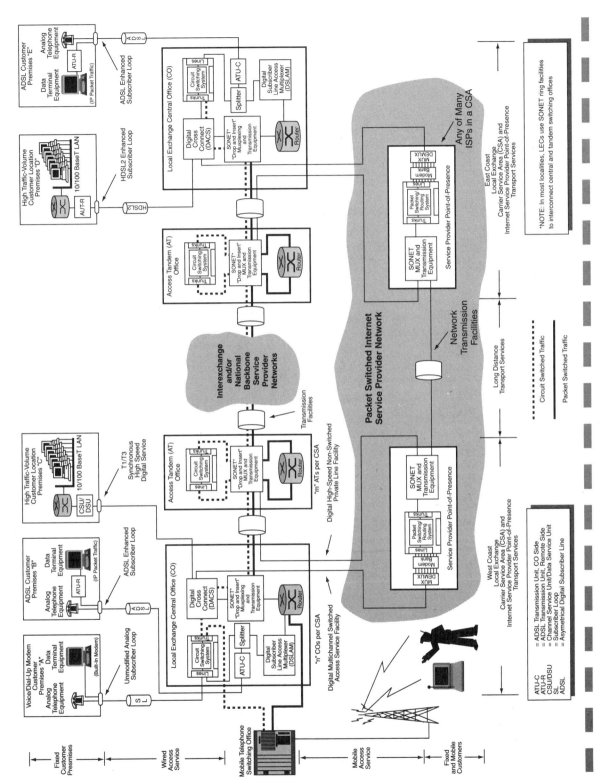

**Figure P4.2** Data network service end-to-end connection example.

long-distance data transport services directly to corporate customers. Corporate customers use these services to implement virtual private data networks, in a manner paralleling the use of voice transport services in virtual private voice networks. However, as explained in Chapter 14, although they benefit from the most modern broadband core network switching and transmission facilities available, neither ISPs nor NSPs offer separate or dedicated transport services directly to Internet subscribers.

Finally, shown at the bottom of Figure P4.2 are East and West Coast examples of local ISP points of presence. The detail reveals that ISPs provision two classes of broadband-access termination facilities. First, they terminate LEC T1/T3 voice transmission facilities that carry multi-channel analog modem Internet subscriber data traffic. These facilities include T1/T3 multiplexers and banks of analog modems to establish individual subscriber-to-ISP data connections. ISPs are also equipped with a variety of broadband digital transmission facilities for the purposes of connecting to both Internet backbone service providers and to local, high traffic-volume business customers. Examples of high traffic-volume user LAN connections are shown at the top of the figure in the premises using T1/T3 or HDSL2 access facilities. Some national service providers also offer local ISP services. In those cases, NSPs may use their own facilities to interconnect their own remote ISP POPs.

In the remaining Part 4 sections, Chapter 13 traces the evolution of premises local area networks (LANs) from the mid 1970s to the most advanced gigabit Ethernet (GE) products being employed today. Chapter 14 presents a similar treatment of early and the most modern LEC/MAN/ISP and IXC/WAN/NSP facilities underlying all relevant data access and transport services. Based on these descriptions, Chapter 15 delineates how providers charge for various services and offers a guide for selecting cost effective approaches that best respond to individual and corporate requirements.

# CHAPTER 13

# Premises (Local Area) Data Network Services

Chapters 13 and 14 furnish service, facility, and technology details of the data services outlined in the taxonomy just presented. Using that taxonomy's organization, Figure 13.1 serves as a roadmap of *premises network services* treated in this chapter. As in the voice-services case, the two major data network services subdivisions are premises access and premises transport services.

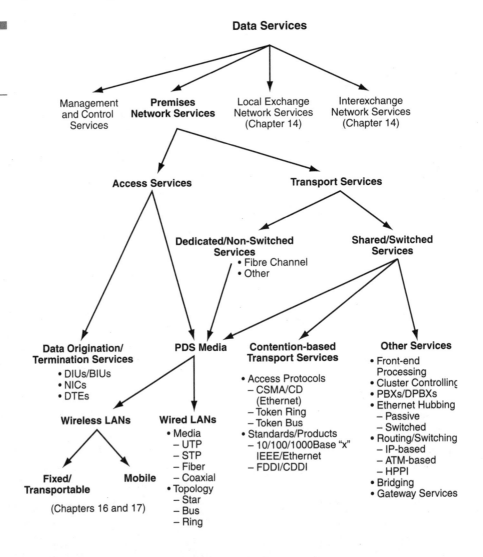

**Figure 13.1**
Taxonomy of data premises network services.

This chapter describes approaches for planning and implementing premises data services. Emphasis is on terminal-to-terminal (peer), terminal-to-server, terminal-to-host processor, and premises-to-metropolitan and wide area network connections. Accordingly, some previous definitions are expanded, and some new terminology is introduced. The chapter briefly traces early technologies, then treats the broad category of local area networks (LANs), which in the last decade have emerged as the premises data services approach of choice, having become a focal point of IEEE and other standards-setting bodies. While the term *local area network* can properly refer to either voice or data networks, the *LAN* acronym is almost universally and exclusively understood to be a premises *data* network.

# Premises Data Access Services

Facilities underlying business and home *premises access services* comprise all manner of terminals, PCs, and workstations designed to directly support operators; remotely controllable or accessible automation and data processing equipment; a growing list of other data-network-compatible information origination/termination *data termination equipment* (DTE); and a premises distribution system (PDS). To treat in any detail the array of applications from word processing and computer aided design to automatic teller machines implemented on DTEs, is beyond the scope of this book. However, due to the success of recent efforts to produce a small set of applications-independent, open-system data communications protocols, understanding communications operations no longer requires an in-depth knowledge of any data-processing application.

For the most part, what needs to be known about protocols that enable DTEs to establish communications across a wide variety of data networks has already been disclosed. Data interface units (DIUs), bus interface units (BIUs), network interface cards (NICs), and certain other hardware and software elements—needed to attach devices to data communications networks—are covered below along with explanations of LAN operations.

As noted, PDS structured wiring, connectors, and cross-connects that conform to EIA/TIA standards are capable of handling both voice and data traffic. Consequently, all PDS components (see Figure 9.3 and its related text) and all the physical, electrical, and electronic industry standards that have an impact on data communications performance, have already been covered in Chapter 9. What remains is to demonstrate how EIA/TIA PDSs are used to support today's popular and high-powered local area data networks.

As Figure 13.1 indicates, PDSs may use wired, wireless, or combinations of those media. Most of the earliest LANs (in the 1970s) employed coaxial cable and IBM-specified shielded twisted pair (STP) cabling for token ring products. By 1986, an unshielded twisted pair (UTP) LAN product was brought to market, about the same time IEEE established its 802.3 UTP LAN working group, later to be called 10BaseT.

Although the IEEE 802.3i fiber optic standard did not appear until 1992, in the mid-70s Robert Metcalfe (Ethernet's inventor) and Eric Rawson, both of Xerox's Palo Alto Research Center, proved that carrier sense multiple access (CSMA—see Chapter 7) signals could successfully be transmitted on fiber optic cable. Consequently, among today's installed LANs, one finds all media types represented. However, the use of coaxial or STP cable is on the wane and most recent and future installations will employ UTP, fiber, wireless, or combinations of those media.

Both Chapters 4 and 9 introduce and describe wireless alternatives to wired PDS media. In certain circumstances, wireless PDS adjuncts are incredibly effective. For example, hand-held devices with product bar-code scanners afford grocery store attendants a highly automated means of coordinating aisle-product pricing with the central databases controlling checkout-register pricing. Similarly useful applications in package delivery, maintenance and repair, and other industries abound. Wireless technologies in data communications are revisited in Chapters 16 and 17.

As noted in Chapter 9, prior to EIA/TIA's efforts to standardize PDS wiring, companies like IBM used over 50 different cable types to interconnect data processing devices in mostly point-to-point, as opposed to wiring closet-based star connections. In that era some form of shielded cable (coaxial or shielded twisted pair) was the predominate choice. By 1994, IBM's token ring LAN design accounted for nearly 50 percent of the installed LAN base. As described below, while devices were logically connected as a ring (like SONET rings cited in Chapter 8), IBM specified that its token ring products be physically star-wired to existing telecommunications closets.

Again, from a wiring topology viewpoint, when the first Ethernets were introduced, one advantage claimed was that a single coaxial "bus-like" cable-run from office-to-office could replace star-based telecommunications closet approaches and thereby simplify wiring. However, limitations on the number and spacing between DTE taps, and the need to re-engineer entire networks for many moves and additions, soon proved that "serpentine" wiring's disadvantages far outweighed its advantages. Today, independent of the cable type selected, hierarchical, wiring closet-based, star-wired topologies are the norm. More discussion of the tradeoffs and design rationale for wired LANs are found below.

# Premises Data Transport Services

It is most often true that completely separate and distinct facilities are used to support the *access* and *transport* parts of *network services.* Although Chapter 10 explains how LEC *transport* services are actually used to furnish users with *access* services to IXC *transport* services, the underlying facilities associated with each category of service remain readily apparent.

In some telecommunications applications, while the notions and the existence of access and transport services remain valid, the underlying facilities supporting each service class may be less apparent, and some are implemented in a single device. Point-to-point two-way radio transceivers are an example. In these transceivers, the radio-power transmitters, the radio-frequency receivers, and the atmosphere are the underlying facilities supporting transport service. In the same transceiver devices, access services use microphone, loudspeaker, and other components.

A similar type of collapsing of facilities supporting access services occurs in *shared medium* LANs. Figure 13.2 illustrates this phenomenon. Part (a) of the figure depicts an example wherein facilities underlying access and transport services in a packet-switched network are readily perceivable. Transport services clearly use switching components of the packet-switch fabric, transmission interface hardware, and connecting transmission facilities. Likewise, access services use session initiation and control and transmission interface DTE components, access transmission and packet-switch input port facilities. Note also in this example, that trunks between interconnected packet switches are *dedicated* to traffic between switch pairs—in spite of the fact that trunk capacity is *shared* among various end-to-end user pairs.

The situation portrayed in Part (b) of the figure is quite different. Today, in a large class of LANs, a common amorphous transmission medium or channel is used to interconnect all attached devices—whether they be powerful packet switches or routers or simple PC-like DTEs. Part (b) generically represents this class of LANs. In early designs, DTEs are attached to a "bus-like" common transmission medium (e.g., a coaxial cable) through standalone BIUs or DIUs. In this case, transport services among DTEs are supported by portions of BIUs and the common transmission facility.

**Figure 13.2**
Data access and transport network services with dedicated and shared transport transmission facilities.

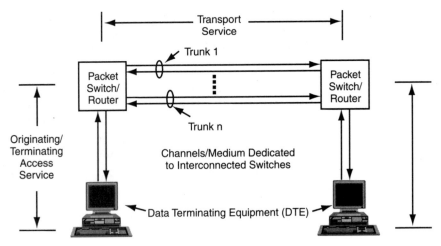

(a) Transport service using dedicated transmission facilities.

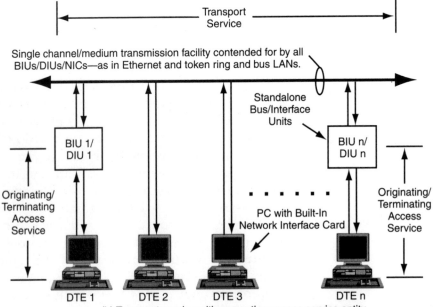

(b) Transport service with competing access service entity "contention" for a single channel/medium transmission facility.

Facilities used to support access services comprise DTE-to-BIU interface portions of both DTEs and BIUs, and office-located connectors and mounting/patch cords, usually limited to ten feet or so in length. In more recent designs, BIU/DIU functions have been consigned to network interface cards (NICs) that can be inserted into DTE (PC) hardware slots, or otherwise absorbed in DTE equipment, further collapsing, or making less distinguishable, facilities used in supporting LAN premises data access and transport services.

In large businesses, one finds that both Figure 13.2 Part (a)- and (b)-type connections are needed. In these cases, PDS systems supply the passive wiring, connectors, and cross-connect facilities supporting both configurations, and consequently PDS facilities can correctly be said to underlie both access and transport services—as indicated in Figure 13.1.

In contrasting modern LANs with longer-established network architectures, we find the most distinguishing feature is the fact that in today's LANs' common channels/media are used to interconnect all attached devices. While in many instances this constitutes a great advantage, it mandates effective multiple-access techniques and protocols that minimize interference and optimize throughput on shared transport facilities. In essence, these techniques arbitrate or otherwise resolve the contention for shared transport facilities among LAN-attached devices. One such technique, dubbed carrier sense multiple access/collision detection (CSMA/CD) is explained in Chapter 7 and addressed further below along with token ring and bus alternatives.

Important contention-resolution and other traditional service alternatives for rendering *shared/switched transport* services, are itemized in Figure 13.1. Implementation examples for each alternative are provided in the remaining parts of this chapter. Although shared/switched transport services are dominant in premises environments, requirements for *dedicated/non-switched*, point-to-point transport services do exist.

For example, ANSI has developed standards for high-speed (133Mb/s-to-1Gb/s) information exchange between mainframes, mass storage devices, workstations, and other computer peripherals. Designed to generate less than one error in $10^{12}$ bits, this technology is dubbed the *Fibre Channel*. Since Fibre Channel is intended for simple point-to-point as opposed to complex networking applications, protocols are rudimentary, implemented mainly in hardware, low in framing and control overhead, and consequently very efficient. Fibre Channel-compatible media include single-mode fiber, multimode fiber, and coaxial cable. Other less exotic paired-device connection requirements can frequently be satisfied by dedicated runs of standard EIA/TIA PDS wiring.

# Premises Data Network Technologies and Products

The rest of this chapter presents a compendium of premises data network technologies and products. Its brief assessment of some predecessor products provides a context for more fully appreciating design and performance advantages that advanced  technologies now make possible. Less than two decades ago, IBM Systems Network Architecture (SNA)-based data network, dominated both local and wide area market segments. In the early 90s, there were still over 20,000 operational SNA networks. By the late 80s, IBM's SNA software involved over two million lines of software code, which at $1,000 per line (typical for that era) represented more than a $2 billion investment in SNA's software development alone.

Since SNA appeared prior to the availability of even modestly powerful personal computers, a major aspect of its design dealt with how to service terminals with virtually no resident processing capabilities. IBM defines a *terminal* as a device, usually equipped with a keyboard, often with a display, capable of sending and receiving data over a communications link. *Dumb terminals* are limited to low-speed operation, do not incorporate local processing, and transmit characters one at a time as they are typed by an operator. Normally hard wired (via dedicated non-switched connections) to a host processor, these terminals cannot support line sharing, polling, or addressing. Polling occurs when a host processor signals a terminal to determine if it is ready to receive or send data.

*Workstations* are input/output devices used by operators that can process data independent of host processors and can be configured to exchange data with other workstations, host processors, or servers. Workstations vary in complexity, ranging from simple personal computers (PCs) to devices with host processor-like capabilities supporting text, database, graphics, imagery, and other data-intensive applications. Workstations can implement multi-layered protocols, as previously described, and therefore can support error control, addressing, and other capabilities to achieve reliable data communications.

In the period following the introduction of PCs, workstations generally connoted standalone processors with capabilities well beyond those of typical PCs. As a consequence of soaring advancements in PC processing power, most new PCs outstrip what had been afforded "workstation" status performance levels just a few years ago. Ever-increasing processor/workstation power and concomitant ability to support end-to-end error free information exchange has been a major factor influencing data network design evolution, as evidenced in the remaining portions of the book. With this background we now proceed with a description of data communications technologies for connecting all manner of terminals, PCs mainframe computers, servers, and other peripheral devices within single buildings or among buildings in campus settings.

# Point-to-Point Premises Networks

A *point-to-point* network consists of a dedicated connection between two pieces of equipment. Figure 13.3 illustrates a point-to-point approach for connecting terminals to host processors. *Cluster controllers* or *multiplexers* can be installed in telecommunications closets to reduce the amount of cabling required between terminals in offices and hosts in central computer rooms. Also, communications front-end processors can be used between hosts and *remote cluster controllers* (cluster controllers in other premises), wherein front-end processors and cluster controllers are connected by separate point-to-point networks.

This approach is typically used with terminals without local processing capability. However, PCs and other workstations can be programmed to emulate IBM's older 3270 and other brands of terminals.

The traditional cabling method for terminals such as IBM's 3270 has been coaxial cable. Premises networks using coaxial cable historically require completely discrete and separately administered premises distribution systems. These installations require make and model compatibility between hosts and terminals. In the past, different equipment required different cable types, so that there was no possibility of standardized, universal wiring. As we saw in Chapter 9, this is no longer the case.

# Integrated Voice/Data PBX and Data PBX Premises Networks

Figure 13.4 illustrates the use of an integrated voice/data PBX for connecting terminals and host processors within premises. In these designs, terminals are connected to PBXs via connectors on digital telephones or special DIUs. Integrated voice/data PBX products use unshielded twisted pair (UTP) for both voice and data signals, and are therefore compatible with EIA/TIA PDS standards. Each PBX port typically supports 64 kb/s.

**Figure 13.3**
Point-to-point
premises data
network.

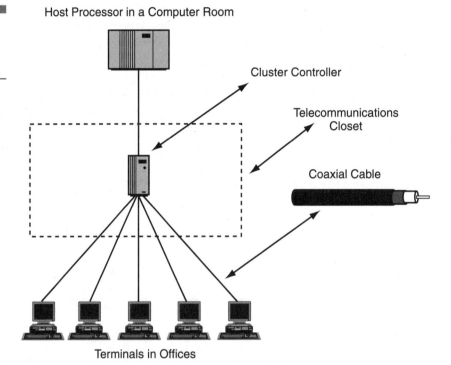

Host Processor in a Computer Room

Cluster Controller

Telecommunications Closet

Coaxial Cable

Terminals in Offices

As described in Chapter 6, high-speed CTI (computer-telephone integration) interfaces provide multichannel PBX-to-computer links so that hundreds of terminal-to-computer connections can be concentrated in a smaller number of PBX-to-computer circuits. Integrated voice/data PBXs, (like older and nearly extinct data-only PBXs) establish connections using circuit switching. Integrated voice/data and data-only PBX approaches have been overshadowed by packet-switching-based LANs described next, primarily because they incur substantially higher per-port costs and because LANs typically offer much higher terminal-to-terminal data communications rates.

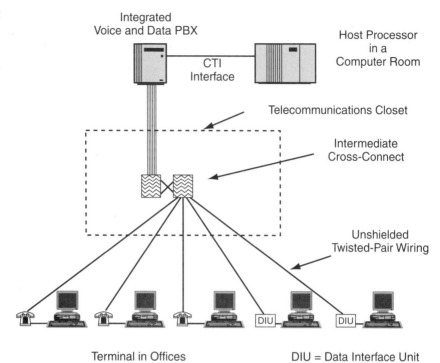

**Figure 13.4**
Integrated voice/data PBX data network.

Terminal in Offices

DIU = Data Interface Unit

# CSMA/CD LAN-Based Premises Networks

In our 1993 first edition we defined a *LAN* as a high-speed (typically in excess of 10 Mb/s) data communications system wherein all segments of the transmission medium (typically coaxial cable, twisted-pair, or optical fiber) are in an office or campus environment. With gigabit LAN products now available some may now quibble with the reference 10 Mb/s as high speed, but otherwise the definition remains valid.

The most popular premises data network approach is the *CSMA/CD LAN* that employs *Carrier Sense Multiple Access with Collision Detection.* CSMA/CD was invented in 1972 by recent Massachusetts Institute of Technology graduate Robert Metcalfe, then a Xerox networking specialist. In 1973, Metcalfe coined the name Ethernet, which he said comes from "luminiferous ether—the omnipresent passive medium once theorized to carry electromagnetic waves through space." Using contention-based "Aloha" radio network concepts postulated by Norman Abramson at the University of Hawaii in the late 1960s, Ethernet efforts have produced the most popular local area networking standards in the world, created new companies, and led many to personal fame and fortune.[1]

Today's Ethernet LANs use *network interface cards* (NICs), a transmission medium to which all NICs are connected, and as options, file servers, print servers, gateway servers, and for larger networks, a network-management system, as shown in Figure 13.5. NICs—printed circuit cards plugged into workstations, PCs, servers or other LAN DTEs—implement LAN protocols and provide medium interfaces (the means for connecting DTEs to LANs), and are generally referred to as *nodes*.

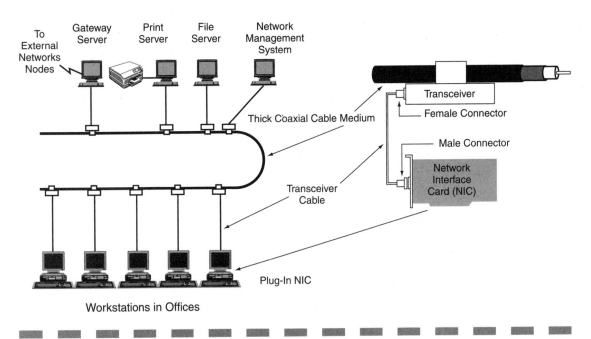

**Figure 13.5**    *CSMA/CD (Ethernet) LAN-based premises network.*

Following Apple-Macintosh and Sun's lead, some Microsoft Windows-compatible PC manufacturers now offer *LAN on motherboard* (LOM) designs in which NIC functions are an integral part of DTE

---

1 For a fascinating and authoritative technical and business history and a detailed explanation of switched and fast Ethernet technologies see *Switched and Fast and Gigabit Ethernet*, by Robert Breyer and Sean Riley, Ziff-Davis Press, 1999.

motherboards, eliminating the need for separate NICs, a practice that demonstrates the popularity and industry acceptance of Ethernet.

*Servers* are devices that provide multiple attached workstations with various host-like services. Examples include storage of and access to centralized data files (file servers), printers, facsimile, and communications gateway services. Servers permit workstations to share resources such as software application programs, computer mass-storage devices, multiplexers, modems, and data communications devices for connecting premises networks to external networks.

A *gateway* is a server that permits client terminal/station access to otherwise incompatible communications networks and/or information systems. A gateway is usually a protocol-translating device that connects a local area network to an external network or any two networks that use different protocols and operating environments.

Figure 13.5 illustrates an Ethernet LAN that uses coaxial cable as the transmission medium. Attachments to the cable are made using transceivers. *Transceiver* is a generic term describing a device that can both transmit and receive. In IEEE 802 LAN standards, a transceiver consists of a transmitter, receiver, power converter and—for CSMA/CD LANs—collision detector and jabber detector capabilities. A *jabber detector* is a timer circuit that protects the LAN from a continuously transmitting terminal. The transmitter receives signals from an attached terminal's NIC and transmits them over the LAN cable medium. The receiver receives signals from the medium and transmits them via transceiver cables and NICs to attached terminals.

Originally, the cable followed a bus or serpentine route from one office to the next. Transceivers located in the ceiling or some other convenient location were connected to NICs by transceiver drop cables (limited to about 150 feet in length). Actual cable connections used *vampire* taps, a means for penetrating coaxial cable cladding and outer metallic sheathing to make contact with its inner conductor. Vampire bolt-on taps eliminate the need to cut coaxial cables when making drop connections. Such wiring plans deviate greatly from telephone twisted-pair universal wiring approaches and as noted have given way to designs that use standard unshielded twisted-pair cabling in conformity with standards described in the Premises Distributions System section of Chapter 9.

As explained in Chapter 7, Ethernet (CSMA/CD) is a local area network contention-based access-control protocol technique by which all devices attached to the network "listen" for transmissions in progress before attempting to transmit themselves, and if two or more begin transmission simultaneously, are able to detect the "collision." In that case each backs off (defers) for a variable period of time (determined by preset "randomizing" algorithms) before again attempting to transmit. All nodes in a shared environment that can listen to and detect collisions are said to be in the same *collision domain*.

Under IEEE rules, collisions must be detectable at all nodes within the first 512 bits of an Ethernet frame transmission. Because the medium and intervening repeaters, hubs, and switching devices introduce a time delay between when transmitters begin a transmission and when receivers detect it, the geometrical dimensions within a single *collision domain* are limited. At 10 Mb/s LAN bit rates, 512 bits translate to 51.2 μsec (a μsec is one millionth of a second) and at 100 Mb/s to 5.12 μsec. This limits maximum *collision domain diameters* to about 1,200 feet for Fast Ethernet 100 Mb/s and 12,000 feet for 10 Mb/s single-segment LANs.

Note that unmodified Ethernet protocols impose a *half-duplex* discipline in that bidirectional transmissions can take place among transmitters and receivers, but not simultaneously (see Chapter 3). As explained below, some *switched Ethernet* designs allow full-duplex communications among pairs of nodes.

The network shown in Figure 13.5 uses a *bus* topology. A bus is a transmission path or channel using a medium with one or more conductors such as are used in Ethernet, where all network nodes listen to all transmissions, selecting certain ones based on address identification. All bus networks employ some sort of contention-control mechanism for accessing the bus transmission medium, such as the CSMA/CD method.

An alternative contention-control method is used in a *token bus,* a LAN access mechanism and topology in which all stations actively attached to a bus listen for a broadcast token or supervisory frame. Stations wishing to transmit must receive a token before doing so; however, the next logical station to transmit might not be the next physical station on the bus. Access is controlled by preassigned priority algorithms.

In a *token ring* LAN, a token or supervisory frame is passed from station to adjacent station sequentially. Stations wishing to transmit must wait for a token to arrive before transmitting data. In a token ring LAN, the start- and end-points of the medium are physically connected, leading to the ring terminology. An example of a token ring LAN diagram is presented below.

Ethernet's architecture has been adapted by the Institute of Electrical and Electronic Engineers (IEEE) and incorporated into IEEE 802.3, one of a family of standards that deals with physical and data-link protocol layers, as defined by the ISO Open System Interconnection Reference Model (OSI—described in Chapter 7). The relationship between the OSI and the IEEE physical partitioning is shown in Figure 13.6.

**Figure 13.6**
LAN protocol physical-layer partitioning and its relation to the ISO OSI model.

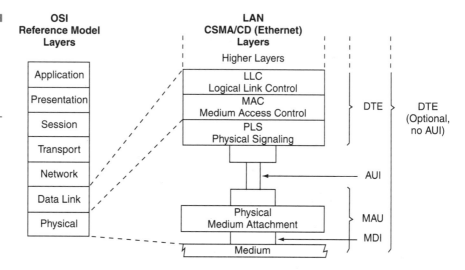

AUI  = Attachment Unit Interface
MAU = Medium Attachment Unit
MDI  = Medium Dependent Interface
PMA = Physical Medium Attachment

Initially IEEE defined two 802.3 coaxial cable CSMA/CD subcategories, namely *10Base5* (known as *thick Ethernet*), and *10Base2* (known as *thin Ethernet* or *cheapernet*). Physically the difference between the two specifications is the diameter of the coaxial cable and the connectors. The thin coaxial cable is much easier to install and less expensive, but sacrifices segment length and supports fewer connections (taps) per segment, as indicated in Figure 13.7.

Next IEEE introduced a third 802.3 CSMA/CD standard, *10BaseT,* that defines LANs using two pairs of unshielded twisted-pair cable (UTP) and supports a 10-Mb/s bit rate. To date there are over 60 different IEEE 802-related standards and committees that define issues ranging from top-level architecture and internetworking to LAN management, to the minutest protocol and physical-layer media and connector details. Components and the operation of 10BaseT and derivative networks are described later in this chapter.

Although bit rates for Ethernet LANs can now be 10, 100, or 1,000 Mb/s, actual throughput rates are less than bit rates. Such reductions in aggregate throughput occur at heavy traffic loads since as traffic increases, there are more chances for collisions which reduce effective throughput. For example, under busy conditions, 10-Mb/s CSMA/CD LANs may only be able to support 4-Mb/s aggregate throughput, (40 percent of the bit rate). Worse yet, if multiple attached DTEs attempt to create aggregate traffic loads in excess of 40 percent, the actual throughput drops below 40 percent. At sufficiently high multiple DTE-offered traffic loads, an Ethernet's traffic-handling ability drops to zero!

**Figure 13.7**
IEEE 802.3 specifications for thick and thin Ethernet.

|  | Thick | Thin |
|---|---|---|
| IEEE 802.3 | 10Base5 | 10Base 2 |
| Cable Type | 802.3 Coax | RG-58 Coax |
| Maximum Segment Length | 500 m (1640 ft.) | 185 m (607 ft.) |
| Network Span | 2500 m (8700 ft.) | 925 m (3035 ft.) |
| Tap Spacing | 2.5 m (8 ft. 3 in.) | 0.5 m (1 ft. 8 in.) |
| Maximum Taps/Segment | 100 | 30 |
| Maximum Stations/Network | 1024 | 1024 |

From an end-user-to-end-user viewpoint, CSMA/CD LANs work best (encounter fewer collisions) under lightly loaded conditions. From a network management viewpoint, multiple aggregate DTE traffic loads beyond 40 percent of bit rates pose the possibility of serious degradation or catastrophic failure of an entire network or network segment, and must be avoided through effective capacity planning.

In decade-old lower-speed PCs and then available NICs and network software, the maximum throughput between two attached DTEs was normally limited to about 200 kb/s (even in light traffic-load conditions) by factors other than bit rate. With today's powerful PCs and high-speed NICs, it is the LAN bit rates and CSMA/CD heavy traffic collision performance that set upper throughput bounds. However, NIC cards can still be purchased that employ Industry Standard Architecture (ISA) bus designs that limit maximum bit rates to about 11 Mb/s. (Don't try to use ISA-based NICs on the Fast or Gigabit LANs discussed below.) NICs employing newer Peripheral Component Interconnect (PCI) buses should support rates exceeding 1 Gb/s. An advantage of token-passing LANs is that under high-traffic conditions, users are afforded aggregate throughput close to 100 percent of supportable bit rate.

# Token Ring LAN-Based Premises Networks

Figure 13.8 shows a token ring LAN. In the early 1980s, IBM marketed token ring LANs using a shielded cable system developed specifically for IBM products. Figure 13.8 illustrates that although wired as a logical ring network, the cable is installed in a star topology—with each DTE connected to a concentrator in a telecommunications closet. IBM token ring networks operate at 4 Mb/s, however, IBM also developed and sold backbone token ring LANs that operate at 16 Mb/s and can be used to interconnect multiple 4 Mb/s rings.

**Figure 13.8**
IEEE 802.5 token ring
LAN-based premises
network.

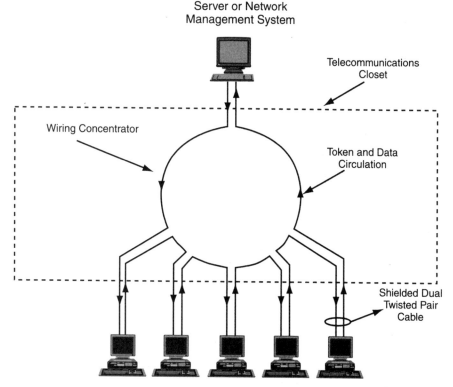

In 1986, IBM issued its type 3 media specification, which allows UTP to be used for horizontal wiring in 4.0-Mb/s token ring segments. A corresponding IEEE specification for unshielded twisted-pair cabling extended bit rates to 16 Mb/s. Even after IBM licensed its token ring integrated circuit chipset to National Semiconductor in 1992 in an attempt to create a low-cost clone industry, the growing popularity of Ethernet soon eclipsed token ring—and even IBM offers built-in Ethernet interfaces in some of its products.

# LAN Components and IEEE 10BaseT Networks

By name and function, LAN components include *transceivers, NICs, repeaters, hubs, bridges, switching hubs, routers,* and *gateways.* As will become evident, a certain amount of functional overlap exists among these components. For example, a switching hub possesses all the capabilities of non-switching hubs. Moreover, vendors often further classify components in terms of where they are deployed in the network. This gives rise to additional component categories such as backbone switches, workgroup switches, desktop switches, and the like. While all LAN switches share basic functionality, in a LAN hierarchy, backbone switching devices at the top level can be expected to possess more functions and higher performance capabilities than desktop switches.

Whereas *workgroup* may commonly refer to a group of people working on a common project, which might include employees working in different buildings, cities, or countries—collaborating in real time via modern telecommunications—when used in a LAN context the word normally refers to a group of people (or rather their workstations) that for one reason or another are attached to a single LAN segment. In LAN parlance then, we treat such descriptors as desktop, workgroup, and backbone as modifiers to the listed categories. To simplify understanding LAN components, it is helpful therefore to concentrate first on basic categories, and subsequently treat modifiers that reflect where they are deployed in LAN hierarchies.

In the above discussion we have already covered transceivers and NICs. With just one additional observation that transceivers can be built in multiport configurations, we continue below beginning with repeater and hub descriptions.

## LAN Repeaters and Hubs

A *repeater,* also known as a *hub* or *concentrator* is a network device that regenerates and forward Ethernet frames. First introduced in two port versions, repeaters were used to extend the reach, or the collision domain diameter, beyond distances that a non-repeated medium would allow. Repeaters are passive or shared components in the sense that they do not logically react to incoming frame control information or any LAN events such as collisions. Repeaters merely receive weak received signals and retransmit regenerated versions of the same signal. Standalone repeater devices are typically available with between 8 and 24 ports at a cost of $30 to $50 per port.

Figure 13.9 depicts a LAN segment using standard, IEEE 10BaseT UTP cabling. In its simplest form, within local area networks a *hub* is a wiring concentrator used in hierarchical star wiring topologies. Those directly connected to terminals or other user devices are often referred to as *local hubs* or *concentrators. Central* or *switching hubs* are those at the highest hierarchical level. 10BaseT hubs are devices used to provide connectivity between DTEs using UTP. Hubs often provide the means for interconnecting 10BaseT, coaxial, or fiber optic cable LAN segments. 10BaseT hubs function so that the LAN performance is identical to that provided by all coaxial cable designs.

**Figure 13.9**
10BaseT standalone workgroup repeater/hub.

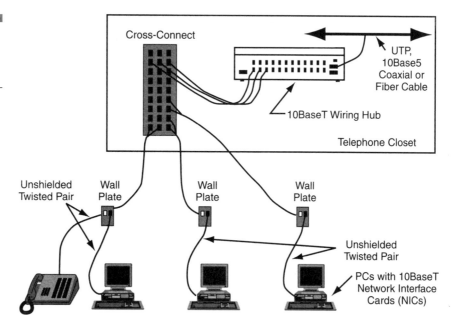

Prior to the advent of *Fast Ethernet* and *Gigabit Ethernet* (100BaseT and 1000BaseT), it was safe to assume that all devices connected to UTP hubs operated at 10-Mb/s bit rates. Today, the same RJ-45 modular connector and wiring scheme and UTP cable are used for 10/100/1000BaseT full- and half-duplex operations. These variations are explained below but for now we want to comment on the *auto-negotiation* or *auto-sensing* capabilities that IEEE has incorporated into its 802.3u standard to help LAN managers, installers, and users to cope with the current seven different operating modes and speeds that may be present on any EIA/TIA PDS link.

As an addition to 10BaseT link-integrity signaling (a feature that allows hub LED displays to reveal the status of each hub port), with auto-sensing equipped LAN components, devices *handshake* with each other when originally connected to determine each other's capability and negotiate the highest-speed performance common operating mode. The process is similar to what was developed for different speed voice-grade modems, wherein modems with higher speed capabilities back off from their maximum rate to accommodate lower-speed connected devices.

As shown in Figure 13.9, the significant advantage of 10BaseT hubs is that all the horizontal wiring can be identical to that used for voice services. This includes the modular phone jacks and UTP cable described in the Premises Distribution System section of chapter 9. Also eliminated are coaxial transceivers and bulky transceiver drop cables. What results is simpler, less costly installation, administration, and maintenance. Recent vendor trends incorporate multiport bridging and network management functions within *intelligent hub* products, functions which are discussed below. Intelligent hubs are available from Cisco, Intel, 3 Comm, Cabletron, and others.

Figure 13.10 looks inside a hub, traces how two UTP pairs support the CSMA/CD operation on a common medium, and relates pair usage to RJ-45 modular jack pin assignments. According to EIA/TIA specifications, pins 1, 2, 3, and 6 are reserved for 10BaseT data traffic. EIA/TIA even specifies wire-insulation colors to specific pin pairs, with pins 1 and 2 using the white/green-green twisted pairs and pins 3 and 6 using white/orange-orange pairs. When not needed for data traffic the other two twisted pairs can be used for voice traffic, assigning pins 3 and 4 (white/blue-blue) and pins 7 and 8 (white/brown-brown). As mentioned before, RJ-11 plugs attached to telephones can be plugged into larger "keyed" RJ-45 sockets and function properly as if plugged into RJ-11 sockets.

**Figure 13.10**
Hub showing internal connections to RJ-45 modular connectors.

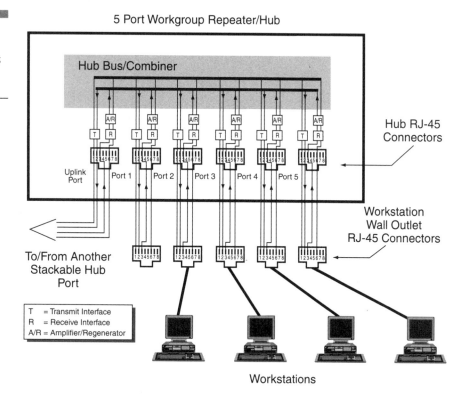

5 Port Workgroup Repeater/Hub

Hub Bus/Combiner

Hub RJ-45 Connectors

Uplink Port   Port 1   Port 2   Port 3   Port 4   Port 5

Workstation Wall Outlet RJ-45 Connectors

To/From Another Stackable Hub Port

T   = Transmit Interface
R   = Receive Interface
A/R = Amplifier/Regenerator

Workstations

Figure 13.10 shows how DTE NIC cards transmit data to a hub using one pair. Hubs then receive, amplify, and regenerate the typically "weak" signals, placing all such signals on a common bus. Hubs provide a bus interface for each attached NIC, and transmit the bus combined signal back to all NICs. Thus, NICs are able to monitor their own, and all other, NIC transmissions and thereby implement the CSMA/CD collision-detection protocol. The figure reveals that the only truly common medium is the very short hub bus. All other parts of the cabling are dedicated to particular attached NICs. It is a design tradeoff whether weak signals are combined first and then amplified and regenerated in common *amplifier/regenerator* (A/R in the figure) components or are separately amplified and regenerated before combination. The latter option is shown in the figure and provides better performance, but requires more A/R equipment.

A hierarchical *hub/concentrator* arrangement, shown in Figure 13.11, illustrates the flexibility afforded by modern multimedia products. UTP can be used where it is most effective, such as for terminal-to-telecommunications closet runs, and fiber optic cable used for longer runs to central equipment rooms, or to "data-hungry" high-powered workstations that demand high data-transmission rates.

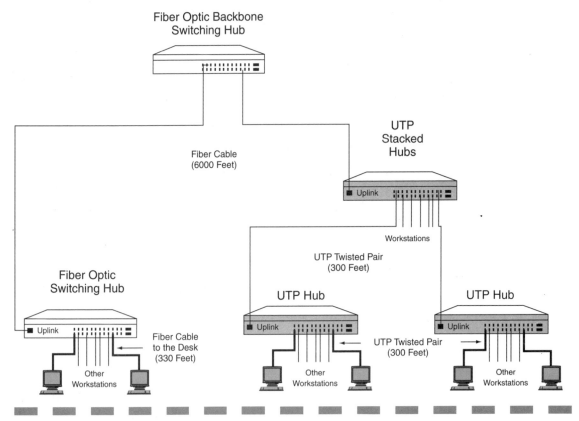

**Figure 13.11**    Ethernet hub hierarchy with stackable, multimedia hubs.

A UTP *stacked hub* example is depicted in the lower right-hand portion of the figure. In most low-end hub designs, except for power, all input and output ports are RJ-45 sockets. An eight-port hub that is stackable would have nine RJ-45 sockets. The ninth socket is usually labeled *uplink*. In stacked configurations, a jumper cable is inserted between the uplink socket on the lower tier hub, and any one of the standard ports on the higher-level hub, as shown in the figure.

Thus, stacking hubs uses up ports that could otherwise serve attached DTEs. In fact, when the uplink port is used on the lower-level hub, one of the DTE ports must be left vacant. In our eight-port example, to connect two hubs together uses up a DTE port on both hubs. In Figure 13.11 with two lower-level hubs connected to another hub, the higher-level hub loses two ports that could otherwise be used for DTE attachment. Again in our eight-port hub example, this means that the two lower-level hubs can support seven DTEs each, and the third hub only six, for a total of 20 DTEs. That is, for this arrangement, three eight-port hubs can serve a maximum of 20, not 24, DTEs.

As illustrated in the figure, multimedia hubs are available that can support various mixes of UTP, fiber, and coaxial cabling. In the figure, the hub in the lower left is designed to accommodate both "fiber-to-the-desktop" and fiber to a central or backbone hub. The higher-level stacked UTP hub on the right side has a "fiber-equipped uplink port." Therefore, either workgroup hub set can be connected to a backbone hub via fiber cable.

As the figure indicates, the workgroup-to-central computer room fiber cable connection allows those two areas to be up to 6,000 feet apart. Moreover, the fiber cable can easily support 100-Mb/s (Fast Ethernet) or even 1,000-Mb/s (Gigabit Ethernet) transmission rates. Other optional workgroup hub embellishments include desktop hub switching, bridging, and full duplex Ethernet operations. Fast and Gigabit Ethernet, switching, bridging, and full-duplex are explained below.

## LAN Bridges

In IEEE 802 LAN standards, *bridges* are devices that connect LANs, or LAN segments, providing the means to extend the LAN environment in terms of numbers of stations, performance, and reliability. An example bridge connection arrangement is shown in Figure 13.12. Bridges operate at the data-link protocol layer (Layer 2) and consequently pass packets without regard to higher-layer protocols. Because of this, bridges can only connect segments of the same LAN types.

Bridges decode Layer 2 *destination addresses* (DAs) encapsulated in incoming frame headers. Comparing detected DAs with those stored in a bridge's database, a bridge decides whether or not to forward packets among connected LAN segments. In this manner, the only way in which traffic in one segment can reach and add to the traffic load in a connected segment is if packets are addressed to that segment.

Bridges perform three basic functions: frame (as opposed to packet) forwarding; learning of station addresses; and resolving of possible loops in the topology (as might occur if bridge 3 in the figure were added) by incorporating an IEEE-defined *spanning tree* algorithm. *Self-learning* bridges construct tables of network addresses by listening to source address information contained in data signal frames, so that no installation or administrative effort is required to configure LAN bridge nodes.

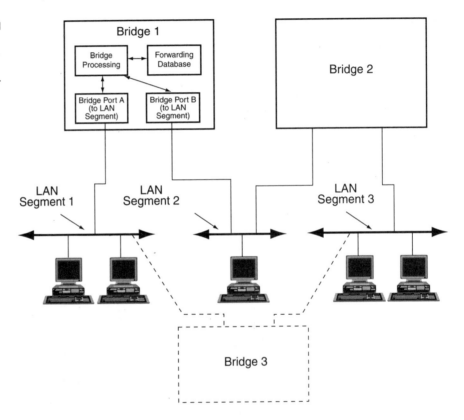

**Figure 13.12**
10BaseT bridge Layer 2 switching.

In connecting network segments as shown in the figure, bridges can filter packets so that traffic originating and terminating within the same segment is not forwarded to other segments. This segmentation of local traffic reduces congestion on each segment. Bridges can also filter traffic from specified terminals for security or other reasons. In short bridges:

■ Regenerate incoming signals—like repeaters.

■ Check for errors and don't forward damaged packets/frames.

- Analyze only Layer 2[2] source and destination addresses to make forwarding decisions.

- Use IEEE spanning tree algorithms to learn, update, and create source and destination address tables automatically and to prevent "loops" from reintroducing already processed traffic.

- Filter or route traffic on the basis of source and destination addresses, security, procedural, quality, and other criteria.

- Use processing and memory components to store packets/frames while making and executing forwarding decisions.

- Expand total network bandwidth/throughput by isolating traffic that originates and terminates in one segment and preventing it from congesting other segments. (Note that if most traffic is between segments then bandwidth expansion will be minimal. Fortunately this is not normally the case.)

- Expand the physical or geometrical size (diameter) of total networks by splitting networks into separate *collision domains*. (Refer to the earlier discussion in this chapter on collision domain definitions and size calculations.)

- Perform speed matching among LAN segments operating at different speeds. (For example, original 10-Mb/s 10BaseT and Fast Ethernet 100-Mb/s 100BaseT segments.)

Some *translating bridges* with limited gateway functions can connect different LAN-type segments; for instance the interconnection of Ethernet workgroup LAN segments to FDDI (defined below) backbone segments. Other bridge functions include collecting and storing network management and control information obtained from traffic monitoring. Clearly bridges implement a type of switching and therefore possess a subset of *switching hub* functions, discussed next.

---

2 A Layer 2 or *media access control* (MAC) address identifies only a station (DTE) on a local area network. Higher-level addresses that direct packets to specific service access points (SAPs) within a DTE or that direct a packet to a destination on external networks like the Internet, are contained in higher *logical link control* (LLC) level or Internet Protocol (IP) headers. As a consequence the complexity of MAC address resolution and required processing power is much less than that required in routers that must contend with global as opposed to local (LAN) addressing fields.

# Switching and Full Duplex Hubs

In packet networks, a *switch* is a component that receives incoming packets on one port, buffers them and retransmits them on other ports. Ethernet *hub switching* is an expansion of Ethernet *bridging* concepts. If it makes sense to connect two LAN segments or subnetworks through a bridge, why not develop a device that connects three, ten, or more networks together. To begin with therefore, LAN switches are able to implement all the bridge functions just delineated. Because switches store incoming frames they may also implement routing functions treated later in this chapter.

Breyer and Riley (see footnote 1 in this chapter) postulate three LAN switching categories, namely workgroup, desktop, and backbone, with the following capabilities:

- **Workgroup Switches**—Also referred to as segment switches, this category congregates multiple shared segments, servers, or high-performance desktop DTEs.

- **Desktop Switches**—A subset of the workgroup category, these devices are specifically designed to connect directly to single LAN DTEs (read nodes).

- **Backbone Switches**—More fully featured and high-performance versions of workgroup switches, these are designed for the highest level in LAN hierarchies. Multiple fiber cable-interconnected backbone switches find application in backbone networks connecting LAN segments or subnetworks within multiple buildings in campus or equivalent settings.

Switches with $n$ ports can typically handle simultaneous and independent communications to compatible ports on $n$ other devices—using dedicated switch-to-device transmission facilities. If Ethernet protocols are used between switches and connected devices (which may be other switches), each connection requires one input and one output switch port. If conventional half-duplex Ethernet CSMA/CD-based protocols are used, since there are only two competing nodes, the only collisions that can occur are those that take place when both the switch and the connected device transmit at the same time. In this situation, throughput rates close to transmission rates, or 10 Mb/s in the case of 10BaseT LANs, can be expected.

However, in such paired-node, switched Ethernet cases, it is possible to suspend normal CSMA/CD protocols and allow both switch and connected devices to utilize full-time the separate bidirectional channels inherent in Ethernet transmission facilities. Because both devices in a 10BaseT design can transmit simultaneously at 10 Mb/s, the effective available transmission throughput rate is counted as 20 Mb/s. Such arrangements are referred to as *full-duplex Ethernet.*

With 10-, 100-, and even 1,000-Mb/s port rates possible, aggregate switching throughput capacity becomes an important switch characteristic. If the switch is "non-blocking" then its theoretical aggregate forwarding rate in half-duplex operations is the product of the number of ports multiplied by the transmission speed, divided by two. A 16-port 100BaseT switch therefore exhibits a maximum theoretical throughput rate of 800 Mb/s. This can be compared with a 100BaseT repeater/hub that supports only 100 Mb/s throughput rates.

With full-duplex features, a 16-port 100BaseT switch can theoretically support 1.6 Gb/s. While it is possible to buy 10 and 100BaseT non-blocking switches, not all gigabit switches have the necessary backplane capacity to support non-blocking service. *Backplanes* are the physical areas (components and/or printed circuit cards) on which major switch fabric components are mounted and to which individual port cards are connected. Backplane capacity is determined by component and circuit board speed limitations and overall switch performance by switch buffer/processor and other design attributes.

An important performance measure in switch and all LAN operations is *dropped* or *blocked* packet rates. In many instances dropped packet rates are a function of packet size. Obviously, manufacturers tend to quote such characteristics based on the packet size that maximizes performance whether or not that packet size is typical in actual operating environments. Thus, along with reliability, mean time to repair, hot-card swapping and other characteristics, blocking, and dropped packet rates are parameters of great interest when making switch procurement decisions.

Figure 13.13 shows a 10BaseT desktop switching hub with separate dedicated connections to individual workstations and servers. The figure indicates that workstations and their NICs are not upgraded or programmed to operate full duplex and therefore are equipped for 10-Mb/s half-duplex service. Individual servers in the *server farm* are full-duplex equipped and therefore able to sustain 20 Mb/s throughput rates.

## Routers

Figure 13.14 portrays a LAN configuration suitable for a large, multi-building campus that includes UTP, coaxial, and fiber cable media,

**Figure 13.13**
Desktop Layer 2 hub
switching example.

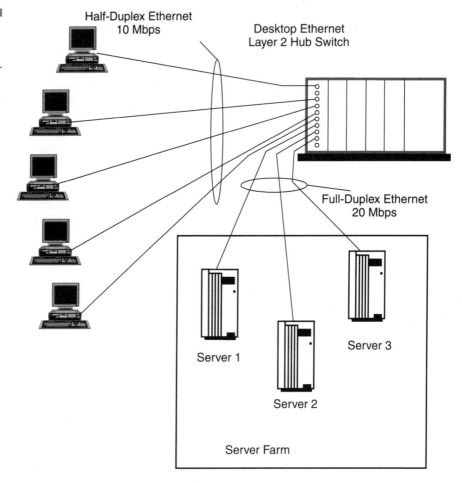

and most of the LAN components previously discussed. It shows the interconnection of hubs to bridges, and bridges connected to routers through multiport transceivers. Although the graphic illustrates how backbone network routers relate to bridge and hub components, the design corresponds to an early-to-mid-90s venue solution and purposely does not include more recent switched Ethernet technologies. A Fast and Gigabit-switched Ethernet configuration is portrayed in Figure 13.16 for contrast.

*Routers* are devices that connect autonomous networks of like and unlike architectures at the network layer (Layer 3). Unlike a bridge, which operates transparently to communicating end-terminals at the data-link layer (Layer 2), a router reacts only to packets addressed to it by either a terminal or another router. Routers perform packet (as opposed to frame) routing and forwarding functions; they can select one of many potential paths based on transit delay, network congestion, or other criteria. How routers perform their functions is largely determined by the protocols implemented in the networks they interconnect.

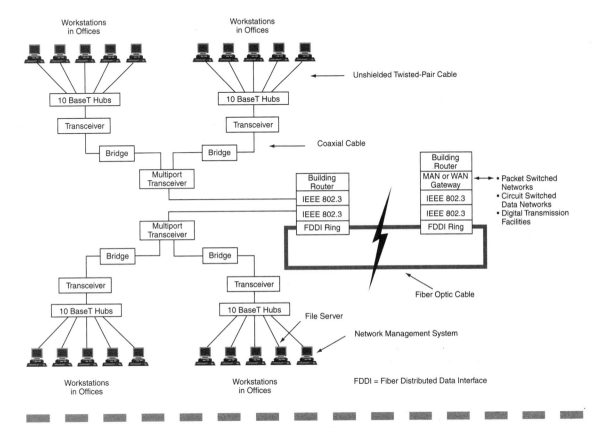

**Figure 13.14** A 10BaseT, coaxial, and fiber cable multi-building premises network.

Routers are often used to interconnect autonomous network segments in multiple buildings or within a campus as shown in Figure 13.14. For these applications routers may be configured to interface with non-802.3 optical fiber backbone networks (e.g., FDDI, defined below) to interconnect buildings, as well as intrabuilding 802.3 LANs. In LAN applications, routers can also be configured to provide gateway services to external networks. *Brouters*, devices that combine the functions of bridges and routers, are also available.

As explained in Chapter 7, routers are the principal switching mechanisms within the Internet, large private intranets, and other packet-switched networks where terabit-per-second throughputs can occur. The router market has been experiencing geometrical growth in the last decade with Cisco capturing the lion's share of the large network market.

Figure 13.15 shows the protocol levels involved with repeater, bridge/Layer 2 switch, router/Layer 3 switch, and gateway devices in terms of the OSI reference model introduced in Chapter 7. To qualify as a router, a device must be equipped to examine the Layer 3 protocol packet headers and route packets according to *network number* and remote network *node destination* addresses.

**Figure 13.15**
Device and functional relationships to the ISO OSI reference model.

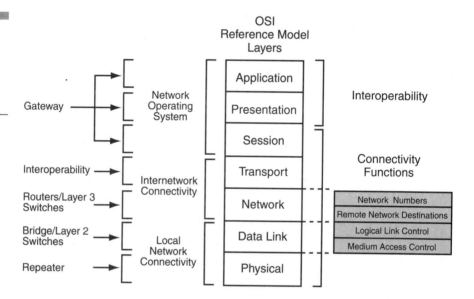

Using the Post Office as an analogy, the Layer 3 address corresponds to the city, state, street number, and zip code used for nationwide mail. In this analogy, Layer 2 addresses are like room numbers in a building. Room numbers are adequate for in-building mail but have no significance elsewhere. In contrast, if employees want to send letters to another building through the U.S. Mail, senders must supply an external *Layer 3* address.

In a similar manner, the local source and destination address information contained in the medium access control (MAC) and logical link control (LLC) Layer 2-level headers are sufficient for delivery to DTE nodes on a local area network. But if an Ethernet frame encapsulates packets for delivery to nodes on the Internet (or any external or autonomous network), then the frame must contain a Layer 3 header that contains external network number and remote node destination address information to enable routers to forward packets properly. In the Internet case, this information is placed in the Layer 3 IP header, as described in Chapter 7.

As it turns out, routers can also (and frequently do) partially implement functions allocated to ISO levels 4 through 7. Routers that perform *traffic filtering, application-based switching, firewall routing,* and other similar functions rely on Layer 4 and higher-level protocols. In yet another example of the overlap among basic LAN components, routers can be imbued with gateway capabilities.

About 16 years ago when the first routers hit the market they exhibited modest (thousands of *packets per second*) throughputs, were used primarily to connect different LANs, and implemented most router functions in software that ran on general-purpose micro-processors. Software-implemented routers have the advantage that they can be programmed to handle multiple protocols and can be easily upgraded as protocols evolve.

However, as data network popularity grew and continues to grow, escalating traffic loads create demands for higher and higher router speeds. As it turns out, the fastest router designs implement most functions in custom designed *application-specific integrated circuits* (ASIC) hardware. While this approach has tremendous speed advantages (today's routers handle more than a million packets per second per port), designing hardware-based multi-protocol routers is expensive, and they cannot be upgraded as easily as software-based machines.

Many high-performance routers actually contain high-speed hardware-based Layer 2 switches to handle all *non-routable* packets and only resort to software-based or other routing techniques to handle routable packets, still another example of functional overlap among basic LAN components. Because it seems to have a more "frontier" technology ring to it, most vendors no longer refer to their router products as routers but rather as Layer 3 switches; according to some experts this may be something of a misnomer.

In summary, whatever you call them, routers operate on the basis of Layer 3 packet header information and find application at network critical junctures in the following backbone tasks:

- Segmenting large networks
- Handling multiple protocols

- Accessing wide-area networks
- Linking different LANs
- Linking major Internet/Intranet subnetworks

# Fast and Gigabit Ethernet LAN-Based Premises Networks

It may very well be that the recent Fast and Gigabit Ethernet developments prove as important and as profound as Robert Metcalfe's basic Ethernet design invention in 1973. The reason for this prognosis is related not just to the 100- and 1,000-Mb/s high speed capabilities, but to the fact that these performance levels are achieved within a common and now-ubiquitous Ethernet architecture, protocol, and basic design envelope.

Ethernet is not just a technology that works and is reliable. Because it is the most affordable approach and because it is amenable to straightforward building-block, plug-and-play-type implementation approaches, Ethernet has become the worldwide LAN technology of choice. By 1998, more than 85 percent of all installed network connections were Ethernet. This represents over 118 million interconnected PCs and other DTEs. In 1998 Ethernet network equipment and technology captured 86 percent of shipments, with Ethernet hub shipments exceeding 48 million ports. By contrast, this means that ATM, FDDI/CDDI, token ring, and all other approaches have dwindled to less than 15 percent of the market.

In this business, success breeds success. As Ethernet sales have mushroomed, countless new companies have jumped into the breach to bring better-performing and less-expensive Ethernet products to market. Ethernet costs have not only spiraled down across the board, but as time passes, the price difference between newer, higher-speed components and lower-speed versions tends to diminish. So the real long-term benefit of Fast and Gigabit Ethernet is not just higher speed but the promise that the economies of scale will apply to these new components because they conform to that single, common design envelope that worked so well at lower speeds.

As an example, Asante has announced a $375 per-port price for its four-port GX4 Gigabit Ethernet switch. According to *Computer Reseller News'* February 28, 2000 issue current industry average costs for desktop gigabit are now $700 compared with $1,200 per port in 1998. Coupled with Asante's announcement, the downward trend for ultra-high-speed Ethernet LAN componentry is well under way. What this suggests is the possibility of doing away with the need to use FDDI or ATM-based backbone technologies in large and/or campus network applications that interconnect enclaves of Ethernet LAN segments. Figure 13.16 depicts what such "pure" Ethernet networks look like.

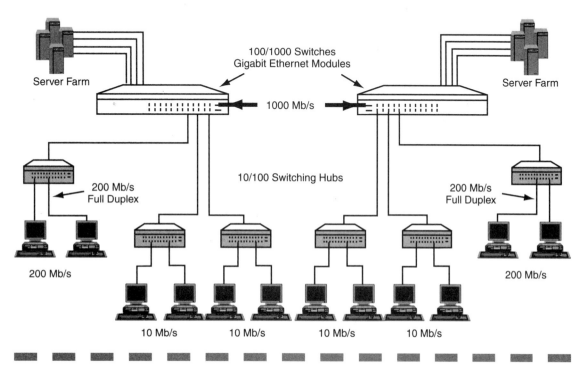

**Figure 13.16**    Fast and Gigabit Ethernets illustrating switching hubs and full-duplex connections.

Spurring these trends and the precursor work are literally hundreds of standards and industry product development efforts. Figure 13.17 lists major IEEE standards efforts and committees related to the LAN technologies covered in this chapter. Because Fast and Gigabit Ethernet employ the same CSMA/CD protocol, the same frame format, and the same frame size as their predecessors, nearly all the earlier standards in updated format remain relevant. No one expects all alternative LAN technologies to disappear overnight. Some LAN alternatives exhibit unique capabilties that will continue to be employed in niche applications. One of those technologies, FDDI, is discussed next.

**Figure 13.17**
Applicable IEEE
standards.

## IEE 802."x" Standards

- 802.1 Architecture and Internetworking
- 802.2 Logical Link Layer
- 802.3 CSMA/CD
- 802.4 Token Bus
- 802.5 Token Ring
- 802.6 Distributed Queue Bus (Switched Multimegabit Digital Service)
- 802.7 Broadband Communications
- 802.8 Technical Advisory Group on Fiber Optic LANs
- 802.9 Voice/Data/Video—Isochronous Trafic (IsoEnet)
- 802.10 Interoperable LAN/WAN Security
- 802.11 Wireless Airwave Standard
- 802.12 100 VG AnyLAN (Demand Priority Access Protocol)
- 802.13 Not Used
- 802.14 100 BaseX—Packet-based Networks

## IEE 10/100/1000Base"x" Standards

| Category | Related 802."x" Standards | Transmission Speed/Mb/s | EAIA/TIA Cable | Pairs Used |
|---|---|---|---|---|
| 10BaseT | 802.3 | 10 | Categories 3, 4, or 5 | 2 |
| 100BaseT | 802.3u | 100 | | |
| 100BaseTX | 802.3u | 100 | Category 5 | 2 |
| 100Base4 | 802.3u | 100 | Categories 3, 4, or 5 | 4 |
| 100BaseFX | 802.3u | 100 | Fiber | |
| 1000BaseT | 802.3ab | 1000 | Category 5 | 4 |
| 1000BaseCX | 802.3z | 1000 | Twinax Shielded | 2 |
| 1000BaseSX | 802.3z | 1000 | Multimode Fiber | 2 Strands |
| 1000BaseLX | 802.3z | 1000 | Single Mode Fiber | 2 Strands |

# Fiber-Distributed Data Interface (FDDI)

FDDI is the ANSI standard X3T9.5 for a 100-Mb/s token ring using an optical fiber medium. It was originally proposed as a packet-switching network with two primary areas of application: first, high-performance interconnection among mainframes, and among mainframes and associated mass storage/peripheral equipment; and second, as a backbone network for interconnecting lower-speed (e.g., 10-Mb/s Ethernet) LANs. The latter application, previously discussed, is shown in Figure 13.14.

The FDDI standard, which basically specifies the two lower protocol layers, is written in four parts: *physical-layer media-dependent* (PMD), *physical-layer protocol* (PHY), *media access control* (MAC), and *station management technology* (SMT). SMT defines functions involved with how the network handles emergencies and how workstations enter or exit the network. Relative to IEEE 802.5 (on which FDDI was originally based), FDDI uses an enhanced *token passing* access protocol which, under light conditions, results in negligible access delay, a property shared with lightly loaded LANs using CSMA/CD (Ethernet) IEEE 802.3 protocols. However, whereas non-switched CSMA/CD data throughput is limited to only 37 percent of the LAN bit rate under heavy loads, FDDI efficiency at heavy loads approaches 90 percent, a feature similar to that of IEEE 802.5 token ring protocols.

Architecturally, an FDDI network employs a dual counter-rotating ring design, depicted in Figure 13.18. Note that when a node is not transmitting its own data frames, signals received from a neighboring node are passed to the next neighboring node. Although connected in a ring topology, physical wiring resembles a star or hub topology. The figure indicates that class A nodes are actually connected to two rings, such that if a cable break (illustrated as an example) or other failure occurs, a back-up ring sustains operation as indicated by the arrows in the figure. Class B nodes have only one connection to a wiring concentrator or hub. Concentrators are equipped to detect failed cable segments or nodes and bypass the failed connection at the hub itself. The dual-ring design greatly enhances service reliability and availability.

In addition to 100-Mb/s throughput and the above-noted advantages, since it uses optical fiber FDDI is superior to metallic cable alternatives in terms of lower noise susceptibility from *radio frequency interference* (RFI) and *electromagnetic interference* (EMI), lower attenuation (signal loss), and other advantages cited in Chapters 4 and 7. In FDDI installations, fiber has a considerably longer span distance (2 kilometers or 1.24 miles using multimode fiber and 40 kilometers or 25 miles with single-mode fiber) before repeaters are required. It is easier to install because of its flexibility, smaller size, and lighter weight (about 30 pounds per kilometer).

Although not completely secure from intrusion, it is substantially more difficult to tap than copper twisted-pair or coaxial cable, and intrusion-detection fiber cable designs are available that meet requirements of U.S. government intelligence agencies. Fiber cables can easily be sourced to meet government TEMPEST (a reference to compromising emanations that may result in unauthorized disclosure of information) specifications.

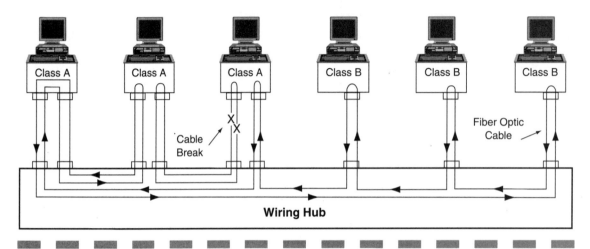

**Figure 13.18**   *Fiber distributed data interface (FDDI) network approach.*

Government agencies and military services have invested heavily in programs such as the Navy's SAFENET (Survivable Adaptable Fiber-Optic Embedded Network) and MFOTS (Military Fiber-Optic Transmission System) to integrate FDDI networks into government communication systems on bases, reservations, and various shipborne, airborne, and space-vehicle environments.

*FDDI II* adds *isochronous* (nonpacket) data-transmission capability, enabling support of voice and video as well as data traffic. Perhaps the principal technical disadvantage of FDDI is the fact that nodes must have operational capabilities to intercept and process signals at 100 Mb/s. When used as a lower-speed LAN interconnection backbone medium, only routers need the high-speed capability. As a competitor for horizontal wiring to the desktop, FDDI imposes high-speed transmission requirements on every workstation or attached device.

For this reason and because the FDDI market has never accelerated like Ethernet's, per-seat and per-port costs are not competitive with Ethernet alternatives. When FDDI first appeared, neither Fast nor Gigabit Ethernet was on the horizon, so network managers eagerly adopted FDDI for backbone and campus inter-building LAN segment connection. Until the mid '90s, FDDI was the only high-speed LAN standard to deliver 100-Mb/s performance. Today, no new products are in the pipeline, and new sales are for existing installations.

# LAN Network Operating Systems

To this point, our LAN discussion has focused on equipment and operational functions. Embedded or installed in the equipment, and implementing most of the functions are *network operating systems* (NOSs), software that controls the execution of network programs and modules. It is the NOS that defines how networks work, specifically how attached nodes communicate within networks, and how communication with external networks takes place. Structurally, networking software is made up of multiple modules, most residing in network servers, but some installed in each terminal/station node accessing network resources.

Peer-to-peer NOSs permit any terminal/station to act as a resource server or a client. In the past IBM and compatible PCs equipped with Microsoft's Disk Operating System (MS-DOS) have been adapted for NOS applications. Since MS-DOS is not designed to run multiple programs or respond to many simultaneous users, most NOSs designed for large networks with dedicated servers/superservers incorporate alternative OSs with extensive multiprocessing and multiuser features.

Although in the 1980s a number of fledging companies competed for the NOS market, Novell (then a small software company in Provo, Utah) managed to capture the lion's share with an introductory product (NetWare) that ran on dedicated PCs and provided simple file and print services to other networked PCs. Perhaps the single most important factor for NetWare's success was that the company realized that to use its NOS product, customers needed to buy networking hardware and began selling Ethernet NICs. In combination, NetWare and Ethernet fast became the world's most popular LAN technology, fueling each other's growth. Later, Novell was one of the first companies to accommodate multiple LAN architectures, such as token ring, stimulating still more growth.

By the mid-90s, NetWare captured 70 percent of the NOS market, which then amounted to 44 million terminations and 400,000 LANs, and supported 30 different LANs and 100 different network adapters. Within the industry as a whole, considerable advances in NOS technology have been made. From humble beginnings like simple password security for a single network have evolved network directory systems (NDSs) that provide a basis for security within today's global network-of-networks environments, along with countless additional operational capabilities.

Likewise, keeping pace with higher microprocessor speeds, multiprocessing architectures, and the migration from 16- to 32- to 64-bit processors, has produced server hardware and software with underlying facilities to power and manage private networks with hundreds of routers, thousands of servers, and hundreds of thousands of users—as well as some of the largest Internet sites.

Overall, among a growing list of heterogeneous multi-protocol and operating system environment capabilities, most advanced NOS products implement the following:

- File server
- Print server
- Applications server
- Database support
- Transaction processing
- Message handling
- Security
- Network management and administration

  - Single and portable platform and application-independent sign-on/log-on
  - Automated desktop software updating
  - Network and desktop configuration and change management

- System diagnostics and fault-tolerance with data protection/data recovery
- Network control center/console support
- Primary server, client, device, and external network drivers

This list of NOS capabilities is representative but not exhaustive, with new productivity-enhancing features being introduced on an ongoing basis.

Among early NOS competitors, Banyan Systems provided many of the same functions as initial NetWare versions, but Banyan's VINES (VIrtual Network Software) products used existing network standards.

Unlike Novell's initial products, which implemented a proprietary transport-layer protocol, VINES runs with TCP/IP, SNA, and other networking protocols.

After IBM had launched its token ring LAN products, it was natural for the company to enter the NOS business. IBM's OS/2 LAN Server is a product that allows users to share directories, printers, and serial devices with other workstations on a LAN. The LAN Server program is installed on a system that has an IBM OS/2 operating system as a base, is optimized for Pentium processors for fast performance, has a true object-oriented user interface to makes it easier to implement and maintain, and doesn't require a dedicated server machine.

For years, Microsoft avoided the networking business. However, when Microsoft acquired 3Com's networking PC LAN technology, it began to integrate networking services into its PC operating system products. The first attempt at this integration, Windows for Workgroups, exhibited some limitations due to its underlying DOS component. When Microsoft designed Windows 95 and Windows NT, it seized the opportunity to integrate core networking services into the platform operating system. Microsoft took two tracks here, pushing peer-to-peer (workgroup) resource sharing via Windows 95 and Windows NT Workstation, and enterprise-scale file, print, and application serving via Windows NT Server.

The introduction of Windows NT heralded the beginning of substantial competition in the LAN NOS marketplace. The merit of the Microsoft strategy to use a single operating system as a platform for all networking services (e.g., LAN servers/Internet servers) has been established in the marketplace. A recent Gartner Group survey reports the following installed base NOS market shares: NetWare 52 percent; Windows NT 27 percent; OS/2 8 percent; Vines 4 percent; DECNET 3 percent; and UNIX 6 percent. While these figures include all NOS applications, they portray an erosion of Novell's early dominance in LAN NOS markets.

Total worldwide NOS software shipments for all applications in fiscal year 1998 were: Windows NT 36 percent; NetWare 24 percent; UNIX 17 perent; Linux 17 percent; OS-2 3 percent; and, others 3 percent.

The PC LAN products offered by all these vendors remain in use today, but the two specific products competing for top billing are Novell's NetWare 5.1 and Microsoft's Windows 2000. Each of these products is offered in 5-connection packages at $1,345 and $999 respectively, but can be purchased in 500-connection increments that support terabytes of disc space and hundreds of thousands of open files for large enterprise and Internet applications.

# LAN Performance Considerations

The operation and maintenance of the first LANs was often characterized by recurring and unpredictable failures that demanded "firefighting" corrective actions by highly qualified staff. One reason for this was that LANs typically incorporated and supported heterogeneous vendor hardware and software, the interoperation of which is not specified or guaranteed by any single source. Consequently, complex device interactions, problem diagnosis, and isolation in mixed vendor equipment environments posed significant challenges.

As a result, in the 80s LANs exhibited chronic failures; LAN operators had little insight regarding design sensitivities, weaknesses, or performance margins in terms of user traffic and utilization. In that era, industry reports stated that in large networks, 25 two- to six-hour outages per year were typical and that administration costs per 100 nodes ranged from $18,000 to $32,000. Worse than just administration and service restoration costs, a 1989 Infonetics study, The *Cost of LAN Downtime,* concluded that the impact on business revenues that results from networks being unavailable dwarfs any other considerations.

One problem is that LANs typically evolve over time, and initially were implemented with minimal front-end requirements analysis, and little formal product development, test, and evaluation. As in any data communications network, response time and throughput are key performance metrics in large heterogeneous networks, but these parameters are difficult to predict based on manufacturers' specifications. Complicating the problem further is the fact that bridge and router performance varies with traffic statistics in different ways for different vendors. Thus, availability, fault tolerance, diagnostics, and service restoration problems reflect the complexity of LAN components and differences in performance characteristics among vendors of apparently plug-compatible components.

Progress in more comprehensive standards development, stricter standards adherence, and effective network management tools and techniques have vastly improved LAN reliability. However, most experts agree that network administration, support, and education expenses still make up more than three quarters of network ownership costs. The need for, and benefits of, modern network management is addressed next.

# ■ ■ LAN Management Considerations

A growing realization among business leaders is that continued viability of their corporations and institutions is inextricably linked to telecommunications operations. For some, the cost impact of service outages is measured in tens, or even hundreds of thousands of dollars per hour.

Beyond a quest for higher intrinsic system reliability, the high-cost impact of outages places a premium on rapid fault detection and isolation, performance, and traffic-monitoring techniques that can detect deteriorating network conditions and alert operators to action before catastrophic failures occur. Centrally controlled reconfiguration and restoration capabilities constitute important aspects of *performance management,* a subset of network management.

Information service users often require ready access to information resources distributed throughout buildings, among buildings on a campus, and at remote locations spanning the nation or the globe. Consequently, the requirement for interconnecting local area, metropolitan, and wide area telecommunications networks—public and private—has become commonplace. Once users employ interconnected networks, they must monitor and manage both individual subnetwork and end-to-end network domains. This task was made more complex by divestiture, which in most instances necessitated the use of multiple-carrier networks.

The operational need, and the market for products that respond to that need, have prompted vigorous efforts from standards-setting bodies and network management product vendors. It is reported that at one time, IBM had assigned over 1,000 people to NetView, SystemView, and other projects related to its network management products. While developing its Enterprise Management Architecture product, Digital Equipment Corporation stated that EMA was the third-largest development project in the company's history. AT&T's Unified Network Management Architecture (UNMA), Hewlett-Packard's Open View, Sun Microsystems' SunNet Manager, and others have made and are making similar levels of investment.

Since many efforts emphasized homogeneous vendor products and systems—such as IBM's NetView, as the management resource for an IBM Systems Network Architecture (SNA) network—one of today's key challenges is achieving integrated, centralized control of distributed systems composed of dissimilar subnetworks and numerous heterogeneous vendor products within those subnetworks. Without integrated management, a service interruption, such as the loss of a main communications line, could trigger a deluge of failure alerts from diverse network components, each measuring different parameters, encoded differently, and transmitted and/or displayed using different protocols and formats.

Recognizing network management's essential role in complex networks and the lack of compatibility among early NM products, in the mid-80s worldwide standards-setting organizations embarked on the development of architectures and frameworks for interoperable telecommunications network management systems.

In the voice telephony arena, the European Telecommunications Network Operators (ETNO), the European Telecommunications Standards Institute (ETSI), the European Conference of Postal and Telecommunications Administration (CEPT), and the European Institute for Research and Strategic Studies in Telecommunications (EURESCOM) are producing architectures and strategic plans incorporating standards-based pan-European integrated network management systems. In particular, the ITU Telecommunications Sector Study Group IV and the ETSI NA4 Technical Subcommittee are completing a set of standards (the M.3010 recommendations) entitled Principles for a Telecommunications Management Network (TMN).

In the data communications arena, there are three principal standards activities. First, the International Standards Organization (ISO) has been working on several Open Systems Interconnection (OSI) network management standards. OSI standards include the Common Management Information Protocol (CMIP), the Common Management Information Service Element (CMISE), and several subsidiary standards.

Next, the Internet Activities Board (IAB) has spearheaded the development of two network management standards, the first called the Simple Network Management Protocol (SNMP, versions v.1, v.2, and v.3), and the second the Common Management Information Services over TCP/IP (CMOT). Lastly, the Institute of Electrical and Electronics Engineers (IEEE) has assumed the lead role in defining management standards for local and metropolitan area networks, and has produced a draft standard entitled LAN/MAN Management. When CMIP is used in conjunction with IEEE standards, such use is often referred to as CMIP over Logical Link Control (CMOL).

Important aspects of these standards and the impact on network management and control technologies are summarized below. Perhaps more than in any other telecommunications industry segment, network management technology's critical value is determined by the degree to which open-system operations are available and supportable in practical and affordable products. For this reason, the remainder of this LAN management discussion focuses on emerging standards-based, interoperable management and control architectures, functional designs, protocols, device-naming, and attribute-specification conventions.

ISO's *Management Framework Standards,* ITU-TS X.700 Recommendations, and the Internet Activities Board's, Requests for Comment (RFCs) characterize management systems as consisting of the following components: a *Structure of Management Information* (SMI), a *Management Information Base* (MIB), and a management protocol such as CMIP or SNMP. Recall from Chapter 7 that protocols are strict procedures implemented in transmitting and receiving devices for the initiation, maintenance, and termination of data communications. Protocols define the syntax (arrangements, formats, and patterns of bits and bytes) and the semantics (system control, information context, or meaning of patterns of bits or bytes) of exchanged data, as well as numerous other characteristics data rates, timing, etc.

ISO/Internet management frameworks are based on the *Agent Process/Manager Process* paradigm, depicted conceptually in Figure 13.19. A *management process* is defined as an application process responsible for management activities. Resources supervised and controlled by network management are called *managed objects.* An *agent process* performs management functions on managed objects.

**Figure 13.19**
Agent
Process/Manager
Process Paradigm.

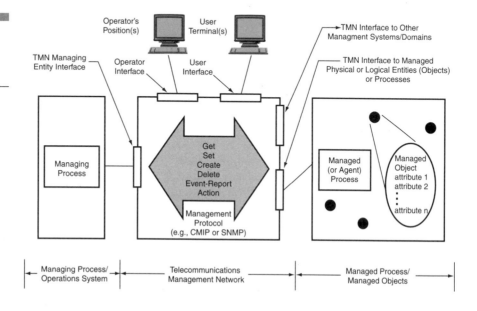

In CMIP/SNMP parlance, a *management domain* may be broken into one or more *managed systems* (sub-domains) and managed systems further broken into managed objects. Managed objects are resources (e.g., modems, T1 multiplexers, bridges, routers, LAN hubs, etc.) to be monitored and controlled by one or more management systems. A *management agent* is hardware and/or software in an object which exchanges management information with *management stations.* CMIP/SNMP establishes a structure for formatting messages and transmitting management information between reporting devices (agents in managed objects) and data collection programs residing in management stations.

Manager processes possess initial and updated global information on whatever physical or logical entity (object) the management system is designed to administer. These entities might be business applications, telecommunications services, physical networks, network elements, or network protocol layers. Managers—implemented in single consoles or within ensembles of distributed consoles—include Graphical User Interfaces (GUIs), databases, and facilities to communicate with the objects they manage.

Consoles enable human managers to access and invoke a variety of software management applications (configuration control, performance monitoring, fault isolation, diagnostics, etc.). GUIs display topologies of managed objects. Typically, operators can retrieve related status and MIB information stored in database repositories, by simply clicking on objects depicted on a GUI display.

*MIBs* define information about managed objects. Within MIBs, managed objects are described in terms of 1) object *attributes* and *characteristics,* 2) operations performed by or on objects, 3) notifications or reports objects can make, and 4) objects' behavior or response to operations performed on them. The SMI identifies information structures describing managed-object attributes, operations associated with attributes (such as get, set, add, remove), as well as operations relating to the managed objects themselves (e.g., read, delete, action).

With hundreds of network management product vendors and even larger numbers of managed network elements, in the absence of object naming, attribute, and communications protocols standards, open system management and control is not possible. In Figure 13.19, the Telecommunications Management Network (TMN) provides communications among managing and managed entities, is ideally logically distinct from managed networks, and where possible, implemented on separate, highly redundant, and reliable facilities. In addition to the managing and managed entity interfaces, the TMN also provides interfaces to "workstation functions" (that is, both operator and user or customer terminals), and an interface to TMNs in other management domains. In networks, management stations or consoles run by an operations staff constitute a network operations center (NOC).

While CMIP theory and structure are sound, its full definition and penetration of traditional local and wide area networks has yet to take place. Filling the gap is SNMP, currently supported by a large contingent of LAN and WAN manufacturers, with hundreds of companies shipping devices that support SNMP agents; the earliest implementations were in bridge and router equipment. Figure 13.20 depicts the generic SNMP architecture in a local area network application.

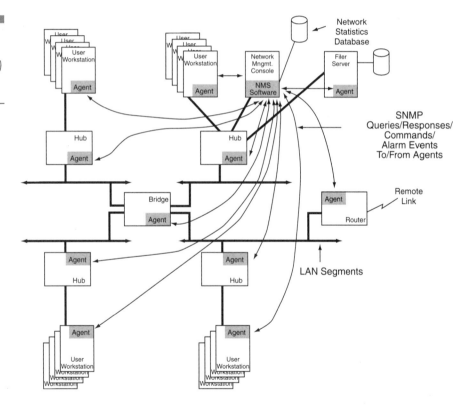

**Figure 13.20**
Simple network management (SNMP) LAN management concept.

Note in Figure 13.20 that the network used by the management console to gather diagnostic information and issue restoration commands is the same network it is managing and controlling. During initial installation and set-up and for major failures, the lack of an independent management network imposes limitations. Also, because

SNMP-based management uses polling, non-secure networks are vulnerable to eavesdropping, tampering, and disruption. Although SNMP v.1 lacked security features altogether, both SNMP v.2 and v.3 and the efforts of the ongoing work of the IP Security (IPSec) working group address provisions for safeguarding packets from unauthorized monitoring or tampering. Chapter 14 explains further IP-based security initiatives and protocols.

Even with some past limitations, the benefits of management products are clear. Gartner Group, for example, estimates that best-of-breed management tools coupled with well-conceived system designs can reduce support costs by $2,000 per year per PC. Since, according to a recent *LAN Times* survey, PCs are increasing by 8 percent and network segments by 71 percent, while network staffs are projected to diminish by 22 percent, such savings could indeed be crucial. Forrester Research projects that desktop costs can be reduced by 50 percent with effective desktop management tools, thereby lowering the ratio of support-to-user personnel from 1 in 50, to 1 in 200.

Although LAN management addresses only a part of enterprise system management, in no other segment has the incidence of heterogeneous equipment interconnection been higher. While no single network management product can represent a complete solution, Novell's Managewise is designed to tie together diverse systems across entire enterprises. Managewise is compatible with HP OpenView, IBM SystemView, and SunNet Manager host-system consoles and various third-party "snap-in" modules to provide custom management capabilities.

As an international company, Novell proves its large-network management mettle by operating its Global Network Operations Centers at Provo, Utah and Capelle, Netherlands using Managewise and its other products. The Global Network Operations Centers are responsible for Novell's 100,000 hits-per-day web site, and global communications between 4,500 employees in offices located in 100 cities using a network that contains 600 routers.

Just as fourth-generation languages are shifting significant software development capabilities directly to end-users, remotely programmable managed objects and advanced management technologies are shifting the ability to "design and build" software-defined complex systems and networks directly into the hands of network managers.

# CHAPTER 14

# Metropolitan and Wide Area Network Services

Today's LANs frequently begin as standalone networks serving small work groups with related job assignments and homogeneous equipment. In large enterprises, however, user requirements for access to company-wide databases and other information services often lead to incorporation of separate LANs into large multibuilding enterprise networks, connecting heterogeneous host and workstation equipment. When individual LANs are unable to meet mounting capacity demands, they are typically split into two or more segments connected by bridges, Layer 2 switches, or routers. As described in the previous chapter, these devices isolate LAN segments by keeping traffic within individual segments, unless it is addressed to terminals on other segments.

Metropolitan and wide area networks, using LAN router/switches or gateways, are the usual means chosen to interconnect LANs in geographically dispersed buildings. IEEE defines a *metropolitan area network* (MAN) as a network in which communications occur within a geographic area of diameter up to 50 kilometers (about 30 miles) at selected data rates at or above 1 Mb/s, consistent with public network transmission rates. In this book, MAN means any network in which communications cross public rights of way and occur within a geographic area of a diameter up to 50 kilometers, independent of data rates. *Wide area networks* (WANs) provide services beyond the distance limitation of MANs.

Under these MAN and WAN definitions, any LEC or IXC voice network discussed thus far could have been referred to as either a metropolitan or a wide area network. Readers might justifiably wonder why this terminology was not applied in previous voice service explanations. The answer lies mainly in a somewhat arbitrary "industry convention." Recall that we mentioned that although PBX-based networks might correctly be referred to as local area networks, by convention when one uses the term LAN, it almost exclusively refers to a local data communications network.

Beyond convention alone, however, it is true that for nearly a century LEC and IXC voice networks have been optimized to deliver call-by-call service to connected telephone instruments. By contrast, data MANs and WANs are optimized to interconnect whole networks in ways that make intervening "connecting" networks transparent to LAN end-users. Ideally, when connected via intervening MANs and/or WANs, users attached to one local area network should be able to access server resources or user DTEs attached to remote LANs with the same speed and operational capabilities that would exist if all LAN users and nodes were connected to a single LAN.

As a context for subsequent discussions of current and emerging MAN and WAN services and technologies, the Introduction to Part 4 summarizes how user requirements and the telecommunications environment evolved over the past two decades.

In today's environment, MAN services are provided mainly by local exchange carriers. This is especially true of the business-to-business data network services. Other competitors, such as Phonescope Communications, Ltd. (PCL), a company that operates Houston, Texas' largest MAN, are not referred to as competitive LECS, but rather simply as MAN service providers, another data service terminology practice that differs from that of the voice arena.

WAN services are provided by interexchange carriers in the case of business services and backbone NSPs (e.g., UUNET, Cable & Wireless, Sprint, AT&T, etc.—many are also ISPs) for Internet applications. See Chapter 2 for a discussion of the data communications industry structure. This chapter explains various types of data services, paralleling Chapter 10's voice-services structure.

# Local Exchange/Metropolitan Area Network Data Services

The taxonomy of MAN data services is shown in Figure 14.1. Access services provide data communications paths between user premises DTEs and switch or DACS ports from which MAN operators deliver transport services. As noted, MAN operators are usually incumbent or competitive local exchange carriers, but may be non-LEC business entities and use non-telco facilities. For example, cable television companies may use cable TV "set-top" modems that deliver data access services over their own cable television facilities. Others, like PCL noted above, have installed fiber cable in urban business districts and among universities or industrial or high-technology parks.

As shown in the figure, MAN operators may offer non-switched transport services using dedicated transmission facilities, or circuit, frame, packet, or cell-switched data transport services. As explained in earlier chapters (Chapter 7 in particular), virtually all data communications involves some form of frame, packet, or cell switching and multilayer protocols among end-user DTEs as well as among intermediate network elements. Customers have two basic transport service options. First, they can procure or lease non-switched transmission

**Figure 14.1**
Taxonomy of
LEC/MAN data
services.

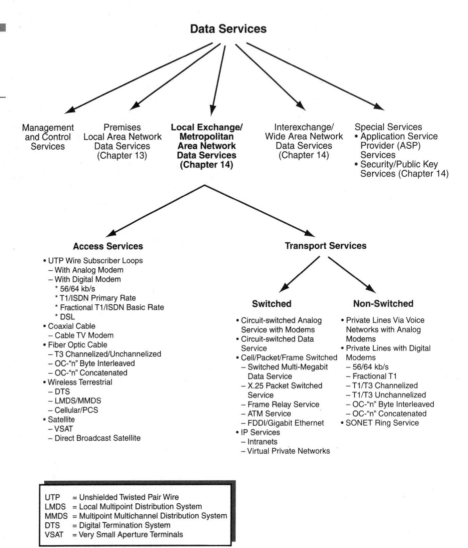

and employ customer-owned switching and protocol implementing
equipment, interfacing with MAN facilities at Layer 1. Second they
can lease switched services to take advantage of the bursty nature of
data traffic and the ecomony of scale of public networks.

In the first case, customers might lease *channelized* or
*unchannelized[1]* bulk T1, T3, or OC-"n" transmission "pipes" to interconnect LAN-attached routers/switches located in multiple customer

---

1 Channelized and unchannelized services are explained further in Chapter 5 and later
in this section.

premises. Note that in these arrangements customers pay fixed, non-usage-dependent charges. For these charges to be cost effective, customers must generate enough data traffic to justify fixed charges. As with voice switched versus dedicate" transport service tradeoffs described in Chapter 12, the impacts of diurnal, weekend, and holiday traffic variations, and other peak and average traffic patterns weigh heavily. Because data traffic is more bursty in nature than voice traffic, peak burst rate-to-average throughput rate is generally of greater consequence in selecting data transport services than busy hour-to-average hour traffic is for voice services. For example, if a plain vanilla dedicated T1 facility is leased to connect two LANs, the maximum burst rate that can be supported is limited to about 1.5 Mb/s.

In the second case, frame relay, packet, or cell-switched transport services, on the other hand, may be procured and paid for on the basis of customer-carrier agreements that specify such parameters as peak burst rates, maximum burst rate intervals, as well as average throughput rates. Under these switched transport carrier service agreements, customers take advantage of economies of scale that accrue in large carrier networks, and the fact that incidences of peak or high burst rate events among multiple carrier customers may tend not to occur at the same instants in time. When this is the case, carriers can offer higher peak burst-rate service than customers would be able to afford using non-switched or dedicated fixed maximum-rate (single-rate) transport services.

As will become evident from our alternative service option explanations below, switched services like frame relay among, for example, four customer nodes can often be substituted for, and provide service otherwise indistinguishable from, six separate dedicated T1 circuits, at lower cost and better peak or burst-rate performance. It will also become apparent that a major discriminator among switched transport alternatives has mainly to do with protocols associated with each alternative.

## MAN Data Access Services

*MAN data access services* typically use the same physical facilities employed for voice access services. If data-service users employ voice network compatible modems (see Chapter 3), any of the LEC voice-access services describe in Chapter 10 may be considered for data service. However, with spiraling demand for high-bandwidth multimedia applications, interest now centers on higher-speed alternatives. The range of possibilities encompasses all media and advanced broadband modulation technologies presented in Chapter 4. While xDSL broadband access via ubiquitous, twisted-pair metallic media remains at the forefront, fiber optic cable is increasingly being brought to high-traffic volume commercial premises by both LEC and non-LEC MAN operators.

Among wireless media alternatives are Digital Termination Systems (wireless T1 service introduced in the 80s), the newer Local Multipoint Distribution Systems/Multipoint Multichannel Distribution Service (LMDS/MMDS), and the direct broadcast satellite (DBS) and very small aperture satellite (VSAT) satellite options described in Chapter 4. Popular services in each of these categories are discussed below.

### UTP WIRE SUBSCRIBER LOOPS WITH ANALOG MODEMS

Historically, twisted-pair metallic wire local loops equipped with voice network-compatible analog modems (see Chapter 3) have been and continue to be the most widely used facilities for delivering data-access services. Since analog modems are designed to be compatible with worldwide local and long-distance voice networks, any facility that can support voice-access service is adequate for data-access service as well—usually at speeds up to 56 kb/s (see Figure 10.2 and related text). Since analog modems connect to the same telco interfaces as standard analog telephones (modular RJ-11 jacks in the United States), the good news is that 120 million U.S. households wired for telephone service are data ready.

For many businesses, the use of modems and ubiquitous worldwide voice networks for data communications is not so straightforward. It is perhaps the industry's greatest irony that one of the greatest advances in voice-service delivery, the appearance of digital PBXs and digital Centrex service, precludes plugging inexpensive analog modems into jacks intended for digital telephone instruments. The reason for this is that to fully benefit from digital PBX and Centrex switching advantages, special telephones employing digital interfaces and digital telephone-to-digital switch line card modulation and signaling techniques must be used. As a consequence, analog modem signals (which are incompatible with digital switch line cards) cannot be switched to analog voice-compatible transport service trunks.

Initially the digital PBX/Centrex "workaround" for this problem was to connect digital DTE signals directly to digital telephone jacks (via either wall outlets or DTE jacks on the digital telephones themselves), and let switches direct those digital DTE signals to banks of modems (modem pools). Modem outputs were then connected to trunks designed to transport analog voice signals. While this approach provides equivalent technical performance to direct modem-to-analog loop connections, for cost, flexibility, and significant performance advantages, businesses with digital PBXs and Centrex service gravitated to one of the data LAN configurations described in Chapter 13 instead.

Once users are attached to LANs, external DTE data communications is accomplished using access and transport services described in this chapter—via LAN gateway servers. Although today most LAN-to-external network traffic is via digital interfaces and data communications facilities, for data traffic that must transit voice network facilities, LAN gateway servers with analog modem pools are invoked.

## UTP WIRE SUBSCRIBER LOOPS WITH DIGITAL MODEMS

In earlier chapters we describe five techniques that enable digital voice signals to be transmitted over twisted-pair wire media. The five techniques relate to:

1. The connection of proprietary digital telephone to digital PBX or Centrex switches.

2. The connection of basic rate, standards-based ISDN digital telephones to ISDN-capable digital PBX and Centrex switches.

3. Synchronous, channelized full-rate (T1/E1 24/30-channel) or "fractional T1" digital carrier connections.

4. Synchronous, primary-rate ISDN connections.

5. Telco subscriber loop carrier (SLC) or *pair-gain* connections.

Recall that the DS1 multiplexers described in Chapter 5 produce 1.544-Mb/s digital signals containing information from 24 separate DS0 signals. T-carrier transmission facilities designed to support such multiplexed signals are referred to as *channelized*, byte-synchronous DS1 facilities. *Unchannelized bit-synchronous* DS1 T-carrier transmission facilities are designed to support 1.544-Mb/s bit rates and establish T1 frame synchronization, but otherwise user-defined data or signaling traffic characteristics remain transparent to telco or other service providers. Among these options in the context of T1 carrier facilities, what remains common are basic modulation and electrical interface characteristics.

Since for either analog-signal or digital-signal applications, the unshielded twisted-pair subscriber loop medium is itself identical, what changes among analog, digital voice, and digital data applications are modem-modulation, transmitter-receiver signaling and other termination equipment interface and protocol-related design features.

Besides analog modem-based use discussed earlier, any of the first four UTP subscriber loop-based techniques designed to support digital voice services listed above may be used to deliver data-access services. The fifth technique, telco SLC facilities, is not made available to customers, but instead is used to extend, or reduce the cost of, UTP-based subscriber loop plant.

When T-carrier facilities are used exclusively for data access, telcos typically offer unchannelized service. In cases where customers wish to use dedicated T1 facilities for both voice and data traffic, channelized service must be chosen. Primarily used in the past to access non-switched dedicated or private line transport services, some LECs now support digital loop protocols (e.g., Bell South's "Fast Packet Options") that enable customer traffic to be directed to a variety of switched data-transport services.

Among larger business customers, T1 service continues to be extremely attractive with 250,000 new lines installed in 1999. Although T1 pricing is highly competitive, small-business and residential customers can rarely afford the cost of bringing two more pairs of UTP to their premises and paying fees on the order of $300 per month. For these users, ISDN and digital subscriber line (DSL)-based service, explained in Chapters 8 and 4 respectively and revisited below, are more likely candidates.

### ISDN-BASED DATA-ACCESS SERVICES

Integrated Services Digital Network (ISDN) is a technology specifically designed to offer switched data services to users over existing local loops. Figure 8.3 shows two kinds of ISDN access services offered today by most LECs. Primary Rate Interface (PRI) service, a variant of T-1 service that provides 23 channels of 64-kb/s service and reserves a single 64-kb/s channel for signaling, is an appropriate selection for customers with high average-bit-rate requirements.

A key PRI feature is the ability to use any combination of 64-kb/s channels (i.e., up to 23) to access different transport services. What makes this capability particularly useful is that these "combinations" can be arranged on a call-by-call, circuit-switched basis. With traditional T1 service, such channel groupings are typically preassigned, or at best can only be modified by "service grooming" operations of channel switching DACS (see Chapter 5 and Figure 5.7).

This promotes more efficient use of access bandwidth. For example, during busy hours, customers may need all PRI channels to handle peak voice traffic loads at acceptable blocking levels. With PRI, during off-peak hours, customers can allocate six channels (384 kb/s) for two or multiparty video conferencing. With conventional T1 access service, six channels would have to be permanently reserved for 384 kb/s access and would therefore not be available to handle busy-hour voice calls.

For lower traffic volume users, ISDN's Basic Rate Interface (BRI) design supports two 64-kb/s customer traffic bearer channels (B-channels) and a 16-kb/s signaling channel (D-channel). This interface is commonly known as the 2B+D interface. To satisfy BRI requirements, telco local loops must support 144-kb/s bit rates. As it turns out, if loading coils are removed and user premises are within 18,000 feet of a central office, then (using 2B1Q modulation, explained in Chapter 5), most existing two-wire local loops are adequate.

As originally conceived, the two B channels are to provide access to circuit-switched voice and/or circuit-switched data transport service while at the same time the 16-kb/s D channel is being used both to satisfy ISDN signaling requirements and to provide customer access to packet-switched data transport service. However, 16 kb/s is no longer considered adequate for data communications.

As a consequence, in most instances today, both B channels are used to establish circuit-switched 128 kb/s connections to data networks (the Internet for example). Since signaling is out of band on the D channel, users may simultaneously initiate or receive voice calls. If a 64-kb/s voice connection is established, one of the two B-channels must service the voice call. In that case the data connection drops back to 64 kb/s for the duration of the voice call. All this B-channel shuffling is transparent to users, except of course for what may be perceptible "slow downs" in data transactions. When voice calls are completed, the data connection resumes 128-kb/s bit operations. Note that ISDN interfaces support full-duplex 128-kb/s communications, which is independent 128-kb/s upstream and 128-kb/s downstream transmission rates, simultaneously.

Despite these apparently attractive characteristics, ISDN BRI market penetration is far below original projections. Part of the reason for this can be traced to the fact that when ISDN first emerged from the standards process, AT&T's monolithic control, which was comparable to foreign PTTs, was being eliminated by deregulation and divestiture. So whereas foreign PTTs in Europe and Japan quickly embraced and implemented ISDN, the U.S. lagged far behind.

Moreover, complications caused by the FCC's network channel terminating equipment (NCTE) order (the NCTE Order), that prevents LECs from owning customer premises-located ISDN interface equipment, further slowed newly independent LEC rollouts. Finally, LECs are not aggressively pricing the service. In the meantime, both ADSL and cable-modem services are becoming available that offer higher downstream bit rates at more attractive prices than ISDN BRI. Applications such as video telephony that require symmetrical bandwidth may keep ISDN services alive, but market forces will ultimately decide its fate.

### DIGITAL SUBSCRIBER LINE-BASED DATA-ACCESS SERVICES

In 1999, a flurry of advertisements—on sides of buses, the radio, and in the printed media—appeared in large cities across the country touting DSL's benefits. Most ads, emphasizing faster access to the Internet, substituted the phrase "worldwide wait" for the meaning of the acronym "www" (World Wide Web), and claimed to shorten the wait. And of course there is much truth in the fact that what is fueling interest in higher-speed data access services is the Internet and residential and small-business users' demand for faster access to it. Three attributes of this demand are driving DSL developments:

- Access services must be inexpensive (comparable to a second telephone line).
- Access services must be deliverable over existing two-wire local loops.
- Downstream traffic volume from the Internet to users is significantly greater than upstream traffic volume from users to the Internet.

From a practical point of view, if one sets out to minimize lifecycle DSL expense while maximizing local loop reach (the first two attributes), then the Internet's asymmetrical traffic flow, the third attribute, favors selecting Asymmetric DSL (ADSL)-based technology over competing alternatives.

As described in Chapter 4 and pictorially presented in Figure P4.2, DSL Access Multiplexers (DSLAMs) in central offices extract data packets from local subscriber loops and route them to ISPs selected by subscribers.

In this design, circuit-switched voice traffic is completely independent from packet-switched data traffic. The data communications path is full period, meaning that if subscriber DTEs are powered on and the communications mode is enabled, the subscriber-to-data network connection remains active. Moreover, data communications has no impact whatsoever on voice network in the sense that no central-office circuit resources are affected at all by data traffic. This means that data traffic in no way contributes to voice network congestion nor the demand on voice networks for dial-tone service. Therefore, the presence or absence of ADSL equipment and operation places no restrictions on and is transparent to telephone users and voice network operations.

Figure 14.2 shows that DSL service providers are generally not the same business entities that provide Internet or other data network services. As the figure implies, in most cases local subscriber loops terminate in an incumbent LEC's wire center, which is usually located in or in close proximity to that LEC's central offices. It is not surprising, then, that the major DSL service providers are incumbent LECs and competitive LEC or MAN operators with collocation agreements to place equipment in incumbent LEC facilities. From an ISP's perspective, it is more economical to connect to a single, centrally located central DSL service provider's router than to place ISP-owned routers in every central office equipped with DSLAMs. The situation today is analogous to what existed with new IXCs soon after divestiture. In that era, connections were made to LEC access tandems until traffic volume justified direct connections to end offices.

Referring again to Figure 14.2, note that connections between DSLAMs and data network or ISP service provider nodes may be implemented over a variety of MAN transport services. These include private lines, frame relay service, ATM service, and IP-based services. Each alternative has advantages and disadvantages. MAN transport service options are described in this chapter and selection rationale in Chapter 15.

The most popular ADSL offering to residential customers to date typically provides 640-kb/s downstream and 64-kb/s upstream capabilities at prices as low as $49 per month. Especially important when comparing DSL to the cable TV modem-based data-access service options discussed below, this capacity is fully available and dedicated to a single subscriber loop and the user DTEs attached to that loop. Also note that DSL service subscribers generally have a wide choice of ISPs, a freedom not normally available from cable TV modem-based service providers.

Business users may opt for higher speeds (up to 7 Mb/s when premises are close enough to central offices) or for symmetrical DSL service at T1 speeds, at correspondingly higher prices.

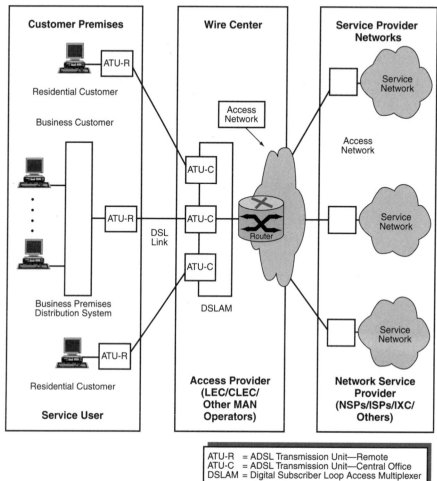

**Figure 14.2**
DSL-based access
services.

With all the advantages, one might reasonably ask why DSL is not more widely installed. There are two parts to the answer to that question. First, the technology is relatively new and many companies planning to offer DSL services are in formative stages. Other reasons are more systemic. At this time, DSL rollout is limited by the availability of qualified local loops and the economics surrounding the cost of removing loading coils, branch taps, and other costs associated with upgrading existing loop plant.

Since 99 percent of all loops terminate at incumbent LEC central offices and currently carry voice service, DSL approaches based on the installation of new or separate subscriber loops are at a grave competitive disadvantage relative to ILEC offerings. Competitive new entrants must vie for space within ILEC facilities to install DSLAMs and router equipment and secure DSL-compliant loops on behalf of prospective customers. Developing rollout strategies under all these uncertainties is complicated at best. The competitive situation has improved, however, since the FCC issued a directive that ordered ILECs to share voice customer loops with competitive LECs to provision DSL service. Accordingly, competitive LECs no longer need to procure separate loops (as unbundled network elements) and cover their cost with DSL revenues.

In time, all DSL-compliant loops will be made available for DSL-based data access service. However, according to current estimates, even after loading coil and branch tap removal, nearly 40 percent of all existing loops may be unsuitable for DSL service.

### CABLE TV MODEM-BASED DATA-ACCESS SERVICES

Cable modem-based data-access services, described in Chapter 4, provide the only metallic wire alternative to the UTP subscriber loop-based access services described above. The most salient attribute of cable TV modem service is that it does not rely on incumbent LEC local loops. Cable TV systems use combinations of coaxial cable and fiber optic cable for their distribution system; both of these are inherently broadband media. (Figure 4.17 illustrates the major components of a cable TV distribution network.) However, this does not mean that cable systems can support access to the Internet, or other data networks, without modification.

Most existing cable systems are designed for one-way (simplex communications) distribution of a common set of *analog* entertainment signals (television, radio, music, etc.) to all subscribers. For this purpose, *tree-like* network architectures are the least expensive to install and maintain. Adapting cable systems for data access requires modifications in two important ways.

First, some analog-signal simplex "head-end," repeater, and customer premises line-terminating equipment must be replaced with full-duplex versions and provisions made for digital data signals generated by routers, DTE digital modems, and other data communications equipment. Second, channels (bandwidth) previously allocated to TV and other entertainment program material, must be reallocated for data transmission. With appropriate modems, single TV channels can support 30-Mb/s data signals. Using CSMA-CD protocols (see Chapters 7 and 13), approximately 400 subscribers with average downstream data rates of 1.5 Mb/s can be accommodated under lightly loaded conditions.

To accommodate much larger numbers of subscribers, there are only two categories of options. First, there are options to add additional analog channels (increasing analog bandwidth) or add separate digital channel capacity, at least between head-end and regional or neighborhood distribution points. In the absence of adding analog bandwidth or digital data throughput capacity, the only other option is to reallocate channels currently being used for TV service and make them available for data access service. Since the number of channels offered is an important cable TV industry marketing factor, any significant drop in the number of offered TV channels is likely to be unacceptable for the entertainment portion of an operator's business.

What this means is that if the data-access business segment becomes enormously successful (a reasonable objective for any new business area), at some time cable TV service providers must add substantially to existing networks. While this is not a surprising conclusion, it is somewhat at odds with the perception that cable TV distribution systems are naturally well positioned to provide residential data access simply because they already bring wires to 60 million households.

Nevertheless, the very existence of this alternate way into homes is attracting the investment necessary to make cable modem Internet access a reality. In fact, cable modem access is currently available to a wider set of users than DSL. Rates for cable TV modem and DSL-based service are comparable.

The major performance differences between cable TV modem and DSL-based services, discussed in more detail in Chapter 4, are related to the fact that subscribers share a common medium using the CSMA/CD protocols to resolve transmit-attempt collisions. Thus in a 400-user segment operating at 30 Mb/s, under peak traffic conditions each user's average throughput share might be as low as 30 kb/s. For today's streaming video Internet traffic such low average throughputs could lead to catastrophic congestion, that is traffic levels that essentially reduce CSMA/CD-based network throughputs to zero.

The second performance consideration is that, unlike DSL installations where a subscriber loop is dedicated to each user, cable modem systems share a common LAN-type medium and each subscriber can listen to all traffic addressed to, or originating from, all other subscribers attached to the same network segment. While encryption and firewall designs mitigate security and privacy risks, networks with no security/privacy features are extremely vulnerable to unauthorized information eavesdropping and malicious tampering. Finally, most cable TV modem-based data-access service providers currently provide connections to just one network or ISP service provider. Generally, the cable modem data-access and ISP service providers are business partners so that there is little hope that subscribers will ever be able independently to select an alternative ISP, as LEC-based DSL subscribers can.

### FIBER OPTIC CABLE-BASED DATA-ACCESS SERVICES

If cost were not a factor, fiber optic cable would be used for virtually all data-access services. As noted in Chapter 4, LECs already have 16 million miles of fiber in the ground, mostly in feeder and distribution segments of subscriber loop plant and among central-office and other LEC switches. LECs often opt to use fiber cable facilities direct to high traffic-volume customer premises—when tariff revenues offset investment outlays. Figure 14.3 lists DS3, OC-3, and OC-12 services that represent typical LEC fiber-based data-access service offerings.

**Figure 14.3**
Summary of LEC access data services.

| Medium | LEC Access Service | Facility |
|--------|--------------------|----------|
| Metallic<br><br>UTP<br>Subscriber<br>Loop | DS0 (56/64 kb/s) | Four Wire UTP |
| | T1/ISDN Primary Rate (1.536 Mb/s) | Four Wire UTP, T1 Bipolar Modulation |
| | Fractional DS1/T1<br>    T-Carrier ("n" x 64 kb/s)<br>    ISDN Basic Rate (128 kb/s) | Four Wire UTP, T1 Bipolar Modulation<br>Two Wire UTP, 2B1Q Modulation |
| | Digital Subscriber Loop (DSL) | Two/Four Wire UTP,<br>"x"DSL Modulation |
| Fiber<br>Optic<br>Cable | T3 (44.210 Mb/s) | 2 Strands Single Mode Fiber<br>DS3 Signal Modulation |
| | OC-3 (149.760 Mb/s) | 2 Strands Single Mode Fiber<br>SONET, OC-3/OC-3c Signal Modulation |
| | OC-12 (599.040 Mb/s) | 2 Strands Single Mode Fiber<br>SONET, OC-12/OC-12c Signal Modulation |
| Wireless | Digital Termination System (DTS) | Point-to-Point, Microwave Radio |
| | LMDS/MMDS | Multipoint, Microwave Radio |

LMDS = Local Multipoint Distribution Systems
MMDS = Multipoint Multichannel Distribution Systems

When used for multiplexed circuit switched voice signals conforming to asynchronous digital hierarchy (ADH) DS "n" structures, SONET OC-3 and higher synchronous payload envelope signals contain byte-interleaved lower rate components, which can be used for byte-synchronous circuit-switched data traffic, or mixed voice and data traffic.

### WIRELESS DATA-ACCESS SERVICES

LECs, cellular telephone system, and other MAN operators offer a number of different wireless data access services. As explained in Chapter 4, some wireless services are intended for truly mobile subscribers, while others are simply wireless equivalents to wired (cabled) access services., In the latter category, *digital termination system* (DTS) equipment employs short-hop, point-to-point microwave radios in the 10.6-GHz band to provide up to DS1 rate service directly to customer premises. DTS services have been available since the 1980s and were primarily intended for data communications, although customers also use them for voice and compressed video. DTS may be employed for data access to LECs or in LEC-bypass applications. For such use, a license must be obtained from the FCC, which refers to the technology as *digital electronic message service* (DEMS).

Local Multipoint Distribution Systems/Multipoint Multichannel Distribution Service (LMDS/MMDS) may be offered by LECs and other service providers. LMDS/MMDS are terrestrial wireless alternatives to DSL and cable modem services, intended for non-mobile users. LMDS/MMDS and a variety of mobile subscriber cellular/PCS data access service, such as cellular digital packet data offerings, are discussed in more detail in Chapters 4, 16, and 17.

In the satellite arena, very small aperture terminals (VSATs) are designed to provide data, voice, and video communications directly from user premises. VSAT technology, operation and sample applications are described in Chapter 4, as is the use of direct broadcast satellite technology as a vehicle for providing data-access service to Internet service providers.

## MAN Data Transport Services

Given their limited geographic extent, MANs lead the way in migrating toward fiber transmission and modern data communications facility deployment. In particular, high-speed data access and transport service offerings first appeared in MANs, driven by several urban-area-related factors. First, highly industrialized cities typically provide homes for multi-location companies, a situation that the fuels demand for high-speed LAN-to-LAN digital data interconnection services. Next, combined with the relatively limited geographic extent (compared to the nationwide dimensions of long distance networks), the high traffic volume and customer density leads to extremely attractive return on investment projections for expenses incurred in installing advanced fiber and data communications equipment facilities.

Because the same "concentration of large corporate customer" arguments apply to voice services as well, it is often the case that voice requirements entice many LECs to embark on major fiber plant expansion programs, long before data service demand begins its geometric growth. The availability of spare capacity in fibers originally installed for voice service, and the ability of DWDM technologies to extend the capacity of existing fiber plant (see Chapter 5) puts LECs in an enviable position to capitalize on escalating urban-area data service demand.

To meet this demand, when appropriate, LECs employ the same facilities used for voice services. This occurs most often in LEC provisioning of circuit-switched-based data service offerings. And, as noted in earlier chapters, it is frequently advantageous to use common transmission facilities for voice and data services independent of service types.

This MAN section focuses on self-healing SONET ring, switched multi-megabit data service (SMDS), fiber distributed data interface (FDDI), and Gigabit Ethernet-based transport services, all of which, today, are unique to metropolitan or local area networks. To eliminate duplication, MAN data-transport services that are nearly identical with WAN data transport services (i.e., circuit switched, X.25 packet switched, frame relay, and asynchronous transfer mode) are described in the last part of this chapter. Circuit-switched and private-line transport services that can carry analog modem data traffic over voice facilities are covered in Chapter 10.

### SELF-HEALING SONET RING DATA-TRANSPORT SERVICES
*Self-healing SONET ring* service is a dedicated, high-capacity network service designed to provide increased reliability, availability, and functionality via ring topologies among multiple customer locations and telco central offices. The service typically provides DS1 (1.544 Mb/s) and DS3 (44.736 Mb/s) channels to customers over primary and alternate paths. The "self-healing" descriptor derives from the fact that service provider facilities continuously monitor service quality, detect and diagnose failures, isolate faults, and switch traffic among alternate paths or initiate other actions that automatically restore service. Figure 14.4 illustrates the major underlying facilities that support this service.

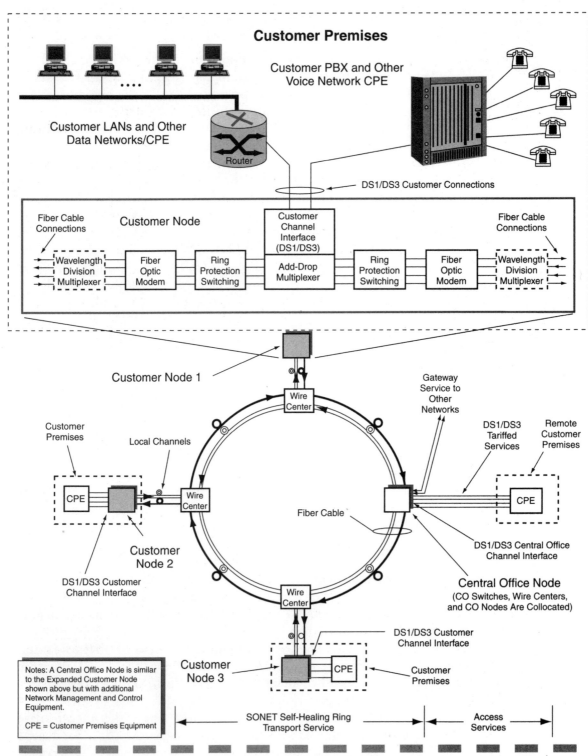

**Figure 14.4** SONET self-healing ring transport service connecting multiple customer locations in a data and voice network.

The operation of these private rings is essentially the same as scaled-down versions of long-distance carrier ring networks. Chapter 8 includes a description of both unidirectional path switched (UPSR) and bi-directional line switched (BLSR) ring architectures and in particular explains how *ring protection switching* is able to restore cable cut failures in 50 milliseconds.

LEC self-healing SONET ring transport service entails the use of multiple *customer* and *central-office nodes.* There must be at least three nodes and at least one customer and one central-office node with remaining nodes of either type. The maximum number of nodes is equipment and facilities dependent. Although constituent elements of ring nodes may vary, Figure 14.4 depicts a representative equipment arrangement. All nodes include add-drop multiplexing, ring protection switching, and fiber optic modem electronic-to-optical signal conversion functionality, as well as fiber cable termination and *customer channel interface* provisions.

Customer nodes are directly connected to a serving central office's wire center via *Local Channels. Alternate Central Office Channels* provide communications paths to alternate central offices. *Interoffice Channels* connect customer and central-office nodes, respectively.

Customer Channel Interfaces (one for each originating or terminating DS3 and/or DS1 tributary) provide channelization at customer nodes. Each node can be configured to allocate 3 DS3, 84 DS1, or any equivalent interface combination. Connectivity at interfaces is specified at the time of order, but may be reallocated subsequent to initial installation.

Since customer nodes represent a large portion of ring transport-service cost, they are only installed on high-traffic volume customer premises. In this case, access service collapses to customer-owned wiring and equipment connecting the LEC installed "customer node" to customer premises equipment (CPE). Lower traffic-volume customer premises can gain access to these private ring transmission networks, however, using one of the MAN access service identified above in this chapter, via central office ring nodes, as shown in the figure.

Central office nodes are configured similar to the exploded customer office node pictured in the figure but with additional network management and control facilities and provisions for gateway interconnections to other DS1/DS3 services.

Figure 14.4 corresponds to applications in which companies with multiple high traffic-generating locations in a single metropolitan area justify connections among remote LANs and PBX voice networks. A representative customer node-to-CPE interface is shown at the top of the figure. In multi-building interconnection applications, most ring nodes will be customer nodes. Figure 14.5 presents an example of how MAN ring transport services may be used by local ISPs to connect to DSL access multiplexers located in central offices throughout a city. In this application, perhaps all but one ring node would be of the central-office node variety.

### SWITCHED MULTIMEGABIT DATA SERVICE (SMDS)

*Connectionless* data services are defined as services that transfer information among service subscribers without the need for end-to-end call-establishment procedures. MAN environments, with their limited geographic extents, are ideally suited to connectionless services. New protocols were developed which extend the traditional LAN protocols to operate over longer distances.

Switched Multimegabit Data Service (SMDS) is a Bellcore-developed LEC service offering public, connectionless, high-speed, fast packet-switched data service throughout metropolitan areas. SMDS provides DS1 or DS3 access to fiber optic-based switched networks. The need for public switched carrier-based versus private MAN service stems from:

- User traffic characteristics.
- The potential need for ad hoc "public" connections between different user organizations.
- The requirement to cross public rights of way.

Regarding user traffic considerations, a key MAN application is the interconnection of LANs supporting bursty, broadband traffic, with short response times or latency.

To be successful, MAN interconnection must be transparent to LAN users. Ideally, these users would enjoy the same throughput and response time performance, whether connected through a single LAN, or via two LANs and an interconnecting MAN. Typically, average LAN-to-MAN-to-LAN traffic may be only 0.5 Mb/s, but critical applications may require peak 10-Mb/s capacities. It is hard to imagine that a user organization would opt to lease a dedicated DS3 (44.7-Mb/s) service for occasional use. Thus, traffic demand and throughput economics increasingly justify switched public service.

SMDS standards are based on a three-level SMDS interface protocol (SIP) stack defined in IEEE 802.6 (covering Layer 1 and parts of Layer 2

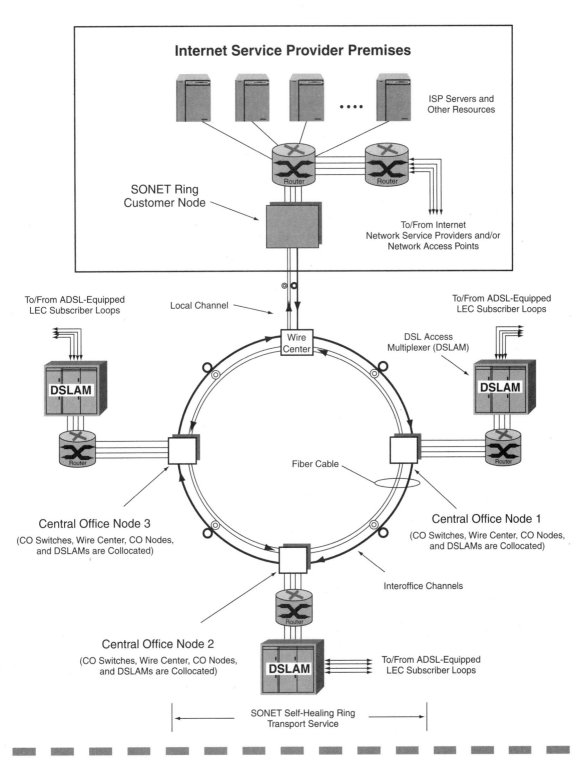

**Figure 14.5** SONET self-healing ring transport service connecting central office DSLAMs to ISP premises.

in the OSI reference model). As depicted in Figure 14.6, the proposed standard calls for a distributed queue dual bus (DQDB) architecture, which defines a high-speed, shared medium access protocol for use over dual, unidirectional, fiber optic bus networks. This connectionless service is similar to other IEEE 802 LANs. As illustrated in the figure, in the literature an SMDS switch is termed a metropolitan switching system (MSS) and the network interface is called a subscriber network interface (SNI).

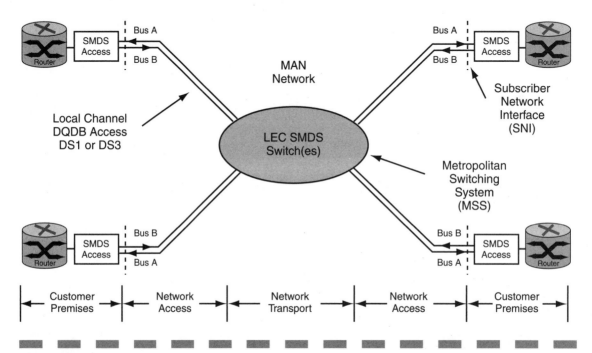

**Figure 14.6**  Switched multi-megabit data service (SMDS).

Distributed queuing is a media access protocol that delivers near-perfect access characteristics. In particular, the protocol enables all of the theoretical "payload" bandwidth (100 percent efficiency) to be used and the average slot access delay approximates that of a perfect scheduler at all traffic loading levels. Payload bandwidth, the bandwidth available for user information-bearing signals, is the channel bit rate—less switching/multiplexing signaling and control overhead.

In the early 1990s, every RBOC was running or planning an SMDS trial, and all were working through Bellcore to set national standards. SMDS was a contender to become the preferred high-speed service in wide area networks as well. However, the increasing popularity and acceptance of IP networks and ATM technology and services has largely eliminated SMDS's WAN role. Today, SMDS remains a MAN transport service offering in some areas, often in low-speed versions known as connectionless data service—which provides throughput rate increments down to 64 kb/s.

## FDDI AND GIGABIT ETHERNET-BASED DATA-TRANSPORT SERVICES

As with SONET self-healing rings, in rendering FDDI and Gigabit Ethernet-based data-transport services, MAN operators establish nodes on high traffic-volume customer premises. In this case, instead of basing these services on a wide area network technology (the WAN long distance self-healing ring structures explained in Chapter 8) MAN providers have adapted technologies originally developed for local area and campus backbone network applications.

Before being acquired by MCI WorldCom, Metropolitan Fiber Systems offered FDDI-based service in over 30 metropolitan areas, calling it Metrofiber Multimegabit Data Service. Until the recent appearance of Gigabit Ethernets, Ethernet technologies had been consigned to local area networks. However, MANs are now beginning to offer Gigabit Ethernet-based data-transport services, with the aforementioned joint Phonescope Communications, Ltd. (PCL) and Cisco venture, an example. For companies with multiple metropolitan area locations, such services are ideally suited for connecting LANs located in different and sometimes widely separated buildings. In that application, such services appear to provide business customers with transport services among company-owned LANs. To the extent that such services provide access to external non-company-owned or wide area networks, they supply data-access services.

With dwindling industry FDDI support and ascending acceptance of Gigabit Ethernet (GE), one would have to predict the latter's future dominance. In purely private, facilities-based GE networks, those with nodes in only one customer's designated facilities, only high-traffic originating/terminating locations can justify installation of gigabit throughput and dedicated fiber cable facilities. Of course, as in SONET self-healing rings, lower traffic-volume customer locations can be afforded access to GE networks via T1, fractional T1 or other lower-capacity facilities.

As an option to facilities-based private GE networks, *virtual private data network* embodiments, discussed again later in this chapter and in

Chapter 15, are possible. As discussed in those sections, in either facilities-based or virtual private networks, it is likely that Layer 1 and 2 GE-based designs may be coupled with higher level Internet IP protocols to form intranets (private networks employing Internet technologies and standards and supporting World Wide Web-compatible browser and other operations).

Two factors make GE/IP networks a likely winning combination. First is the almost seamless transition between *switched* gigabit-rate Ethernet network segments, and *CSMA/CD-based* Ethernet segments that currently support hundreds of millions of existing lower rate Ethernet users. Chapter 13 explains why this is true.

Second, intranets use network-layer protocol, browser and other operating technologies that have proven their mettle in the most stringent of all real-world test scenarios—the public World Wide Web. Worldwide use and user familiarity with the Internet has created perhaps the largest-ever base of competent users, each able to manipulate and effectively employ its common browser and other operational features. Why develop, and then have to train people to become familiar with, new proprietary private network operations when most employees are already familiar with proven and always evolving Internet techniques?

What we are witnessing with GE-based intranets is a marriage between the world's most successful physical layer design and the world's most popular networking and operating technology.

# Interexchange/Wide Area Network Data Services

Technology has made its impact on telecommunications felt nowhere more than in the realm of wide area network data services. As late as 1990, there was almost a complete lack of viable, high-speed public data network services. Packet switches were too slow and limited in capacity to accommodate the large number of subscribers that public services would attract. Private networks were the usual alternative for users with high bit-rate and traffic-volume requirements, and what public networks existed were based on X.25 protocols operating at a maximum speed of 56 kb/s. Deployment of high-speed switching and transmission facilities in the early 1990s, combined with standard protocols that satisfied diverse performance applications, produced a groundbreaking set of services that have withstood the test of time until now, and promise to be the foundation for even more dramatic data networking capabilities in the future.

Figure 14.7 shows the taxonomy of extant WAN data services. Since IXC/WAN access and transport data services using analog modems are delivered via unmodified voice network facilities, Chapter 10's IXC voice services discussions apply and will not be repeated here.

## Interexchange/Wide Area Network Data-Access Services

Access services provide data communications paths between user premises DTEs and switch or DACS ports from which WAN operators deliver transport services. IXC data-access services employ both *user premises-to-LEC premises,* and *LEC premises-to-IXC premises* facilities. As with IXC voice access services, the LEC premises-to-IXC premises portion of IXC access services are delivered using LEC transport services, both switched and non-switched. For the user premises-to-LEC premises portion, any of the user-to-LEC services and underlying facilities listed in Figure 14.1 may be used. For simplicity, user-to-LEC services are not shown on Figure 14.7.

In particular, access to ISP WAN points of presence may be via analog modem connections to LEC switches and LEC voice circuit-switched transport services to ISP POP-located modem banks, as shown in Figure P4.2. But any of the other access service possibilities listed in Figure 14.1, such as the DSL or cable TV modem alternatives discussed above in the LEC/MAN part of this chapter, may also be used.

Because it is simple, available to all telephone service subscribers, and relatively inexpensive, modem- and voice network-based dial-in access is the choice of most residential users to reach ISPs. But it is equally true that dial-in access to the Internet makes horribly inefficient use of local circuit-switched telco facilities. The main reason for this is that voice networks are sized for voice calls that average about three minutes in duration, while the average length of an Internet call grew to 30 minutes in 1998, an increase of 55 percent over 1997.

Recently, a heavy snow closed the federal government and most businesses in Washington, DC. The large number of workers at home dialing into their workplaces and the Internet caused widespread telephone system disruption. Throughout the entire workday, lack of dial

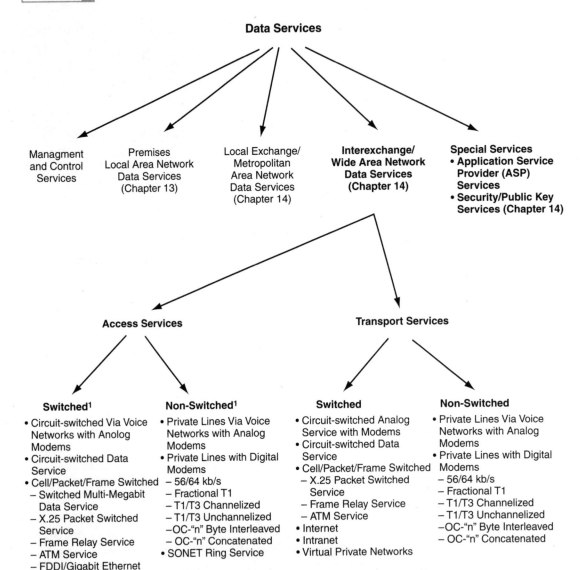

**Figure 14.7**   Taxonomy of interexchange/WAN data services.

tone and busy signals in the voice network were commonplace. What is disheartening about this situation is the fact that Internet-access line use is typically very low. In spite of the fact that little or no packet traffic is exchanged for most of the time that a user is connected, voice-network switching and transmission capacity remains unavailable for other subscribers for the entire duration of Internet access calls.

As noted, DSL-based Internet access techniques need make no demands on voice networks at all. Cable TV modem approaches that do not require analog modem voice-network connections for upstream traffic (some do), likewise place no burden on voice networks. Choosing any of the other user premises-to-LEC access methods listed in Figures 14.1 and 14.7 not only avoids inefficient voice-network use but results in dramatic Internet-user performance enhancements.

While it is true that residential and small-business users account for the lion's share of the more than 150 million devices now accessing the Internet (most of which can benefit from DSL and other "single-user" Internet access technologies), large corporations often must provide Internet access to hundreds or even thousands of employees. Company LANs with thousands of attached stations are not uncommon, and increasingly there is a need to provide all stations with at least occasional Internet connectivity.

Organizations with such large-scale Internet access requirements must obviously select high bit-rate access facilities such as those supporting T1, T3, and even greater capacity OC-n access services. Precisely what service is needed is determined using traffic engineering-based network design methods discussed in Chapter 11. Chapter 15 provides a guide to selecting both access and transport data services.

Although the number of ISPs offering customers dedicated T1 access is growing, only large ISPs or Network Service Providers (NSPs) that are also ISPs can handle some large traffic-volume customers.

Many companies now employ server-based applications to conduct e-business with customers via the Internet. Some of these installations, like that of Amazon's book and other resale operations, must be sized to handle hundreds of thousands of transactions per day. So another high-capacity access requirement that has emerged is the need to connect *superservers* to the Internet. The last section of this chapter treats application service providers (ASPs), a new business arena set up specifically to provide direct-to-fiber backbone superserver-to-Internet connections.

Referring to Figure 14.7, all LEC/MAN transport service alternatives (used to provision IXC/WAN access service) except circuit-switched data service, X.25, frame relay, ATM, and IP services have already been discussed. Since these LEC/MAN transport services are nearly identical with IXC/WAN counterparts, they are treated below.

# Interexchange/Wide Area Network Data-Transport Services

### CIRCUIT-SWITCHED DATA-TRANSPORT SERVICES

Although today's switched voice services are often perceived as analog in nature, the underlying switching and transmission facilities are nearly all digital. In most cases, the only segments of end-user-to-end-user connection paths where analog signals exist are in local sub-scriber loops. Thus, there is a natural evolutionary progression from circuit-switched voice services to what are most often referred to as *circuit-switched data services* (CSDS).

Like circuit-switched voice networks, CSDS networks employ data traffic-independent signaling to establish and disconnect call connections. Once the connection is made, intervening switch and transmission systems remain transparent to carried data traffic, and only physical or Layer 1 protocols apply at network interfaces. CSDS networks are designed to use, unmodified, the same asynchronous digital hierarchy (ADH—see Chapter 5) multiplexers that voice networks use and, as a consequence, generally offer service in 64-kb/s transmission-rate increments.

For higher than 64-kb/s rates, *unchannelized* transport services at specified rates are typically offered. Network providers offer such services by using adjacent channels on underlying ADH multiplexer infrastructures. If LECs offer user access services with these capabilities (e.g., ISDN Basic Rate Interface or Primary Rate Interface access), then multiple-64-kb/s rate service can be provided on a call-by-call basis between end-users.

If ISDN access is not available, business users can employ less-flexible means of accessing CSDS. A standard T1 access line can be used if the appropriate number of adjacent channels are permanently assigned to the CSDS service and an extra 64-kb/s channel is reserved for signaling. In this configuration, the access channels used for CSDS cannot be used for regular voice calls, a capability that the ISDN Primary Rate Interface provides. Another method of implementing high-rate CSDS without ISDN access is to use multiple subscriber loops and 64-kb/s transport service in conjunction with customer premises-located *inverse multiplexers* (IMUXs). The IMUX operating principle is as follows. If, for example, a 384-kb/s channel is needed to support video conferencing, the 384-kb/s signal is fed to an IMUX that *inverse-multiplexes* it into six 64-kb/s signals. At the customer premises, six separate 64-kb/s CSDS calls are established to the destination customer premises. At the destination, the six separate 64-kb/s signals are *re-multiplexed* into the original 384-kb/s format by a companion IMUX. IMUX operations restore and compensate for timing or synchronization anomalies among individual calls.

While standalone switched data services exist, the most common service delivery method is via virtual private network service options. For instance, AT&T provides, as part of its Software Defined Network service, a Software Defined Data Network (SDDN) option. Switched data rates of 64 kb/s, 128 kb/s, and 384 kb/s are supported. Far from being random or arbitrarily selected, these rates correspond to the coder-decoder (codec) output rates that are used in popular video conferencing systems. In 1990, as part of its video conferencing H-series, the ITU-TSS established $P \times 64$ standards, where p is an integer.

As it turns out, $P \times 64$ standards and CSDS service are ideally suited for video conferencing, which is by far CSDS's most common application. The reason is that circuit switching and ADH multiplexing are specifically designed to deliver high QoS (that is, constant and low-latency performance) for voice signal transport, an attribute of equal importance to video signal transport. $P \times 2$ or 128-kb/s access rates are supported by ISDN Basic Rate services. Higher access rates require ISDN Primary Rate services or IMUX operation. Since data rates can be changed on a call-by-call basis, PRI provides true *bandwidth on demand* functionality, although at maximum rates far below those envisioned for ATM and other packet-switched bandwidth-on-demand technologies.

Figure 14.8 illustrates LEC and IXC facilities arranged to deliver CSDS services. PRI access to IXC CSDS-based switched transport services is generally provisioned by dedicated access T-1 service groomed for PRI service by LEC DACS equipment. BRI access is provisioned through BRI-equipped local subscriber loops and LEC CSDS-based switched transport service.

## X.25 PACKET SWITCHED DATA-TRANSPORT SERVICES

The X.25 protocols developed in 1976 by the then CCITT provided the first public data network standards. X.25 networks proliferated and dominated the data transport scene in most parts of the world. In the U.S. however, most corporations continued to choose the already entrenched proprietary IBM Systems Network Architecture (SNA)

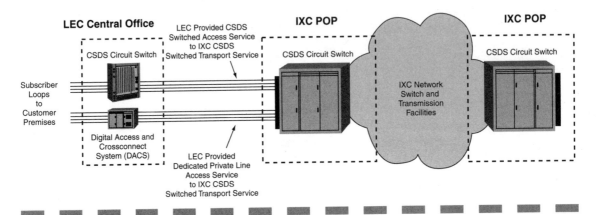

**Figure 14.8**   Circuit-switched data services.

protocols to power private data networks. Not only could IBM offer over two million lines of proven, nearly defect-free software code (representing some $2 billion in development costs), but the code itself was designed for compatibility with IBM computers, which commanded the lion's share of the mainframe market.

In addition, U.S. military, research, and education institutions were hard at work developing TCP/IP protocols for use in their packet-switched networks. Nevertheless, nationwide X.25 packet switched WAN services were offered by AT&T's ACCUNET Packet Service, British Telecom's Tymnet, CompuServe Inc.'s CompuServe Network Services, General Electric's GE Information Network Services, and Sprint International's Sprintnet (formerly Telenet).

As new transmission system bit-rates increased and bit error rates plummeted (relative to the earlier transmission systems for which X.25 protocols are optimized), X.25 performance limitations increasingly became a major issue. Although X.25 networks generally operate in virtual circuit modes, assuring delivery of packets in the proper order, X.25 has no flow control mechanism and there is no means to guarantee throughput. In addition, the processing requirements to implement the full complement of X.25 protocols at every switch node limited throughput to 56-kb/s rates.

Figure 14.9 depicts major facility components underlying WAN X.25 service. Access to WAN X.25 services is usually via dedicated circuits. Customers so connected are said to be *on-net* and attached CPE may include router or other LAN gateways or mainframe computer front-end processor equipment, but at a minimum include DTEs with packet assembler/disassembler (PAD) and other X.25 interface provisions. *Off-net* customers may gain access to X.25 networks via dial-in public-switched voice networks and analog modems or via CSDS circuit-switched data services.

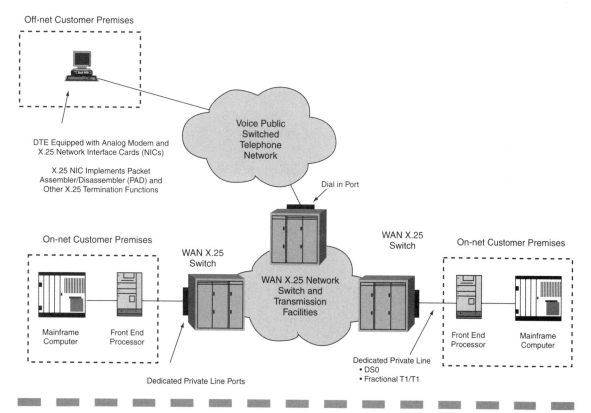

**Figure 14.9** X.25 packet-switched services.

In areas where LECs offer X.25 metropolitan area service, MAN X.25 services can be used to access WAN X.25 services, but such internetworking arrangements are rare. Internetworking among WAN X.25 networks is more common. In the mid-1980s, prior to the general availability of the Internet, where throughput was not an overriding requirement, X.25 service became the dominant choice among users requiring universal, public connectivity, largely due to the fact that X.25 networks offered large numbers of network nodes and standard interfaces.

Today, in the U.S., X.25 service is largely consigned to legacy installations being supplanted in private corporate networks by frame relay (where higher guaranteed throughputs are required), and by Internet services where universal connectivity is a necessity. While X.25 services remain in greater use in Europe and other foreign markets, ultimately the Internet is destined to dominate public data network service delivery.

### FRAME RELAY DATA-TRANSPORT SERVICES

Frame relay is a high-speed switching technology that achieves 10 times the packet throughput of existing X.25 networks by eliminating two-thirds of the X.25 protocol complexity and adding out-of-band signaling. The details of the frame relay protocol are presented in Chapter 7. Frame relay service is based on transparent delivery of frames defined at the data-link layer (Layer 2) of the OSI reference model. It provides either switched or permanent virtual circuit connection-oriented service. Although *switched virtual circuit* (SVC) operation was standardized by the Frame Relay Forum in 1997, most carrier offerings remain limited to *permanent virtual circuit* (PVC) service. PVC addresses are loaded into frame relay switches at service initiation. From then on, unless the service is modified, only loaded PVC addresses can be interconnected. As a consequence, PVC-based frame relay networks are normally restricted to closed or virtual private network applications, and as discussed below, are chiefly marketed as alternatives to dedicated private-line services.

In contrast, SVC-based service is endowed with the ability to establish virtual circuits among any pair of nodes on a corporate frame relay network at the time at which connections are established.

For transporting today's bursty LAN traffic (i.e., traffic that has high data-rate requirements during peak times followed by longer periods of inactivity), frame relay has many advantages over competing services. By operating at Layer 2—as illustrated in Figure 7.14—it can encapsulate a variety of network-layer protocols. Its low processing overhead supports high throughput rates and its connection-oriented flow control processes make it possible for service providers to guarantee throughput rates. Finally, because switches only respond when there is input traffic to route, frame relay is inherently more efficient than dedicated circuit networks in handling bursty traffic.

Figure 14.10 compares frame relay (Part (a) in the figure) with dedicated private-line service (Part (b) in the figure) in a four-node corporate data communications network. Part (b) depicts the *dedicated private lines* needed to establish connections between all pairs in a four-node network. In this example, three 512-kb/s access service chan-

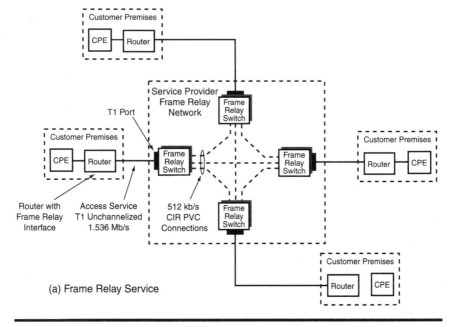

**Figure 14.10**
Frame relay service and dedicated transmission service compared.

(a) Frame Relay Service

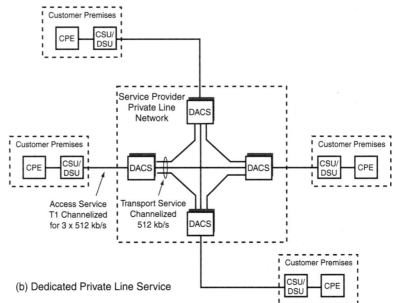

(b) Dedicated Private Line Service

nels are provisioned using a single T1 facility between customer and service provider premises. Each 512-kb/s access channel is permanently connected to one of the six 512-kb/s interswitch transport channels. Because all access and transport service is *channelized,* bit rates between any two points on the network can never exceed 512 kb/s. As a consequence, dedicated networks must be engineered for *peak* throughput requirements, which in bursty applications can be many times average throughput requirements. You buy much more capacity than you need—on average!

As shown in Part (a) of the figure, access to *frame relay networks* is usually provisioned using dedicated unchannelized T1 access facilities between customer premises and network operator service delivery "ports." Customers must have routers equipped with frame relay interface cards or some other termination equipment that maps LAN and other CPE signals to the frame relay formats described in Chapter 7. Port speeds are matched to access-service rates which in most cases are limited to T1 rates, with some providers offering fractional T3 or full T3 capabilities. Transport service between any two ports on a frame relay network is arranged at the time of service initiation by ordering permanent virtual circuits (PVCs) and specifying connection service bit rates.

As implied in Figure 14.10(a), multiple PVCs can be defined at any given port. Frames arriving at a port via access channels are routed to appropriate remote PVC-related ports using data-link control identifier (DLCI) frame header address information, as described in Chapter 7. Since frame traffic is generated only when users have information to transmit, not all PVCs emanating from a port are likely to be active simultaneously. Because of this fact, service providers permit the sum of the individual transmission rates of all PVCs assigned to a port to exceed that port's maximum throughput rate. This is known as "oversubscription" and to the extent that good engineering practice in specific applications allows, it can result in substantial performance and cost benefits compared with dedicated private-line alternatives.

Studying the network configuration illustrated in Part (a) helps one to understand the factors behind frame relay's flexibility and efficiency. To begin with, PVCs are fundamental components of frame relay service-provider tariffs. Hence, total service charges are largely determined by the quantity and characteristics of PVCs that customers order. PVCs are characterized by *Committed Information Rate* (CIR), *Committed Burst Size* ($B_c$) and *Committed Rate Measurement Interval* ($T_c$) traffic parameters, often referred to as service descriptors when used to specify particular frame relay services. CIRs represent average data transmission rates that users commit to pay for—whether or not actual average rates are less than or equal to agreed-to CIRs. CIRs may be anywhere between 0 percent and 100 percent of access-line rates. Conversely, service providers guarantee to deliver $B_c$

bits over $T_c$ intervals of time. Lastly, $T_c$ is computed as $B_c$ divided by CIR.

Under these agreements, users may transmit above subscribed CIR rates (up to port speeds if no other PVCs are vying for capacity), a process known as *bursting*. If the burst is short enough so as not to exceed $B_c$ bits over a $T_c$ time interval, no action is taken by the network. Otherwise, once $B_c$ thresholds are exceeded, future frames are marked as *Discard Eligible* (DE). These frames are carried on a *best-effort* basis only, which means that during periods of network congestion DE frames are discarded first.

When frame relay services were first introduced and network traffic low, to minimize bills, customers specified understated CIRs, anticipating that not much data would be discarded under DE rules. As the service became more popular, customers opted for more realistic CIRs.

Clearly, these features reduce costs and enhance burst-traffic handling flexibility. It should also be clear that the burst size parameter is an important indication of frame relay's ability to improve burst traffic-handling performance. Unfortunately, this parameter is rarely publicized and may not be specifiable on a per-subscriber basis. Users are urged to acquire details of the frame relay service provider's flow-control operations before selecting a service.

Frame relay is the most pervasive data network service supporting today's corporate networks. Currently it is the only data service being included in multi-service special contracts between carriers and customers. Until SVC-based services emerge, because frame relay is connection oriented, supporting only permanent virtual circuits, its primary application will remain in single-customer private networks. Internetworking among various corporate entities is largely accomplished with IP services in the U.S. and IP and X.25 services elsewhere.

### ASYNCHRONOUS TRANSFER MODE DATA-TRANSPORT SERVICES

What frame relay is to data users, Asynchronous Transfer Mode (ATM) is, and more, to multimedia users. As explained in Chapter 7, ATM-based telecommunications systems support multiple classes of service simultaneously over common transmission and switching resources. ATM was also designed from scratch to support high throughput rates, with connection rates starting at T3 (45 Mb/s) and going all the way to OC-48 (2.4 Gb/s). These impressive characteristics make it clear that from the outset ATM was designed to be a core-network carrier-class technology, resulting in uniform network infrastructures to efficiently and economically accommodate all types of traffic.

While ATM's original objectives remain valid, widespread deployment in core networks has not yet occurred and end-user applications themselves have become more bandwidth intensive. Desktop computers and corporate servers are doubling in speed every 18 months and local area networks within business-customer premises typically operate at 100 Mb/s with speeds of 1 Gb/s just around the corner (see Chapter 13 for details). New applications include ultra high-speed data transfer, streaming isochronous compressed video and audio signals, high-definition TV, and CD-quality entertainment program material, high-resolution graphics, and numerous others.

Both MAN and WAN service providers now offer ATM-based services, with components similar to frame relay's services (e.g., ports, PVCs, SVCs with various attributes). From a service-delivery perspective, ATM's main advantages over frame relay offerings are higher throughput rates and a wider range of service classes. However, in an attempt to widen the appeal of both technologies, frame relay operators are beginning to offer higher-rate ports and PVCs (in the fractional and full T3 range), while some ATM operators now offer T1 ports and PVCs extending down to fractional T1 rates.

Because of the relatively high cost of ATM equipment, frame relay is normally the best choice for non-real-time data communications applications, and ATM the choice when high QoS transport performance is crucial. Recall from Chapter 7's discussions that the short 53-byte ATM cell size was chosen to endow ATM networks with small and constant latency and delay characteristics—essential for satisfactory video and voice service delivery—while allowing them to remain viable for both bursty transaction and large file-transfer data communications.

The remainder of this discussion is devoted to describing extant ATM service class offerings, traffic parameters, or service descriptors associated with service classes, and other components used to define specific service offerings and determine service-provider charges.

Figure 14.11 depicts a representative customer premises-to-ATM service provider point-of-presence access service arrangement. Customers normally use high-speed routers equipped with ATM interface cards or small ATM edge switches to perform *ATM adaptation layer* (AAL) and other ATM termination equipment functions. ATM service provider (ASP) switch ports are provisioned at speeds of 1.536, 44.210, 149.210, and 599.040 Mb/s. Currently, PVCs are offered that support four of the five classes of service defined in Chapter 7, namely:

- Class 1—Constant Bit Rate (CBR)
- Class 2—Variable Bit Rate—Real Time (VBR-RT)
- Class 3—Variable Bit Rate—Non-Real Time (VBR-NRT)
- Class 4—Unspecified Bit Rate (UBR)

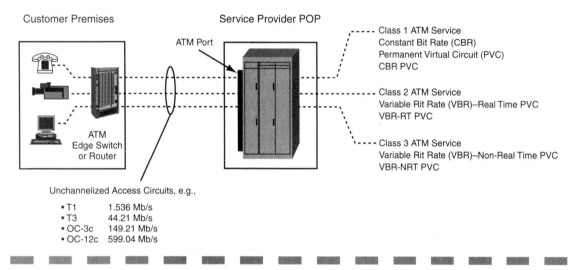

**Figure 14.11**   Asynchronous transfer mode (ATM) service.

As the figure implies, Class 1 CBR access service is appropriate for constant bit rate isochronous digital voice (standard voice network digital formats) and video signals that require low and constant latency and preserve application sensitive timing.

Class 2 service supports variable bit-rate and timing-sensitive applications. While uncompressed and most compressed video codecs generate constant bit-rate outputs and are therefore candidates for CBR service, more recent variable bit-rate compressed video codecs tend to generate bursty output signals. These codecs reduce average output rates at no sacrifice or loss in quality and are ideally suited to *packet video* transport, but remain sensitive to latency and timing (roughly 150 msec end-to-end delay with no more than 30 msec variance). One of ATM's advantages is that complex buffering and rate smoothing equipment that would be needed to transport variable-rate codec signals over synchronous time division multiplexer-based services, is eliminated. This is an excellent example of how ATM was designed from the ground up to accommodate multiple application classes, as opposed to synchronous TDM-based approaches that forced applications to fit networks.

Variable bit-rate non-real-time Class 3 service is appropriate for the vast majority of data communications applications for which no precise delivery time or synchronized signal timing requirements exists. Class 4 unspecified bit-rate service is a connectionless best-effort service with no guarantees of either throughput rate or cell loss performance. Class 4 service is similar to IP-based services. Because of that, it is used in IP traffic over ATM applications. A fifth class of service, available bit-rate (ABR) service, although defined in ATM specifications, is yet to be offered by service providers.

Note in Figure 14.11 that mixed classes of traffic can be accommodated over a single ATM access service. The figure lists the range of commonly employed of access facilities. What follows is an examination of traffic parameters (service descriptors) that may be used in conjunction with service classes to describe specific service offerings and related tariff charges.

ATM service specifications define a set of *traffic parameters* that relate to the characteristics of information being transported by PVCs, and also specify service levels guaranteed by PVC flow control mechanisms. These traffic parameters are:

- **Peak Cell Rate (PCR)**—The PCR, in cells per second, is the maximum source traffic rates that can be transported over PVCs. This parameter applies to both CBR and VBR classes (it is the only parameter applying to the CBR class).

- **Sustainable Cell Rate (SCR)**—SCRs, in cells per second, are guaranteed average cell rates of PVCs over time. It is roughly equivalent to the Committed Information Rate defined for frame relay PVCs. This parameter applies only to the VBR classes.

- **Maximum Burst Size (MBS)**—MBS is the maximum number of consecutive cells that may be transmitted at the peak cell rate. It is roughly equivalent to the burst size defined for frame relay PVCs. This parameter applies only to the VBR classes.

When procuring PVCs, the above traffic parameters are specified by customers. The relationship between PVC traffic parameters and *equivalent bandwidth* numerics, used by some service providers to describe service offerings, depends on service classes. Rounded-up equivalent bandwidth and other traffic parameters then become the basis for ordering services and *traffic contracts* between users and providers. As long as user traffic *conforms* to agreed-upon traffic parameters, network operators guarantee to transport customer data cells over specified PVCs. When actual traffic exceeds parameter bounds, priority levels in all subsequent cell headers are lowered. From that point on, until actual traffic once again conforms to parameter bounds, lower priority-marked cells are more likely to be discarded during periods of network congestion. The following paragraphs examine relationships between traffic parameters and equivalent bandwidth.

PVCs for non-UBR service classes are ordered and priced per Mb/s of *equivalent bandwidth* (or per 64 kb/s of equivalent bandwidth for PVCs less than or equal to 1.536 Mb/s). Equivalent bandwidth is a capacity utilization measure related to the above traffic parameters. For the CBR class, the only parameter specified is the peak cell rate. In this case, equivalent bandwidth is obtained by multiplying the PCR by 0.000424. The 0.000424 constant comes from the fact that there are 53 bytes in a cell and 8 bits in a byte. ($53 \times 8 = 424$ and if 424 is multiplied by the peak cell rate and divided by 1 million, then the PCR is transformed into a quantity expressed in Mb/s.)

If the bandwidth is greater than 1.536 Mb/s, it is rounded up to the next integer number of 1-Mb/s increments. If the bandwidth is equal to or less than 1.536 Mb/s, the bandwidth is divided by 0.064 and the resulting number rounded up to the next integer number of 64 kb/s increments.

For the VBR classes, all three traffic parameters are specified. The maximum burst size for the VBR-RT class is 32 cells and for the VBR-NRT class is 100 cells. For the VBR-RT class, the PCR is usually fixed at twice the SCR. In this case, the equivalent bandwidth depends on the assumed percentage of time the user spends at the SCR, the PCR, and rates in between.

According to one ATM service provider (BellSouth), the SCR is multiplied by 0.000512 to obtain the equivalent bandwidth. The number of increments of bandwidth are determined by rounding as described above. For the VBR-NRT class, the PCR is usually fixed at four times the SCR. In this case, according to Bell South specifications, equivalent bandwidth is obtained by multiplying the SCR by 0.000804. In some cases, customers may specify PCRs and SCRs separately for VBR-NRT service. The BellSouth formula for equivalent bandwidth is then:

$$BW = PCR \times 0.000133 + SCR \times 0.000286$$

Unlike the 0.000424 constant, there is no obvious basis for these latter constants. They are the result of BellSouth's analysis of the relative amount of time customers spend bursting above their subscribed SCR. This is BellSouth's way of translating customer-ordered services into charges for those services. Other service providers substitute the use of extensive tables to translate a wide range of PCR and SCR parameters, in many combinations, to related charges.

It should be obvious that the characterization of user applications, the selection of the appropriate ATM service classes, and the engineering of PVC sizes is a complex process. To create efficient networks that meet all service goals, ATM service design requires intense customer and service-provider cooperation. The end result, however, can be the holy grail of unified network architecture and infrastructure that can be managed as a single entity, satisfying the gamut of user requirements.

We hasten to add, however, that since only PVC-based ATM services are currently being offered by service providers, the holy grail depiction applies only to private networks. Since no switched virtual circuit (SVC)-based services are yet offered, achievement of the original ATM design objective, to develop a unified network architecture and infrastructure that can underlie all public and private network services, is not yet possible. As a consequence, today's ATM offerings, like those of frame relay-based networks, are essentially alternatives to service provider dedicated private-line offerings.

To reiterate, compared to frame relay, the principal ATM performance advantages are the ability to handle a larger range of service classes and higher operating rates. Since underlying ATM facilities are more expensive than frame relay facilities, ATM's principal disadvantage is that it is generally a more costly service option.

## IP-BASED DATA-TRANSPORT SERVICES (FROM THE INTERNET TO EXTRANETS)

The subject of IP-based network services covers a wide range of topics and environments. On one level, IP is just another protocol on which customers can build private networks or carriers can build *closed* networks rendering dedicated services to single customers. These implementations certainly exist, but their influence on the world of networking pales when compared to Internet services. Internet's history and evolution are covered in Chapter 2. Its success stems from the very precepts that created it: it was designed to be a network-of-networks, a loose federation of locally controlled autonomous networks. The whole Internet, relying on standards now implemented in virtually every computer in the world, is pervasive and robust, changing itself to adapt to generations of application demands. This chapter concentrates on services offered to users by ISPs in their many forms.

Most single-user access to the Internet uses dial-up services and analog modems. By contrast, large businesses access ISPs with dedicated circuits at T1 and greater speeds. Corporations deploying servers for information dissemination or electronic commerce require very high-speed connections, starting at T3 (45 Mb/s) and working up to OC-12 (620 Mb/s) and beyond. Finally, the very largest content providers collocate servers at the very heart of the Internet backbone, within the data fortresses of *Application Service Providers* (ASPs), new enterprises that owe their existence to phenomenal Internet growth. ASPs are revisited in the last section of this chapter.

Beyond IP protocols and standards that make possible seamless interconnection of networks around the globe, the popularity of the Internet is in large measure traceable to two additional factors. First, since this is a public utility with virtually no access restrictions anywhere in the world, the number of potential users is huge. Nearly every computer, whether owned by an individual or an enterprise, is a candidate for Internet connection. The second factor is the Internet's dominant application, one that allows people and organizations wanting to share information to interact with other people and organizations seeking information. Open access policies and easy-to-use browsers and search engines have made interactions among these two classes of users today's most popular Internet application.

To this point in time, IP-based networks have been implemented according to three different networking/application paradigms. The Internet's paradigm is consistent with the above operational capabilities, and is therefore characterized as follows:

- Networks designed to connect very large numbers of small client computers with relatively smaller numbers of superserver sites.

- Asymmetrical upstream- and downstream-traffic throughput rates.

- Networks employing IP protocols and standards (browsers, file transfer, directory naming, network management, and numerous other standards and practices).

- Unrestricted worldwide public-network access.

By comparison, *intranets* are optimized to satisfy corporate networking requirements. Although definitional variations exist, intranets are commonly understood to be IP-based private networks employing Internet technologies and standards and supporting user interfaces and capabilities nearly identical to those of the Internet. Intranets are often used to give employees easy access to company information. Intranet networking/application paradigms or models exhibit the following characteristics:

- Networks designed to support server-to-server connections at typically higher speeds than client-to-server connections.
- Symmetrical traffic throughput rates.
- Network connectivity limited exclusively to specified business sites.
- Networks employing IP protocols and standards (browsers, file transfer, directory naming, network management, and numerous other standards and practices).
- Access limited to employees, customers, or suppliers.

Intranets may be implemented using either facilities-based or virtual private network designs, options similar to the facilities-based and virtual or software-defined voice network alternatives described in Chapters 10 and 12.

Software-defined implementations usually use the ever-present Internet as their underlying transport infrastructure. In these cases, the networks are referred to as Virtual Private Networks (VPNs). Note that this term has already been used in the industry to refer to software-defined voice and non-IP based data networks. Readers must use context to understand how the term is being used in specific instances. However, today, VPN most often refers to data network implementations being discussed here.

The third IP-based networking/application paradigm in use today is employed in what are now called extranets. *Extranets* are intranets to which access is provided via the Internet. Extranets capitalize on the Internet's global-in-extent, public services to provide remote access to corporate intranets for employees (away from their offices), and customers—wherever in the world they may be located. Since employees are often granted privileges to access and manipulate data not granted to customers, employee access normally involves some form of *password* and *firewall* security protection methods.

Password and firewall security protection *tunneling* methods are described in more detail below, but for now we want to emphasize the benefits of using established Internet communications privacy and security technologies to gain access to intranets by examining what is required if they are not used. If employees at home or on travel require off-net access to company intranets, one solution that works worldwide is to use ubiquitous public-switched voice-network services. Under this approach, intranet nodes are equipped with banks of modems to establish connections with employee transportable PCs. If traveling employees require access to sensitive company data, then again under this approach, companies must implement their own password, firewall, and other communications security provisions.

By contrast, Internet-based extranet access to intranets, using established Internet VPN technology security safeguards, eliminates complexities involved with multichannel dial-in voice-network connections and expensive independently owned and operated password and encrypted authentication provisions.

As useful and popular as IP-based networks are, it is important to understand current IP-based transport service limitations. Unlike frame relay and ATM-based transport services that employ virtual circuits and flow-control mechanisms that allow service providers to "guarantee" throughput rates and quality of service performance, today's IP-based networks lack those facilities. With that said, two qualifying statements should be added. First, the enormous growth and volume of traffic successfully transported via the Internet today leads one to the irrefutable conclusion that there are vast quantities of traffic that don't require flow control, throughput and delivery guarantees, low and constant latency, real-time and timed response, etc. And even when today's Internet is used to carry streaming voice and video signals, the results may not live up to high-fidelity, high-resolution, fast-motion tracking expectations of program-quality purists, but nevertheless such applications remain highly useful and in great demand.

The second point is that vigorous efforts are now under way to develop the means to endow IP-based transport with improved voice, video, and other QoS capabilities. These efforts include the *multiprotocol label switching* (MPLS), *differentiated services* (Diffserv), *tag switching*, *bandwidth reservation*, and *digital wrapper* techniques described in Chapters 7 and 8.

It may have occurred to readers that, although the most modern and high-speed core network facilities in the entire telecommunications realm support Internet transport services, these services are not offered directly to Internet service subscribers. Internet subscribers cannot specify transport service characteristics or performance; neither are they directly charged for such services. ISPs use a portion of the fees they collect from subscribers to pay for Internet transport services.

The next section continues the discussion of how large corporations gain access to and use IP-based services and the key aspects of virtual private network and IP security (IPSec) protocols.

# Special Services

## Application Service Provider (ASP) Services

Corporations providing content on the Internet do so using host computers known as servers. These computers run the application software that provides all the functions of a Web site. These functions can be split into three functional tiers: providing graphical interfaces to the client-computer Web browsers; running site business-function application software; and storing and retrieving business-function data. At high-volume sites, these functions are typically distributed among many server computers at each layer. Front-end load balancers direct customers to least-busy servers. This multiple server architecture makes Web sites scalable. As demand increases, more servers are added, capacity increases linearly and complicated transitions to larger computers are avoided. It is not unusual for *server farms* to contain as many as 50 multiprocessor servers and hundreds of gigabytes of database storage.

Maintaining large installations reliably is a challenging undertaking for business organizations. Providing space, security, power, backup power, and high-speed access to the Internet is both expensive and complex. *Applications service providers* (ASPs) are providing an increasingly popular solution to these problems. ASPs construct large buildings expressly fitted to house and supply HVAC, power, and other environmental support for client Web servers.

Racks may be located in secure or nonsecure compartments equipped with preinstalled structured cabling to distribution frames used to make connections to local Internet backbone transport service points of termination. Battery backup power is provided to prevent transient fluctuations in utility-provided power from disrupting operations and motor-alternator sets for long-term power outages.

ASP sites are located directly on high-capacity Internet backbone junction points of major National Service Providers (NSPs). As a consequence, ASP clients are less susceptible to Internet access congestion than they would be if connected to smaller ISP sites. Clients are free to install company-owned server hardware and software on ASP premises or obtain those facilities from ASPs or third-party vendors. Figure 14.12 illustrates a distributed server architecture along with photographic views within one of AboveNet Communications ASP facilities. The main server room, the carrier-class Cisco 12000 routers

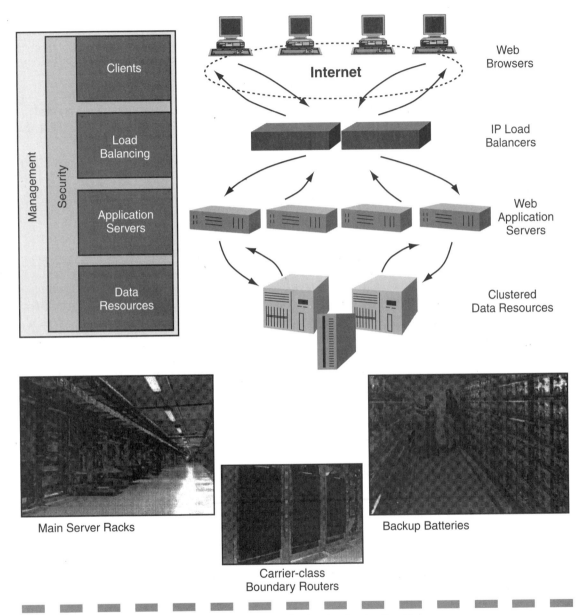

**Figure 14.12** Distributed server architecture at an Application Service Provider.

used to connect to the Internet, and the room containing backup batteries are shown.

The largest content providers take the distribution of functions across a multiple server paradigm within a single ASP facility one step further. As reliable as ASP facilities are designed to be, they may fail or become victims of fire, storms, floods, or other acts of God. Accordingly, some large content providers establish redundant facilities in ASP locations across the country and even around the world. Distributed-server facility architectures result in improved operational performance as well as higher reliability and availability. Distributed or "mirrored" database designs do, however, impose strict concurrency requirements. Synchronized and near-instantaneous updates of all distributed databases are required when the content in any one of them is changed.

## Virtual Private Networks and Security/Public Key Encryption Services

Virtual private networks (VPNs) employ a combination of technologies that allows users to transmit traffic over the Internet with information privacy and security assurances equal to what can be expected from facilities-based private networks. Motivation to use the Internet as opposed to private facilities derives from the fact that, due to the Internet's size and pervasiveness, opting for the Internet alternative saves money and greatly enhances the number and geographic dispersion of connected end-point locations. Simply put, due to its size and extent, the Internet offers the same economy-of-scale efficiency and flexibility benefits that accrue to other large public voice and non-IP-based data networks. Since so many company customers and suppliers are already connected to the Internet, granting them access to corporate intranets is as easy as giving them the correct Internet universal resource location address (URL).

The downside of the Internet's reach is that a world of unwanted intruders has physical connectivity to the same VPN resources that authorized users employ. Without adequate communications security, most companies could not risk the havoc that hackers could wreak on their intranets, and facilities-based private networks would be the only option. The major security functions needed fall into three categories.

The first is a reliable method to identify and authenticate users seeking to gain intranet access. Clearly when authentication provisions fail, not only does an unauthorized party have access to a company's intranet, but he is erroneously viewed as "trusted." The second functional capability needed is one that protects sensitive information content from being revealed or compromised by intentional or unintentional eavesdroppers. Lastly, some means must be available to prevent malicious data tampering, and in particular undetected data manipulation.

The technology at the heart of modern authentication and information protection systems is a branch of mathematics known as *cryptography*. An in-depth treatment of cryptography and cryptographic systems is beyond the scope of this book. However, the basic operating principle powering reliable communications security systems is that digitally encoded transmitted signals can only be decoded with a secret "key," a key known only to message transmitting and authorized message receiving parties. Encoding a message using a secret key is a process called *encryption*. Ideally, encrypted messages can only be recovered by a decoding process that employs the identical secret key, a process known as *decryption*.

We are now ready to explore important aspects of all VPN implementations by tracing the steps involved in establishing secure connections through the Internet. The first step in the process is user *authentication*, a responsibility of the company operating the intranet to which a user seeks access. Historically, the simplest form of authentication is the familiar login ID and password approach. One drawback of this approach is that user passwords may be transmitted as clear text during logon and be stolen by intruders monitoring the connection. Safer systems require passwords plus a number from a "token" (a separate device providing numbers that change every minute). Host computers run user-unique algorithms that recreate the number and authenticate the combination of user ID, password, and token number. Eavesdroppers monitoring such logon transactions may decipher the token number, but because it is changed every minute, it cannot be used after that period.

Other approaches establish an encrypted link using a shared secret key tied to users' accounts. Once an encrypted link using the shared key is established, user equipment securely forwards a password to server equipment authenticating the user, and user and server equipment exchange a new key to be used only for the current session. All subsequent user-to-server data exchanged during that session is encrypted and decrypted using the newly agreed-upon key. Encrypted data streams become the payload for all packets exchanged between user and server equipment. Intervening Internet routers and other equipment are denied access to the plain text (unencrypted) content embedded in the packets. The end-user and server systems are said to have created a *tunnel* between them.

*Tunneling* is at the heart of all VPN implementations. There are currently three tunneling protocols used in most VPNs: *IP Security* (IPSec), *Layer 2 Tunneling Protocol* (L2TP), and the *Point-to-Point Tunneling Protocol* (PPTP). PPTP is the most widely used protocol to date due to Microsoft's support of the protocol within Windows. L2TP evolved from a Cisco-developed protocol. IPSec, a product of the Internet Engineering Task Force, is destined to become a major contender for use in VPNs.

Once a user has been authenticated and an encrypted tunnel established, the next step is to invoke *access control* provisions. These are designed to limit any particular user's access to only those resources for which he has been granted usage permission. Special "firewall" processors filter packets based on source address, destination address, and packet type, effectively partitioning networks into restricted segments. Access control can apply down to file levels in operating systems implementing user-file system security (e.g., UNIX and Windows NT).

Encryption is obviously an important part of all VPN facets. Secure and economical cryptographic key distribution has been a major communications security objective for years. A breakthrough that resolves the problem occurred with the introduction of *public-key cryptography*, a system using a pair of keys with the following property. Messages encoded with one key can only be decoded with the other key. One key is held in secret within a particular user's equipment (a *private key*); the other key, a *public key*, is shared and can be stored in any other user's equipment.

The way the technology works is as follows. If I know a user's public key, I can encrypt a plain-text message with it and send the encrypted data stream to him. The encrypted data stream can then only be decoded with that intended user's private key. If the user knows my public key, I can append text to my plain text message encrypted with my private key. By decoding this text with my public key, the user authenticates that the message in fact came from me. This process is known as a *digital signature.*

Public-key systems are very powerful, but require distribution of public keys and users must have a high level of confidence that public keys really belong to alleged owners. One option, though not very practical, is for users personally to distribute their public keys to everyone with whom they wish to communicate. Fortunately, public-key cryptography properties support a more efficient solution. A trusted entity, known as a *Certificate Authority* (CA), verifies user identities and issues private/public key pairs to users. Public keys are placed in files known as *digital certificates*. These files contain the digital signature of the certificate authority (i.e., material encoded with the CA's private key).

To establish secure communications with an application on a server or another user, communications-initiating users send CA digital certificates electronically to any application (or other user) that requires them. The authenticity of the certificate is determined by verifying the CA's digital signature. The public keys of well-known CAs are built into all major Internet client and server applications. Internally, corporations can act as their own CAs if they install their CA certificate in the appropriate application software. The entire process of issuing, using, and validating certificates and the entities involved in the process is known as a *Public Key Infrastructure* (PKI). The X.509 standards developed by the IETF are widely accepted as the basis for such infrastructures.

Figure 14.13 illustrates a typical VPN implemented by the ABC Corporation, a company engaged in electronic commerce. The figure emphasizes the use of VPNs as a means of providing access to ABC's e-commerce Web server from ABC business partners. It also reveals how VPNs can provide mobile and telecommuter employees access to any resource connected to ABC's intranet. *Firewall* gateways protects servers from unintentional modification or malicious tampering of ABC's information processing resources by unauthorized users.

As shown, secure remote access for telecommuters and mobile workers is accomplished using VPN tunnels through the Internet (tunnels 2 and 3 in the figure). Once authenticated, employees accessing the Web server via VPN tunnels have the same privileges as if they were directly connected to corporate intranets.

Given the global extent of the Internet, ABC's business partners can also be given secure privileges on the corporate network, regardless of where they are located. Selected individuals in partner companies use VPN technology to access special servers acting as a gateways to ABC's intranet. An encrypted tunnel is established through the Internet (tunnel 1 in the figure) and business-partner users are granted whatever special (or restricted) privileges are allocated to them. Note that the configuration shown is also a graphical example of an *extranet*.

The illustrated capabilities are not easily implemented with any other business networking technology thus far described. In particular, consolidated Internet VPN access to intranets has major advantages over traditional dial-in methods. As mentioned earlier, the costs associated with modem pools, company-owned and operated communications security provisions, and local and long-distance calls can all

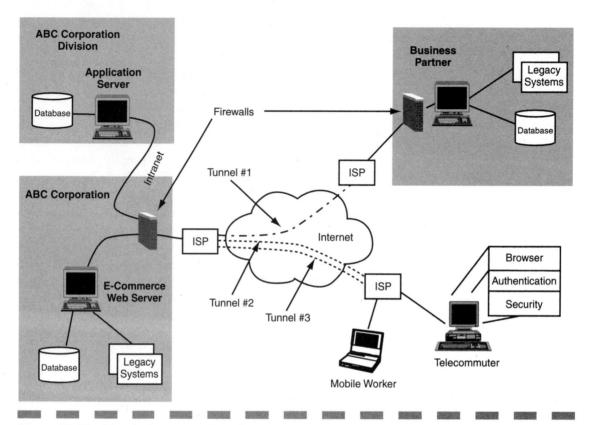

**Figure 14.13**   Virtual private network configuration.

be avoided. In addition, VPNs may be the only practical access method possible for off-site employees equipped only with DSL, cable modem, or wireless data access to the Internet. Small remote corporate sites can also take advantage of VPN service in situations where it is too expensive to extend corporate intranet connectivity to those sites.

VPNs and the establishment of a national and global PKI are crucial to unlocking the gates to the rich world of electronic commerce. Figure 14.14 illustrates the vast scale of the Internet economy, which currently rivals major standard industries in size. By any measure, the impact of the Internet and its IP-based services on commerce and society is undeniable.

**Figure 14.14**
Electronic commerce
statistics (1999).

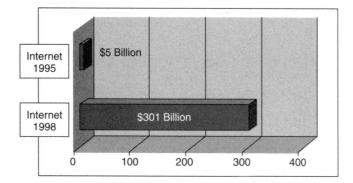

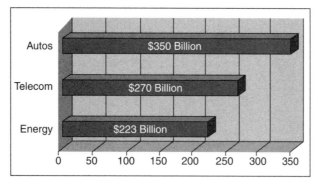

- Infrastructure (e.g, Modems, ISPs) — $114.9B
- Applications (e.g., Browsers, Search Engines) — $56.3B
- Intermediary (e.g., Portals, Brokerages) — $58.2B
- Commerce (e.g., e-tailers, Subscriptions, Sales, Services) — $101.9B

$301.3B

# Selecting Data
# Network Services

MAN and WAN data network design objectives are similar to those of voice networks—to provide specified service and quality levels using the least-expensive approaches. The principles of traffic aggregation to achieve economies of scale, explained in Chapter 11, also apply to data networks. Finally, as noted in Chapter 14, the evolution of public data networks toward digital, high-performance architectures has now made public network services an attractive alternative to private data networks. The migration from private to public data networks is now following the same path taken by voice systems in the late 1980s.

The main reasons for the emergence of public data networks are cost effectiveness, scalability, ease of management, and protection from technical obsolescence. As with voice networks, the economy of scale of a public network dwarfs that of any single organization's network. The large number of points of presence of a WAN carrier provides more efficient access to a network backbone than the relatively few backbone switches in a private corporate network. Finally, carriers are continually upgrading and refreshing network infrastructures with the latest hardware and software technology in order to reduce costs, increase performance, and compete with rivals in the marketplace. Individual using organizations do not have the capital assets available to make these kinds of investments in their private networks, nor can such investments be justified from a business viewpoint. Public data services allow the carriers to make the investments and users to reap the benefits.

The primary difference between selecting voice services and selecting data services is the nature of the traffic itself. Voice is a true commodity in the marketplace today; all carriers provide essentially the same quality and performance and services are mainly differentiated by price (and features). Data services, on the other hand, vary widely in characteristics and in the quality of service they provide. Various data networking technologies have evolved to support specific user requirements and traffic characteristics in the most cost-effective manner. Requirements such as connectivity, bandwidth, and quality of service are driving factors behind selection of a particular service, while price is often used as the discriminator between different providers of that same service. Figure 15.1 shows the characteristics of the major network services (identified and described in Chapter 14) in today's marketplace. *Ubiquitous connectivity* means the ability to communicate with terminals outside a user's organization, including, in some cases, the general public. *High bandwidth* refers to services supporting data rates above 1.536 Mb/s. *Performance guarantees* refer to throughput rate and QoS parameters and provide a minimum level of service on a point-to-point basis. *Multiple classes of service* support applications with varied performance requirements (QoS, constant bit rate, variable bit rate, etc.) on the same network simultaneously. Finally, some services are tolerant of *bursty traffic* and are economical solutions for carrying such traffic.

**Figure 15.1**
Characteristics of
data transport
services.

| | Ubiquitous Connectivity | High Bandwidth | Performance Guarantees | Multiple Classes of Service | Bursty Traffic |
|---|---|---|---|---|---|
| **Dedicated Circuits** | | X | X | | |
| **X.25 Service** | X | | | | X |
| **SMDS** | MAN Only | X | | | X |
| **Internet** | X | | | | X |
| **IP Intranet** | | X | | | X |
| **Frame Relay** | | | X | | X |
| **ATM** | | X | X | X | X |

The following sections describe LEC/MAN and IXC/WAN data services, their characteristics and rate structures, and make comparisons among services where appropriate.

# Local Exchange/Metropolitan Area Network Data Services

MAN data services are described in Chapter 14. This section focuses on the user characteristics that determine the right selection of access and transport services. In most cases, access and transport selections are independent; access types can be used with multiple transport services. Access-service selection is discussed below, followed by the discussion of transport services.

## MAN Data-Access Services

MAN data-access services are various forms of local loops from the customer's premises to the first point of switching at service provider premises. User requirement characteristics that affect access service selection are:

- Amount of connect time required per day
- Amount of data transmitted per session (transmission utilization)
- Burstiness of the data transmitted
- Bit rate required during transmission
- Upstream/downstream traffic-flow symmetry requirements

As an example, consider the Web browser employed by a typical residential user. The connect time is usually only a few hours per day and the amount of data transmitted per session is relatively low and bursty (i.e., there is considerable "think time" between periods of actual data transmission). The bit rate required is a function of user tolerance, but currently varies within the range from 20 kb/s to a few hundred kb/s. The traffic is highly asymmetrical; the user sends short queries and receives whole Web pages in return, most of which are laden with graphics and images that require a large number of bits to render. Contrast this with a video teleconferencing application which transmits a steady stream of data per session and requires high bit rates to be viable. Traffic requirements are also symmetrical in the latter case. In the sections below, the access types described in Chapter 14 are matched with appropriate applications. Figure 15.2 summarizes the matching of application characteristics to access technologies.

### SUBSCRIBER LOOPS WITH ANALOG MODEMS

As stated previously, analog modems allow the access types used for voice traffic also to be used for data applications. Since voice-transport services accessed with analog modems are inherently switched services, this access method supports traffic types suited for dial-up service. This implies that the connect time required per day is low and required maximum bit rates are 56 kb/s or less. Dial-up connections are full duplex, so symmetrical traffic is supported. These characteristics are normally associated with residential users or business users who are on the road and require occasional connectivity to their corporate offices. While telecommuters can also employ this access method, today's business applications require more bandwidth to operate efficiently and integrate the telecommuter into the office environment. Hence, some of the digital access methods discussed below are better choices for telecommuter applications.

Analog subscriber loop costs are discussed in Chapter 12; these same services can be used with analog modems at the same cost as they are for voice applications.

▄▄▄  ▄▄▄  ▄▄

**Figure 15.2**
*Access technologies matched to application characteristics.*

| | Connect Time Per Day | Transmission Utilization | Burstiness | Bit Rate | Traffic Symmetry |
|---|---|---|---|---|---|
| **Loop with Analog Modems** | Low | High | Low | Low | Symmetric |
| **Dedicated Loops with Digital Modems** | High | High | Low | High | Symmetric |
| **ISDN Services** | Low | High | Low | Medium | Symmetric |
| **Asymmetric DSL** | Medium | High | Low | High | Asymmetric |
| **Symmetric DSL** | Medium | High | Low | High | Symmetric |
| **Cable TV Modems** | Medium | Low | High | High | Asymmetric |
| **Wireless Loops** | High | High | Low | High | Either |

## SUBSCRIBER LOOPS WITH DIGITAL MODEMS

Digital subscriber loop services fall into three general categories. Traditional digital loops provide full-duplex, synchronous transmission over multiple twisted wire pairs or over fiber optic cable at a wide range of speeds from 56/64 kb/s to 600 Mb/s. Traditionally, these loops have been used to access private-line transport services. With the advent of LEC/MAN switched data transport services, digital loops also provide access to these services to match the demands of the business customer. These demands are characterized by traffic generated from hundreds or thousands of client and server computers connected to corporate LANs. Here, bandwidth requirements are usually high and traffic flow is symmetrical in nature. Traffic engineering techniques, discussed in Chapter 11, are used to determine appropriate access services. Digital loops are priced on a monthly recurring basis, with a nonrecurring charge at installation time. Prices vary by bandwidth and the term commitment that a customer is willing to make (i.e., the minimum length of time the customer commits to using the service). If a customer's central office does not offer the MAN switched service being accessed, customers must also procure loop extensions, which are priced by mileage to the closest central office that does provide the service. Figure 15.3 shows typical charges for digital loops and representative 20—50 mile loop extensions, taken from BellSouth's Broadband Exchange Line Service tariff.

**Figure 15.3**
Typical charges for digital local loops.

| | Nonrecurring Charge | Month to Month | 12 to 36 Months | 37 to 60 Months |
|---|---|---|---|---|
| **Broadband Exchange Line** | | | | |
| 56 kb/s | $450 | $70 | $61 | $51 |
| 128 kb/s (2B1Q) | $450 | $105 | $92 | $77 |
| 1.536 Mb/s | $465 | $155 | $146 | $136 |
| 44.210 Mb/s | $1,000 | $1,500 | $1,400 | $1,300 |
| 149.760 Mbp/ | $1,800 | $2,550 | $2,200 | $2,000 |
| 599.040 Mb/s | $3,600 | $5,100 | $4,335 | $3,900 |
| **Broadband Extension 20–50 Miles** | | | | |
| 56 kb/s | $75 | $30 | $23 | $15 |
| 1.536 Mb/s | $120 | $280 | $210 | $140 |
| 44.210 Mb/s | $350 | $2,600 | $2,480 | $2,350 |
| 149.760 Mb/s | $750 | $6,785 | $6,250 | $5,900 |
| 599.040 Mb/s | $1,500 | $14,890 | $13,725 | $12,950 |

ISDN-based and Digital Subscriber Line-based access services are finding application in the home-office worker and small-business community where higher performance than analog modems can provide is required, but the cost of dedicated digital loops is prohibitive. These digital loop alternatives are discussed next.

### ISDN-BASED DATA-ACCESS SERVICES
Chapters 8 and 14 give readers the technical background and rich historical saga behind ISDN. At one time, the ISDN Basic Rate Interface (BRI) was destined to become the standard access method for accessing "bandwidth-on-demand" transport services from residential or small-business locations. The long implementation cycle, the lack of aggres-

sive pricing in the United States, and the emergence of new technologies have blunted the penetration of ISDN, but the service remains a viable option to analog modem service where it is available. ISDN BRI is sold as a loop service coupled with switched transport service. For residential users, many LECs apply a per-minute usage charge for all local calls while others specify a threshold (typically 12,000 minutes per month) above which usage charges apply. The basic service charge is normally slightly higher than two analog lines (BRI provides two "B" channels, which may be used to make two calls simultaneously). However, the installation charges are steep and ISDN modems are considerably more expensive than analog modems (by a factor of five to ten times the cost). This is balanced by the fact that symmetrical bandwidth can be obtained at more than double the rate of analog modems, enough to support a viable teleconferencing application or a collaborative virtual meeting.

The ultimate usefulness of ISDN BRI for the telecommuter will be the support for ISDN dial-up service at the destination end of calls. Corporations can arrange to provide this capability through their dial-in modem pool equipment. To be used for higher speed Internet access, the Internet Service Provider chosen by the customer must support these types of connections and must allow the combining of the two "B" channels if full 128-kb/s throughput is to be realized. ISPs that cater to business customers typically offer such connections, but many "bargain basement" providers of low-cost Internet access do not provide this capability. Potential users are advised to investigate all operational attributes before purchasing ISDN (or any advanced access service).

For large-business users, the ISDN Primary Rate Interface access service is the best choice for the most flexible use of access bandwidth. This access service is *channelized* into twenty-three 64-kb/s "B" channels in contrast to other unchannelized 1.536 Mb/s digital modem access service described in Chapter 14. The difference is that channelized ISDN PRI is used to access circuit-switched transport services as opposed to packet/frame-switched transport services, which are better served by unchannelized access services. PRI's flexibility comes with the ability to combine several of the access channels into a single higher-speed channel on a per-call basis, an important capability if video calls are being mixed with voice calls over the same PRI facility. The cost of a PRI access circuit is greater than a standard T1 local loop (typically more than $500 per month), as is to be expected given the amount of processing involved in implementing advanced PRI call-setup features. Nevertheless, PRI is a very popular access service among large customers, especially those employing sophisticated PBXs on their premises.

### DIGITAL SUBSCRIBER LINE-BASED DATA-ACCESS SERVICES

Digital Subscriber Line (DSL) services have exploded on the access scene as something of a phenomenon in the past year. As described in Chapters 4 and 14, DSL provides high-bandwidth connectivity over existing two-wire facilities at very attractive prices. In most applications, DSL outperforms ISDN at lower cost. While current implementation rates are slower than expected (see Chapter 14), technical problems are being resolved and most local carriers, incumbent and competitive LECs alike, have aggressive DSL service-buildout plans.

DSL services come in many flavors, each with its own personality in terms of performance characteristics and cost. The high-bandwidth versions (e.g., HDSL, HDSL2, SDSL) are symmetrical in nature and are suited to business customers, where lower cost makes them an attractive alternative to traditional T1 service. The impetus behind DSL demand, however, is its application as a high-speed Internet access method for residential and small-business users. The asymmetric version of this technology (ADSL) is most suited for this application and can be delivered as a service at a price point competitive with a second telephone line. For Internet applications, ADSL provides higher performance than ISDN at a lower monthly cost.

The main factor in choosing DSL-based access service is its availability at customer premises. Serving central offices must be equipped with DSL Access Multiplexer (DSLAM) equipment (see Chapter 14) and the customer local loops must be on metallic facilities all the way to central offices and be within prescribed distances from those offices. If a customer premises is served by a remote terminal using fiber or other carrier facilities to reach the central office, the DSLAM equipment has to be placed in the remote terminal location. This configuration is rare today, but it will become more common as new fiber nodes are deployed.

Since most local loops are owned by incumbent LECs, it is no surprise that they provide the bulk of the DSL service in use today. However, competitive LECs are using Telecommunications Act of 1996 provisions to place their own DSLAMs in the incumbents' central offices and provide DSL service over local loops procured as *unbundled network elements*. Many CLECs do not provide voice service at all and are aggressively pursuing DSL markets as their only focus. The November 1999 FCC ruling on line sharing gives CLECs access to local loops where incumbent LECs are already providing voice service (the CLEC accesses only the upper frequencies on the same loop from which the incumbent is deriving voice revenue). This results in a dramatic decrease in the cost to CLECs to provide DSL service and will increase competition.

An ADSL service providing 640-kb/s downstream bandwidth is priced as low as $49 per month, with Internet access bundled in for an additional $10 per month. At the end of 1999, there were 500,000 DSL lines in service, with 77 percent provided by incumbent LECs, 22% by competitive LECs, and 1 percent by IXCs.

## CABLE TV MODEM-BASED DATA-ACCESS SERVICES

Cable modems currently represent the only high-speed alternative to telco-based access services. The nature of the cable TV transmission plant lends itself to high downstream bandwidth, but restricted upstream capability. This paradigm is sufficient for residential Internet users, which is the most significant market for cable modem service. At the end of 1999, cable modem service was available to 43 million homes, about 40 percent of all homes passed by cable. The service had 1.8 million subscribers, almost four times as many as DSL services in the same time period.

As described in Chapters 4 and 12, users on a cable modem segment share a 30-Mb/s downstream channel. While this is sufficient for individuals with bursty, Internet browsing applications, it does not scale well to more sophisticated uses such as video teleconferencing, LAN interconnection where many users share a single connection, or the connection of a server to a cable modem (which may overload low-speed upstream channels). With present architectures, then, cable modem service will remain primarily a residential service alternative.

Pricewise, cable modem service is very competitive with DSL, typically offering 1.5-Mb/s downstream bandwidth for less than $50 per month. Of course, cable modem bandwidth is shared and DSL bandwidth is dedicated, but at the present loading levels per neighborhood, cable system performance should be adequate. An additional consideration, when comparing this service to DSL service from an ISP, is the bandwidth of the ISP's upstream connection to the Internet. Because DSL provides dedicated bandwidth on local loops, unless ISPs significantly upgrade their connection bandwidth to the Internet backbone, these connections will become bottlenecks as the number of DSL subscribers grows (a situation very much like the cable modem environment). In both cases, rigorous traffic engineering capacity management actions must be taken by ISP service providers to ensure that quality of service objectives are being met.

Large investments are being made in the infrastructure to support cable modem services by non-LEC companies, who see this avenue as the only competitive alternative to local-loop access services. This will lead to expanded and totally new service offerings from this industry segment in the near future. Business users should remain current on changes as they occur and evaluate the applicability of these changes to their needs.

### WIRELESS DATA-ACCESS SERVICES

The next emerging alternative for high-speed data access will be fixed wireless services such as Local Multipoint Distribution System (LMDS) and Multipoint Multichannel Distribution Service (MMDS). Although both systems have application to downstream video applications, they are also being considered for two-way, high-speed data applications. The bandwidth available in each LMDS "cell" can support 75 downstream video channels as well as 1.5 Gb/s for data channels. These data channels can be configured at the service providers' option and may very well depend on the mix of business and residential users in the cell coverage area. In a business district, a smaller number of symmetrical channels may be appropriate (e.g., 650 full-duplex T1 channels if the total bandwidth is used for data) while in residential cells, a larger number of asymmetrical channels competitive with ADSL can be offered (e.g., 2,000 channels at 640 kb/s downstream and 64 kb/s upstream plus the 75 video channels). The interfaces deployed on the customer premises in LMDS prototype systems include Ethernet, ATM, and T1.

Projected price points for LMDS systems at this time look very competitive with current alternatives, but deployment of the technology is still years away. Market forces will ultimately determine the mix of services that providers will offer. Current systems are in the prototype stage and significant technical concerns remain, many involving the reliability of the propagation over a wide range of terrain, foliage, and weather conditions. Nevertheless, LMDS remains one of the more exciting prospects for high-bandwidth, multimedia residential and business service delivery.

# MAN Data-Transport Services

Like Chapter 14, this section focuses on transport services unique to metropolitan areas: SONET ring data-transport services and high-speed connectionless data services (i.e., SMDS, Gigabit Ethernet, and FDDI).

These services capitalize on the wide availability of fiber optic transmission close to customer premises and the need for very high-speed data communications between corporate entities located in a distributed campus setting within a single metropolitan area. SONET underlies most fiber optic implementations today, but is rarely offered as a service directly to end-users.

The main selection criteria for the MAN services noted above are the traditional tradeoffs between dedicated and shared resources: connectivity and utilization. Dedicated (non-switched) services reserve transmission resources whether the user is transmitting data or not and, hence, are priced accordingly. In addition, since switching is not involved on either a circuit or a packet basis, transmission capacity must be reserved individually and permanently between each pair of end-points ("permanently" in this context means until changed by a network management action as opposed to changed on a per-call basis). For this type of service to be cost effective, point-to-point traffic must be well known, stable, and large enough between any two points to result in high utilization of the dedicated bandwidth (greater than 50 percent is usually required and 70 percent is desirable). Standard tariffed private-line services offer point-to-point dedicated circuits at fixed monthly rates, but the SONET ring service discussed below provides a much higher level of flexibility and reliability for users with the traffic volume and reliability requirements to justify the high cost.

On the other side of the spectrum, connectionless data services are well suited for applications where all user locations require connectivity to all other locations in a bursty, unpredictable manner. These services are the metropolitan area equivalent of the local area networks discussed in Chapter 13. The pricing structure for these services is based on the total bandwidth at a location and not on the specific requirements between any two points. Later in this chapter, this same theme reappears when connection-oriented data services are discussed with pricing structures representing a mixture of dedicated and connectionless elements.

## SONET RING DATA-TRANSPORT SERVICES

SONET ring service is more than the sum of individual circuits between end-points that require connectivity. It represents an entire dedicated transmission architecture for all locations in a metropolitan area. Since the minimum offered capacity of SONET ring service is at OC-3 rates (155 Mb/s of aggregate throughput), with OC-12 rates (620

Mb/s of throughput) more common, this service is for customers with very large volumes of traffic among sites within a restricted geographical area. The absolute need for reliable service is also required to justify the high cost of the SONET infrastructure. In these applications, the business cost to the customer of a telecommunications outage must exceed the extra cost of SONET reliability. A prime example is the interconnectivity in lower Manhattan that connects traders and brokers associated with stock and bond exchanges. An hour of outage in this community can result in billion-dollar losses stemming from missed opportunities. Similar, though less extreme, situations exist in many industries and metropolitan areas. Often, requirements for SONET ring service come from organizations representing large numbers of individual users, no one of which justifies the expense of a SONET ring. An example of this application is the SONET ring service provided by the Department of Defense in Washington, DC for use by all DoD components in the Washington metropolitan area.

The topological structure of a typical SONET ring service was explained in Chapter 14. The pricing structure and design criteria for the service are discussed below.

The rate elements for service are:

- **Customer Nodes**—Ring switching capabilities on customer premises.
- **Central Office Nodes**—Ring switching capabilities at central offices.
- **Interoffice Channels**—Fiber optic paths between central-office nodes and between central-office nodes and the serving wire centers of customer nodes.
- **Local Channels**—Fiber optic paths from customer nodes to their serving wire centers.
- **Alternate Central Office Channels**—Fiber optic paths from customer nodes to alternate serving wire centers providing diverse paths from local channels.
- **Customer Channel Interfaces**—Individual DS1 and DS3 interfaces at customer nodes.
- **Central Office Channel Interfaces**—Individual DS1 and DS3 interfaces at central office nodes, where they may be connected via tariffed services to remote customer locations.

Both nonrecurring and monthly recurring charges apply to all rate elements. The recurring charges are discounted based on the length of the time commitment a customer is willing to make for the service.

Figure 15.4 shows the rate elements from BellSouth's tariff for OC-12 SONET rings.

**Figure 15.4**
SONET ring service rate elements.

| | Nonrecurring Charge | Month to Month | 12 to 36 Months | 37 to 60 Months |
|---|---|---|---|---|
| **Local Channel** | | | | |
| Fixed | $525 | | | |
| Per Quarter Mile | | $105 | $90 | $80 |
| **Alternate CO Channel** | | | | |
| Fixed | $525 | | | |
| Per Quarter Mile | | $685 | $360 | $295 |
| **Interoffice Channel OC-12** | | | | |
| Fixed | $200 | | | |
| Per Quarter Mile | | $55 | $50 | $45 |
| **Customer Node OC-12** | $460 | $3,590 | $3,090 | $2,865 |
| **Customer Channel Interface** | | | | |
| DS3 | $235 | $170 | $135 | $310 |
| **Central Office Node OC-12** | $460 | $2,680 | $2,280 | $2,080 |
| **CO Channel Interface** | | | | |
| DS3 | $290 | $115 | $90 | $85 |
| DS3/1 Multiplexer | $285 | $700 | $600 | $550 |
| DS1 | $265 | $18 | $14 | $12 |

The major decisions for this service are how many SONET ring nodes to order, which should be customer nodes, and where the nodes should be located. Customer nodes serve requirements only at the node location itself (remote locations must access the ring at central office nodes) and use fiber optic local channels and alternate central-office channels to access the ring. Hence, they must have significant data throughput requirements and a requirement for diversity to justify the expense of establishing a node. Customer nodes are normally located at major computer centers that are the primary sources and sinks of corporate traffic. In contrast, central-office nodes serve as aggregation points for smaller remote customer locations. These locations access the ring using other (non-SONET ring, see Figure 14.4) tariffed services from their premises to the central-office nodes. This remote access *may* be diverse (i.e., connect to two separate central-office nodes over physically separated facilities), but it is usually *not* diverse because of the high cost. Central office nodes are located near concentrations of customer sites to minimize this access cost. Node location is a significant network design problem that must be addressed and solved, usually with the assistance of the service provider, before proceeding with the rest of the design.

Once nodes are established, the remaining configuration decisions are straightforward. Figure 15.5 shows the rate elements included in a four-node OC-12 SONET ring. Two of the nodes are customer nodes located at major data centers. These data centers require high-bandwidth connectivity with each other for real-time data synchronization. The remaining two nodes are central-office nodes, which act as collection points for other customer remote nodes in the area. The remote nodes in this example are connected to the ring without diverse access, but enjoy the additional reliability provided by the ring for the transport portion of their connectivity. The customer nodes require four DS3s between them; each remote node requires 28 DS1s to each of the customer nodes (as shown in the circuit requirements box in Figure 15.5). The remote nodes use DS1 tail circuits to access the central office nodes. Two DS3/1 multiplexers at each central office node combine these circuits into two DS3 circuits for delivery to the data centers. Assuming this distribution of traffic and a 48-month term commitment, the total cost of the service (excluding tail circuits) breaks down as follows:

| Element | Quantity | Nonrecurring Cost | Recurring Cost |
|---|---|---|---|
| Customer nodes | 2 | $920 | $6,180 |
| Central-office nodes | 2 | $920 | $4,560 |
| Interoffice channels—3 miles each | 4 | $800 | $2,400 |
| Local channels—1/2 mile each | 2 | $1,050 | $360 |
| Alternate central-office channels—1 mile each | 2 | $1,050 | $2,880 |
| Customer channel interfaces—DS3 | 12 | $2,820 | $1,620 |
| DS3/1 multiplexers | 4 | $1,140 | $2,400 |
| Central-office channel interfaces—DS1 | 112 | $29,680 | $1,568 |
| **Total cost** | | $38,380 | $21,968 |

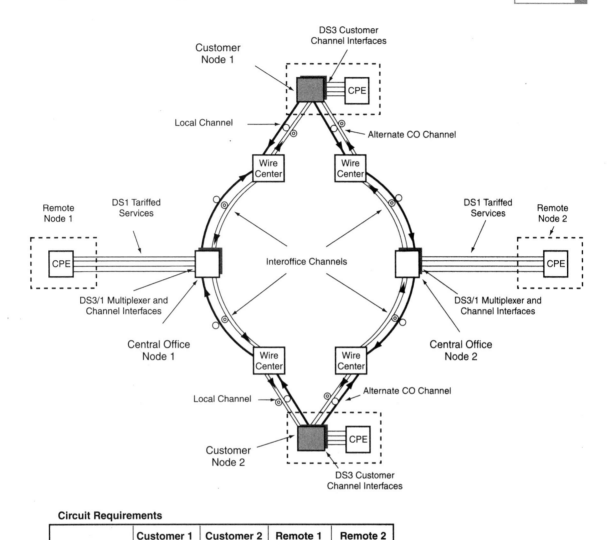

**Circuit Requirements**

|        | Customer 1 | Customer 2 | Remote 1 | Remote 2 |
|--------|-----------|-----------|----------|----------|
| Cust 1 |           | 4 DS3     | 28 DS1   | 28 DS1   |
| Cust 2 |           |           | 28 DS1   | 28 DS1   |

**Figure 15.5**  Example SONET ring configuration.

Based on a fair allocation of costs between DS1 and DS3 connectivities, the recurring cost of reliable transport across this network is $98 per month per DS1 and $2,746 per month per DS3. Using the tariff rate of $210 for a 20—50 mile DS1 Broadband Extension as a benchmark, the SONET ring is highly cost effective in this example and provides greater reliability than conventional private lines. The key to the cost effectiveness demonstrated here is high utilization of SONET ring capacity (a total of eight DS3s of capacity was used on the ring, resulting in almost 70 percent utilization).

### SWITCHED MULTIMEGABIT DATA SERVICE (SMDS)

SMDS is an example of a high-speed, connectionless data-transport service offered by LECs in a metropolitan area. Like all connectionless services, the pricing structure is simple and consists of access charges and a *connection charge* which depends only on bandwidth. Detailed billing is not provided and there are discounts for term commitments. Figure 15.6 shows the rate structure for BellSouth's SMDS offering (designated Connectionless Data Service by BellSouth). Access to SMDS consists of a digital local loop configured for switched services of the appropriate bandwidth. Traffic engineering is simplified with this service, requiring only that the aggregate bandwidth of each location be known.

**Figure 15.6**
SMDS rate elements.

| | Nonrecurring Charge | Month to Month | 12 to 36 Months | 37 to 60 Months |
|---|---|---|---|---|
| 56/64 kb/s | $400 | $80 | $69 | $49 |
| 112 kb/s | $460 | $120 | $104 | $74 |
| 128 kb/s | $460 | $120 | $104 | $74 |
| 192 kb/s | $460 | $190 | $165 | $125 |
| 256 kb/s | $460 | $240 | $207 | $147 |
| 320 kb/s | $460 | $300 | $259 | $184 |
| 384 kb/s | $525 | $410 | $369 | $319 |
| 448 kb/s | $525 | $410 | $369 | $319 |
| 512 kb/s | $525 | $410 | $369 | $319 |
| 576 kb/s | $525 | $410 | $369 | $319 |
| 640 kb/s | $525 | $410 | $369 | $319 |
| 704 kb/s | $525 | $410 | $369 | $319 |
| 768 kb/s | $525 | $410 | $369 | $319 |
| 1024 kb/s | $525 | $410 | $369 | $319 |
| 1152 kb/s | $525 | $410 | $369 | $319 |
| 1.536 Mb/s | $525 | $410 | $369 | $319 |
| 44.210 Mb/s | $1,225 | $3,500 | $3,250 | $3,000 |

The main drawback of connectionless data services is the lack of guaranteed performance levels across the network. The quality of service provided is adequate for data applications, but may not be suitable for delay-sensitive applications such as voice and video. Both performance and cost assessments must be performed in making the selection of SMDS for MAN applications.

Other connectionless MAN alternatives have similar pricing structures and characteristics. FDDI-based services were more popular in the past, but are now fading from the scene. In the near future, MAN services based on Gigabit Ethernet will gain in popularity and will likely be offered by new service providers as well as incumbent LECs.

# Interexchange/Wide Area Network Data Services

In today's increasingly electronic economy, the corporate wide area data network is an important strategic asset where low cost is not the only criterion for judgment of its effectiveness. High performance, high reliability, scalability, and ease of management become important when the public (i.e., paying customers) as well as employees interact with corporate information resources. Wide area data services exist today to address the varying needs of business applications (refer to Figure 15.1 for a summary). In the previous section on MAN services, the essential tradeoff between dedicated (non-switched) services and connectionless services was discussed. For WAN applications, the option of building totally private networks with customer-owned switches and dedicated transmission services connecting the switches is rapidly disappearing in the wake of capable and cost-effective public service offerings (paralleling the disappearance of private voice networks in the 1980s). The expense, management difficulty, and inflexibility of private networks does not match customer needs in the fast-moving world of modern data communications. Technology refreshment of owned equipment alone would be an intractable and expensive task given the 18-month life-cycle of most major data communications equipment.

Given this environment, the major service selection issue for the WAN is between connectionless public services and connection-oriented public services. The connection-oriented services (e.g., frame relay and ATM) retain some of the characteristics of dedicated services; there is some form of performance guarantee across the network, usually in the form of a guaranteed average throughput over a measurement interval. This guarantee is implemented in software by the service provider and, since capacity is not used when a user is not actually transmitting, the carrier can gain efficiencies that result in lower service prices. Chapter 14 describes in detail the advantages in flexibility that are gained by using frame-switched permanent virtual circuits as opposed to dedicated circuits; the following sections will describe the price structures of WAN data services and give illustrative examples of their application.

## Interexchange/Wide Area Network Data-Access Services

As with voice services, the major decision to make in accessing WAN transport services is between dedicated access and switched access. In the data arena, however, "switched" access can take many forms. The traditional form of switched access, circuit-switched calls, is implemented in the form of dial-up services provided by most WAN service providers. Once a call is established, the same data protocols are used over the connection as would be used over a dedicated access circuit. The disadvantages of dial-in access are limited bandwidth due to the use of switched-voice facilities to complete the call, and inefficient use of resources if the required connection times are long. Dial-up voice calls to packet-switched WAN data services are typically local calls and do not incur usage charges for the calls. For access to circuit-switched data services, LEC switched digital services are used between the customer premises and the WAN provider's POP (usually an IXC POP for these services). LEC switched digital services do incur usage charges, so the choice between switched and dedicated access is an economic one. LEC switched digital services typically cost four cents per minute for each 64 kb/s of switched bandwidth; this cost times the expected usage in a month is traded off against the cost of dedicated circuits of appropriate speeds.

For connection-oriented WAN services, the above choices are the most prevalent, with dedicated access being the most popular for corporate locations and dial-up access for mobile workers and telecommuters. For IP services, however, the alternative access services discussed above (i.e., DSL, cable modem, and wireless) can be used for access to WAN services as well. Internet Service Providers maintain high-speed connections to LEC and cable system networks to gather aggregated IP traffic from many end-users. There are no additional charges for this access beyond what the customer pays for the DSL or cable modem service; the cost for using the LEC or cable operator network is built into WAN provider transport charges.

## Interexchange/Wide Area Network Data-Transport Services

### CIRCUIT-SWITCHED DATA-TRANSPORT SERVICES

Circuit-switched data services (CSDS) are delivered under the same price structures as switched voice services. Since interexchange switched data transport is mainly a business service, the most common method of utilizing this capability is in conjunction with a virtual private network service or special contract service. Typically, large volumes of voice traffic justify virtual networks and dedicated access connections to large customer sites. If video or switched data applications are present, the customer orders ISDN PRI access connections in place of standard T1 access connections. The appropriate bandwidth is then configured through the network on a call-by-call basis.

CSDS transport is priced using separate price schedules under a virtual network tariff or special contract. The most common bit rates supported are 64 kb/s, 128 kb/s, and 384 kb/s. Figure 15.7 shows on-net-to-on-net schedules for both 64-kb/s and 384-kb/s service. The prices are mileage based and time-of-day sensitive. The ability to make a digital data call to an off-net location is a new feature of virtual network services and reflects the availability of digital switched access in LEC infrastructures.

Circuit-switched data are suited to applications where connectivity is required on a part-time basis only, but use of the circuit is high once the connection is established. Video teleconferencing is the major application for CSDS and switched 384 kb/s is the most popular switched bit rate for this purpose.

### FRAME RELAY DATA-TRANSPORT SERVICES

Of all the WAN data-transport services used for corporate networks in the U.S., frame relay service is the most popular and fastest growing. It has virtually replaced X.25 service as the standards-based method of connecting business computers and servers. Because it is a Layer 2 proto-

Switched 64 kb/s Service

| Mileage | Initial 18 Seconds | | | Additional 6 Seconds | | |
|---|---|---|---|---|---|---|
| | Day | Eve | Night | Day | Eve | Night |
| 0–55 | .0147 | .0147 | .0147 | .0049 | .0049 | .0049 |
| 56–292 | .0156 | .0156 | .0156 | .0052 | .0052 | .0052 |
| 293–430 | .0243 | .0243 | .0243 | .0081 | .0081 | .0081 |
| 431–925 | .0276 | .0276 | .0276 | .0092 | .0092 | .0092 |
| 926–1910 | .0309 | .0309 | .0309 | .0103 | .0103 | .0103 |
| 1911–3000 | .0336 | .0336 | .0336 | .0112 | .0112 | .0112 |
| 3001–4250 | .0402 | .0402 | .0402 | .0134 | .0134 | .0134 |
| 4251–5750 | .0402 | .0402 | .0402 | .0134 | .0134 | .0134 |

Switched 384 kb/s Service

| Mileage | Initial 18 Seconds | | | Additional 6 Seconds | | |
|---|---|---|---|---|---|---|
| | Day | Eve | Night | Day | Eve | Night |
| 0–55 | .0639 | .0385 | .0322 | .0213 | .0171 | .0143 |
| 56–292 | .0851 | .0506 | .0423 | .0230 | .0200 | .0188 |
| 293–430 | .0970 | .0574 | .0479 | .0323 | .0255 | .0203 |
| 431–925 | .1112 | .0689 | .0574 | .0371 | .0306 | .0255 |
| 926–1910 | .1238 | .0756 | .0630 | .0413 | .0336 | .0280 |
| 1911–3000 | .1348 | .0803 | .0671 | .0449 | .0357 | .0298 |
| 3001–4250 | .1505 | .0905 | .0754 | .0502 | .0402 | .0335 |
| 4251–5750 | .1568 | .0938 | .0783 | .0523 | .0417 | .0348 |

col, frame relay can provide transport for any and all network-layer
protocols in use throughout a corporate network. These may include
IP, IPX (the network-layer protocol for Novell networks), SNA, and
even X.25 itself. As a connection-oriented protocol, frame relay provides

guarantees of average throughput between specific end-points by establishing *virtual circuits* between them and by associating a *committed information rate* (CIR) with each virtual circuit (see Chapters 7 and 14).

Since frame relay combines attributes of a shared service (frame relay switches are shared among customers) with aspects of a dedicated service, its pricing structure reflects elements of both. Users access frame relay service through *ports.* A port terminates an access circuit and usually matches the access circuit's bit rate. As described in the example in Chapter 14, many virtual circuits can be carried through the port, each with its own destination and CIR. The sum of the CIRs may exceed the port bandwidth, since not all virtual circuits may be active at the same time. However, the total bit rate at any one time may not exceed the port capacity. It is not surprising, then, that a port charge is one of the pricing elements of frame relay service, and that the charge depends on the port bandwidth.

For connectionless data services, this port charge is the only charge. The additional functionality that frame relay provides on each permanent virtual circuit (the guarantee of average throughput equal to the purchased CIR) is paid for through a PVC charge that depends on the CIR ordered. In contrast with dedicated circuits, the prices for PVCs are flat rated and do not depend on the mileage between end-points or the location of the end-points. Since they are implemented in software, PVCs can be offered in finer steps and at bit rates lower than 64 kb/s, making them much easier to match to requirements than dedicated transport services. As noted previously, a PVC is also much more flexible in allowing the actual bit rate to fluctuate above or below the CIR.

Figure 15.8 shows a price schedule for frame relay service taken from the federal government's FTS2001 contract, a large multiservice contract providing a wide range of services to all federal agencies. The PVCs in this example are full-duplex virtual circuits (the CIR is guaranteed in both directions over the PVC). In some tariffs, PVCs are specified and priced as simplex PVCs (the CIR is guaranteed in only one direction). A separate PVC of a different CIR could be provisioned between the same two points in the opposite direction.

Figure 14.10 provides a direct comparison between a frame relay and a dedicated-circuit solution to a four-node customer configuration. The additional capabilities provided by the frame relay service are discussed there. The table below compares the total monthly costs of the two solutions using dedicated-circuit prices from FTS2001 and assuming 600 miles between the network POPs. In the dedicated-

■■■ ■■■ ■■■
**Figure 15.8**
Frame relay service
rate elements.

| Ports | Monthly Charge |
|---|---|
| 64 kb/s | $93.64 |
| 128 kb/s | $159.29 |
| 256 kb/s | $230.01 |
| 384 kb/s | $289.83 |
| 512 kb/s | $357.25 |
| 768 kb/s | $440.20 |
| 1024 kb/s | $687.33 |
| 1536 kb/s | $769.03 |

| Permanent Virtual Circuits | Monthly Charge |
|---|---|
| 16 kb/s CIR | $15.35 |
| 32 kb/s CIR | $28.18 |
| 48 kb/s CIR | $38.17 |
| 64 kb/s CIR | $42.17 |
| 128 kb/s CIR | $84.82 |
| 192 kb/s CIR | $135.72 |
| 256 kb/s CIR | $169.64 |
| 320 kb/s CIR | $238.31 |
| 384 kb/s CIR | $271.43 |
| 448 kb/s CIR | $333.64 |
| 512 kb/s CIR | $381.30 |
| 576 kb/s CIR | $428.96 |
| 640 kb/s CIR | $476.63 |
| 704 kb/s CIR | $524.29 |
| 768 kb/s CIR | $593.76 |
| 832 kb/s CIR | $619.62 |
| 896 kb/s CIR | $667.28 |
| 960 kb/s CIR | $714.94 |
| 1024 kb/s CIR | $848.23 |
| 1152 kb/s CIR | $935.92 |
| 1280 kb/s CIR | $1,039.91 |
| 1408 kb/s CIR | $1,143.90 |
| 1536 kb/s CIR | $1,247.89 |

circuit solution, a T1 multiplexer is needed on the customer premises to derive the separate 512-kb/s channels, a function the router performs using frame switching.

In this example, the frame relay solution is not only more flexible, it is also less costly.

| Element | Quantity | Frame Relay Cost | Dedicated Circuit Cost |
|---|---|---|---|
| T1 router w/frame relay interface | 4 | $142.46 | |
| T1 multiplexer w/three CSU/DSUs | 4 | | $368.60 |
| T1 Access circuits | 4 | $1,002.93 | $1,002.93 |
| T1 Frame relay ports | 4 | $3,076.12 | |
| 512-kb/s frame relay PVCs | 6 | $2,287.81 | |
| 512-kb/s dedicated-circuit transport (600 miles) | 6 | | $7,490.45 |
| **Total cost** | | **$6,509.32** | **$8,861.98** |

The traffic engineering needed to configure a frame relay network is only slightly more complex than a connectionless service. The access and port speed depend only on the aggregate bandwidth requirement at a location, much as with connectionless service. In order to specify the correct CIRs for the PVCs, an estimate must be made of the point-to-point traffic. However, this estimate need not be exact since the traffic will be carried on a best-effort basis even above the contracted level. CIRs may then be adjusted after analyzing monthly PVC performance data; a new CIR on an existing PVC can often be in effect within minutes of placing a change order.

For data-only requirements at speeds up to 1.5 Mb/s, frame relay offers the ideal blend of cost effectiveness and performance. If higher speeds are required or if a mixture of voice, video, and data is carried on the same infrastructure, then ATM service is recommended.

## ASYNCHRONOUS TRANSFER MODE DATA-TRANSPORT SERVICES

Pricing structures for ATM services are very similar to those for frame relay except for the added complexity of multiple service classes. Access lines and ports are priced in the same manner, based only on bandwidth. ATM ports are available at higher speeds than frame relay ports are, up to OC-12 (599.040 Mb/s) for MAN ATM services, and are also generally available at speeds as low as T1. Ports are engineered in the same manner as frame relay ports; the speed required is based on the sum of the speeds of the PVCs utilizing the port. Also, as in frame relay, ports may be oversubscribed based on a knowledge of the traffic utilization on the PVCs using the port.

PVC charges are based on the service class specified and on the *traffic parameters* discussed in Chapter 14 (peak cell rate, sustainable cell rate, and maximum burst size). Different ATM service providers use different price structures for PVCs. BellSouth computes an equivalent bandwidth based on the service class and traffic parameters for a given PVC (these formulas were presented in Chapter 14 as an example). PVC pricing then depends on service class and equivalent bandwidth. Figure 15.9 shows the BellSouth ATM prices for ports (called "customer connections" by BellSouth) and PVCs. In the BellSouth tariff, PVCs are priced as segments and two must be ordered to implement an end-to-end virtual circuit. In this example, PVCs for different service classes have the same price. This is not generally the case; constant bit rate PVCs are typically more expensive than variable bit rate PVCs.

**Figure 15.9**
Metropolitan area ATM service rate elements.

| | Nonrecurring Charge | Month to Month | 12 to 36 Months | 37 to 60 Months |
|---|---|---|---|---|
| **Customer Connection** | | | | |
| 1.536 Mb/s | $595 | $550 | $450 | $415 |
| 44.210 Mb/s | $1,225 | $3,500 | $2,800 | $2,550 |
| 149.760 Mb/s | | | | |
| Fixed | $2,175 | $5,580 | $4,650 | $4,200 |
| Per Mile | | $140 | $132 | $130 |
| 599.040 Mb/s | | | | |
| Fixed | $4,750 | $14,550 | $12,650 | $11,500 |
| Per Mile | | $205 | $195 | $190 |
| **Permanent Virtual Circuit (CBR, VBR-RT, VBR-NRT)** | | | | |
| Per PVC Segment | $70 | $5 | $5 | $5 |
| Per Mb/s | | $40 | $40 | $40 |
| Per 64 kb/s | | $2.60 | $2.60 | $2.60 |
| **Permanent Virtual Circuit (UBR)** | | | | |
| Per PVC Segment | $70 | $5 | $5 | $5 |
| 1.536 Mb/s Activation | | $10 | $10 | $10 |
| 44.210 Mb/s Activation | | $250 | $250 | $250 |
| 149.760 Mb/s Activation | | $500 | $500 | $500 |
| 599.040 Mb/s Activation | | $1,000 | $1,000 | $1,000 |

In some cases, carriers charge explicitly for various combinations of peak and sustainable cell rates for VBR PVCs. Figure 15.10 lists a portion of the MCI WorldCom PVC price table from the FTS2001 contract. Note that for a fixed sustainable cell rate, the price increases as the peak cell rate increases. A higher PCR causes more uncertainty in the actual traffic flow and more resources need to be allocated to maintain the performance guarantees on the virtual circuit. Inspection of the rates for a fixed PCR and variable SCR reveals an interesting pattern. This carrier has optimized his rates (and presumably his network) around a peak to sustainable ratio of 2:1. Other ratios are allowed, but are priced higher for the same peak rate.

**Figure 15.10**

ATM PVC price examples based on peak cell rates and sustainable cell rates.

Variable Bit Rate–Non Real Time PVCs

| Peak Cell Rate | Sustainable Cell Rate | Monthly Charge |
|---|---|---|
| 2048 kb/s | 1536 kb/s | $990.37 |
| 3088 kb/s | 1536 kb/s | $1,002.56 |
| 4096 kb/s | 1536 kb/s | $1,435.28 |
| 4632 kb/s | 1536 kb/s | $1,605.57 |
| 6144 kb/s | 1536 kb/s | $2,672.42 |

| Peak Cell Rate | Sustainable Cell Rate | Monthly Charge |
|---|---|---|
| 4096 kb/s | 1024 kb/s | $1,963.16 |
| 4096 kb/s | 1088 kb/s | $1,943.73 |
| 4096 kb/s | 1170 kb/s | $1,868.96 |
| 4096 kb/s | 1216 kb/s | $1,797.09 |
| 4096 kb/s | 1280 kb/s | $1,727.97 |
| 4096 kb/s | 1344 kb/s | $1,661.50 |
| 4096 kb/s | 1408 kb/s | $1,582.39 |
| 4096 kb/s | 1472 kb/s | $1,507.03 |
| 4096 kb/s | 1536 kb/s | $1,435.28 |
| 4096 kb/s | 1786 kb/s | $1,366.93 |
| 4096 kb/s | 2048 kb/s | $1,301.83 |
| 4096 kb/s | 2381 kb/s | $1,497.11 |
| 4096 kb/s | 3072 kb/s | $1,721.68 |

To be cost effective, ATM services must be priced as a whole network with a mixture of traffic classes or with a significant number of very high-speed connectivity requirements (45 Mb/s or higher). ATM efficiencies occur at the edge of the network where ATM premises switches combine several classes of traffic from different sources. The ATM flow control mechanism allows cells from all applications to share the available bandwidth while providing each application with its guaranteed quality of service. Since ATM equipment is relatively expensive, savings from reduced transmission costs must be realized through traffic engineering.

## IP-BASED DATA-TRANSPORT SERVICES

IP transport services are the primary connectionless data services for WAN applications. Like all connectionless services, IP charges are limited to port charges based on bandwidth and not on the number of packets actually transmitted. This concept of flat-rate pricing was pioneered by the Internet service providers and is generically referred to as *Internet pricing*. The most common version of this pricing is the dial-up account used by most residential customers to access the Internet. A fixed charge, usually less than $20 per month, allows almost unlimited usage at speeds up to 56 kb/s.

For business users, ISPs offer a spectrum of dedicated ports and access technologies. Charges for the access lines are sometimes bundled with the port costs. Access technologies include symmetrical DSL, T1, and fiber optic transmission at T3 or higher speeds. Innovative pricing schemes include ports that are accessed with a full T1 circuit, but are charged based on the use actually encountered during the month 95 percent of the time (the highest 5 percent of the utilization samples are discarded). The most cost-effective port configuration depends on the traffic profile at the customer site.

Because of their current lack of end-to-end QoS controls, IP services are rarely used for satisfying all corporate networking requirements. They are used by businesses to access the Internet for applications where interaction with the public or with other businesses is required. A special case of this situation is the e-commerce Web site, which requires extremely high performance and reliability from the IP network serving the hosts. Chapter 14 discusses the configuration where ASPs provided hosting services for these large sites with access directly to an Internet backbone network. Hosting services are typically priced in terms of rack space, security features, and network connections speed. Not surprisingly, major national backbone network providers are also entering the hosting business, building large hosting facilities collocated with major backbone nodes. These providers include UUNET, Cable & Wireless, and Qwest Communications.

When IP services are used to satisfy corporate WAN requirements, they generally take the form of special contracts between ISPs and customers. These contracts are written in terms of *service level agreements* (SLAs). These specify performance levels to be monitored by ISPs and results provided to customers on a monthly basis. Carrier/service provider failure to meet performance levels results in credits to the customer. This paradigm provides incentives for ISPs to engineer their backbones and customer-access arrangements for optimum performance. In some cases, backbones used for corporate services are completely separate from those used to carry general Internet traffic, allowing a higher performance standard to be maintained. As methods for implementing differentiated services in IP networks become established, new offerings based on multiple classes of service will appear and compete with frame relay and ATM for high-end corporate WANs.

# 5

# Wireless Services

I n just the last two decades, wireless services have catapulted from the practically nonexistent to over 86 million U.S. and more than 200 million subscribers worldwide. In the U.S., a new wireless subscriber is added every two seconds, about 42,000 per day. Within the telecommunications industry, only the Internet's growth is in the same league as wireless service growth. It is generally estimated that there will be more mobile devices than PCs connected to the Internet by 2003.

Some trace the origins of today's booming wireless market to Edwin Howard Armstrong's 1935 breakthrough discovery that frequency modulation (FM), as opposed to amplitude modulation (AM) in broadcast radio applications, greatly enhances speech and audio quality. While FM and even more exotic modulation and multiple access techniques (Chapters 4 and 5) remain crucial in modern cellular telephone systems, the middle 80s' appearance of inexpensive, lightweight, and low-power microcomputers to automate complex subscriber set call handling, power control, and other signaling and supervisory functions, makes microcomputer electronics an equally important enabling technology.

But if radio modulation, multiple-access techniques and microcomputer electronics stand as twin technology pillars, wireless' spectacular growth in the United States must be attributed in no small part to the open and competitive environment in which the cellular telephone industry found itself in its formative years. In May 1981, a Federal Communications Commission (FCC) Order mandated that there be two licenses granted in each of 306 urban metropolitan statistical areas (MSAs—75 percent of the population and 20 percent of the landmass), and in each of 428 rural service areas (RSAs—25 percent of the population and 80 percent of the landmass)—one to be granted to incumbent local "wireline" carriers, and the other to competitive non-wireline companies.

As a measure of the FCC Order's impact on fostering competition, the Cellular Telecommunications Industry Association states that in 1999 "wireless competition has accelerated to the point that 238.7 million Americans can now choose between 3 and 7 wireless service providers. More than 87.9 million Americans can choose from among 6 or more wireless providers, and 87.7 million Americans can choose among 5 wireless providers." It is informative to contrast this situation with what might have occurred if AT&T had maintained its virtual monopoly position on telecommunications in the United States.

Recall from Chapter 1 that based on a late-1970s Bell Labs market-research study that predicted a mobile-subscriber base of only 800,000 by 2000, AT&T chose not to pursue the business vigorously. Thus, in the absence of competition, growth in the U.S. wireless industry might have been severely stunted. Fortunately, the FCC broke with its traditional stance that telecommunications services in the United States are best rendered by regulated monopolies. Because of that decision, other aggressive, entrepreneurial companies—that envisioned an enormous market potential—were given the opportunity to pursue, and did indeed develop a new, affordable mobile telephony service now considered indispensable.

Moreover, for the first time, local exchange carriers found themselves vying with competitors in the marketplace for customers and profits, as opposed merely to appealing to regulatory commissions for product and service approvals—and guaranteed rates of return. It is safe to assume that the general public's positive reaction to competition in the cellular telephone arena emboldened the FCC in its 1996 Telecommunications Act deliberations to expand competition to nearly all facets of local telecommunications service. Since telecommunications services are crucial in most businesses, the salutary benefits to the economy and society in general may make the FCC's mobile telephone rulings one of the most fortuitous yet enlightened and invaluable regulatory actions ever taken.

The variety of currently available wireless services is extensive, with new industry and standards groups activities paving the way for even more options. As explained in Chapter 4, in 1988, the FCC authorized a class of service allowing telephone companies to use digital radios instead of copper wires when cost effective. These and similar services are designed for stationary or fixed-equipment use. So we have fixed and mobile classes of wireless service, with gradations within the mobile category that distinguish between services intended for fast moving-while-operating user equipment (automobile, boat, or airplane located), pedestrian, and movable but stationary-while-operating applications.

From a radio-frequency spectrum point of view, the FCC and the World Administrative Radio Conference (WARC) assign both bandwidth and spectral location to specific wireless service classes. Within those assignments, wireless facilities are used for voice, data, video, entertainment, and numerous special services such as message paging, voice paging, and two-way acknowledgment paging. Finally, wireless services may be classified as either terrestrial or satellite-based.

Again, as noted in Chapter 4, services and facilities intended for one class of service are often adapted for others. Cellular subscribers are free to take mobile telephones indoors and use them as a substitute for fixed wireless or wireline services. Universal mobile telephone service (UMTS) handsets promise to provide access to both cordless indoor and cellular outdoor services. Overall, the trend toward integrated, multimedia (voice, data, other) service provision within wired systems is occurring in the wireless world as well.

In countries with antiquated or nonexistent wireline facilities, wireless technologies are particularly effective for rapidly implementing modern telephone service. For example, the Moscow Telephone System still employs paper insulated loop wiring (some of which was installed around the turn of the 20th century), and wireline switching systems that predate stored-program computer control designs and modern signaling systems (like SS7). Of course, these technologies make possible the kind of modern telecommunications that most people in industrialized western countries now take for granted. By 1993, Moscow already had several analog cellular telephone systems to alleviate the problem, and in that year one of the authors participated in the design of an advanced digital cellular system that was implemented in less than a year.

As mentioned in Chapter 1, beyond rapid deployment advantages, declining per-subscriber wireless access costs are now often less expensive than wired alternatives. To place the wired local loop versus wireless alternative in perspective, consider the following. Sixty-three percent of US West's central offices serve rural customers. Statistically this amounts to only 32 subscribers per square mile, a figure that may be compared with the national average of 168 subscribers per square mile. In some situations US West has expended as much as $50,000 per subscriber line and some lines could cost $200,000. Today, the average cost of a 1,500-foot wired subscriber loop ranges between $1,000 and $1,500 per subscriber.

According to CTIA, at the end of 1999, all cellular service providers included in their annual survey had made cumulative capital investments of $71.265 billion. With 86 million revenue-generating subscribers, this works out to $828.22 per subscriber. Understanding that this cellular capital investment includes switching and all other facilities, and not just an amount allocable to replacing local loops, if the above "back-of-the-envelope" analysis even approximates reality, it is a powerful factor in any service provider's subscriber access service investment business case. It may be one reason why in these "competitive-LEC days," established carriers considering future upgrades to aging wire facilities are often heard saying, "no more copper!"

While FCC decisions to permit and encourage competition in the cellular telephone market sector resulted in explosive growth, one of the competition's downsides was a lack of compatibility among separately developed systems. As evidenced in the following sections, the most serious incompatibility problems in the U.S. had to do with the technical ability to support users "roaming" outside home-service provider areas, prior to the development of workable inter-service provider billing mechanisms.

Our treatment of wireless services and underlying customer premises equipment (CPE) and network facilities that support them begins with an exposition of cellular telephone services and system operations. The key cellular operational principle involves frequency reuse. Since radio spectrum is limited, and the potential exists for multiple users to interfere with one another, to accommodate ever-growing numbers of wireless subscribers necessitates some method of reusing spectrum on a noninterfering basis. This involves being able to assign the same frequencies to multiple users by ensuring that they are geographically separated by specified distances and controlling transmitted power levels. The technical details of how this was accomplished in the mid-1980s to support mobile telephone service is presented first, but the basic technology is at the heart of most emerging capabilities such as the broadband, multimedia local multipoint distribution system (LMDS) and other wireless services discussed here and elsewhere in this book.

# CHAPTER 16

# Terrestrial Wireless Services

# Cellular Systems and Design Principles

Mobile cellular telephone, first marketed in the early 1980s, has proven to be the impetus for an ever-expanding class of wireless mobile, transportable, and personal communications products and services. As noted, the key idea behind cellular service, which makes it and the other wireless devices practicable, is that the same frequency channels can be systematically reused. Unlike previous radiotelephone designs that used a single set of frequencies to cover an entire city, severely limiting the maximum number of simultaneous users, cellular uses low-powered transmitters covering geographic cells (usually less than 8—10 miles in radius) as shown in Figure 16.1a. Frequency assignments are made in patterns that minimize interference (i.e., the same frequencies—see footnote in the figure—are not assigned to adjacent cells).

Because the frequency spectrum is a limited resource, without a systematic way to reuse frequencies, there would be no way to accommodate ever-increasing user requirements. Implicit in the cellular concept is the capability to shrink cell sizes to accommodate growth in the number of subscribers. A cellular system may begin with 10 cells, but faced with increased demand, find that number inadequate. If the initial cell pattern has been properly laid out, it is possible (but often expensive) to split existing cells in high-usage areas, such as downtown business areas, into three to six new cells. Various cell sizes can thus be used in metropolitan areas in accordance with geographical traffic patterns.

Figure 16.1 depicts three categories of equipment present in all cellular telephone systems, mobile telephone or transceiver sets, cell sites or base transceiver stations, and mobile telephone switching offices (MTSOs). Radio links connect mobile telephones to fixed base station-cell sites that, in turn, are connected to MTSOs. Functional attributes and operations of these components are explained below.

Cellular design imposes the requirement that as an active mobile subscriber moves through an area, his call has to be *handed off* between cell-site transmitters. This creates a control problem, which had to wait until microcomputer and digital switch technology made it practical and affordable to achieve hand-off without losing call connections or being audibly perceptible to subscribers.

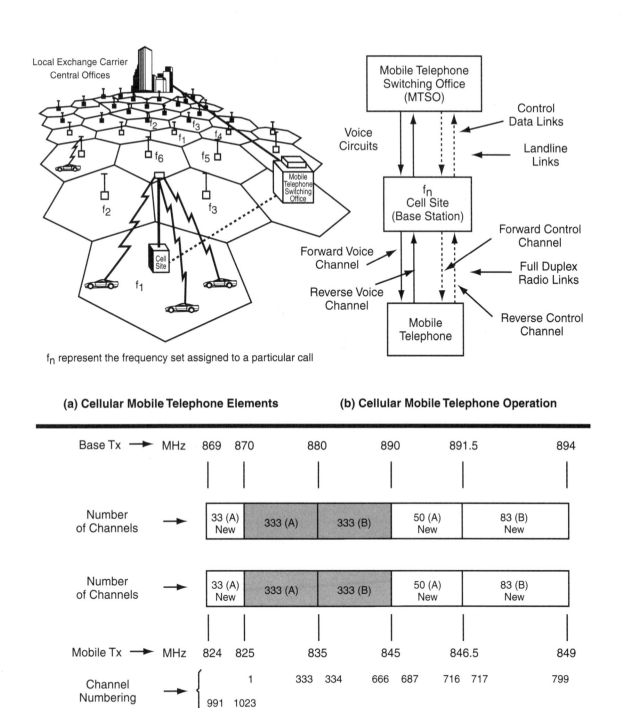

**Figure 16.1** Cellular telephone system design and AMPS frequency allocation.

Handing-off uses computers at three levels—microprocessors in *mobile transceivers,* in *mobile control units* (MCUs) at cell sites, and in *mobile telephone switching offices.* MTSOs, controlling cell sites, normally provide connectivity to nationwide telephone networks. L.M. Ericsson, AT&T, and Nortel are leading cellular switch vendors.

Figure 16.1b shows that separate transmit and receive channels (full-duplex operation) are assigned for both voice and control functions. Base station-to-mobile set transmission links are called *forward channels* and mobile set-to-base station links *reverse channels. Control channels* exchange various types of signals that allow base stations to "sense" the presence of "powered-on" mobile telephone sets, to *page* or notify mobile telephone sets of incoming calls, to transfer dialing instructions for outgoing calls, and accomplish hand-off and numerous other call monitoring and supervisory functions.

A sequence of electronic identification numbers is assigned to both cellular carrier systems and each individual mobile telephone. The numbers, appended to control signals exchanged between base stations and mobile telephones, allow carriers to locate and forward incoming calls to specific telephones, and to verify that outgoing calls originate from telephone sets duly registered with some carrier. The set of numbers is used to assign call charges to subscribers directly, or to other carriers in cases where subscribers seek service from other than their "home location" carrier. The series of numbers and their use are summarized as follows:

| Mnemonic | Description |
|----------|-------------|
| MIN1 | Mobile identification number 1—A 24-bit number corresponding to the 7-digit telephone number assigned by a carrier to a particular telephone. |
| MIN2 | Mobile identification number 2—A 10-bit number corresponding to the area code assigned by a carrier to a particular telephone. |
| ESN | Electronic serial number—A 32-bit number assigned by manufacturers that uniquely identifies a particular telephone set. |
| SID | System identification number—A 5-bit number uniquely associated with a particular cellular carrier. |

# Advanced Mobile Telephone System (AMPS)

Figure 16.1c depicts FCC cellular spectrum allocations in the 800 MHz band. The technical and operational design attributes that the FCC envisioned for use in its spectrum allocations were defined in 1979 at Bell Laboratories and codified under Bell Labs' *Advanced Mobile Telephone Service* (AMPS) descriptions. AMPS uses a frequency division multiple access (FDMA) technique to share cellular frequency allocations among constantly changing sets of users. Within each FDMA channel, frequency modulation (FM) is used to modulate FCC-assigned cellular radio frequency (RF) carriers (channel center-frequencies) with baseband analog user information (voice or voice-compatible modem or facsimile) signals.

In 1983, 666 channels were allocated to cellular service in each serving area. Of these; 333 are assigned to incumbent wireline carriers and 333 to independent enterprises—originally selected by lottery. In Figure 16.1c, bands A and B represent allocations to the two service providers, with B blocks assigned to wireline carriers. In those initial allocations, 21 channels were used for control signal traffic, and 333 minus 21, or 312 channels were used for voice traffic. Later, new FCC allocations boosted the total number of channels to 832. The 166 new channels are all used for voice traffic and again are divided equally among two service providers, so that each now has 395 voice traffic-bearing channels.

The bottom part of Figure 16.1c shows mobile telephone and the top portion base station transmitting channel frequency assignments. Channels are assigned numbers from 1 to 1,023, noting that channel numbers 800 through 990 are not used. Channel center-frequencies are spaced 30 kHz apart. The relation between channel numbers and mobile transmitter center frequencies, in MHz, is given by the equation:

$$f_{mobile} = \begin{cases} 0.003n + 825.00 & 1 \leq n \leq 799 \\ 0.003(n - 1023) + 825.00 & 990 \leq n \leq 1023 \end{cases}$$

The relation between channel numbers and base station transmitter center frequencies, in MHz, is given by the equation:

$$f_{base\ station} = \begin{cases} 0.003n + 870.00 & 1 \leq n \leq 799 \\ 0.003(n - 1023) + 870.00 & 990 \leq n \leq 1023 \end{cases}$$

In the design of cellular networks, all of Chapter 11's traffic engineering principles and practices may be used to optimize network resources in terms of projected user traffic. However, cellular network design and optimization involves factors not encountered in wired or cabled telecommunications systems. For example, while both wired and wireless networks experience diurnal, day-of-week, and seasonally driven peak traffic conditions, geographical factors have further impact on wireless network traffic. Rush-hour vehicular traffic jams, particularly at bridges, tunnels, and other bottlenecks, create much higher concentrations of offered user telephone traffic than occur on the average in other urban or suburban cellular coverage areas. Such factors directly drive both base station location and cell sizing decisions.

Other factors affecting cellular system operational viability include physical line-of-sight radio wave transmission obstructions from either natural (terrain undulation) or man-made (building) causes. Cellular network design tasks typically begin with smooth terrain, uniform cell morphology baseline assumptions—to be refined by design iterations that invoke ever more detailed terrain-driven cell morphologies. In most cases, calculations are verified with sample field measurements. Still other performance-influencing factors include the degree to which providers can offer in-building and underground (e.g., in-subway) service; interference among user operations within and between cells; probabilities of satisfactory connection boundaries; and installed-system call-blocking and dropped-call probabilities.

In view of these complexities and considering the added synchronization difficulties that subscriber sets moving at relatively high speeds pose, reliable cellular system design and engineering is a formidable task. Fortunately, the adage "practice makes perfect" applies and competent designers have developed powerful *computer-aided design* (CAD) tools to cope with the complexities. Even so, to this day, users express quality-of-service complaints, and service providers are constantly refining hardware and software designs based on actual performance data gathered daily from ever-increasing numbers of user calls. As an indication of how rapidly baseline design changes can occur, some sales agents advise customers to return their telephones on a monthly basis for the most recent software updates.

## Other Cellular Systems, Characteristics, and Operational Considerations

Although the cellular concept was advanced by Bell Labs in 1947, it was not until 1979 that the first automatic cellular system commenced operation in Japan, and in 1981 in Nordic countries. In the United States, the first commercial AMPS cellular wireless system was placed in service in Chicago in October 1983. Early North and South American cellular systems conform to AMPS standards, but in the rest of the world, a number of different types of analog cellular systems emerged.

With as many as five different and incompatible air interfaces in Europe, in 1982 the Conference of European Posts and Telecommunications (CEPT) established the *Groupe Spéciale Mobile* (GSM—now commonly understood to mean Global System for Mobile Communications). By 1987, through a remarkably cooperative effort, all developing parties agreed to a compatibility specification that supported hybrid FDMA (analog) and time division multiple access (TDMA) technologies. Today, 215 service providers in over 100 countries support GSM specifications which are now the responsibility of the European Telecommunications Standards Institute (ETSI).

Figure 16.2 lists cellular systems employing various standards developed since 1980, along with some key technical attributes. The first five columns, from AMPS to TACS, are all referred to as *first-generation (1G)* cellular systems. 1G designs use FDMA, analog voice formats, and analog frequency modulation (FM) in voice channels, a process that varies transmitted carrier frequencies above and below assigned FCC cellular channel center frequencies in direct proportion to the amplitude of voice or other user information signals.

Even 1G cellular systems employed digital frequency shift key (FSK) modulation for control channel traffic. In the U. S., since 1988, 1G control channels conformed to initial versions of IS-41 (Interim Standard-41). IS-410 (read IS-41 "zero"), the initial version of IS-41, defined handoff procedures, with later versions addressing MIN/ESN-checking pre-call validation; roaming operations; the correlation of telephone numbers with signaling system 7 (SS7) network nodes; means for supporting short message service (SMS); and other advanced features like caller ID and enhanced emergency 911 services.

Readers who used mobile services in the mid-80s probably recall how difficult it was to make calls when "roaming." The term *roaming* describes a mobile customer's ability to use mobile telephone outside the "home" service area of the operator with which the customer established mobile service. In those days, cellular growth was so rapid that although mobile telephone service was available in many cities, network operators had not yet established procedures and standards for assessing charges and collecting fees from one another for providing service to subscribers roaming outside of their home locations.

## Cellular Mobile Telephone Systems

| System \ Parameter | AMPS | MCS-L1 MCS-L2 | NMT | C450 | TACS | D-AMPS/ IS-54/IS-136 | PCS | GSM[3] | IS-95[4] |
|---|---|---|---|---|---|---|---|---|---|
| Transmit Frequency (Mhz) Base Mobile | 869–894 824–849 | 870–885 925–940 | 935–960 890–915 | 461–466 451–456 | 935–960 890–915 | 869–894 824–849 | See Figure 16.4 | 935–960 890–915 | 869–894 824–849 |
| Multiple Access Method | FDMA | FDMA | FDMA | FDMA | FDMA | TDMA/FDMA (Dual Mode) | TDMA | TDMA | CDMA |
| Channel Bandwidth (Khz) | 30 | 25 12.5 | 12.5 | 20 10 | 25 | 30 | 200 | 200 | 1500 |
| Traffic Channels per RF Channel | 1 | 1 | 1 | 1 | 1 | 3 | 16 | 8 | 55 |
| Total Number of Traffic Channels | 832 | 600 1200 | 1999 | 222 | 1000 | 832 × 3 | 375 × 16 | 125 × 8 | — |
| Voice Format/Encoding | Analog | Analog | Analog | Analog | Analog | VSELP | RELP | RELP | QCELP |
| RF Modulation Control Channel Voice Channel | Digital FSK Analog FM | Digital FSK Analog FM | Digital FFSK Analog FM | Digital FSK Analog FM | Digital FSK Analog FM | Digital PSK Digital PSK | Digital GMSK Digital GMSK | Digital GMSK Digital GMSK | Digital PSK Digital PSK |

---

### Cordless Telephone Systems

CT1

CT2

CT3

Digital European Cordless Telephone (DECT)

Universal Mobile Telecommunications System (UMTS)[1]

---

### Acronyms

| | |
|---|---|
| AMPS | Advanced Mobile Phone Service |
| MCS | Mobile Communication System |
| NMT | Nordic Mobile Telephone |
| C450 | Nordic–1 |
| TACS | Total Access Communications System |
| GSM | Global System for Mobile Communications |
| PCS | Personal Communications System |
| IS-54/ IS-136 | EIA/TIA/IS-54—A Standard (March 1991) (1994) |
| CSMA | Code Division Multiple Access[2] |
| TDMA | Time Division Multiple Access[2] |
| FDMA | Frequency Division Multiple Access[2] |
| FFSK | Fast Frequency Shift Key Modulation[2] |
| FSK | Frequency Shift Key |
| GMSK | Gaussian Minimum Shift Key Modulation[2] |
| RELP | Residual Excited Linear Predictive Coding[2] |
| VSELP | Vector Sum Excited Linear Predictive Coding[2] |
| QCELP | Qualcomm Code Excited Linear Prediditive Coding[2] 8 and 13 kb/s and Enhanced Variable Rate Vocoder (EVRC) |

---

### Notes

1. Registered trademark of ETSI.

2. See Chapters 3 and 4 for modulation, multiple access, and voice coding discussions.

3. GSM also operates in the 1900 MHz band in North America and in the 1800 MHz band in Europe.

4. IS-95 CDMA operations in the 1900 MHz band are also approved in the United States.

**Figure 16.2**   Cellular systems introduced since 1979.

Collaborative efforts among industry participants, nurtured by the *Cellular Telecommunications Industry Association* (CTIA) and savvy entrepreneurs who sensed an enormous business opportunity, changed that situation. Together they developed TIA/EIA-41 Intersystem Operations and other standards, a number of agreed-upon formats for compiling and exchanging data—such as TIA/EIA-124's Call Detail and Billing Record Exchange Formats, and the Cellular Intercarrier Billing Exchange Record (CIBER) billing record format.

Many new service providers were initially ill-equipped to cope with the management and administrative complexities of operating a local telephone company, much less national, and now international, requirements to enter into business relationships with hundreds of other companies. Cibernet Corporation, which began offering financial and administrative intercarrier settlement services to mobile service providers in 1988, now lists 200 client carriers in over 50 countries.

It should become evident from ensuing discussions that, as in the broader wireline telecommunications business segment, since all service providers have access to the same technologies, over time, technical performance and quality of service diminish as competitive discriminators. Among competent service providers, the costs related to high-quality telephone service delivery also tend to equalize. As a result, over the long term the market discriminators that ultimately determine winners and losers become those business and administrative capabilities that lead to superior customer service, response and satisfaction.

Returning now to technical considerations, cellular systems listed in the last four columns of Figure 16.2, when introduced, represented the first appearances of *second-generation (2G)* capabilities. While there are no rigid generational definitions, 2G designs are typically credited with employing either time division or code division multiple access techniques (TDMA/CDMA); some form of narrowband digital voice encoding; and digital control and voice channel radio frequency (RF) carrier-modulation techniques.

2G designs, so to speak, "pull out the stops" in terms of equipping users with as many telephony features as might be found in the most advanced desktop wired telephone sets. More important, digital coding, modulation, and multiple access technologies not only enhanced service quality and reliability, but also greatly increased the number of subscribers who can be served per Hertz of FCC-allocated channel bandwidth.

The importance of "bandwidth efficiency," the number of subscribers who can be served per Hertz, is hard to exaggerate. From a natural resource perspective, since available spectrum is finite, efficient spectral usage is a prerequisite to being able to provide mobile telephones to all the people who want them. From a service provider's perspective, finding ways of packing more subscribers into a fixed and limited FCC spectral allocation is the only way to accommodate more subscribers continually and thereby increase revenue.

Defining characteristics and features of *third-generation (3G)* designs are the object of much industry and standards-body work. The ITU-T initiative for the development of *International Mobile Telecommunications for the year 2000* (IMT-2000) is an umbrella effort directed at 3G specifications. While still evolving, 3G capabilities will include a full complement of telephony features: numeric, text, and voice paging; two-way data communications; Internet access and Web browsing; and even video conferencing. Prototype subscriber handsets incorporate color liquid crystal flat displays, touch-panel controls, and miniature and back-lit *qwerty* (standard computer or typewriter) keyboards for outgoing message composition. Handsets and services supporting some 3G features are already on the market.

Whereas most subscriber equipment is designed (1) to operate in a single frequency band, (2) to support either analog or digital information formats and/or modulation, and (3) to possess a single telephone number and be registered in only one service provider or carrier system, multifunction telephones are available. For example, some dual-band telephones operate in both 800- and 1900-MHz bands and *dual-mode* telephones support both analog and digital operations. Finally *numerical assignment modules* (NAMs) in each device relate MINs (telephone numbers) and ESNs (manufacturer-assigned serial numbers) to a particular service provider. Some telephones have dual NAM capabilities, which means they can have two telephone numbers, registered to the same or two different service provider systems.

The top row in Figure 16.2 indicates nominal base-station and mobile-set transmission frequencies for the different systems illustrated. Note that around the world, GSM, PCS and IS-95 systems, all of which are discussed separately below, may each be implemented in different frequency bands. Other rows in the figure indicate the channel bandwidth; the number of traffic channels per RF channel; total numbers of traffic channels not considering frequency reuse; voice format/encoding and control; and voice signal RF modulation techniques.

In Figure 16.2, CT1, CT2, CT3, and DECT (Digital European Cordless Telephone) are European cordless telephone specifications. The use of cordless telephones designed to these specifications in Europe is markedly different from the typical use of cordless telephones in the U. S. In the U.S., most cordless telephones in residences and businesses communicate with customer-owned base stations (see Chapter 4), and are merely "untethered" replacements for corded customer premises telephone equipment. In Europe, however, cordless telephones designed in accordance with the above specifications are intended to be replacements for coin-operated public telephones, which in many localities are in short supply. In Europe, these devices are sometimes referred to as telepoint telephones. As with all wireless technologies, if one waits long enough, differences among various manifestations tend to disappear, and universal devices blur what had once been clearly defined operational distinctions.

## Digital AMPS, IS-54, and IS-136 Cellular Systems

Within a few years after analog cellular service was introduced in 1983, it became clear that higher-capacity, more-reliable, and lower-cost systems were needed to meet growing demand. To stimulate technology development, in 1987 the FCC declared that cellular licensees could employ alternative technologies in 800-MHz bands, provided that they did not interfere with existing services. Responding to predictions that system capacity would be exhausted by the early 1990s, CTIA established a subcommittee in 1988 to identify both improved service and capacity requirements, and solution alternatives.

Among postulated requirements were a tenfold increase in capacity relative to AMPS; dual-mode, simultaneous analog AMPS and digital operations; new data features including facsimile and short message service; and enhanced service-availability by 1991. The Telecommunications Industry Association was asked to create a specification to satisfy those requirements and released Interim Standard-54 in 1991. IS-54 is an approach that permits older analog AMPS and newer digital mobile telephones to share the same base stations—with each operating in designated channels that previously were all committed to AMPS service.

IS-54 operating principles are illustrated in Figure 16.3. Along the top of the figure is a mobile transmitter frequency profile corresponding to the FCC channel allocations first depicted in Figure 16.1. As Figure 16.3 indicates, some of the 416 channels assigned to a service

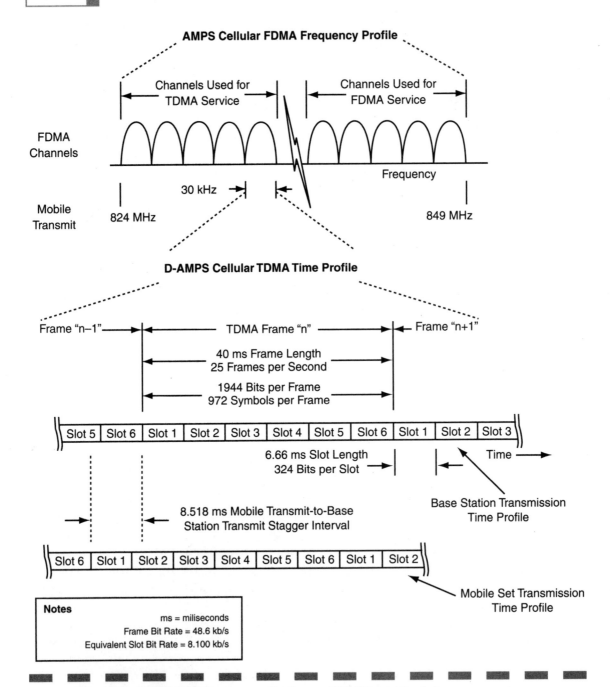

**Figure 16.3** Digital AMPS FDMA frequency utilization and TDMA frame structure.

provider are designated for older AMPS analog FDMA service and continue to operate in the fashion described above. The remaining channels are used to support digital TDMA service implemented using time division multiplexing (TDM) operating principles explained in Chapter 5.

The bottom portion of the figure portrays the TDMA frame time profile. Each of the 30-kHz-wide frequency division channels designated for digital TDMA service contains TDM signals with frame lengths of 40 milliseconds (ms)—equivalent to 25 frames per second. Each frame consists of six time slots used to carry both user traffic and TDMA signaling and control signals. The digital signal bit rate is 48.6 kbps, which means that there are 1,944 bits per frame. This translates to 324 bits per time slot and an equivalent per-slot throughput rate of 8.1 kbps. Note that there is an offset in time between mobile set and base station transmission of 8.158 ms. The reason for the offset is that a mobile set that transmits in time slot 1, for example, completes its transmission before the set needs to receive the corresponding time slot 1 transmission from a base station. In this way, mobile sets need not be capable of transmitting and receiving at the same time.

In voice user-traffic cases, the information placed in each time slot corresponds to "digital voice samples," which means that some form of analog-to-digital voice encoding must be used. IS-54 specifies the use of *vector sum excited linear predictive coding* (VSELP—see Chapter 3 for a discussion of alternative digital voice processing techniques) that produces an output bit rate of 13 kbps. Since, as noted above, the maximum bit rate that can be supported by a single time slot is 8.1 kbps, in IS-54 conforming designs, two time slots per frame must be used to support 13-kbps VSELP output signals. Because two time slots can support 16.2 kbps, this leaves 16.2 minus 13 kbps of capacity for signaling and control purposes.

To squeeze IS-54's 48.6-kbps transmission bit rates into AMPS 30-kHz channels, a 2B1Q (2 Binary, 1 Quaternary) line encoder is used (see Chapter 4 for basic coding and modulation techniques and rationale explanations). In 2B1Q encoder/modulators, successive pairs of user data bits are mapped into one of four (quaternary or 4-ary) symbols. Thus transition rates for 48.6-kbps IS-54 signals are limited to a maximum of 24.3-kbps symbol rates, a rate that fits comfortably in 30 kHz AMPS channels.

The RF modulation technique selected for IS-54 is known as π/4-shift, differentially encoded quaternary (4-ary) phase shift keying (π/4-QPSK). The name π/4-QPSK derives from the fact that carrier phase shifts on successive two-bit symbols are restricted to $\pm\pi/4$ (that is $\pm$ 45 degrees) or $\pm 3\pi/4$ (that is $\pm$ 135 degrees) instead of 0, $\pm\pi/2$, or $\pm\pi$, (that is 0 degrees, $\pm$ 90 degrees, or $\pm$ 180 degrees) for conventional QPSK. When used with 2B1Q encoders, like conventional QPSK modulation, $\pi/4$-QPSK limits symbol transition rates to one half input bit rates, but exhibits superior adjacent-channel interference performance and is compatible with analog FM demodulators, a characteristic important in dual mode cellular radio receivers.

TDMA IS-136, also called Digital AMPS, is a revision of TDMA IS-54 that was published in 1994. Note in Figure 16.2 that the maximum number of voice channels that can be supported in either IS-54 or IS-136 cellular systems, excluding frequency reuse considerations, is three times the 832 channels that AMPS supports. Thus, movement to TDMA technology has to some extent satisfied the requirement to expand capacity beyond original AMPS capabilities.

The fact that both analog FDMA and digital TDMA operations can coexist means that the enormous number of people who purchased analog mobile telephones need not immediately give them up. In fact, according to CTIA, at the end of calendar year 1999, of 86,047,003 revenue-generating wireless subscriber sets, 45,604,912 were still analog, while 40,442,091 were of some digital design. Relative to other parts of the world, the incredibly fast growth in numbers of analog subscriber sets just after cellular service began in the U.S., has always been an impediment to rapid upgrades and changes in network technologies.

Thus, one of the greatest advantages of IS-136 is that it provides immediate improvements in capacity and performance, while accommodating a more gradual phase-out of customer-owned analog subscriber sets. The principal advantage of IS-136 compared to IS-54 is that the new specification takes into account advanced hardware, software, system, and control technologies that were not available when IS-54 was formulated.

## GSM Cellular Systems

Like IS-54/IS-136 efforts in the U.S., GSM was developed in Europe as a means of improving the capacity and performance of existing analog cellular systems. Unlike the United States where the FCC was attempting to move the industry from a single analog standard to a new generation of *competing* digital standards, ITU-T began with a half-dozen incompatible analog cellular system designs and embarked on a course to define a single, international open, nonproprietary digital cellular standard, initially in the 900-MHz band.

Whereas U.S. efforts were constrained by "grandfather clause" requirements to preserve existing mobile subscriber set and compatible cell site operations, GSM was in essence a "clean slate" design effort. To their collective credit, European standards and industry engineers thought of just about everything needed to produce a high-capacity, high-quality, and high-reliability cellular system that fully anticipated operational, business, management, and administrative requirements that truly international operations demand.

As just one example, each GSM telephone contains a *subscriber identification module* (SIM) receptacle, into which small plastic SIM cards can be inserted. Each SIM *smart card* contains a microprocessor with 8 kilobytes of memory that contains all the information needed to identify a particular subscriber (person not set), and an algorithm used to encrypt that subscriber's voice or other information transmitted over the airwaves by GSM conforming mobile sets. Although SIM cards are used in other PCS (discussed next) mobile sets, GSM devices won't work at all without SIM cards.

Any telephone into which a SIM card has been inserted, whether it be owned by a subscriber or mounted in a taxi cab for public use, becomes uniquely identified with the particular subscriber who owns the card. Subscribers owning cards thus become responsible for any calls placed or received by whatever telephone contains their card.

Clearly SIM cards are a fundamental building-block for a universal and international roaming chargeback system. But beyond mere "after-the-fact" call-charge tracking, GSM operations use worldwide SS7 networks (the signaling and supervisory networks built for wireline telecommunications systems) to provide real-time pre-call authentication and validation for outgoing calls and roaming-subscriber location determination for incoming calls. Attached to every *mobile services switching center* (MSC—MSC is GSM terminology for U.S. MTSOs) are both *home location registers* (HLRs) and *visitor location registers* (VLRs), databases that facilitate roaming and local GSM operations.

Figure 16.4 summarizes GSM's architecture. Notice that although the topology resembles U.S. AMPS systems depicted in Figure 16.1, the terminology for some components is slightly different. The *equipment identity register* (EIR), for example, is a database that contains what we earlier referred to as *electronic serial numbers* (ESNs). In acquiring a working knowledge of GSM operations, it is helpful to become familiar with the GSM components pictured in Figure 16.4. Note that the seven numbers in circles correspond to the International Consultative Committee for Radio's (CCIR's) Future Land Mobile Telecommunications System functional references model interface numbers, one indication of how thorough the GSM specification is.

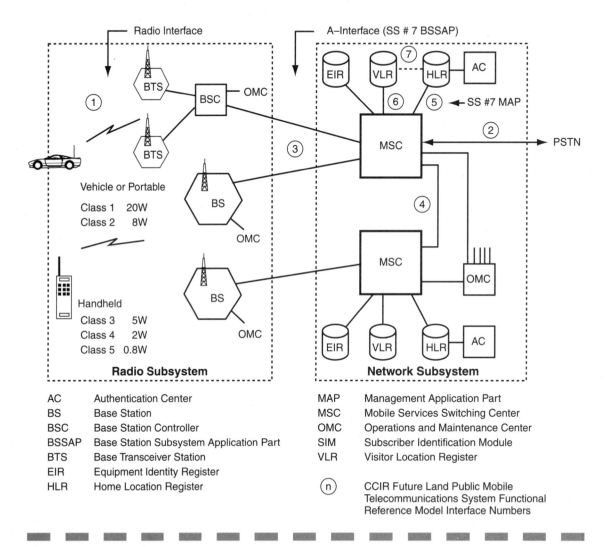

| | | | |
|---|---|---|---|
| AC | Authentication Center | MAP | Management Application Part |
| BS | Base Station | MSC | Mobile Services Switching Center |
| BSC | Base Station Controller | OMC | Operations and Maintenance Center |
| BSSAP | Base Station Subsystem Application Part | SIM | Subscriber Identification Module |
| BTS | Base Transceiver Station | VLR | Visitor Location Register |
| EIR | Equipment Identity Register | | |
| HLR | Home Location Register | (n) | CCIR Future Land Public Mobile Telecommunications System Functional Reference Model Interface Numbers |

**Figure 16.4**   *Global System for Mobile Communications (GSM) system architecture.*

European GSM telephones operate in the 900- and 1800-MHz bands. In the U.S., GSM telephones operate in the 1900-MHz band so only dual-band sets (e.g., the Bosch model 718) can be used in both locations.

However, SIM smart cards can be used anywhere. Thus, U.S. travelers in Europe and Asia may rent European-version telephone sets and be "called" wherever they may travel and automatically have accumulated outgoing call charges transferred to their U.S. service provider accounts.

Companies offering GSM services in the U.S. include Omnipoint, Pacific Bell, Aerial Communications, Powertel, and Voicestream. GSM wireless services are now available in most major U.S. cities. Omnipoint, for example provides service in over 3,500 cities in the U.S. and Canada, and in over 65 countries around the world. Nokia 5190 and 6190 mobile telephones support both digital 1900-MHz band GSM and analog AMPS 800-MHz operation. Modern GSM telephones and networks are feature laden and designed to support a broad range of voice, paging, data, Internet access, and other services.

Figure 16.5 reveals the breadth of mobile network-to-wireline network and supporting business-institution connectivity encompassed in GSM standards and actual designs. It also reveals an extensive and rich assortment of technical, administrative, marketing, security, and business-management functions embedded in modern systems.

## PCS Cellular Systems

Many industry experts believe that wireless communications for voice, data, and other services is becoming the most significant telecommunications development since fiber optics. In the early 1990s when the term first came into vogue, *personal communications service* (PCS) meant different things to different people. To some it was a cellular system with smaller cells and lower-power (and therefore longer-battery life) mobile telephone sets. For others it was an outgrowth of cordless telephone technologies. Still others emphasized wireless LAN and PBX applications.

More recently the FCC defined PCS broadly as a family of mobile or portable radio services that can be used to provide service to individuals and businesses, and can be integrated with a variety of competing networks. It anticipated that PCS licenses would be used to provide such new services as advanced voice paging, two-way acknowledgment paging, data, and other services.

Today, PCS is a term that encompasses a wide range of wireless mobile services, chiefly two-way paging and cellular-like calling services that are transmitted at lower power and typically higher frequencies (1850 to 1950 MHz) than analog cellular services. Although PCS systems use cellular structures, in some circles, "cellular services" now

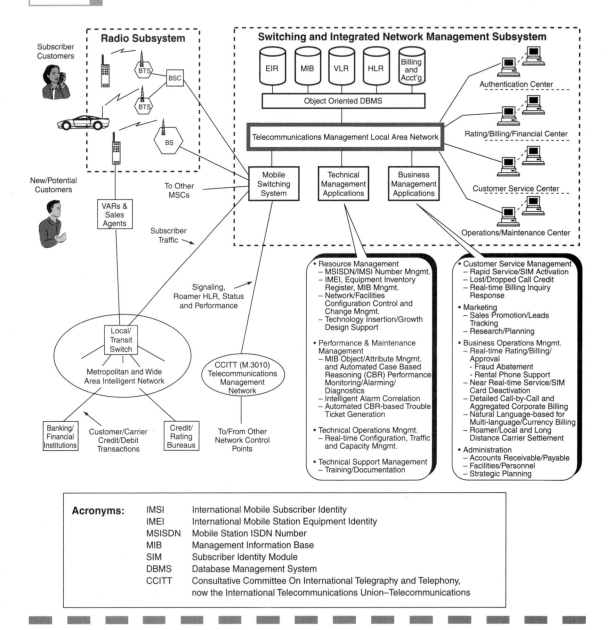

**Figure 16.5** *GSM architecture illustrating network interfaces and representative technical, administrative, marketing, and business-management functions.*

connote only those offered in the original AMPS 824-to-894 MHz band. In fact, as just noted, GSM encompasses all services envisioned for PCS, and many GSM sets operate in the 900-MHz band. Thus, PCS is more accurately defined as capabilities to support a far-reaching variety of personal communications services, without regard to transmission frequency bands.

Although the notion of PCS-like service features was not new, what was new in 1993 was the decision to auction licenses in a part of the electromagnetic spectrum that the FCC had allocated for emerging communications technologies. In October 1993, the FCC concluded that a combination of *major trading areas* (MTAs) and *basic trading areas* (BTAs) would, "promote the rapid deployment and ubiquitous coverage of PCS and a variety of services and providers." MTAs are large geographic areas defined in the Rand McNally Statistical Atlas.

The spectrum was divided among three PCS service categories: narrowband PCS in the 900—901-MHz, 930—931-MHz, and 940—941-MHz bands; broadband PCS in the 1850—1990-MHz band; and, an unlicensed portion of spectrum at 1910—1930 MHz. Broadband PCS allocations are detailed in Figure 16.6. Each MTA license extends over 30 Megahertz of radio spectrum, 15 MHz each for mobile and cell site base station transmission.

In 1994, the FCC began a series of license auctions intended to foster competition, to raise money for the U.S. Treasury, and to increase the number of service providers in each market from as few as two to as many as eight. Under auction rules, wireless providers were eligible to bid for new radio spectrum in MTAs not covered by their existing licenses. In 1995, the FCC granted 99 licenses in 51 MTAs in the largest auction held to date, raising $7 billion dollars. Three more auctions followed, as the FCC offered additional C, D, E, and F block PCS licenses covering BTA-based markets.

When the FCC declared in 1987 that cellular licensees could employ alternative technologies in 800-MHz bands to stimulate technology development, it signaled a policy decision to decouple spectral allocations from specific radio technologies. As a consequence, no single, modulation, multiple access, or physical network design or technology can be uniquely associated with FCC PCS allocations. To date in North America, three digital cellular phone technologies have been implemented in the new bands, namely GSM 1900, CDMA IS-95, and TDMA IS-136. With the basics of GSM and IS-136 already covered, we next address CDMA IS-95.

## CDMA Cellular Systems

As explained in Chapter 5's discussion of multiplexing techniques, frequency division multiplexing (FDM) divides a transmission circuit's frequency spectrum into subbands, each supporting single, full-time communications channels on a noninterfering basis. In time division multiplexing (TDM), a transmission facility is shared in "time" rather than "frequency." Unlike either FDM or TDM, with *code division multiplexing* (CDM), individual channel signals are modulated with special, *orthogonal coding signals* in such a way that multiple signals can be transmitted in the same frequency band and at the same time without significantly interfering with each other.

In real systems, to approach ideal results, the bandwidth occupancy of coding signals is often hundreds of times as large as the input signal's. For example, when 10-kbps voice information input signals are modulated by coding signals; a modulator output signal is produced that occupies a bandwidth equal to the coding signal's bandwidth; typically it will be in excess of 1 Mb/s. This bandwidth expanding process is referred to as "spread-spectrum" modulation.

*Spread-spectrum* technology first appeared in military systems in the 1940s because transmitted spread-spectrum signals can be virtually immune from intentional enemy jamming signals. The 1970s and 80s saw growing commercial interest and in the late 80s and early 90s, Qualcomm proposed and developed a CDMA-based cellular system. In 1993, the Telecommunications Industry Association completed IS-95, a CDMA cellular specification based on Qualcomm's design.

Since then, industry proponents formed the *CDMA Developers Group* (CDG), an organization that promotes cellular CDMA technology and standards. Operator and customer support have been spectacular. From the first commercial system launch in 1995, by the ends of calendar years 1997, 1998, and 1999, worldwide CDMA subscribers reached 7.8, 23.0, and 50.1 million respectively. Like TDMA IS-136, CDMA operates in both the 800- and the 1900-MHz bands.

Major U.S. carriers include AirTouch, Bell Atlantic Mobile, GTE, Primeco (a consortium of AirTouch, Bell Atlantic, and US West), and Sprint. Sprint markets its CDMA offering as Sprint PCS and claims to operate the largest 100 percent digital, 100 percent PCS wireless network in the United States. Serving the majority of the nation's metropolitan areas including 4,000 cities and communities across the country, Sprint PCS has licensed coverage of nearly 270 million people in all 50 states, Puerto Rico, and the U.S. Virgin Islands.

Key CDMA advantages include the ability to (1) serve more subscribers than FDMA or TDMA in a given spectral allocation (like the 800-MHz AMPS and 1900-MHz PCS bands); (2) provide higher quality and more reliable service (fewer dropped calls); and (3) enhance interference immunity from other subscriber signals or noise from other sources. All these advantages are in large part due to the spread-spectrum nature of CDMA's transmitted signals. A complete explanation of why CDMA spread-spectrum cellular systems exhibit these advantages can literally fill another book. For those readers needing the details, Jhong Sam Lee and Leonard E. Miller's 1,228-page book published in 1998, *CDMA Systems Engineering Handbook*, is a highly recommendable and comprehensive reference.

While those details may be beyond the interests of most readers, to comprehend even top-level operational, performance, business, and market implications of this new cellular technology requires some appreciation of how and why it works.

Orthogonal coding or spread-spectrum signals specified in IS-95 are generated in part by using what are known as *pseudo-noise* (PN) code generators. The highest chip-rate PN code specified in IS-95, and the one determining the spread-spectrum bandwidth of transmitted signals, is 1.2288 million chips per second (Mcps). Chip rates are roughly equivalent to bit rates, defined in Chapter 3, except that because PN code generators produce random digital signals, the rate that codes transition from binary "1" to binary "0" signal states also occurs randomly. For PN codes, therefore, the chip rate refers to the maximum rate at which PN code generators can change state.

RF carrier modulation in IS-95-conforming systems employs a special form of quaternary phase shift key (QPSK) modulation. Unlike IS-54/IS-136 QPSK modulation, wherein pairs of data bits are encoded into a single phase position, in IS-95 the four modulation phases are used to enhance detection quality and to minimize interference among competing user transmissions and other noise sources, rather than to conserve transmission signal bandwidth. (See Chapter 4's DSL modulation discussion for background.)

As a simple example of how CDMA allows multiple signals to be transmitted in the same frequency band at the same time, and still provide the means to recover each information signal separately, consider the following. Assume a situation in which we have ten equally powered CDMA telephones using 10-kbps digital voice encoders that employ 1-Mb/s CDMA spread-spectrum modulation. Next, assume that a particular receiver receives all ten spread-spectrum signals with equal power density. Then, as measured in the "spread" bandwidth, any particular desired signal level is roughly only one tenth of the power level of all transmitted signals. Another way of stating this is that in the spread bandwidth, the ratio of desired signal power to total competing signal power is negative—that is, desired signals exist at a ten-to-one, negative signal-to-noise ratio.

In the process of detecting or recovering a particular desired signal from the ensemble of signals, if the receiver has a local version of the identical PN code used in the transmitting station, when that code is used in the demodulation process, a remarkable result occurs. What happens in a properly designed demodulator is that a replica of the narrowband transmitted digital voice signal is recreated at the receiver, and it is roughly ten times the noise-power level produced as a result of the presence of competing subscriber set signals in the common or shared radio frequency spread-bandwidth channel.

Basically, in CDMA systems, this relative spread-spectrum signal-to-noise level improvement factor is called *processing gain* and is given by the ratio of the spread bandwidth (or chip rate) to the information signal bandwidth or bit rate. In the above example this ratio is 1 Mcps-to-10 kbps or 100. So mixing ten equal-powered signals in a common spread-spectrum channel places desired signals at a disadvantage, or negative desired signal-to-competing signal ratio of 10. But demodulation boosts the relative power of a desired signal by a factor of 100, to produce a net positive desired signal-to-competing-signal power ratio of 10.

In this example, note that ten subscriber signals are carried in 1 MHz of common or shared RF bandwidth. The bandwidth efficiency is 10 users per 1 MHz or 100 kHz of bandwidth per subscriber. If satisfactory voice quality can be obtained with a demodulator output signal-to-noise ratio of less than a factor of 10, then more than ten subscriber signals can be packed into the same 1 MHz of RF spectrum, and the bandwidth efficiency might rise to 15 subscriber signals per 1 MHz of RF spectrum. If 10 MHz of RF spectrum is available, then instead of packing 10 or 15 subscribers in a single 1 MHz spectrum slot, 100 or 150 subscribers can be placed in the 10 MHz slot. Note however that bandwidth utilization efficiency remains the same.

In addition to CDMA processing gain, there are other important design techniques that allow even more subscribers to be accommodated in a given RF spectrum. Techniques such as power control, variable voice encoding rate, and ability to capitalize on speech pauses, directional antenna-based sectorization, and frequency reuse efficiency are identified and explained below.

In a mobile telephone system cell, some subscriber sets may be closer to base stations than others. Since radio signal transmitter-to-receiver power loss is inversely proportional to the square of the distance separating them, subscribers who are much closer than others have an enormous received power advantage over others, and can "jam" distant weaker signals beyond recovery. To counter this possibility, IS-95-conforming sets use control channels between subscriber-set and base-station transceivers to adjust transmitter power levels so that all signals received by either subscriber sets or base stations are nominally equal.

Another feature that makes IS-95 CDMA systems more tolerant of network congestion and geographically induced signal degradation is the ability to reduce the output rate of the digital voice encoders embedded in mobile telephones. As a numerical example, assume that a vocoder can produce digital voice output signals at both 10 kbps and 1 kb/s. For a constant 1-MHz RF bandwidth, as in the example above, dropping the digital voice rate to 1 kbps raises spread-spectrum processing gain from a factor of 100 to a factor of 1000. What this means is that now instead of ten subscribers per 1 MHz of RF bandwidth, 100 users can be accommodated.

The penalty paid for this increased capacity is reduced voice quality. Vocoders operating at 1 kbps produce voice with severe intelligibility and speaker-recognition problems. Whereas a constant 1-kbps vocoder rate would be unacceptable, momentary operations at such rates are acceptable. A few seconds of degraded voice quality is normally a better alternative to "no service" at all, or a "dropped call." Since subscribers originate and terminate calls at random intervals and vehicles enter and leave cells at random times, traffic congestion within cells is often a dynamic phenomenon.

It is important to recognize that although the ability to substitute lower data-rate voice temporarily within existing calls, as an alternative to service denial or dropped calls, is relatively easy to support in CDMA systems, there are no straightforward FDMA- or TDMA-based design equivalents. As indicated in Figure 16.2, in IS-95 systems, analog voice signals are converted to digital format using a Qualcomm-invented form of code excited linear predictive coding (QCELP). QCELP vocoders produce two fixed output digital bit rates at 8 and 13 kb/s. Qualcomm vocoders can also produce variable output bit rates at 1.2, 2.4, and 9.6 kb/s.

Another traffic congestion management technique possible in CDMA systems is the ability to reduce transmitted power during speech pauses. In full-duplex voice communications, unless two parties talk at the same time, either the forward or the reverse channel is idle. Even when one person is talking, there are significant pauses between what are known as *talk spurts.* As described in Chapter 3, both *digital speech interpolation* (DSI) and analog-based *time-assigned speech interpolation* (TASI) have been used for decades on transoceanic transmission to essentially double traffic-carrying capacity.

Again, in cellular systems, reducing transmitter power during speech pauses has immediate congestion relieving capacity-expanding benefits in CDMA designs. By contrast, within today's FDMA/TDMA cellular designs, reducing transmitter power during speech pauses has no direct impact on traffic capacity at all.

Using complex analysis, and taking all these factors into account, it can be established that the CDMA systems can support 106 users in 1.2288 MHz, and FDMA systems 60 users in 12.5 MHz of spectrum. Thus, CDMA system designs can support about 18 times more users than FDMA in the same spectral bandwidth. In comparing CDMA to TDMA spectral utilization, CDMA systems are about 1.7 times more efficient than TDMA systems.

Finally, CDMA offers two additional design options not intrinsically supported in FDMA/TDMA approaches. One non-communications application for spread-spectrum signals is in the field of precise distance measuring, position location, and navigation equipment. Without getting into details, when modulated electronic signals are used in position measurement apparatus, resolution, or the degree of accuracy obtainable, is directly proportional to modulation bandwidth. Thus, CDMA's spread-spectrum signals offer the possibility of precision distance (location) measurement, while simultaneously being used for voice or data information communications.

Lastly, the use of complex PN codes greatly enhances privacy and security against unsophisticated eavesdroppers. Moreover, if nonlinear PN codes are used, a form of transmission security can be achieved that poses a significant code-breaking challenge.

Clearly, the laundry list of CDMA-based design options has much to do with the fact that cellular system operators appear to be rushing to embrace the technology. As noted, however, at the end of 1999 more than half of U.S. cellular subscriber sets were still of analog design. Consequently, a complete CDMA rout is not anywhere on the immediate horizon. Then too, FDMA-TDMA proponents are not rolling over. We can look for more vigorous competition, a characteristic that has been the hallmark of the cellular market since its inception.

## Specialized Mobile Radio (SMR) and Enhanced SMR Systems

Originally, SMR referred to a service based on large cells of up to 50 miles in radius. Also called *trunk mobile radio*, SMR employs a CB radio-like push-to-talk operation that finds application in police, taxicab, fleet delivery, and similar industries. Accordingly, it is often called *radio dispatch service*. In older systems, users selected from a plurality of FDM channels, listened for traffic activity, and if none was heard, pushed microphone buttons to talk and transmit voice signals to a high-elevation signal-repeating base station. Originally, SMR base stations simply translated signals received on one set of frequencies to pair-wise related transmit frequencies.

Used in 1921 by the Detroit Police Department, the first commercial service was offered by AT&T in 1946 in the city of St. Louis. By the early 1990s, there were 5,000 operators and the FCC had granted over 900,000 licenses. FCC allocations for SMR service are in the 806—821-, 929—932-, and 935—941-MHz bands. Gradually SMR operators converted from analog to digital technologies and added cellular-like automation features.

*Enhanced SMR* preserves dispatch-like push-to-talk operations, but adds cellular mobile telephone dialing features and ESMR-to-PSTN wireline network connectivity, allowing SMR operators to compete directly with any of the cellular carriers described above. Nextel and Geotek Communications Inc. were instrumental in developing ESMR, and although Geotek filed for bankruptcy in 1998, Nextel has established a nationwide network serving 92 of the top 100 U.S. markets and thousands of communities across the United States. In addition to the mobile cellular telephone features discussed above, Nextel offers what it calls Direct Connection®, a dispatch push-to-talk service among paired or groups of users. According to Nextel, while cellular service is available nationwide, Direct Connection® service is only available in home markets.

On April 3, 2000, Nextel announced Nextel Worldwide®, a digital wireless service in the U.S. and more than 65 other countries offering service in 18 of the world's 25-largest cities. Outside Nextel SMR-based serving areas, subscribers use GSM cellular service. Nextel offers dual-mode Motorola i2000® iDEN® telephones (both Motorola trademarks) to customers, equipping them with one-telephone, one-number service in any of Nextel Worldwide's coverage areas.[1]

Motorola's SMR series of iDEN products are designed to support dispatch, standard mobile cellular telephone dial-up connectivity, data, and paging services when operated in networks that support those services. Although the i2000 telephone weighs just 5.9 ounces and is only 5.3 inches high, its features include Nextel's Direct Connect two-way radio dispatch feature; built-in speaker phone for hands-free conference calling; text/numeric paging; Caller ID; voicemail; one-touch dialing for eight preprogrammed numbers; an ability to add phone numbers and private IDs to personal electronic directories; last ten numbers received/sent list; a missed-call indicator; 3-way calling; and a "vibration" silent notification. Users may select from among English, French, Spanish, and Portuguese as languages in which menu and other information appear on a four-line display.

The fact that it now offers wireless Internet service (using i500plus®, i700plus®, or i1000plus® Motorola telephones), means that Nextel (and other ESMR providers) are in a position to compete with any of the cellular telephone companies operating in the 800-MHz AMPS or 1900-MHz PCS bands, across the entire gamut of mobile services.

## Terrestrial Wireless Data Services

A broad context for wireless service technologies is established in Chapter 4. In particular, Figure 4.19 depicts relationships between spectral allocation, bandwidth (data rate), and terminal mobility. Within that context, the merits of today's *wireless data network* (WDN) services can best be presented and understood by first segmenting the growing number of service-provider offerings into a smaller number of categories. Accordingly, as Figure 4.19 suggests, some WDNs are designed to accommodate high-mobility terminals (those mounted in automobiles or airplanes), hand-carried or personal slow-moving or merely portable (fixed while operating) terminals, and lastly fixed terminals (those that can be equipped with permanently installed antennas).

---

1 Overseas, the European Telecommunications Standards Institute's (ETSI) standard for Terrestrial Trunked Radio (TETRA) combines the features of mobile cellular telephony, dispatch radio, and 28.8 kb/s packet radio data transfer that supports short message service (SMS) and both numeric- and text-based messaging. The TETRA forum, formed in 1994, includes over 60 organizations from dozens of countries.

Next, WDNs may employ either packet or circuit switching, which, as in fixed-data networks, produce data services with decidedly different traffic-related characteristics and efficiencies. The third category that helps one to sort through the array of wireless data offering possibilities has to do with the specific technical approach and standards used to design the facilities that deliver the services. The ensuing discussion reviews important, popular data services in terms of whether underlying facilities conform to open system standards or use proprietary designs and protocols; and whether or not they rely on established FDMA, TDMA, CDMA modulation and multiple access technologies (discussed above), or some other vendor/provider-specific technique.

Clearly, within these categories, some approaches are better suited than others in supporting e-mail, short message service, access to news, financial, travel, sports information, general access to the Internet and entertainment sources, access to corporate LANs and WANs, and business data processing applications (file transfer, decision support, etc.), paging and numerous other services. Paging is addressed separately in a section below.

It is technically possible to transmit data through mobile cellular networks using standard voice-network wireline modems (e.g., V90 modems described in Chapter 3), and some cellular telephones provide jacks to do so. An advantage of this approach is that with cellular-to-wireline network connections, wireless data users have ubiquitous worldwide access to any PSTN recipient equipped with a compatible modem. There is no need for protocol-converting, cellular-to-wireline gateway connections. The disadvantage is that because mobile connection reliability is so poor (due to cellular hand-off operations and other cellular system transport vagaries like dropped calls), the approach has virtually been abandoned and will not be further addressed.

### CELLULAR DIGITAL PACKET DATA NETWORKS

*Cellular Digital Packet Data* (CDPD) is an IP-compatible packet switching data protocol designed to work over AMPS analog FDMA (800-MHz) facilities. The packet-switching and handling parts of the protocol can also be used in digital TDMA PCS (1900-MHz) systems. In AMPS systems, data are transmitted over idle (no voice traffic) FDMA channels. In heavy traffic, CDPD designs "squeeze" data traffic onto channels being released from call connections, prior to reassigning them for voice traffic. Data traffic may also be temporarily placed in FDMA channels as mobile voice terminals are handed off among different base stations.

CDPD providers across the nation offer service to users in home serving areas and to roamers as well. GoAmerica Communications Corporation, for instance, is a national wireless Internet service provider company founded in 1995 to offer mobile access to e-mail and corporate data, using information appliances such as 2-way messengers, *personal digital assistants* (PDAs), handheld, and laptop computing devices. One mobile terminal hardware arrangement involves plugging business card-sized wireless PC cards (for example the Sierra Wireless Inc. Air Card 300 PC Card) into laptop Type 2, *Personal Computer Memory Card International Association* (PCMCIA) slots. Such cards contain AMPS-compatible FDMA transceivers designed to automatically establish and maintain connections with any CDPD-capable cellular service provider.

GoAmerica offers service on CDPD networks throughout the continental United States and in Hawaii. In the northeast, Bell Atlantic Mobile (known as Verizon Wireless since its merger with Vodaphone Air Touch PLC and PrimeCo on April 4, 2000) provides CDPD facilities. Other companies providing facilities for GoAmerica subscribers include BellSouth Wireless Data and American Mobile Satellite Corporation. American Mobile Satellite Corporation changed its name to Motient Corporation on April 25, 2000.

### MOBITEX® AND ARDIS NETWORKS

BellSouth uses Mobitex networks to deliver wireless data services to GoAmerica and other subscribers. There are now 29 networks in 22 countries built to the Mobitex Interface Specification. A Mobitex Operators Association exists to foster interoperability and roaming operations. Originated by Ericsson, Mobitex networks work in the 80, 800 and 900 MHz bands. In American networks, mobile terminals transmit at 896—902 MHz and receive signals in the 935—941-MHz band, using a type of carrier sense multiple access (CSMA—see Chapter 13) and supporting user data rates at 8 kbps. Mobitex networks can connect to customer hosts or LANs using wireline networks and X.25 protocols. Research in Motion manufactures and sells its RIM Wireless PC Card, which like the Sierra Air Card 300 plugs into laptop Type 2, PCMCIA card slots.

Another WDN option for GoAmerica and other subscribers needing wireless data service is the ARDIS network. This was originally developed by IBM and Motorola for use by IBM field engineers and service technicians in IBM customer facilities. As odd as it may seem, in the early 1980s, IBM—a company whose image is often equated to computing—held more FCC radio licenses than any other single organization except the U.S. Park Service. In March 1998, Motient Corporation—then American Mobile Satellite Corporation—acquired the ARDIS network, which today is the country's most extensive wide area packet-data network, with 1,700 base stations in 430 metropolitan areas and 10,000 cities across the United States, Puerto Rico, and the Virgin Islands.

What differentiates packet-data networks from other networks is that users are charged only for the data they actually exchange, and not for the time connected. This allows terminal equipment to be constantly turned on, and always ready to send or receive information. GoAmerica, for example, offers a pay-per-use price plan at $9.95 per month for the first 25 kilobytes and an additional $0.10 per kilobyte thereafter. An unlimited usage plan is also available for a monthly fee of $59.95. In either case, there are no additional roaming charges.

### GSM-BASED MOBILE DATA SERVICES
GSM's TDMA mode of operation (and that of any TDMA-based wireless cellular network) makes it well suited for data service. Today's GSM data transfer rate of 9,600 bps adequately supports e-mail, short message, facsimile, and other services not demanding higher rates, and is achievable using only a single TDM time slot in an eight-slot GSM TDM frame. But, like ISDN-based and other wireline *circuit switched data services* (CSDS) described in Chapter 14, the GSM design can intrinsically be adapted to assign multiple time slots to single calls resulting in 19.2-, 28.8-, and even 64-kb/s end-to-end data communications services. Identified in GSM specifications as *high-speed circuit-switched data* (HSCSD), a version that provides bandwidth-on-demand, variable-rate data service is also cited.

GSM specifications also provide for *General Packet Radio Service* (GPRS), the European standard digital cellular packet-switched data service. GPRS is a significant enhancement to GSM cellular networks, providing an ideal capability for Internet browsing and other bursty data applications. In November 1999, BT Cellnet, the UK's first mobile telephony and Internet provider, completed the world's first live GPRS data transfer call over a GSM network.

### IS-95 CDMA-BASED MOBILE DATA SERVICES
Like GSM and all other TDMA-based wireless cellular networks, as explained above, CDMA networks require that analog voice signals be converted to digital format for transmission. Being able to transport digital voice signals imbues these networks with intrinsic capabilities

to transport digital data traffic as well. IS-95B, released in 1988, specifies high-speed data operation using up to eight parallel codes (the equivalent of eight voice channels) to deliver a maximum bit rate of 115.2 kb/s. Beyond this second-generation (2G) CDMA capability, Part 6's *Outlooks for the Future,* identifies 3G wideband CDMA efforts that target user information rates up to 2 Mb/s.

### PROPRIETARY WIRELESS DATA SERVICES

A number of alternative wireless data services are appearing. Because most enterprises offering these services are start-up companies, current coverage is generally limited, with significant planned expansion. In November 1999, MCI WorldCom and Vulcan Ventures each purchased $300 million worth of preferred stock in Metricom, a public company since 1992. Metricom currently offers a wireless data service called *Ricochet* in the San Francisco Bay Area, Seattle, Washington D.C., and ten national airports. Ricochet is based on a digital packet-switched network using frequency-hopping spread-spectrum radio frequency modulation. The network consists of shoebox-sized radio transceivers, also called *microcell radios,* typically mounted atop street lights or utility poles, strategically placed at roughly half-mile intervals.

Within 20 square miles, containing about 100 microcell radios, Metricom installs wired access points, equipment that converts wireless network packets to formats for transmission via wireline frame relay networks to IP-based backbone networks. Metricom projects that today's 28.8-kb/s service will grow to a 128-kb/s capability available in 12 major markets by mid-2000, and cover 100 million people by 2001. Metricom has changed its business model and will no longer sell directly to end-users, but instead through authorized Ricochet service providers, the first one, of course, being MCI WorldCom.

# Local Multipoint Distribution Service/Microwave Multipoint Distribution Service

*Local multipoint distribution service* (LMDS) is a regulatory designation for broadband fixed-wireless systems that in the United States operate in licensed allocations in the 28- and 31-GHz bands. On March 25, 1998, after 128 rounds, the FCC closed an auction that began on February 18,

1998, awarding 864 licenses in 493 Basic Trading Areas (BTAs). In each BTA, an "A" Block license (consisting of 1,150 MHz of spectrum) and a "B" Block license (consisting of 150 MHz of spectrum) were auctioned.

The auction raised $578,663,029 and represented the largest amount of spectrum the Commission had auctioned to date. At that time, FCC Chairman William E. Kennard announced that the marketplace had 104 new LMDS players with the potential for offering real competition in the local loop business segment. In keeping with the trend to foster the maximum amount of competition and to capitalize on the inventive ingenuity of each licensee, a policy the FCC adopted with its cellular telephone allocations, the FCC leaves LMDS technology and service description details up to licensees.

Overall it is anticipated that, topologically, LMDS systems will resemble the architecture of mobile cellular telephone systems and be designed for line-of-sight communications between customer premises and base stations within cells of 1.5-to-4 miles in radius. Unlike cellular telephone systems, since both base stations and customer premises are fixed, high-gain, directional, and cross-polarized antennas at each location can be used. This minimizes interference among different base station-to-customer premises communication links, a factor that increases the number of subscribers that can be served in each cell, and boosts frequency reuse efficiencies among cells. Early estimates are that up to 80,000 customers can be provided wide-bandwidth, high data-rate service in each cell.

Like all the other telecommunications services discussed, LDMS service can be viewed as consisting of access and transport components, with end-to-end connections requiring both customer premises and network facilities. Customer premises LDMS equipment comprises an antenna or rooftop unit and a network interface unit (NIU). NIUs provide an interface between external LDMS networks and customer LANs or in-house premises distribution systems (PDSs). Chapters 4 and 9, and in particular the text discussions surrounding Figures 4.21, 9.10, and 9.11, provide insight into the functional and physical aspects of LDMS NIU implementation.

Facilities underlying LDMS network access service include NIU-to-base station/node and node-to-backbone network facilities. At this point, a key LDMS design objective is to avoid the single traffic type characteristics that most of today's backbone networks exhibit. Thus, ideally, LDMS transport services will support with equal ease IP-based and, at least an equivalent to ATM's multiple quality and class-of-service traffic-handling capabilities. Capitalizing on traffic-independent backbone transport service capabilities imposes the same type of requirements on facilities underlying access services.

Chapter 4 provides an overall context for wireless LDMS/MMDS services and relates U.S. developments to *European International Mobile Telecommunications-2000* (IMT-2000) and *Mobile Broadband System* (MBS) programs and concepts. Chapters 14 and 15 indicate how LMDS services relate to more traditional existing offerings and predict probable top-level service characteristics.

*Microwave multipoint distribution systems* (MMDS) is a 2-GHz mid-70s technology whose initial application was a wireless alternative to cable TV systems. But just as today's cable TV systems have been adapted to provide high-speed downstream Internet access along with primary entertainment traffic (see the cable TV discussions in Chapters 14 and 15), this possibility also exists within MMDS systems. A first-quarter 2000 decision by Sprint and MCI WorldCom to purchase most of the major MMDS operators and convert existing plants to broadband data service delivery, as opposed to existing analog television programming indicates that adapting MMDS to Internet/multimedia service delivery is underway (reported in Wireless America, March 2000).

# Paging Services

Originally, pagers were small analog wireless receivers that alerted users to incoming messages via audible tones or mechanical vibration. Now referred to as numeric pagers, these simple devices display and store a quantity of 20-digit call-back telephone numbers. As the paging industry made the transition from analog to digital signaling and modulation technologies in the 1970s, it also began a trend toward feature-laden smart pagers.

The next rung up the pager technology ladder adds capabilities to receive and display alphanumeric financial, traffic, weather, sports, general news update, and other subscription-dependent broadcast information, as well as personal messages from Web-connected PCs, two-way pagers, and telephones. Callers with access to PCs establish Internet-based connections to service-provider Web sites where they key in pager telephone numbers and text messages. Pagers equipped with up to 500,000 bytes of memory can typically store 25 separate personal messages along with whatever broadcast information pager owners decide to retain.

As an alternative to PC-generated messages, some service providers maintain call centers wherein callers dictate to operators who then transcribe short messages, usually up to 80 characters in length, to be sent to pagers. When caller identification information is available, alphanumeric pager networks automatically append calling-party telephone numbers to text messages. Other modern pager features include data and time clocks, alarm functions, directory, and note book storage for discretionary use by pager owners.

Unlike the numeric and alphanumeric pagers just described, which can only receive radio frequency transmissions from base stations, two-way alphanumeric pagers are able both to receive messages from, and send messages to service-provider message centers, using full-duplex pager-to-base station radio frequency communications links. Full-duplex pagers exhibit two principal advantages over simplex, receive-only devices. First, pager owners need not seek telephones or other means to reply. They simply key in reply messages using miniature "qwerty" keyboards. This not only saves the time needed to locate and activate alternative communications devices, but often a short message obviates longer-than-necessary voice calls.

The second important two-way pager advantage is *assured message delivery*. With the availability of pager-to-base station radio communications, service provider message centers can store messages for pagers until pagers send "successful message receipt" acknowledgment reply messages. In this way, if pagers are not powered on, or if they are outside coverage areas, message centers can store and retransmit messages at later times until an acknowledgment reply is received. Assured message service providers guarantee 99.9 percent-reliable delivery and store undelivered messages up to five days.

The most recent pager technology to come to market is *voice paging*. PageNet VoiceNow service, available in Dallas-Fort Worth, Chicago and Atlanta, permits users to store up to six 30-second voice messages using special pagers manufactured by Motorola. Stored voice messages can be listened to as many times as desired, and if a pager's memory is full or not powered on, the VoiceNow network holds subsequent messages until a pager is ready to receive them. Motorola likens pager-based voice messaging service to a wireless, portable voice answering machine.

Pagers operate in a large number of FCC-assigned very high frequency (VHF—i.e., 30—300-MHz) and ultra high frequency (UHF—i.e., 300—3000-MHz) spectral bands with modern digital pagers more likely to be found in the 900-MHz and PCS (see Figure 16.6) parts of the spectrum. A wide variety of local and national service plans are available starting at about $10 and $30 per month respectively. Numeric pagers can be purchased for as little as $30. High-end two-way, qwerty-keyboard models run as high as $360, but as in the cellular telephone sector, pager service has become a commodity and many dealers waive activation fees and sell pagers at a loss to attract new annual or longer subscription-period customers.

**Figure 16.6**
Broadband Personal Communications Services (PCS) FCC frequency allocation plan.

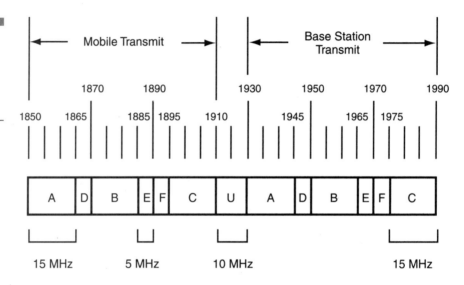

Since many cellular telephone service providers now offer numeric and text paging, short message service, and even full-blown Internet browser and e-mail service—all in a single instrument not much larger than a pager—readers might pose the question: why would anyone choose to subscribe to both mobile cellular telephone and pager services and carry two different instruments? In particular, why would customers opt for pager-based voice messaging when most mobile cellular services now include voice-mail features?

Only the market can answer those questions, but there is a substantial difference in pager and cell phone service delivery. Whereas mobile cellular voice service demands real-time, full-duplex communications between two or more parties, pager-based services only operate in non-real time, using store-and-forward technologies.

# Personal Area Networks and the Bluetooth Initiative

Harald II ("Bluetooth"), born around 950 A.D. is credited with consolidating the Danish realm and reigning as king of a unified Denmark. In our wireless context, *Bluetooth* was selected as a code name for a personal area wireless networking project and standard, originally envisioned as producing a single, unified wireless replacement for cables used to interconnect desktop PCs, laptops, printers, facsimile machines, pagers, personal digital assistants (PDAs), and cell phones. Objectives include synchronizing operations and stored data (personal information management), file transfer, and general-purpose data transmission.

In April 1998, Intel and Microsoft established a consortium with IBM, Toshiba, Nokia, Ericsson, and Puma Technology to create a single, digital wireless protocol and define end-user requirements for interconnecting various mobile, transportable, and desktop devices. There are over 1,000 members of the Bluetooth Special Interest Group (SIG), and development of Version 1.0 of the standard was announced in July 1999.

Bluetooth-conforming systems operate in the 2.5-GHz *Industrial, Scientific, and Medical* (ISM) frequency bands providing license-free operation in the United States, Japan, and Europe. The protocol supports packet and circuit switching, making it suitable for both voice and data traffic. Each synchronous voice channel supports 64-kbps digital signals. Full-duplex asynchronous data channels support either asymmetric data transmission (721 kb/s in one direction and 57.6 kb/s in the reverse direction) or symmetrical 432.6-kb/s transmission in both directions. The nominal connection range is up to 30 feet but can be extended with higher power to 300 feet.

Although it was originally envisioned as only a cable replacement technology, a broader view sees it as creating *personal area networks* (PANs) where all the devices a person owns work seamlessly together. Under this perception, PANs can be created anywhere, in homes, offices, hotel rooms, and even automobiles. In automobiles, for example, headsets could establish wireless connections to Bluetooth-conforming cellular telephones enabling hands-free operation. In hotels, Bluetooth-capable PDAs or laptops could connect with local access points or Bluetooth-compatible cellular telephones to access e-mail or to browse the Internet.

Motorola, a Bluetooth SIG Early Adopter "envisions a global portfolio of compatible, user-friendly devices to facilitate and maximize the use of Personal Area Networks," and already markets PC Card Adapter, USB (universal serial bus) Adapter, and Hands-free Car Kit products. Clearly, this initiative is bound to enhance future wireless operations and products. One might even suspect a marriage of universal mobile telephone service (UMTS—see Chapters 1 and 4) and Bluetooth concepts. A proliferation of microcell-based Bluetooth access points in urban areas and public buildings might materially shape the wireless world to come.

# Wireless Market Trends

In the mid-1980s, unit prices for cellular telephones hovered in the $2,000—$3,000 range, and monthly subscription charges averaged $125—$150, resulting in lackluster growth in demand for units and services. By 1987, unit prices broke the $1,000 barrier and with steadily declining usage costs touched off an annual growth rate of 40 percent as depicted in Figure 16.7. According to CTIA's Semi-Annual Wireless Survey, 1985—1999, between December 1987 and December 1998, average monthly subscriber bills steadily dropped from $96.83 to $39.43.

By December 1999, the average had risen back to $41.24. In fact the monthly expenses may continue to rise. But from an individual's overall telecommunications budget, total monthly costs could be declining. The reason is that with recent mobile telephone plans like $70 per month for 700 minutes for either local or long-distance calls, many subscribers may be reducing wireline long-distance costs.

Originally, mobile telephones could be purchased in three different formats: *mobile* (permanently installed in vehicles), *transportable* (bag phones or hard-case models fitting in briefcases), and *handheld* (some as small as wallets). The latter two styles have self-contained batteries. Although some tote bag models are still available, nearly all cellular telephones are manufactured as handheld devices with car

**Figure 16.7**
Wireless subscriber growth.

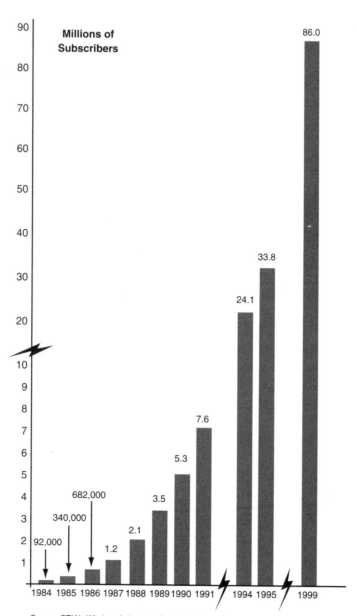

Source: CTIA's Wireless Industry Indices Report, 2000.

mounting and hands-free operating kits sold as accessories. Mobile telephone service has reached commodity status with activation fees waived and a wide range of telephones often sold at a loss to new subscribers who are willing to agree to one-year or longer service contracts. Compared with 1990 mobile telephone instruments, even low-end $50 models are more feature laden. The following list of features indicates that today, hand-held cellular phones rival those found in the most expensive desk sets:

- Auto redial
- Calculator
- Calendar/event reminder
- Caller ID, call waiting caller ID
- Call waiting
- Call forwarding
- Clock with alarm
- Dialed, received, missed calls
- Digital data
- Keypad lock
- Multiple language support
- On-screen battery indicator
- One-touch credit card dialing
- Phone book
- Profile settings
- Restricting calls, authentication
- Ring tones, silent ringing, vibration alert
- Silent keypad
- Three-way calling
- Voice calling
- Voicemail capability
- Short message service (SMS—numeric paging and text messaging)
- E-mail, Internet access, and browsing
- Dual band (e.g., 800MHz/1900Mhz)
- Tri band (e.g., 800MHz/1800MHZ/1900Mhz)
- Dual mode (e.g. AMPS analog/PCS digital, U.S. SMR/GSM)

The list is by no means complete but it does indicate a trend to include virtually any feature that can technically be supported. Excluding the last five features (SMS, text messaging, dual mode, dual band, tri-band, and e-mail/Internet browsing capabilities) telephones with all other features can be purchased (without plan commitments) in the $50—$300 range. The top-of-the-line Qualcomm pdQ model with HTML Web-surfing capability carries a no contract price of $899. Generally, features most likely to run up the price are dual mode, dual or tri-band, and data communications and Web-surfing capabilities. Even so, the Motorola Timeport model P8167 is a dual-mode (AMPS analog and Sprint PCS CDMA) and dual-band (800 and 1900 MHz) model that permits browsing the wireless Web and can be purchased for $299.

As in other electronics sectors, low-cost powerful microprocessors eventually eradicate feature-driven hardware cost factors. Moreover, as an incentive to sign on for long-term $60-per-month Internet access plans, once these plans become popular, service providers can be expected to underwrite or discount whatever cost differences remain between Internet-capable and non-Internet-capable mobile telephones.

From the service-provider point of view, the trend is overwhelmingly toward mergers and the eventual emergence of only a handful of large national operators. In the mid-1980s, when commercial cellular mobile telephone started, most customers were satisfied with local service. In our highly mobile society, as subscribers begin to take mobile service for granted, they more and more expect to be able to use a single telephone wherever they happen to travel. By 1993 there were more than 1,000 cellular companies, but since that time we have witnessed one merger after another, with each resulting new organization claiming to offer seamless nationwide service.

As mentioned, on April 4, 2000 Bell Atlantic Mobile and Vodaphone AirTouch PLC and PrimeCo completed their merger and formed a new company called Verizon Wireless, and Bell Atlantic indicated plans to complete the purchase of GTE Corporation by mid year. That combination of companies will have 24 million wireless subscribers, with a likely increase to 28 million by year-end. On April 6, 2000, SBC Communications and BellSouth Corporation announced their own merger, making them the nation's second-largest wireless carrier with 16 million subscribers. AT&T will then be the third-largest company, with 12.2 million wireless subscribers. Three companies will have more than 50 percent of all mobile subscribers as customers.

In response to the question about why the spate of mergers is occurring now, Patrick J. Devlin, president of Verizon's Mid-Atlantic region stated, "It's happening now because to really succeed in this business you need to be very large. It's a very capital-intensive business. We announced we are spending $3.3 billion on networks. You can only do that if you have a large scale and scope . . . Customers demand high quality. They want it to be almost home-like and office-like. There's only one way to do that, and that's to be big."

The word *verizon* is said to be derived from the Latin word *veritas*, which means certainty and reliability, and *horizon*. One would have to agree that technical integrity and performance can better be achieved in single, well-integrated networks, as opposed to the interconnection of a plurality of heterogeneous networks. However, competition has been a major factor in rapidly developing the remarkable capability that we all enjoy today. With an understanding of technical issues surrounding national and international mobile network design and operation, readers may draw their own conclusions regarding the motivation behind the mergers. Beyond technical merits, however, the aggregation of many profitable businesses clearly is one way to expand profitability. Time will reveal the overall results, but for now the trend to "large" is undeniable, and wireless business potential has never been greater.

The estimate that there will be more mobile devices than PCs connected to the Internet by 2003, if true, means that the two fastest-growing telecommunications segments may "stoke each other's fire." We made the statement earlier and it bears repeating. Many industry experts believe that wireless communications for voice, data, and other services is becoming the most significant telecommunications development since fiber optics.

# CHAPTER 17

# Satellite Wireless
# Services

# Fundamentals of Satellite Communications

Chapter 4 introduces wireless communications and key operating principles. The discussion surrounding Figures 4.7 through 4.9 and 4.18 through 4.21 is particularly relevant to this chapter's explanations of wireless satellite communication. With the background as well of additional treatments of wireless communications in later chapters, (and Chapter 16's treatment of terrestrial systems in particular), we are now prepared to focus on what makes satellite and terrestrial wireless communications different.

In all radio communications we speak of the *propagation* of signals from transmitters to receivers, where by propagation we mean the movement or transmission of a wave in a medium or in free space.[1] In traveling from transmitters to receivers all line-of-sight radio signals experience what is known as *propagation loss,* that is, a loss of signal power level proportional to the square of the distance separating transmitters and receivers. What this means is that if one receiver is ten times further than another receiver from a given transmitter— other factors being equal—its received signal power is one hundredth that of the closer receiver.

In all terrestrial wireless systems discussed so far, the distance separating transmitters and receivers has been limited to 20—30 miles for carrier microwave transmission systems, and even shorter distances in most cellular networks. Recall from Chapter 4 that *geosynchronous earth orbiting* (GEO) satellites, that is, satellites in orbits that make them appear stationary from any position on earth, are at an altitude of about 22,300 statute miles. Even *low earth orbiting*[2] (LEO) satellites may be hundreds of miles high. Because propagation loss is proportional to the path distance squared, satellite-link transmitter power levels and budgets are markedly different from those for terrestrial links.

---

1 *Communications Standard Dictionary*, Martin H. Weik, D.Sc., Van Nostrand Reinhold Inc.

2 Low earth orbit (LEO) is a term that typically refers to satellites in orbits ranging from 500 to 1,500 miles in altitude above the earth. Medium earth orbits (MEOs) are those in which satellites are in the 6,000 to 13,000 mile altitude range. Little LEOs generally refer to LEO satellites that operate at 800MHz or less. Big LEOs operate above 2 GHz and Mega LEOs from 20 GHz to 30 GHz.

For line-of-sight radio transmission, another factor affecting propagation loss is a radio signal's wavelength, which, as Chapter 5 explains, is inversely related to its frequency. Here too, it can be shown that propagation loss is also directly proportional to the transmitted radio frequency—squared. Conversely, antennas—such as the *parabolic dishes* sometimes seen gazin" upwards at orbiting satellites—exhibit higher gains at higher frequencies than whip or stick antennas used on most cellular telephones. Simple whip antennas have unity gain. That means that they normally neither enhance nor degrade received signal power levels.

Since the gain of any particular physically sized parabolic antenna can be shown to be proportional to transmitted frequency squared, one could conclude that added propagation loss incurred by raising transmitter frequencies can be offset with increased antenna gain. While this is generally true for *high-gain* or *highly directional* antennas (when transmitting and receiving antennas are perfectly aligned, as in fixed satellite terminals), misaligned directional antennas (antennas not pointing at each other) may actually reduce power levels beneath what is achieved with unity-gain whip transmitting and receiving antennas.

Since in most cases, the need to "aim" mobile cellular telephone antennas (other than that they be generally perpendicular to the earth) imposes untenable user requirements, high-gain or directional antennas are not a viable means of compensating for propagation loss in cellular systems. Thus, network designers have had to find other means to deal with increased propagation loss to distant satellites at transmitting frequencies high enough to penetrate the ionosphere. Note also that the need to have transmitting and receiving antennas "staring" at each other means that such arrangements are unsuitable for most broadcast, one-to-many, applications. (A brief explanation of ionospheric impacts on radio transmission is given below.)

One other geosynchronous satellite communications characteristic, noticeable in two-way voice conversations, is information signal delay. With 22,300 miles separating earth terminals and satellites, round-trip voice signal delays of about half a second are incurred. For simplex (one-way) communications applications like broadcast radio or television, such delays are of no consequence. In real-time interactive conversations, however, a half second is long enough that often, not hearing responses from other parties, persons will begin talking simultaneously and, after detecting the interference, must pause and begin again. This phenomenon alone motivates the use of LEO orbits for conversational voice applications.

The long propagation delays also have an impact on data traffic and Internet IP traffic in particular. To eliminate the possibility of undetected transmission errors, transmitters generally store already transmitted data until they obtain "correct message received" acknowledgment messages from remote receivers. At multimegabit-per-second data transmission rates, storing data for a half second necessitates a great deal of memory and a revamping of most protocols designed for terrestrial data communications. Work to enable efficient transmission of IP and ATM traffic via satellites is an ongoing technical effort receiving much attention.

As a prelude to delving into specific satellite services and networks, we present in Figure 17.1 a summary of radio spectrum terminology used in those discussions. Lower radio frequencies, up to about those in the high-frequency (HF) band, are essentially reflected back to earth by the ionosphere, the region of the atmosphere that extends from about 30 to 250 miles in altitude. Subject to atmospheric conditions, radio signals in the lower parts of the spectrum may use the ionosphere and the earth's surface as something like a radio waveguide, so that global communications is often possible.

Since higher frequencies are not as affected by ionospheric effects, satellite communications systems generally use frequencies at or above the UHF band. At the higher bands the effects of rain, fog, and thermal variations all have an impact on propagation and must, to the extent possible, be accounted for in link-power budget margins. Illustrations of satellite systems operating in the bands listed in Figure 17.1 are described below.

# Satellite Systems and Design Principles

Like terrestrial wireless systems, some satellite systems are designed to accommodate fast-moving high-mobility terminals whereas others are better (or only) suited for fixed or slow-moving earth terminal applications. Some of the first communications satellites were experimental, and these were soon followed by military and commercial versions. Hundreds of different satellites and satellite systems have been deployed since the first telecommunications satellite was launched December 18, 1958. Because not all of these can be dealt with separate-

▬▬  ▬▬  ▬▬

**Figure 17.1**
Frequency band
descriptors and
related spectral
occupancy.

| Band Descriptor | Spectral Occupancy | Comments |
|---|---|---|
| ELF-band | 30–300 Hz | ELF = Extremely Low Frequency |
| VLF-band | 3–30 KHz | VLF = Very Low Frequency |
| LF-band | 30–300 KHz | LF = Low Frequency |
| HF-band | 3–30 MHz | HF = High Frequency |
| VHF-band | 30–300 MHz | VHF = Very High Frquency |
| UHF-band | 300–3000 MHz | UHF = Ultra High Frequency |
| P-band | 0.230–1.000 GHz | |
| L-band | 1.530–2.700 GHz | |
| SHF-band | 3–30 GHz | SHF = Super High Frequency |
| S-band | 2.700–3.500 GHz | |
| C-band | Downlink 3.700–4.200 GHz<br><br>Uplink 5.925–6.425 GHz | |
| X-band | Downlink 7.250–7.745 GHz<br><br>Uplink 7.900–8.395 GHz | |
| Ku-band | Downlink<br><br>FSS 10.700–11.700 GHz<br><br>DBS 11.700–12.200 GHz<br><br>Uplink<br><br>FSS 14.00–14.500 GHz<br><br>DBS 17.300–18.100 GHz | FSS = Fixed Satellite System<br><br>DBS = Direct Broadcast Satellite |
| Ka-band | 18–31 GHz | |
| EHF-band | 30–300 Ghz | EHF = Extremely High Frequency |

ly, our discussion emphasizes examples depicted in Figure 17.2, selected to illustrate the complete range of applications now supported.

Considering the large number of countries and companies that
either are operating or have plans to operate satellite systems (just as in
terrestrial wireless systems), and the finite amount of usable radio

| Category | System/ Satellite Name | Orbit/ Number of Satellites | Band | Comments |
|---|---|---|---|---|
| **Military Satellite Systems** | | | | |
| Military/ Strategic (33 inch to 30 foot Diameter Antennas) | DSCS (Defense Satellite Communications Systems Phase III) | GEO/ 9 Total (5 Active, 4 Reserve) | X-band UHF | 6 at 50–85 MHz Bw in X-band Single Channel Trasponder in the UHF-band for Higher Authority National Command Use |
| Military/ Tactical (Small Antenna Terminals) | FLTSATCOM (Fleet Satellite Communications Systems) | GEO/ (5 Active) | UHF-band | 23 Communications Channels 10 for Navy, 12 for Air Force and 1 for Higher Authority National Command Use |
| **Broadband, High-Capacity—Large Earth Terminal Satellite Systems** | | | | |
| Large Terminals | Intelsat/(IP 5 to 9) International Operation | GEO/ 17 Satellites | C-band  Ku-band | 38 at 36 MHz Bw  6 at 36 MHz Bw |
| Large Terminals | GE Americom/GE-4 Private Company Operation | GEO/ 12 Planned | Ku-band  C-band | 24 at 36 MHz Bw 4 at 72 MHz Bw 24 at 36 MHz Bw |
| **Broadband, High-Capacity—Small Earth Terminal Satellite Systems** | | | | |
| VSAT Terminals | GE-4 (GE Americom) | GEO/ 12 Planned | Ku-band  C-band | 24 at 36 MHz Bw 4 at 72 MHz Bw 24 at 36 MHz Bw |
| **Personal Communications—Small or Handheld Earth Terminal Satellite Systems** | | | | |
| Portable Terminals | Inmarsat/ Inmarsat-1, 2, 3 International Operation | GEO/9 Satellites | L-band | See the text for service descriptions |
| Hand-held Terminals | Globalstar/ Globalstar | Low Earth Orbit (LEO) 48 Active at about 900 Miles Altitude | L-band Subscriber Uplink 1.61–1.63 GHz S-band Subscriber Downlink 2.48–2.5 GHz C-band Fixed Uplink 5.09–5.25 GHz C-band Fixed Downlink 6.70–7.08 GHz | 16 Spot-beams per Satellite 13 Spread Spectrum Channels at 1.2288 MHz per Channel Complete Earth Coverage Except at Latitudes > 70° Variable Rate Vocoder Digital Voice (1.2–13 kb/s) and 1.2, 2.4, 4.8, and 9.6 kb/s Data |
| Hand-held Terminals | Skycell/MobileSat Motient the U.S. Operator; Cable, and Wireless Optus Pty. Ltd. in Australia | GEO 2 Satellites Serving Australia | L-band | 500 Full Duplex Optus MobileSat Channels Supporting Fixed Voice, Facsimile, Data, and Paging Services |
| Hand-held Terminals | ICO global Over 70 U.S. and non-U.S. Investor | MEO 10 Satellites at about 6000 Miles Altitude | L-band x Subscriber Uplink 2.170–2.200 GHz L-band Subscriber Downlink 1.980–2.100 GHz C-band Fixed Uplink 5.0–5.2 GHz C-band Fixed Downlink 6.8–7.0 GHz | 160 Spot-beams with a Four-Cell Frequency Reuse Pattern Full Duplex Digital Voice at 4.8 kb/s, and Data and Facsimile at 2.4 kb/s |
| **Broadband Communications—Small Earth Terminal Satellite Systems** | | | | |
| Small Terminals 10-inch Antennas | Teledesic | LEO 288 | Ka-band Uplink 28.6–29.1 GHz Downlink 18.8–19.3 GHz | Asymmetrical User Data Rates, 2 Mbps Uplink and 64 Mb/s Downlink with "Bandwidth-on-Demand" Options |
| Small Terminals 24-inch Antennas | Celestri | 9 GEO 63 LEO | Ka-band | |
| Small Portable Terminals | Skybridge | LEO 64–85 | Ku-band | Asymmetrical User Data Rates, 2 Mbps Uplink and 60 Mb/s Downlink with "Bandwidth-on-Demand" Options |

**Figure 17.2** Representative high-capacity and personal communications military and commercial satellite systems.

spectrum, efficient mechanisms for frequency reuse are essential. To accommodate as many operators as possible, GEO satellites are now spaced only two degrees apart.

Although not drawn to scale, Figure 17.3 illustrates how large satellite communications earth-terminals, with high antenna gains (narrow beams), permit the reuse of exactly the same uplink and downlink frequencies for satellites in equatorial, geosynchronous orbits spaced only two degrees apart. What the figure implies is that the narrow "pencil-beam" signal radiating from the pictured earth terminal, illuminates the target satellite receiver antenna, providing it with a signal strengthened by both transmitting and receiving high-gain antennas. Conversely, because satellites to either side of the target satellite are in "side lobes" of the transmitting earth terminal, their antennas receive only greatly attenuated levels of incident power. While the two-degree spacing limits to 180 the number of GEO satellites transmitting at the same up- and down-link frequencies, additional satellites can be placed in GEO orbits if they operate at different frequencies. Currently there are some 2,200 satellites in orbit. By country, the U.S. operates about 660, the CIS (formerly the Soviet Union) about 1,300, Japan 55, France 24, the United Kingdom 18, Germany 15. The rest are scattered among other countries; some 59 are operated by international organizations.

Figure 17.3 also illustrates the operation of so-called satellite *earth-coverage* and *spot-beam* antennas. An earth-coverage satellite antenna is one that has a beam width wide enough to illuminate the maximum area on the earth "visible" from a particular satellite. Note that an antenna beam width wider than what is necessary would not encompass any greater earth surface area, but would reduce available signal-power levels. Earth coverage from equatorial GEO satellites is illustrated on the figure by the light-colored dashed elliptical line. GEO earth coverage satellites reach north and south to about 70 degrees latitude, leaving polar-cap areas uncovered.

Some satellites are equipped with spot beams, illustrated in the figure by a small light-colored elliptical dashed line. Satellite spot beams permit frequency reuse within a single satellite. Earth terminals transmitting at exactly the same frequency but in different spot-beam antenna patterns, can be detected separately on the satellite. The operation is similar to the frequency reuse among cells sufficiently separated in terrestrial cellular mobile telephone systems as explained in Chapter 16. Satellite spot-beam antennas are also useful in reducing the effectiveness of intentional enemy jamming in wartime conflict situations. In these instances, enemy jammers outside of spot-beam illumination areas are reduced by the high-gain factor of the directional spot-beam antenna.

**Figure 17.3**
Geosynchronous
earth orbit (GEO)
satellite systems
illustrating typical
satellite and Earth
terminal antenna
proper ties.

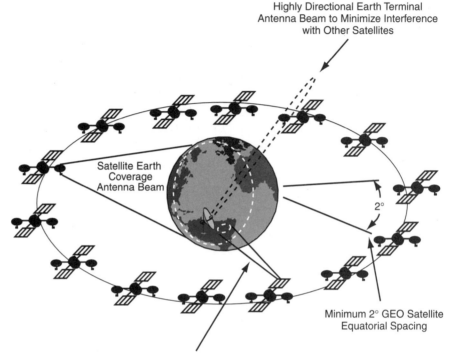

Highly Directional Earth Terminal
Antenna Beam to Minimize Interference
with Other Satellites

Satellite Earth
Coverage
Antenna Beam

2°

Minimum 2° GEO Satellite
Equatorial Spacing

Highly Directional Satellite "Spot Beam" Antenna to
Increase Earth Terminal and Satellite-Received Signal
Power, to Permit Frequency Reuse, and to Reduce
Intentional (Enemy Jamming) or Unintentional
Interference from Other Ground Emitters

## Military Satellite Systems

The military was quick to recognize satellite communication's potential in both strategic global warfare and tactical battlefield operations. Even initial models of *Defense Satellite Communications Systems* (DSCS, see Figure 17.2) satellites exhibited impressive capacity and anti-jam communications capabilities. There are a total of nine DSCS satellites in geosynchronous orbit, five active and four in reserve, with expected lifetimes of ten years. Current satellite models carry six transponders with 50 to 85 MHz of bandwidth, whose signals can be received by antennas that range in size from 33 inches to 60 feet.

In the absence of enemy jamming, these transponders provide a means for establishing secure, multichannel voice, high-speed data, and video-capable links between stateside higher authority command centers and military command posts and battlefield units anywhere in the world. Under enemy jamming conditions, the wide-bandwidth transponder capabilities can be allocated to anti-jam modulation operation. Coupled with agile and highly directional spot-beam antennas that can place enemy-jammer transmitters in antenna-null positions, these capabilities result in essentially undeniable communications for the military's most critical communications needs.

While physically large earth terminals are normally employed, the Department of Defense (DoD) has developed truck-borne and ship-borne terminals that enable "go-anywhere" capabilities. Moreover, electronic tracking conformal antenna arrays (that is electronically "steerable," high-gain antennas that have no moving parts and conform to an aircraft's or ship's exterior structural envelope) are now available that extend applications even to advanced airborne command centers.

In the tactical arena, the DoD uses *Fleet Satellite Communications* (FLTSATCOM) for U.S. Navy and other tactical military operations. A FLTSATCOM advantage is that high-gain antennas and the need to point such antennas at satellites are not required. Of course this means that per-channel bandwidth is limited and links are susceptible to enemy jamming. Nevertheless, over the years the capability has proven critically valuable in military situations.

## Broadband, High-Capacity—Large Earth Terminal Satellite Systems

Some of the earliest high-capacity, commercial communications satellites were designed and launched by the *INternational TELecommunications SATellite* (Intelsat) organization, now counting more than 140 member countries and 40 investing entities.

Originally, Intelsat's focus was on transoceanic and transcountry transport among private (U.S.) or government-owned (Post, Telephone, and Telegraph—PTT—administrations overseas) public telecommunications networks. Today the active Intelsat satellite system comprises 17 geostationary satellites. Current and planned services include:

- Public and private voice and data networks

- Internet and intranet traffic
- Synchronous digital hierarchy (SDH) and asynchronous transfer mode (ATM) traffic
- Broadcast and direct-to-home video
- Other broadband applications such as high data-rate trunking, telemedicine
- Tele-education, interactive multimedia/video services

Mirroring Intelsat's *high-capacity, few relatively large earth-terminal* business and technical models are similar capabilities operated world-wide by private companies. GE Americom's service, listed in Figure 17.2 and described further in Chapter 4 is just one of many competitive offerings based on a new breed of advanced satellite technology designs.

Whether operated by international or privately held companies, the wide-bandwidth or high-capacity class of service just described is best accessed using large earth terminals. In most cases, service providers employ terrestrial facilities to ultimate end-users of such services. Examples include the use of terrestrial public switched telephone networks for telephony services and television and radio broadcast facilities for entertainment services.

## Broadband, High-Capacity—Small Earth Terminal Satellite Systems

Other categories of wide bandwidth, high-capacity satellite services that employ *small earth terminals* and typically deliver service directly to end-users include *direct broadcast satellite* (DBS) and *very small aperture terminal* (VSAT) services, both of which are explained in Chapter 4. As depicted in Figure 17.1, DBS service, introduced in the United States in 1994, is supported in the Ku-band.

## Personal Communications—Small or Handheld Earth Terminal Satellite Systems

We discuss now a variety of satellite-based services intended to compete with, or at least provide services similar to, terrestrial mobile cellular service in areas of the world where terrestrial services are not now available. The first such system to become operational was Iridium, a design proposed by Motorola in 1987. By 1991 Motorola incorporated Iridium LLC as a separate company, eventually owned by an international consortium of 17 invester organizations. By 1998 Iridium completed the launch of a 66 LEO satellite constellation at altitudes of about 500 miles.

Iridium employed three types of communications links: L-band links between mobile-user terminals to satellites (uplinks and downlinks at 1.62—1.63 GHz.); Ka-band between fixed earth stations and satellites (uplinks at 19.4—19.6 GHz, downlinks at 29.1—29.3 GHz); and Ka-band intersatellite links at 23.2—23.4 GHz. Inter-satellite links made it possible for subscriber calls originating on one side of the world, to be transmitted to satellites on the opposite side of the world and directly to remote mobile subscribers, without the need of passing through terrestrial long-distance networks. Fixed earth stations, on the other hand, served as gateways to worldwide PSTNs so that Iridium subscribers could originate or receive calls from wireline telephones, anywhere in the world.

While Iridium must be regarded as one of the world's most complex system design and deployment feats, on March 17, 2000 Iridium LLC announced it was terminating service and that it intended to decommission and de-orbit Iridium satellites. Although speculation abounds about the details of why this happened, one fact is certain—the service never attracted enough customers to make it a viable business.

Given that it played important roles in Kosovo refugee and Central American hurricane Mitch disaster relief operations, from a humanitarian viewpoint Iridium's demise is certainly regrettable. Among theories advanced to explain what happened is the belief that, when it was postulated in 1987, no one could have predicted the rate at which terrestrial systems have been established worldwide, even in underdeveloped third-world countries. Chapter 16's terrestrial cellular growth statistics lend credibility to the belief that rapid terrestrial cellular system deployment certainly reduced the number of prospective Iridium customers.

Other experts cite $2 to $4 cost-per-minute charges as reasons why the service did not attract enough affluent customers and was not affordable for the world's poor. Still others cite the inability to complete calls from within buildings (standard use-instructions include directions to walk outdoors to areas where neither buildings nor trees obscure one's view to the horizon). The fact that handheld telephones weighed about a pound and contained bulky antennas sensitive to vertical alignment (as compared to terrestrial cellular models that weigh as little as 3.6 ounces and are relatively insensitive to antenna alignment) may also have been relevant factors.

Whatever the actual reasons, Iridium's failure has not yet damped the hopes and activities of some dozen similar satellite-based personal communications enterprises. Figure 17.2 lists examples of interest to readers attempting to acquire an appreciation of important technical and business trade-offs. Each is described below.

### INMARSAT

*INternational MARitime SATellite* (Inmarsat) service grew out of a United Nations-sponsored organization created on July 16, 1979. Originally based on six leased satellites from Marisat, Marecs, and Intelsat, the service proved so successful that in 1983 Inmarsat decided to operate its own satellites.

Currently, the Inmarsat system comprises nine GEO satellites and 40 land earth stations in 31 countries. Land earth stations act as gateways connecting Inmarsat traffic to appropriate terrestrial wireline telephony and data communications networks. Inmarsat terminals operate in L-band with uplinks at 1.6265—1.6605 GHz and downlinks at 1.525—1.559 GHz.

Because Inmarsat services have been offered for so long, they have become a baseline against which newer services are compared. Figure 17.4 characterizes Inmarsat portable terminals and voice, data, and facsimile capabilities. The terminal that can be most closely compared with terrestrial cellular telephones is the Mini-M. Worldphone; a version of this terminal manufactured by Nera ASA is housed in a case that is approximately $2 \times 10 \times 10$ inches and contains both an antenna and a separate telephone. Worldphones can be purchased for about $2,900 and Inmarsat calls cost between $2.50 and $3.50 per minute plus long-distance charges.

Note in Figure 17.4 that Inmarsat-C does not support voice services. Inmarsat-C does support personal two-way messaging, automated, or polled data-reporting messaging, and position-reporting messaging—when integrated with position data derived from *Long Range Aid to Navigation* (LORAN), *Global Positioning System* (GPS), the *GLobal NAvigation Satellite System* (Glonass—a Russian 24 MEO satellite navigation system), or certain other navigation systems.

Not listed in Figure 17.4 are Inmarsat-E and D+ services. The Inmarsat-E system supports global maritime distress alerting using *Emergency Position Indicating Radio Beacons* (EPIRBs). EPIRBs are required by the International Maritime Organization's *Global Distress and Safety System* (GMDSS) in all vessels exceeding 300 gross tones and all passenger vessel engaged in international voyages. EPIRBs are also available for small craft, enabling them to transmit distress alerts, which are an integral part of *Maritime Rescue Coordination Centre* (MRCC) search and rescue operations.

**Figure 17.4**
Inmarsat land portable terminals and services.

| Service<br>Feature | Inmarsat-A | Inmarsat-B | Inmarsat-M | Inmarsat<br>Mini-M | Inmarsat-C | Inmarsat-M4 |
|---|---|---|---|---|---|---|
| Service Type | Analog | Digital | Digital | Digital | Digital | Digital |
| Voice Type | FM | 16 kb/s | 4.8 kb/s | 4.8 kb/s | N/A | 4.8/64 kb/s |
| Data Rate | 64 kb/s | 64 kb/s | 2.4 kb/s | 2.4 kb/s | 600 bps | 64 kb/s |
| Fax Rate | 9.6 kb/s | 9.6 kb/s | 2.4 kb/s | 2.4 kb/s | 600 bps | 2.4/14.4 kb/s |
| Antenna Type/<br>Size | Dish<br>36 inches | Flat or Dish<br>38 × 30<br>inches | Flat<br>18 × 15<br>inches | Flat<br>10 × 8<br>inches | Omni<br>10 × 6<br>inches | Flat<br>16 × 29<br>inches |
| Equipment<br>(H × W × D) | 10 × 13 × 17<br>inches | 9 × 18 × 21<br>inches | 4 × 15 × 18<br>inches | 2 × 10 × 8<br>inches | 10 × 8 × 5<br>inches | 2 × 8 × 8<br>inches |

Inmarsat-D+ offers global two-way data communications utilizing equipment no larger than personal CD players, targeted at the rapidly expanding mobile and fixed asset-tracking industry, a market vigorously pursued by Orbcomm (see below). The modular transport industry is especially interested in tracking containers and monitoring the condition of high-value contents during shipborne and overland transit. As a consequence, the global coverage of Inmarsat and other satellite communications systems makes them ideal candidates for carrying position-determination-supporting payloads. Satellite system operators have apparently concluded that the incremental capital required to address asset-position-determination and monitoring business segments is well worth the investment. It is interesting to note that Iridium backers proposed to offer *radio determination satellite services* (RDSS) in the 1.610—1.626.5 GHz band, a portion of the spectrum set aside for RDSS service, provided it could simultaneously be used to for Iridium's other two-way digital voice and data services.

## GLOBALSTAR

Globalstar is a consortium of international telecommunications companies originally established in 1991 by Qualcomm of San Diego and Loral Aerospace in Newport Beach to provide mobile and fixed satellite-based telephony services. Globalstar's current constellation consists of 48 LEO satellites at altitudes of about 900 miles, and they officially commenced operations on October 11, 1999. The total cost of the system is estimated at $4 billion, which may be compared with Iridium's $7 billion figure.

A significant technical design and business structure difference between Globalstar and Iridium is that Globalstar essentially provides only satellite-based *originating and terminating access service* to earth-bound wireless terminals. Whereas Iridium, with its inter-satellite links, could provide transport services among satellites on opposite sides of the earth, and originating and terminating access services to mobile and fixed earth-bound terminals in those remote locations, Globalstar satellites, having no inter-satellite cross-links, merely provide satellite-based access to upwards of 200 fixed earth stations. These stations include mobile switching office (MSO) capabilities similar to MSOs in terrestrial cellular systems, and act as gateways to other fixed and cellular telephone networks in more than 100 countries on six continents.

Thus Globalstar, relying on local consortium member-provided transport services for end-to-end call completion, acts as a wholesaler selling access to its system to regional and local telecommunications service providers around the world. This more inclusive business arrangement may be pivotal in Globalstar's ability to avert an Iridium-like financial failure. Many experts believe that Iridium's ability to compete with transport services offered by connecting fixed and mobile carriers undermined those carriers' best long-term interests and consequently their motivation to help make Iridium a success.

Of course, the fact that Globalstar satellites are of a hard-limiting, bent-pipe design, as opposed to Iridium's more complex signal processing and inter-satellite link-capable designs, makes them far simpler and less expensive—accounting for some of the difference in the two systems' investment costs. Globalstar upgrade plans call for increasing the number of LEO satellites to 64 and adding 4 GEO satellites.

Subscribers use L-band 1.61—1.63 GHz uplink and 2.48—2.5 GHz downlink radio frequency signals that may be placed in any of 16 spot beams, per satellite. The radio frequency modulation technique basically complies with IS-95 spread-spectrum, CDMA specifications. Accordingly, up- and downlink L-band channels are divided into thirteen 1.2288 MHz subchannels. In each of these channels, multiple 1.2288 mega-chip per second (Mcps) pseudo-random codes are used to modulate digital vocoder voice and other user digital data information signals (as explained in Chapter 16) so that each user's signal may be individually recovered by receivers.

Just as in terrestrial CDMA systems, during speech pauses, vocoder output rates and transmitter power levels are reduced to minimize interference with active-speaker signals, helping to maximize the total number of users that can be accommodated. In industry jargon, adaptive communications systems are those that in real-time "sense" actual user and overall propagation conditions, modifying modulation and other transmitter and receiver configuration parameters in ways that optimize individual link and/or system-wide performance.

Qualcomm's CDMA wireless system designs accomplish these objectives with a degree of success that could only be dreamed about just two decades ago. In that era, the theory and advantages of rake receivers (four tines in IS-95) was well known. *Rake receivers* employ a plurality of redundant receivers to detect and track multiple signals. For example, radio signals may follow multiple transmitter-to-receiver paths. Some signals may follow a direct path, others may be reflected off buildings or the earth. In conventional receivers, sometimes the signals received from multiple paths add constructively, but it is usually equally likely that they may combine in a way that decreases performance. Rake receivers have the ability to detect multiple-received signals separately, and by appropriate signal processing combine or select only the combination that maximizes performance.

Although adaptive, rake-receiver operation has been understood and demonstrated for some time, it took the appearance of advanced digital signal processing techniques, implemented in tiny, powerful microprocessors, to make it a reality in small, low-cost handheld terminals—such as those used in CDMA-based wireless systems. As we noted above, such capabilities make possible the ability to translate voice inactivity to expanded maximum numbers of subscribers that wireless networks can support. The techniques also markedly improve overall QoS and hand-off procedures—in the satellite case when user-signals and user-control are switched from spot-beam to spot-beam or from satellite to satellite.

Satellite-based mobile systems can be compared to terrestrial counterparts by simply imagining that both subscriber terminals and base stations may be moving. In CDMA satellite networks, before a subscriber signal in a spot-beam of a particular satellite is handed off to either another spot-beam or another satellite, that subscriber's signal is detected and received in the new spot-beam while the original connection is still maintained. This redundant or double coverage is continued until QoS thresholds in succeeding spot-beam channels exceed specified thresholds. This *make-before-break* or *soft-handoff* procedure greatly enhances communications reliability and reduces the probability of *dropped calls*.

In spite of all of these functional capabilities, the Qualcomm GSP 1600 handheld mobile telephone weighs 13 ounces and operates in three different modes. Besides the CDMA satellite mode, GSP 1600 phones support both AMPS analog and CDMA operation in 800-MHz band terrestrial cellular networks. Users in range of terrestrial service may opt for either terrestrial or satellite service with just one telephone.

Globalstar telephone instruments are still considerably larger than typical terrestrial cellular counterparts and contain relatively bulky, orientation-sensitive antennas that must be used in satellite operation. Globalstar's instructions for satellite use include directions to go outdoors and seek building- and tree-free visibility to the horizon, a limitation some experts say contributed to Iridium's business failure. Weatherproof Globalstar terminals may be mounted on rooftops or at other outdoor locations, with wire connections to standard desktop telephones, making indoor satellite-based calling possible. Globalstar telephones are available today at about $1,200 with service provider plans, and $1,500 without plan commitments.

### SKYCELL/MOBILESAT

Skycell and MobileSat are names for satellite-based mobile services offered respectively in the U.S. by Motient and in Australia and some southeast Asia and western Pacific regions by Cable and Wireless Optus Pty. Service is rendered via 500 full-duplex channel, GEO satellites with L-band uplinks at 1.6455—1.660.5 GHz and downlinks at 1.545—1.559 GHz. User terminals like Westinghouse's S3000 and NEC's S2 are usually mounted in vehicles and support push-to-talk voice, facsimile, data, and paging services. The service is based on spot-beams and coverage of the continental U.S., Alaska, Hawaii, and the Caribbean is provided by Motient. All of Australia and up to about 120 miles out-to-sea coverage is offered by Cable and Wireless Optus Pty. In Australia, subscribers are charged $1.80 peak and $1.65 off-peak per minute for calls.

### ICO GLOBAL

ICO, with over 70 international investors, is designed to include ten MEO satellites at about 6,000 miles altitude, with L-band uplinks at 2.170—2.200 GHz and downlinks at 1.980—2.010 GHz. Orbits are selected to provide continuous coverage of the entire globe and high elevation angles (averaging 40—50 degrees) to all users, a fact that minimizes antenna-pointing difficulties. Satellites have 160 spot-beams each arranged in four-cell frequency reuse patterns to maximize the number of subscribers that can be supported.

### ORBCOMM

ORBCOMM is currently a 35-LEO satellite system. It is designed to enable businesses to track remote and mobile assets such as trailers, railcars, locomotives, and heavy equipment; monitor remote utility meters and oil and gas storage tanks, wells, and pipelines; collect environmental and industrial (*supervisory control data acquisition*—SCADA) data; and stay in touch with remote workers anywhere on the globe. ORBCOMM features data-only packet switching and does not support voice communications. Store-and-forward capabilities ensure that messages are delivered regardless of a user's location at the time a message is sent.

Besides the space segment, ORBCOMM includes gateway control centers, gateway earth stations (GESs), a network control center, and subscriber communicators (SCs)—that is, hand-held devices for personal messaging, as well as fixed and mobile units for remote monitoring and tracking applications. ORBCOMM uses 137—138 MHz VHF and 400-MHz UHF frequencies for downlink transmission to mobile or fixed data communications devices and 148—150-MHz VHF frequencies on uplinks. Uplink VHF transmissions from SCs are at 2.4-kb/s data rates. Downlink transmissions to SCs are at 4.8 kb/s using VHF/UHF frequencies, with 9.6-kb/s service planned for the future.

Messages generated by SCs are relayed by satellites to GESs, where they are forwarded to end-user destinations through the Internet or other terrestrial networks. Messages to remote SCs can be initiated from computers, using the public Internet or ORBCOMM's Private Frame Relay or Virtual Private Networks service offerings to access GESs. The low 500—600-mile high orbits, and low VHF/UHF radio frequency operations, mean that smaller, lighter-weight, lower-power satellite and user equipment can be used. The estimated cost of the system is $330 million.

## Broadband Communications—Small Earth Terminal Satellite Systems

If the four satellite-based system examples just discussed correspond to today's terrestrial mobile cellular service described in Chapter 16, the last three example satellite systems listed in Figure 17.2 correspond to local multipoint distribution service (LMDS), also described in Chapter 16. The Teledesic, Celestri, and Skybridge programs all have as objectives providing *broadband access service to users with relatively small (10—24-inch) antennas,* anywhere in the world.

Teledesic, the most talked-about program, calls for 288 LEO satellites with Ka-band uplinks at 28.6—29.1 GHz and downlinks at 18.8—19.3 GHz. The Teledesic design places satellites at about 800 miles in altitude, a height that permits all users with continuous, nearly directly-overhead visibility. Again, greater than 40-degree elevation angles simplify ground antenna pointing procedures and dramatically reduce the possibility of obstruction from surrounding buildings, trees, and terrain undulations that can prevent service delivery.

Proponents project asymmetrical uplink and downlink data rates at greater than 2 and 64 Mb/s respectively. They also view user terminals as Teledesic Network *edge devices* that interface with a wide variety of standard network protocols including IP, ATM, ISDN, and others. With these capabilities, Teledesic applications include Internet, corporate intranets, LAN interconnection, and other broadband, high traffic-volume services. Since developmental and deployment investment costs are estimated at greater than $9 billion, embarking on this project is a somewhat daunting undertaking. Its backers include Bill Gates, Craig McCaw, and the Boeing Company. Beyond cost, accomplishing the launching of 288 satellites in the planned 18—24-month time span itself poses a considerable challenge.

Though less ambitious in terms of numbers of satellites, the Celestri (sponsored by Motorola) and Skybridge programs target the same broadband services and applications for customers equipped with small and inexpensive terminals. In terms of high-rate data communications, 20-millisecond roundtrip delays to LEO-altitude satellites compare favorably with terrestrial system latency and contribute to the achievement of seamless satellite and terrestrial system interoperation—using existing protocols.

The aggregate capacity of proposed broadband systems takes on truly astonishing proportions. In its 64-satellite configuration, Skybridge expects to be able to support 10—20 million simultaneously connected broadband end-users.

Despite the promise of the above systems, perhaps the toughest lesson that Iridium's demise taught is that in public telecommunications systems, it is not enough to conceive, design, develop, and deploy technological marvels, no matter how impressive they may be. What customers buy is basically service, and if they can obtain equivalent service at less cost from networks that are far less grandiose or elegant than more expensive super-high technology alternatives, they'll take the cheaper approach every time.

# Outlooks for the Future

Y ou are approaching the end of the *McGraw-Hill Telecom Factbook*. If you look back, you should now be able clearly to discern the book's building-block structure and its focus on relationships between advanced technology and modern business needs. Part 1 begins with basic definitions, terminology, and background, providing a basis for the more complex explanations of modern telecommunications technologies, services, and operations that follow. It then moves to discussions of historical and current regulatory environments and how they have shaped the business and technical infrastructures that now deliver telecommunications services.

Part 2 presents fundamentals of transmission, multiplexing, circuit switching, packet communications, and advanced network concepts. These chapters provided an in-depth, "under the covers" look at technologies and techniques supporting today's advanced telecommunications services, many of which are completely new since our book's first edition. These include the Internet, a relatively unknown network for research and education communities before its debut as the World Wide Web in 1994, frame relay and ATM data services, and ubiquitous wireless services.

Part 3 explains voice communications services from several viewpoints. Premises services are discussed, as are important distinctions between *access* and *transport* services. This part continues with descriptions of tariff structures, fundamentals of traffic engineering, voice network design principles, and characteristics of cost-effective business networks. The final chapter in Part 3 relates price structures to popular voice services and contains a tutorial guide on how to become an informed business telecommunications product and service consumer.

In a similar manner, Part 4 examines local, metropolitan, and wide-area data communications network developments and technologies, discussing the great variety of data services now offered and their relation to business-critical applications.

Part 5 tackles the fast-moving field of wireless communications, encompassing terrestrial cellular networks and satellite-based systems—revealing myriads of new technologies and innovative and highly competitive business enterprises. Narrowband and broadband, fixed and mobile subscriber, voice, paging, e-mail, Internet browsing, and other data, video, and imaging applications are described.

Virtually all important services and technologies are treated in these chapters, and a special effort made not simply to define and explain individual topics, but to illustrate how technologies, products, and services relate to one another, and how they reflect industry patterns and trends.

Throughout these chapters, we identify telecommunications technology and business initiatives—some with enormous capital investment backing—with the potential to sustain and possibly even accelerate beyond the last decade's phenomenal industry growth rate. Even more happily, these initiatives promise to further enhance ways in which telecommunications improves business operations and personal endeavors. While some possibilities appear to be sure bets, the marketplace ultimately separates winners from losers. Who, for example, would have predicted that following a nearly flawless execution of one of the world's most significant and impressive technical developments, that the $7 billion investment in Iridium would simply be abandoned—and over 60 satellites incinerated in space as they are de-orbited?

In spite of the hazards of doing so, at this point in books like this, it is always tempting to wax omniscient, confidently predicting future developments. In fact, however, if you, the reader, have grasped the implications of material already presented, your prognostications are as reliable as ours, since we now possess a common view of historical events and technological capabilities, and are faced with the same marketplace uncertainties. Nevertheless, based on lessons learned we offer some thoughts about technology and business developments that deserve watching. We structure our thoughts along the same lines as the rest of the book, that is, we present basic technology outlooks first, followed by impacts that technology developments are likely to have on service delivery. We cite the defining role that standards and consumer-price thresholds have played in past "killer" technology/application successes—as at least partial justification for our observations.

# Technology Outlooks

Though rarely of interest or visible to end-users, technological progress is at the core of the last decade's sweeping changes. Not only do technological advances provide more bandwidth for applications, they drive costs down to keep application advances affordable. Both business and residential users are super-sensitive to price. They want 10 to 100 times the capacity and capability of today's systems, but are willing to pay only 2 to 3 times the price. Three technology areas appear to be pivotal to future developments: 1) advanced modulation technologies and the digital signal processing and high-speed, low-power integrated circuit chips that make them possible; 2) fiber optic cable and photonic-based multiplexing/transmission system technologies; and 3) ultra—high-speed and -capacity switching/routing technologies.

## Advanced Modulation Technologies

The development of practical and affordable hardware—implementing advanced modulation and optimum matched filter receiver techniques—is a key high-technology area where utilizing bandwidth-limited media is the only option access service providers have to meet growing user demand. The two most prominent incidences where this technology has already produced breakthrough results include terrestrial and satellite-based wireless mobile communications (where nature limits the amount of useful spectrum); and, the need to better utilize copper twisted-pair wiring brought to nearly all small-business and residential premises in the United States. As Chapters 16 and 17 state, although advanced modulation and optimum matched filter receiver theories have been understood for decades, it is only in the last half-decade that *application specific integrated circuit* (ASIC) and general purpose *digital signal processing* (DSP) technologies have made it possible, practical, and affordable to apply these theories in small hand-held low-cost terminals.

In 1965, Gordon Moore, an Intel cofounder, postulated that advanced integrated circuit chip complexity and capacity would double every eighteen months. History has proven his hypotheses correct. Moore predicts this phenomenon will continue for another decade. Consult Chapter 16 for the incredible difference that DSP-enabled code division multiple access (CDMA) technology has made in both terrestrial and satellite-based mobile telecommunications system performance. In the wireline arena, advanced modulation techniques embedded in current digital subscriber line (DSL—see Chapter 4) hardware has already increased downlink throughput rates by factors of 10 to 30, and newer *very high speed DSL* (VDSL) promises another factor of 10 to 50 improvement.

The next breakthrough application of advanced modulation techniques will be broadband, fixed wireless local loops. These systems, based on LMDS/MMDS paradigms, offer true multi-application service alternatives to traditional guided media (wired or cabled) facilities; may have unique advantages for hard-to-reach customers in rural areas; and may be delivered by either terrestrial or satellite service providers.

# Fiber Optic Cable and Photonic-based Multiplexing/Transmission System Technologies

The introduction and widespread deployment of fiber optic cable-based technologies has already had a tremendous impact on network economics. Rather than leveling off, each year, using dense wavelength division multiplexing (DWDM—see Chapter 5), "clearer" fiber that exhibits a more uniform loss characteristic over a wider range of wavelengths, dispersion shifted fiber, Solitons, and other technologies, the capacity of fiber transmission systems continues to grow unabated.

You might say that fiber-based transmission system capacity is following a geometric progression similar to Moore's Law for integrated circuits. This leads to a very important probable result. Whereas in the past, the cost of most transport services has been proportional to throughput rates, speed, or bandwidth, fiber-based transmission system capacity increases may eventually eliminate such factors from "cost-to-deliver-service" equations. In the near future, transmission systems will carry terabits per second of data on single strands (one terabit = one million million bits). Given the high fixed costs of establishing transmission routes, regardless of capacity, these technologies combine to drive down unit transmission costs. Recall from Chapter 8 that bandwidth is "so close to being free" that in high-reliability SONET-ring networks 50 percent of the available capacity is held in reserve simply to recover from *possible* cable cuts or other failures.

At some low cost per bit per second, it is conceivable that service providers may prefer to deliver higher rate ports—if in doing so the length-of-session or actual network usage times can be shortened. Then too, higher maximum throughput rates can reduce latency problems arising from the need to transport a mix of isochronous voice and video and other traffic types. Certainly users, especially those engaged in interactive file transfers, will opt for the highest affordable throughput rates, making most downloads appear instantaneous.

Another important technology trend is the gradual substitution of photonic, as opposed to electronic, technologies in multiplexing, switching, and other transmission equipment. In the first fiber optic cable installations, fiber optic modems were used to convert electronic signals to photonic signals on one end of a fiber cable run, and another modem to convert photonic signals to electronic signals at the other end. Chapter 8 and Figure 8.15 portray a transition from these "conversion-intensive" initial installations to what ultimately may be *all-optical* networks. While serious limitations prevent that ultimate achievement now, *optical add/drop multiplexers* (OADMs) and *optical cross-connects* (OXCs) are already revolutionizing today's transmission systems.

These advancements coupled with photonic erbium-doped fiber amplifier-based products (EDFA—also discussed in Chapters 8 and 4), mean that amplifier/regenerator equipment can now be spaced 2,000 miles apart—as opposed to 125-mile electronic amplifier separation. In combination with the new technologies mentioned above, EDFAs and other photonic equipment are producing additional transmission system economies—further driving down the cost of delivering transport services.

## Ultra—High-Speed and -Capacity Switching/Routing Technologies

But high transmission rates are only beneficial if the capacity can be efficiently shared among large user populations, a situation that fosters high utilization of facilities and low unit costs. Recall that switching, in its many forms, is primarily a mechanism for sharing transmission resources. Chapter 11 demonstrates that in all communications networks there are economies of scale. For the same grade of service, aggregating more traffic leads to higher facility utilization. Large systems are more efficient and cost effective than small systems; single large switches and routers utilize bandwidth more efficiently than multiple smaller switches and smaller bandwidth transmission facilities. Hence, major Internet backbone providers are deploying routers capable of switching packets at 10 Gb/s (directly driving OC-192c transmission channels—9.95328 Gb/s); each router driving a wavelength on a DWDM system. This combination keeps backbone transport costs low and capacity high in anticipation of new broadband access service demands that threaten to open traffic volume floodgates.

Of course, these are not the only areas where technology plays a major role. But, as seen below, even these three enabling technologies materially affect transport and access services, and will therefore eventually be felt by all users—residential and business.

# Transport Services

Transport services are *network switching, transmission and related services that support information transfer between originating and terminating access facilities.* Network transport facilities used to deliver voice services have evolved from manual plug-boards and copper wire voice-frequency carrier systems to today's automatic stored-program-controlled circuit switches and modern SONET-based transmission systems. As much as today's circuit-switched and synchronous digital hierarchy-based transport facilities may justifiably be regarded as technological marvels, as Chapter 8 explains, many experts believe *IP-over photon* designs may not only be preferable for data communications traffic, but ultimately (with MPLS, differentiated services, digital wrappers, protocol tags or other techniques) the best approach in all-application/all-service, backbone transport networks.

In the past, separate transport networks (and, therefore, separate transport services) for each application were developed. Each evolved to satisfy specific application requirements (e.g., circuit-switched networks for voice, inexpensive packet-switched networks for low-speed data, and expensive fiber-optic router-based networks for high-speed data). ISDN, the first attempt to achieve economies through integrated digital network approaches to multiservice applications, failed because its implementation stretched over decades and original objectives fail to meet even current requirements.

ATM/SONET was postulated as ISDN's successor in pursuit of the holy grail of service integration. Again, slow rollout, expensive equipment, and ATM's inherent telephony orientation makes it vulnerable to a competing approach, the ubiquitous Internet with its TCP/IP protocols and router-based networks. With all these technologies in play, it is not a question of whether *convergence*[1] of voice, video, and data applications and underlying integrated, all-service digital network designs will take place, but only a question of when and under which network design paradigms.

---

1 *Convergence* is the name applied to a concept for satisfying all voice, data, video, imagery, and other applications and all access, transport, and other service requirements (all-applications/all-services) over single telecommunications facilities. It does not imply that a single network supports both access and transport services. It does imply, for example, that a single transport network facility accommodates all voice, data, video, imagery, and other services and related applications. It might also imply that a single access network facility accommodates all voice, data, video, imagery, and other services and related applications.

ATM networks surely have a head start with existing capabilities to support multiple service classes, but TCP/IP networks may duplicate this capability within a year with the introduction of the aforementioned Multiprotocol Label Switching (MPLS) and Differentiated Services. Initially, MPLS IP traffic may be routed over existing ATM infrastructures, but some new service providers will skip ATM/SONET layers altogether and transmit IP packets directly over DWDM-enhanced fiber optic transmission systems as described in Chapter 8. Technological innovation and competition among service providers will benefit end-users by offering them more options and lower prices, regardless of the eventual technological outcome.

Likely future technology impact on *transport services* can be inferred from some changes already taking place. In the past customers procured and paid for specified transport services and carriers delivered guaranteed quality and classes of services. For long-distance service, even residential customers selected a *primary interexchange carrier* (PIC) and received bills that detailed IXC long-distance charges. Compare that situation with the offerings of many mobile carriers that today provide a specified number of call minutes for fixed monthly charges, independent of whether or not subscribers make local or long distance calls. With these offerings, subscribers no longer have to dial "1" and mobile carriers subsume the obligation to render long distance service and pay any IXCs (other than those associated with the mobile carrier) that may be needed to complete calls. This is a voice service example of how decreasing transport costs may lead to their amalgamation into access service provider charges and how transport services may gradually disappear as separately procurable entities.

A similar change in data transport services is materializing. As noted in Chapter 14, corporations may procure advanced frame relay and ATM-based services able to support multiple classes of service and receive service delivery guarantees from carriers. What was also explained in Chapter 14 is the fact that today, such services are limited to virtual circuit offerings, and therefore are essentially replacements for more expensive leased private-line facilities. This means that unlike public voice and data network services (where anyone can communicate with anyone else connected to public networks), in corporate networks using only virtual circuit service, connectivity is limited to the number of virtual circuits purchased (usually "in-house" or specified customer or supplier locations).

Another observation made in Chapter 14 is that except for the Internet, there are no public data communication networks offered by carriers that dwarf the size of existing private corporate data networks. Recall that the immense size of today's public voice networks, and the economies of scale that they yield, means that carriers can offer corporations *virtual private voice network* services. Although voice VPN services are implemented using public network facilities, they render services that are indistinguishable from those of yesterday's facilities-based corporate private networks. The advantage is, of course, that today's VPN voice services cost much less than the same services obtained via private facilities-based voice networks.

In contrast to this voice network situation, as just noted, only the Internet, and all the backbone transport networks that support it, are of a size that dwarfs private corporate data networks. As a consequence it is only the emerging Internet-based *virtual private data network* services that hold forth the promise of delivering virtual private data network services to corporations at substantial cost savings. Moreover, Internet-based VPN services offer the possibility of universal, worldwide connectivity to anyone connected to the Internet.

Again in the case of Internet VPN service, corporations cannot now obtain multiple service class, guaranteed end-to-end throughput rates, latency, or other quality of service guarantees that can be specified in most frame relay and ATM-based transport services. Balancing this is the fact that corporations do not procure or pay for separate long-distance Internet-based data transport services. Instead, corporations contract with ISPs or NSPs (Internet Service Providers or National Service Providers) and pay monthly access charges based on port speeds. So again, in the case of Internet one detects the disappearance of transport services as separately procurable entities.

As another example of transport service trends, remember from Chapter 16 that unlike Iridium's satellite-based access and transport services offered on a global basis, that newer Globalstar offerings provide only access service to "partnering" regional wireline or mobile terrestrial carriers. Globalstar's business model relies on connecting service providers to deliver transport services necessary for end-to-end call completion. Thus, while Globalstar provides its subscribers with nationwide and global long-distance transport service through partnering, as we have seen, other mobile communications companies have pursued mergers among local and long-distance service providers.

As a final example of the changing transport service marketplace recall from Chapter 2 that although transoceanic cables were once the purview of monopolistic telcos that financed undersea cables for their own use, that is no longer the case. Responding to market projections and profit potential, Fiber Line Around the Globe (FLAG), an independent submarine cable venture, began offering service to multiple carriers in 1997. Currently FLAG provides 10 Gb/s of capacity to Sprint, AT&T, Cable and Wireless, China Telecom, Deutsche Telekom, MCI WorldCom and others. Global Crossing and CTR Group (Project Oxygen Network) are two other similar ventures, which in combination with FLAG, constitute about $14 billion in recent transoceanic cable investment providing several hundred landing points in hundreds of countries.

Here is an example in which some transport service providers sell no services directly to end-users, but rather have become carriers for other carriers. Note that increasingly, the actual facilities underlying end-to-end transport services in fact comprise an agglomeration of interconnected networks, owned and operated by multiple service providers. Interestingly enough, today's ultra-broadband Internet backbone transport networks came into being, and remain, agglomerations of separate carriers' carrier networks whose services are generally only available through ISPs or NSPs that also offer local ISP access services.

Examining recent transport service trends leads one to the conclusion that the opportunity for either corporate or residential customers to procure transport services as separately specifiable entities may be vanishing. This conclusion is consistent with regulatory tendencies to allow regional Bell holding companies to enter long-distance transport service markets, and to allow IXCs (or anyone else) to compete for local telephone access service markets and ever-decreasing per-unit bandwidth long-distance costs. It is also consistent with the trend to build all-application/all-service (voice, data, video, imagery, etc.) network facilities with the ability to support all access and transport services. If these trends continue, ultimately we may be able to establish single contracts to deliver voice, data, video, imagery, Internet access, and any other service on a local, regional, national, or global basis.

## Access Services

Likely future technology impact on *access services* can also be inferred from some changes already taking place. In the wireline arena, unshielded twisted-pair local loops that for years supported only voice access services are now being equipped with one of several varieties of digital subscriber line (DSL) technologies to provide access services to the Internet and other data communications networks.

Another example of convergence is found within Cable TV and direct broadcast satellite (DBS) access facilities. Again, until recently, these access facilities supported only entertainment program material but now are used at least for downlink Internet access. Of course, since the Internet itself is rapidly being modified to accommodate voice, data, video, imagery, and other traffic, such arrangements take on the appearance of being all-application access service facilities.

Perhaps the most all-encompassing examples of access facility application and service convergence are found in today's terrestrial cellular and satellite-based wireless telecommunications systems. While some data service quality and performance levels must be improved, it is nevertheless true that many of today's mobile subscriber sets (you can hardly call them just telephones anymore) support voice, paging, short message service (SMS), e-mail, Internet access and browsing, personal digital assistant, emergency, and other applications. Expanding wireless application and service possibilities even further are Bluetooth initiative personal area networks (PANs) introduced in Chapter 16 that many view as a single, universal means of providing individuals with access to all telecommunications networks and devices.

On the tethered or wireline front, future convergent applications are likely to require more bandwidth than the nominal 1.5 Mb/s capabilties currently available from DSL and cable modem systems. As video origination and server applications in homes or home-offices become more prevalent, the need for symmetrical uplink and downlink bandwidth will follow. Next-generation wired systems will rely heavily on fiber optic cable-based feeder and distribution local loop segments, minimizing copper UTP cable run-lengths to customer premises. This will enable providers to deliver higher speed service to customer premises without completely replacing the existing local loop outside plant. Current activities include specifications for simpler and less expensive Passive Optical Networks[2] (PONs) as alternatives to more complex SONET-based feeder networks.

---

2 Passive optical networks are point-to-multipoint access networks that consist of optical line termination (OLT) equipment at service provider locations and optical network units (ONUs) in neighborhoods, in buildings, or in customer premises, that are connected via fiber optic cables, couplers, splitters, and other passive components. Depending upon ONU placement, PONs provide fiber-to-the-curb (FTTC), fiber-to-the-building (FTTB), or fiber-to-the-home (FTTH) service.

In conjunction with very high speed DSL (VDSL) hardware, service providers will be able to deliver 15 Mb/s service over one-mile copper loops and up to 58 Mb/s service over short (<1000 ft) loops. These throughput rates will be allocable to upstream and downstream traffic, thus supporting either symmetrical or asymmetrical services. Wireless local loop systems, based on terrestrial LMDS/MMDS (see Chapter 16) or satellite-based (the Teledesic, Celestri, and Skybridge initiatives noted in Chapter 17) technologies will provide similar broadband, multi-application/ services direct to user premises.

# Wireless Services

Beyond the wireless trends pertaining to both mobile and fixed user applications already treated here and in Chapters 16 and 17, we mention two additional important developments. Emerging third-generation mobile systems will not only provide an increasing number of voice features, but are slated to support much higher data rates and as a consequence more multimedia and Internet applications. In extensive research efforts in Europe, the U.S., Japan, and Korea, *wideband code division multiple access* (WCDMA) is the leading candidate for powering third-generation (3G) wireless personal communications systems. Now being harmonized in a joint effort known as the Global Third Generation (G3G) project, introduction of 3G mobile networks will have to progress in a way that maintains backward compatibility with current systems.

International Mobile Telecommunications System 2000 (IMT-2000) 3G network requirements include the ability to support full coverage and mobility service at 144—384-kbps user information rates, and limited coverage and mobility service at 2 Mb/s rates. Eventual designs are likely to be derived from existing IS-95 CDMA specifications to promote backward compatibility. Unlike IS-95's "air-interface" 1.25-MHz channel bandwidths and 1.2288-Mcps (million chip per second—see Chapter 16) modulation chip rates, the nominal bandwidth in all 3G WCDMA proposals is 5 MHz, with 3.6864-Mcps (three times IS-95's 1.2288 Mcps rate) and 4.096-Mcps spread-spectrum modulation rates currently under consideration.

High data throughput rates will extend the all-application/all-service convergence trends in other sectors to mobile subscriber markets. Despite air interface and subscriber equipment size issues, wireless

Internet applications are poised to thrive in the near future. As noted earlier, experts project that by 2003 more people will access the Internet via wireless devices than with PCs, totaling over 830 million wireless connections. This merger of wireless and Internet technologies is moving faster than either trend on its own and will profoundly change the face of communications as we know it.

Other new mobile wireless applications and services relate to public safety. Mobile telephones are already crucial in life-saving emergency assistance situations. The ability to reach 911 emergency call centers is one of the most-often voiced reasons Americans give for purchasing wireless phones. In fact, between 1996 and 1998, wireless 911 calls grew from 60,000 to 100,000 per day. While current capabilities are significant, Enhanced 911 (E911) services now being implemented across the country promise even greater utility.

Although advanced 911 systems automatically pinpoint *wired* subscriber locations, the location of *wireless* callers is currently only generally known. Precious time is wasted determining the location of wireless callers in distress, even if their location is known by the callers, and additional time is lost in determining the location of the closest emergency response teams. As a consequence, the FCC has mandated that automatic location identification (ALI) features be added to all public wireless carrier systems.

When the FCC released its Order in 1996, the only subscriber position-location solution envisioned was network based. Under this approach, multiple base stations receiving subscriber set signals can use triangulation methods to pinpoint subscriber set locations. Since that time, advancements in subscriber equipment-based position-location alternatives have materialized. One such approach uses mobile telephones with global positioning satellite (GPS) receiver capabilities. In this approach GPS-determined location information is automatically forwarded to E911 call centers.

The growing popularity of CDMA systems strengthens handset-based solution arguments. First of all, powerful processors must be embedded in handsets just to handle CDMA's enormously complex signal-processing requirements. Slightly more processing power to add position-location capabilities does not impose a significant processing load increment. As explained in Chapter 16, CDMA handsets employ power-control features to reduce transmitted power as handsets move closer to a base station. While power control is essential to efficient CDMA operation, reducing power when in close proximity to one base station reduces the likelihood that multiple other base stations could monitor its signal, as required by network-based triangulation solutions. As a result, the FCC modified its Order to allow both network and handset-based solution approaches to satisfy its requirements.[3]

On October 9, 1999, the Cellular Telecommunications Industry Association (CTIA) issued a request for proposals soliciting industry solutions to FCC E911 requirements as they apply in the Washington, D.C. metropolitan area. Whatever solution is adopted must be workable with any of the extant mobile telecommunications systems discussed in Chapter 16. Since the Washington, D.C. area is served by five different carriers operating AMPS analog and digital (TDMA, CDMA, and GSM) networks in both the 800 and 1900 MHz bands, it is an excellent locality in which to seek and test solution approaches.

# Standards, Pricing-Thresholds, and Universal Access Impacts on "Killer" Application and Product Market Status

Beyond carrier abilities to offer wide varieties of convergent services, and the underlying technologies that make them possible, standards, pricing thresholds, and universal access are three additional key factors crucial to past, present, and future "killer" application and product market successes.

Looking at standards first, it is useful to consider the fax market development. Although the concept was demonstrated in London in 1850 by F.C. Bakewell, it was not until 1934 that dry recording processes and suitable papers appeared to make commercial facsimile products possible. Even after that, although the technological basis for practical products existed, the fax market potential remained dormant. At first, when few organizations possessed fax equipment, new

---

3 For network-based solutions, the requirement is location determination within 100 meters for 67 percent of the calls, 300 meters for 95 percent of the calls. For handset-based solutions, the requirement is 50 meters for 67 percent of the calls, 150 meters for 95 percent of the calls. The tighter accuracy standard for handset approaches compensates for the fact that these approaches will generally require a longer phase-in period.

users had to justify investment based on "in-house" demand. Following adoption of Group III standards by fax equipment manufacturers, things changed rapidly. As ever-increasing numbers of Group III machines were sold, the application broke free of in-house limitations and soon, faxing between independent enterprises became commonplace and expected.

Growth was fed by regenerative forces—as more units were sold, more buyers felt compelled to purchase. Where at first, most small businesses considered fax machines optional, today small enterprises and even individuals view fax equipment ownership as essential.

Looking back, it is easy to spot the critical ingredients in the fax success story. First, Group III standards permitted interoperation between fax machines made by different manufacturers, essential for the market to encompass inter-company applications through universal connectivity. Moreover, universal connectivity depended upon the ability of Group III machines to operate reliably over the ubiquitous voice telephone network.

A final factor contributing to fax's growth involves pricing. All potential telecommunications applications have threshold price levels. No matter how appealing or novel a technology may be, there is a market sustaining price level above which only a small number of consumers—either for status or special operational reasons—will participate. Prices which elevate applications to "killer" status are usually significantly below market sustaining prices. We witnessed this in the consumer VCR market. Above $1,000, sales languished and only relatively affluent, technology-oriented users entered the market. At $200-300, a large percentage of television set owners purchased one or more units. In the fax case, the fact that any manufacturer's equipment works with any other's Group III compliant model unleashed a fierce competition for market share that sent prices plummeting.

We also witnessed the pricing-threshold effect in the cellular telephone market. In the mid-1980s, unit prices for cellular telephones hovered in the $2,000–$3,000 range, and monthly subscription charges averaged $125–$150, resulting in lackluster growth in demand for units and services. By 1987, unit prices broke the $1,000 barrier and with steadily declining usage costs touched off an annual growth rate of 40 percent (as depicted in Figure 16.7). That geometrical growth rate continues to this day. Of course the fact that cellular systems from the outset have been interconnected to the worldwide public voice networks satisfies the ubiquitous access requirement. There is little argument that in breaking price barriers, VCRs, fax machines and cellular telephones overcame the last obstacle to achieving key application status. Note the interdependency among the pricing, standards, and universal connectivity factors cited above.

Two methods lead to standards that profoundly influence our marketplace today. The first is based on international treaties, officially sanctioned working groups, and a hierarchy of national standards bodies. This is embodied in the International Telecommunication Union (ITU)[4], whose telecommunication standardization sector (designated ITU-T) adopts recommendations with the goal of establishing standards on a worldwide basis. Most of the standards-based services we discuss throughout the book emanate from ITU activities (ISDN, X.25 packet switching, ATM, frame relay, to name a few examples). ITU continues to move forward in attempts to create standard frameworks within which industry can innovate and define service implementations.

The second method by which standards responsible for cosmic growth may be put in place is the adoption of de facto industry standards, sometimes established with no official standards group participation. In the 1970s while ITU (then CCITT) was adopting standards (mostly related to telephony—the major existing network infrastructure at the time), another completely different approach was being pursued by a relatively closed military, research, and education community. As noted in Chapter 7, a set of Internet standards built around TCP/IP protocols evolved through an informal, collaborative process whose participants were volunteers from affected communities.

Despite attempts by the "official" standards community to replace TCP/IP, the Internet proliferated to such an extent that TCP/IP reigns as the once and future king of data communications protocols. Even as it becomes the essential building block of national systems for electronic commerce, TCP/IP's standards are still determined outside official international channels by members of the Internet Engineering Task Force (IETF). Proponents of IETF cite the rapidity with which the IETF has reacted to changing market requirements in adopting new standards for use in the Internet as a reason for it to continue in its role; the slow-moving ITU is deemed unsuitable for the task. Some see an eventual confrontation over jurisdiction over the precious resources of the Internet and its evolution, but more likely is a cooperative arrangement that will assimilate IETF standards into ITU rec-

---

4 See Appendix A for comprehensive and historical dissertation on U.S. and worldwide standards setting.

ommendations and allow ITU frameworks to influence IETF deliberations. This process is already under way with new ITU "Y" series recommendations covering IP topics.

Regardless of the method by which Internet standards were established, without them there would be no single network of the Internet's proportions. There might be hundreds of incompatible data networks with large numbers of users, but there could never be a single network through which 150 million devices can establish reliable communications; and by which hundreds of millions of users are able to peruse with equal abandon over 850 million different Web pages. Quite simply, it's the de facto industry standards that make Internet's killer application/service status a reality.

# Summary

It should be apparent that future developments are driven by the many interrelated factors just discussed. To help sort through them we provide the following list of six principal industry trends, most of which manifest themselves as some form of convergence of what historically have been separately pursued service or network design objectives.

1. **Application convergence.** The Internet forced the recognition that neither individuals nor enterprises have separable voice, data, paging, video, imagery, e-mail, browsing, or other application requirements. We have need for and are steadily moving toward integrated multi-application services and networks that can support them.

2. **Convergence toward single-protocol integrated digital networks.** Efficiently supporting multi-application services means that separate voice and data network designs of the past must eventually yield to or converge upon single-protocol integrated digital networks. By integrated digital networks we don't just mean networks that can transport voice, data, and other traffic, but a more systemic integration in which switching, multiplexing, and other functions converge economically in single, multifunctional equipment, a fundamental ISDN precept. This means extending what began years ago when PBXs and central-office switches absorbed multiplexing functions once accomplished in

separate boxes, and continued with the evolution of DACS and ATM hardware.

By single-protocol networks we mean ultimately the abolition of separate switching, multiplexing, and transmission systems for voice, data, video, and other traffic. At the present time the most likely unified approach is IP-based packet-switched networks embellished with MPLS, diffserv, tag switching, digital wrappers, and other techniques aimed at enabling the Internet to support all classes of services. Interestingly enough, these packet-switching embellishments in essence imbue packet networks with the ability to establish switched virtual circuits "on the fly," a characteristic of circuit-switched networks. Thus, while it is relatively easy to make packet-switched networks act like circuit switched networks, the converse is not true.

Until just a few years ago voice traffic dominated the telecommunications landscape. The thought of unraveling or abandoning circuit-switched voice networks in favor of IP-based networks would have been looked upon with great suspicion if not disdain. Now with other forms of traffic gaining parity and promising to dominate in the near future, the trend toward a single network paradigm not only appears reasonable, but we already see large-carrier movement in that direction. Of course before the benefits of single-protocol integrated digital networks can be fully realized we need to figure out what to do with perhaps a billion or so very workable and reliable telephones that are essentially operable only over existing circuit-switched networks.

Interim "gatekeepers" joining remnants of old and new networks will work for a while, but in the end if we do progress to all-IP networks we will need to develop a substitute for that omnipresent telephone. And the device, or appliance, will have to be as inexpensive and just as easy to use as today's telephones—not a trivial requirement by any measure.

3. **Trend toward "carrier's carrier" transport services.** As unit transport costs spiral downward due to constantly increasing fiber optic/DWDM generated transmission bandwidth availability, as non-telcos enter Internet backbone and international transport markets, and as mergers among access and transport service providers accelerate, transport costs will continue to be absorbed "invisibly" in Internet and other access service provider customer charges.

4. **Local and long-distance company ownership convergence.** As competitive provisions of the Telecommunications Act of 1996

become fully implemented and as Judge Green's modification of final judgement (MFJ) strictures affecting LATA and inter-LATA operations abate, the distinction between local and long-distance services as separate entities will diminish, and even more mergers of currently separate local, regional, and long-distance service providers can be expected.

5. **Wired and wireless company ownership and operational convergence.** Chapter 4 highlights the convergence now occurring among wired and wireless operations. Existing products support the "twinning" of mobile and corded telephones such that either can be used and both respond to the same listed directory number. Future UMTS telephone capabilities to operate with either public cellular systems or privately owned base stations and the requirements for highly interconnected wired and wireless telephone networks all reflect the rationale behind the number of wired-wireless company mergers that have already taken place.

6. **Convergence among telecommunications and program material/intellectual property service providers.** With equally good engineers and technicians and access to virtually identical technologies, ultimately the discriminators that distinguish one telecommunications service provider from another become cost, customer service, and other business management attributes. While competition on these grounds can be rewarding, aggressive entrepreneurs are always interested in using existing investments and market positions in more profitable enterprises.

When Alexander Graham Bell discovered that a good deal of money could be made by essentially "charging people for electron activity," he unleashed the unbridled pursuit of telecommunications business ventures in all its current forms. More recently businessmen have discovered a new enterprise that may ultimately dwarf Bell's monumental discovery, namely the ability to charge people for the privilege of accessing and viewing program material, ideas, knowledge, and other intellectual property, an even more ephemeral commodity than electron activity.

While the language here is somewhat "tongue-in-cheek," the facts are that businesses and individuals have an insatiable appetite and need for intellectual property in all its varied forms. Supplying that need may yet dictate the corporate composition of tomorrow's telecommunications service providers. America Online is certainly a markedly different company from ISPs content with merely providing users with electronic access to the Internet. In the long term the truly appealing aspect of combining telecommunications and intellectual property or program material service delivery is that both services appeal to the highest and most noble of all human activities, man's ability to think and his ability to communicate and exchange ideas with other men. It's not the kind of service that's likely to be replaced by some new technology.

# Future Paradigms

In a discussion of service innovation trends, a relatively new "active network" paradigm needs to be mentioned. Whereas most networks are merely passive transport mechanisms for data bits, *active networks* allow individual users, or groups of users, to inject customized software programs into network nodes where nodal processors execute the program's instructions. This paradigm goes beyond simple downloading of computer software programs for later installation on network-embedded or standalone computers. In *encapsulated active network approaches,* the programs are integrated into every packet of data transported over the network.

Active networks are particularly germane to this discussion of trends and new services since one of their key objectives is the ability to accelerate innovation by making it easier to deploy new network services. Active networking is not a totally new concept. Firewall packet filtering, wherein routers decide which packets should be blocked, is a rudimentary example of the type of operations that active networks support. What is new is the ability to deploy programs dynamically to nodal processing engines.

As we close not only this section but the book itself, we acknowledge that the outlooks discussed in this section are conservative; they are possible, even probable, directions for the near future. However, the truly significant innovations of the next decade may be those that no one predicts. Just as surely, they will evolve from today's capabilities, which may be the best reason why one should acquire a thorough knowledge of existing telecommunications technologies and networks.

Robert Lucky, Corporate Vice President of Telecordia Technologies (formerly Bellcore) and a well-known prognosticator of the future, admits that "forecasting isn't what it used to be, it's pretty much become impossible." Lucky's top ten telecommunications trends mirror many of the topics stressed in these pages: broadband access, quality of service, DWDM, next-generation convergent networks, and the wireless revolution. But he reserves his number-one trend for appliance networks, having everything connected to networks. If readers of this book contribute to creating an environment so rich that anything conceived can be realized, then we have succeeded mightily.

Thanks for the reading!

# A

# Standards Setting in the United States

# Introduction and Scope

Successful provisioning of facilities for telecommunication services, whether public or private, is largely dependent on the *interconnectivity* and *interoperability* of those facilities. Interconnectivity and interoperability are cardinal objectives of standards-setting processes. The subject of telecommunications standards is many faceted, often surrounded by confusion and misunderstanding. This appendix examines the "what, how, when, and why" of standards. From an historical perspective, the telecommunications industry grew up around a standards-based approach.

In the past several decades the data processing industry and more particularly the discipline of data communications—itself an intersection of the data processing and telecommunications worlds—has greatly complicated the traditional standard-setting proclivity of the telecommunications industry. To gain an appreciation of the role of standards in telecommunications and data communications, some background is necessary. Data processing standards are addressed only insofar as they affect data communications.

# Background

Telecommunications standards date back to the 1880s with the founding of the *International Telegraph Union* (ITU) based in Geneva, Switzerland, as the driving force to achieve interconnectability and interoperability among European telegraph systems—then mostly government appendages to railways or postal administrations. The ITU constituted a collegial process for establishing recommendations, which while not binding on member telegraph administrations, were considered as treaties by foreign ministries. The Transatlantic Telegraph Cable linking North America (essentially the United States) with European international networks greatly accelerated the significance of ITU "standards" (actually recommendations). Thus, the U.S., through the State Department, became a signatory to ITU recommendations, an arrangement which persists to this day.

With the advent of the telephone, which used wire media much as telegraphy did, and with the appearance of wireless radio communication, the ITU split into two principal bodies and a third offshoot body. The first two bodies divided the telecommunications domain into Radio and Telephone and Telegraph with acronymic "names" derived from the initials of their formal names in French, the CCIR—the International Consultative Committee for Radio, and the CCITT—the International Consultative Committee for Telephone and Telegraph. The radio domain also merited a third entity—the IFRB—the International Frequency Registration Board—one of the first attempts at fostering international cooperation between often hostile, and sometimes warring states. In 1993, a new ITU—the International Telecommunication Union was reborn and reorganized to focus on the needs of modern telecommunications. Three new sectors were created:

- **ITU-R**—The radiocommunication sector combining the functions of CCIR and IFRB.
- **ITU-T**—The telecommunication standardization sector absorbing the functions of CCITT.
- **ITU-D**—The development sector facilitating and enhancing telecommunication development worldwide.

The U.S. and most other nations are members of the ITU out of necessity, since electromagnetic waves respect neither geographic nor political boundaries. Today the ITU is one of the few really working parts of the U.N. and remains based in Geneva.

Early on, the U.S. domestic scene, essentially private-enterprise driven, did not see the need for direct government involvement in standards setting, other than the State Department's continuing involvement in treaty making through the ITU. The rise of the Bell System colossus under the leadership of Theodore Vail gave the U.S. a standards-making engine of awesome speed and power. Until its demise at divestiture in 1984, the old Bell System drove standards for telephony and also to some extent for telegraphy—despite the fact that telegraphy was controlled by Western Union and individual private rail companies.

The Bell System established the world's best and lowest-cost universal telephone service—albeit in the guise of a monopoly, courtesy of the 1916 Kingsburg agreement between the U.S. Government and AT&T (the proprietor of the Bell System). Prior to divestiture, government impact on standards-making took the form of FCC initiatives loosely termed "deregulation." There were also some true standards making bodies such as the IEEE (Institute of Electrical and Electronic Engineers—a U.S.-based professional society), the EIA (Electronic

Industry Association a trade group[1]), and ANSI (American National Standards Institute). The FCC codified existing standards developed by the Bell System into Rules and Regulations. Two principal standard areas were Customer Premise Equipment (CPE)—the purpose of Part 68 Rules, and interconnection of "Other Common Carriers" as a consequence of deregulation.

# ■ ■ Impact of U.S. Developments

From the late 1960s, telecommunications standards have been greatly affected by federal judicial and regulatory initiatives directed toward the Bell System and its workings. These activities can be grouped under two categories; *deregulation*—the regulatory domain; and *divestiture*—the consequence of the U.S. Department of Justice antitrust suit against the old AT&T. The combined effect of these initiatives drastically altered the way standards are established and administered in the U.S., and to some extent how the ITU operates, reflecting U.S. influence on its activities.

Pre-divestiture, U.S. telecommunications was run under the rubric of the Bell Systems—a benign monolith—but a monolith nevertheless. On the plus side, this made for controlled introduction of new technology in a manner that made standardization a built-in objective of the process. Undeniably, this approach led to the establishment of the world's finest, large-scale domestic telecommunications system.

Also undeniable was the cost-effectiveness of the service provided by the Bell System. This was attributable to the lack of competitive pressure on the financial structure of the Bell System, which conformed only to a regulatory "rate-of-return" philosophy. The negative side of this arrangement, however, was that standardization was also used to delay introduction of new technology having cost and/or performance advantages.

Early federal initiatives were designed to mitigate the negative aspects of standardization, increasingly viewed as skewed to AT&T's financial advantage. Furthermore, Bell System standards were not visible to the community at large except for those affecting interfaces with CPE, and those visible by virtue of the Bell System's relationship

---

1 In March 1998 EIA, the Electronic Industries Association changed its name to Electronic Industries Alliance.

with independent telephone (i.e., non-Bell affiliated) companies, represented then by the United States Independent Telephone Association (USITA).

Joint Bell System (through Bell Labs) and USITA issuance of "compatibility specifications" was a major standards accomplishment. Their influence on the engineering and operation of the physical plant of the Bell System were of inestimable value. They made possible the levels of both customer and competitor interconnection that ultimately led to today's competitive telecommunications environment.

Divestiture was the instrument causing the dissolution of the Bell System and its unilateral standard-setting activities. The loss of the concept of "system" in the U.S. telecommunications infrastructure has turned what was heretofore a Prussian General Staff approach to standards into a more democratic, collegial system, much like the ITU in Europe before its replacement by CEPT (the Conference of European Posts and Telecommunications—a grouping of administrations, usually associated with governments, striving for consensus).

# Current Status of Standards Setting

U.S. Telecommunications standards (including data communications) today are no longer dominated by any single entity—including the U.S. government. To the extent the Department of State participates in ITU standards recommendations and that the resulting "treaties" are ratified by Congress, the international standards-setting process has changed little from two decades ago, except for a quickening of pace. The ITU carries out its work in *Study Groups* which meet over a four-year period culminating in a *Plenary* session, that adopts recommendations in categories identified by letters, as shown in Figures A.1 and A.2. Major representation changes, however, have resulted from the replacement of the Bell System by various federal agencies and a variety of quasi-nongovernmental organizations such as ANSI, IEEE, and EIA/TIA.

The international arena is bifurcated between the ITU and its constituents with its standards cast as recommendations, together with the International Organization for Standardization (ISO). The ISO is much more active in data communications than telecommunications. A third entity, the International Electrotechnical Commission (IEC) works largely through the ISO. The designated U.S. representative to

**Figure A.1**
ITU-T
recommendations in
force.

| Recommendation Series | Title |
| --- | --- |
| A | Organization of the work of the ITU-T |
| B | Definitions, symbols, classification |
| C | General telecommunications statistics |
| D | General tariff principles |
| E | Overall network operation and service operation |
| F | Non-telephone telecommunication services |
| G | Transmission systems and media, digital systems |
| H | Audiovisual and multimedia systems |
| I | Integrated services digital network (ISDN) |
| J | Transmission of TV, sound, and multimedia signals |
| K | Protection against interference |
| L | Construction, installation, and protection of cables |
| M | TMN and network maintenance |
| N | Maintenance of international sound and TV circuits |
| O | Specification of measuring equipment |
| P | Telephone transmission quality |
| Q | Switching and signaling |
| R | Telegraph transmission |
| S | Telegraph services terminal equipment |
| T | Terminals for telematic services |
| U | Telegraph switching |
| V | Data communication over the telephone network |
| X | Data networks and open system communications |
| Y | Global information infrastructure and IP aspects |
| Z | Languages and general software aspects |

ISO is ANSI which, through its accreditation of the Alliance for Telecommunications Industry Solutions (ATIS), has made it the primary vehicle for telecommunications, and to a limited extent data communications, standardization in the U.S. today. The formal standards-related entity of ATIS is known as the T1 (for telecommunications—not T carrier!) Committee composed of interested parties from LEC and IXC service providers, equipment manufacturers, the government, and end-users when they elect to participate. While under the ISO aegis, through a loose coupling with ITU, many ANSI initiatives are also considered in ITU proceedings.

**Figure A.2**
ITU-R recommendations in force.

| Recommendation Series | Title |
|---|---|
| BO | Broadcasting—satellite service (sound and TV) |
| BR | Sound and television recording |
| BS | Broadcasting service (sound) |
| BT | Broadcasting service (television) |
| F | Fixed service |
| IS | Inter-service sharing and compatibility |
| M | Mobile, radiodetermination, amateur, and related satellite services |
| P | Radiowave propagation |
| RA | Radioastronomy |
| S | Fixed satellite service |
| SA | Space applications and meteorology |
| SF | Frequency sharing between fixed satellite service and fixed service |
| SM | Spectrum management |
| SNG | Satellite news gathering |
| TF | Time signals and frequency standards emissions |
| V | Vocabulary and related subjects |

U.S. data communications standards are influenced by the CBEMA (Computer and Business Equipment Manufacturer Association) working on the ISO X3 Committee in the areas of data communications and Open System Interconnectivity (OSI)—a major initiative of the ISO. Other entities such as the IEEE and the EIA/TIA (Electronics Industry Association/Telecommunications Industry Association) also produce standards, which are often adopted by ANSI. Figure A.3 depicts these relationships and indicates the presence of the major U.S. federal government players—the FCC and the Department of State. Most federal input to the standards bodies comes from the Department of Commerce's National Institute of Science and Technology (NIST)—the old National Bureau of Standards and the National

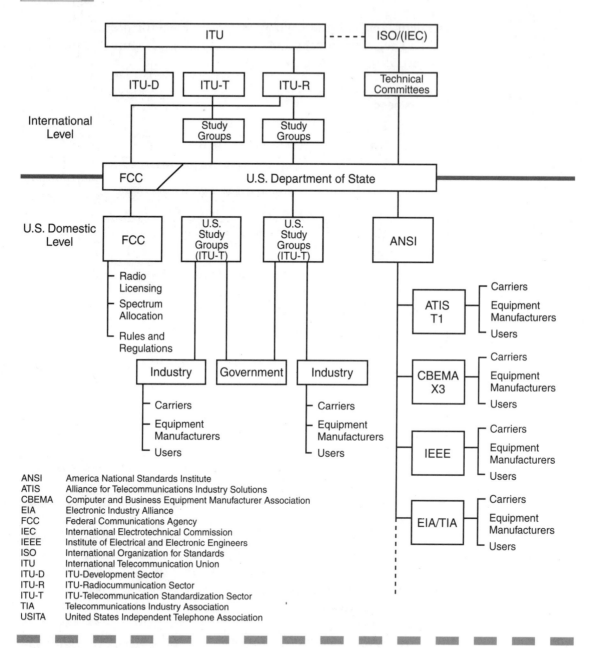

**Figure A.3**    Organization for telecommunications standards setting in the U.S.

Telecommunications and Information Agency (NTIA). Other government participants include components of the Department of Defense,

and other interested agencies via the National Communications System (NCS)—an Executive Branch agency operating at the White House level.

# ISDN Standards Activities in the United States

The foregoing generally describes telecommunications standards setting both at international levels and within the United States. The present section illustrates how lengthy and complex the standards-setting process can be, particularly when the intersection of several different ITU recommendation series must be dealt with. It does this by way of example, tracing historically how those processes and market forces combined to delay U.S. migration toward Integrated Services Digital Network (ISDN), relative to ISDN's more rapid adoption in Europe and other industrial countries. As it turns out, because superior technological alternatives have materialized so quickly, the rapid overseas ISDN buildout, and its expense, is now regretted by many PTTs.

Because the need for setting ISDN standards in the U.S. coincided with deregulation and divestiture, some pundits cite extended ISDN standards deliberation as evidence that somehow the FCC actions supporting deregulation and divestiture were ill advised, and that U.S. telecommunications might have been better served by continuing the Bell System monopoly. But, as evidenced throughout this book, the results of the FCC's "dual provider" cellular telephone decisions, its 1996 Telecommunications Act provisions, and the appearance of new long-distance competitors and rate reductions directly related to deregulation and divestiture, clearly establish the merits of greater levels of competition in all aspects of the U.S. telecommunications market. Nevertheless it is true that the existence of large numbers of competitors, each with a stake in the outcome of proposed standards, itself greatly complicates standards setting. So although the following overview focuses on ISDN standards, it also depicts, in general, the complex way in which international standards setting interacts with U.S. market realities. The example serves a second purpose in that it provides deeper insight into the number of steps involved in ITU deliberations, as well as the number of different committees and separate recommendations that must taken into account when concepts and paradigms as broad as ISDN are encountered.

The concept of an ISDN was first formulated in the councils of the CCITT before the 1976 Plenary Session. While the concept incorporated significant contributions from the U.S., it can fairly be said to have reflected a European vision of telecommunications. In the early 1970s European telecommunication administrations, and to some extent Japan, viewed evolving digitization of their telecommunication infrastructure, initiated by U.S.-based activities, as needing a standards framework to ensure future interoperability. In those days of analog technology, switching and transmission systems were seen as independent network segments. Digitization of the switching and transmission plants made possible the integration of many functions, particularly multiplexing, and led to the "Integrated Digital Network" (IDN) concept. But this technological improvement could not reach customers until a systematic provision for access was established. The concept of user access to the IDN underlies the Integrated Services Digital Network.

By the 1976 CCITT Plenary Session, the concept had gelled sufficiently to warrant creation of recommendations within the "I" Series heading (coinciding with ISDN) in the "Green" Book issued by the Plenary Session. The pattern for evolution was set, and has persisted through four successive Plenary Sessions—1980 (Yellow), 1984 (Red), 1988 (Blue), and 1992. Extensive coverage of ISDN appears in not only the I Series but also the G (Transmission), Q (Switching and Signaling) Series, and to a lesser extent in the X (Data Communication Networks) and E (Tariffs) Series. The topic was developed from a "recommendations" (the polite term for standards ) perspective, and ISDN proceeded rapidly in Europe and Japan—but not in the U.S. Why?

To answer this question, we must look to what makes telecommunications and data communications in the U.S. different from other industrialized nations. Part of the answer is that an integrated, monolithic telecommunications administration corresponding to foreign PTTs no longer exists in the U.S. to rapidly define and enforce standards. Aside from the loss of this standards-setting capability, other events impeded U.S. developments to support access to an IDN.

Underlying the concept of ISDN is a basic notion of the NT—the Network Termination. As originally conceived, NT is service provider-owned equipment on customer premises that establishes the point of telecommunications service delivery by a carrier (U.S.A.) or an administration (most of the rest of the world). This would have left the details of how the NT and central office equipment would be designed entirely in the hands of service providers, to adapt classic two-wire loops to customer premises so that they are able to support basic and primary rate signals, Thus, basic and primary rate access to CPE is standardized at the customer side of the NT—identified as the S or T reference points in I Series parlance.

The ISDN primary rate interface for the U.S. reflects the four-wire DS1 level of the digital signal hierarchy. Provisioning for basic rate interfaces reflects the great amount of two-wire plastic insulated copper (PIC) twisted-pair facilities composing local telephone distribution in this country. These decisions represented the status of CCITT recommendations at the time of the 1980 Yellow Book. During the four-year period leading to the 1984 Red Book, deregulation and divestiture changes took place in the U.S. These changes are not reflected in the Red Book, which largely followed and expanded upon the Yellow Book.

The FCC's computer Inquiry II decision in 1979 mandated the separation of any and all forms of terminal equipment at a customer premises from the service provided by a carrier. By the early 1980s, the FCC was confronted with a highly disputed proceeding with a formal title under Docket 81-216 variously styled as the "NCTE Order" or "NCTE Deregulation." The Commission's decision, rendered in 1983, took the position that all NCTE was to be treated as CPE (the Computer Inquiry II term for terminal equipment subject to Part 68 of the FCC Rules and Regulations). This decision reflected strong arguments, primarily from the data processing industry, that "separation of NCTE from service provision will not retard network innovation."

Following the "NCTE Order" decision, and the January 1, 1984 Bell System divestiture date, the U.S. embarked on serious, complex standards-setting activity to define what is now termed the "U" Reference Point involving an army of interested parties under the auspices of the Exchange Carriers Standards Association (ECSA) T1 Committee. This activity was to develop a standard for U.S. applications that bore some resemblance to the CCITT ISDN definition in the 1984 Red Book. The result, which reflects significant communication and semiconductor technology advancement, arrived in 1988 as ANSI T1.601-1988, entitled "ISDN-Basic Access Interface." This document in an earlier draft form was submitted in time for inclusion in the 1988 Blue Book and now appears in a similar form in G.960 and G.961 of the G Series of Recommendations.

Developers of CPE containing an NT could finally incorporate standards into product developments with a reasonable level of confidence. However, five to six years were expended in the process. The basis for the complexity of standards-setting activities for ISDN access is graphically portrayed in Figure A.4. The figure illustrates the tremendous amount of coordination that must take place among activities developing related standards and the participating international organizations. Undoubtedly, this complexity at least partially accounts for the fact that an informal group of interested users chose to develop the TCP/IP protocols that now dominate the worldwide Internet outside of cumbersome international standards setting processes.

**Figure A.4**
ISDN access
standards setting
relationships.

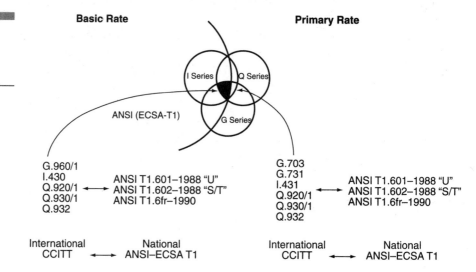

APPENDIX **B**

# Letter Symbols for Units of Measurement

While disparity often exists in the way units and prefixes are abbreviated, good technical practice calls for standardized symbols to ensure a common understanding of their meaning. To this end, the American National Standard Letter Symbols for Units of Measurement (ANSI/IEEE Std 260.1-1993) was developed as a basis for all units and abbreviations of units, and applies throughout this book. The table below lists relevant telecommunications industry prefixes and units, gives the appropriate standard symbol, and also provides alternate nonstandard symbols that sometimes appear in the literature.

| Name/Prefix | Standard | Nonstandard |
|---|---|---|
| tera | T | |
| giga | G | |
| mega | M | |
| kilo | k | K |
| milli | m | |
| micro | $\mu$ | m |
| nano | n | |
| ampere | A | |
| bit | b | |
| bit per second | b/s | bps |
| byte | B | |
| cycle per second | Hz | cps |
| decibel | dB | |
| gigahertz | GHz | |
| gigabyte | GB | |
| gigabit per second | Gb/s | Gbps |
| hertz | Hz | |
| kilohertz | kHz | KHz |
| kilobyte | kB | KB |
| kilobit per second | kb/s | Kb/s, kbps, Kbps |

*continued on next page*

| Name/Prefix | Standard | Nonstandard |
|---|---|---|
| megahertz | MHz | |
| megabyte | MB | |
| megabit per second | Mb/s | Mbps |
| minute | min | |
| month | mo | |
| second | s | sec |
| terabyte | TB | |

# GLOSSARY

**2B1Q** A modulation scheme in which successive pairs of user information data bits are mapped into one of four (quaternary or 4-ary) symbols. See Chapter 4 for examples.

**access** All facilities and related services needed to support information transfer to or from an information source to some network's transport service.

**access connection** See *central office connection*.

**access coordination fees** Small surcharges added onto underlying LEC charges when access is ordered by interexchange carriers on a customer's behalf.

**access service** Specified sets of information transfer capabilities furnished to users at telecommunications network points of termination (POTs) to provide access to network transport services.

**access tandem (AT)** A LEC switching system that performs concentration and distribution functions for inter-LATA traffic originating or terminating within a LATA.

**acknowledgment** A message sent to the originator of a data communications packet or frame indicating that the packet or frame was correctly received.

**active networks** Networks allowing individual users, or groups of users, to inject customized software programs into network nodes where nodal processors execute the program's instructions. In encapsulated active-network approaches, the programs are integrated into every packet of data transported over the network.

**address signals** Signals conveying destination information such as a called 4-digit extension number, central-office code, and when required, area code and serving IXC carrier code. These signals may be generated by station equipment or by a switching system.

**agent process** A process that performs management functions on managed objects.

**alternate mark inversion (AMI)** See *bipolar signals*.

**alternate space inversion (ASI)** See *bipolar signals*.

**ampere** In electrical circuits, the flow of electrons, called electric current, is measured in *amperes*. When a constant voltage (non-time-varying) of one volt amplitude is applied to a resistor with a resistance value of one ohm, it produces a steady or constant current of one ampere.

**amplitude shift key modulation** A type of modulation in which carrier signals are modulated in amplitude by information-bearing signals. In on-off key amplitude-shift key modulators driven by binary signals, for example, digital baseband signal levels corresponding to logical "1" values produce modulator output bursts of carrier signal sinusoidal cycles. Baseband signal logical "0s" inhibit modulator output signals altogether.

**analog carrier system** A transmission system that uses repeaters that compensate for analog medium impairments, and produce output signals that are linear-scaled versions of input signals. Analog carrier systems can carry speech, data, video, and supervisory signals, although they are best suited for speech signals.

**analog signal** A continuous signal that varies in some direct correlation with an impressed phenomenon, stimulus, or event that typically bears intelligence. Sound waves and their electrical analogs are characterized by loudness (a quantity proportional to amplitude) and pitch. Analog signals can assume any of an infinite number of amplitude values or states within a specified range, in accordance with (analogous to) an impressed stimulus. *Pitch* refers to how many times per second the signal swings between high and low amplitudes, i.e., its frequency.

**application specific integrated circuit (ASIC)** Integrated circuits designed for a specific application or function.

**area code** The first three digits of a telephone number in North America (also known as the *numbering plan area*).

**arrival rate** In a queuing system, how often users arrive or request resources.

**asymmetrical digital subscriber line (DSL)** Services or equipment that support higher downstream than upstream data rates.

**asynchronous** A condition among signals or signal components derived from different frequency or timing reference sources and hence having different frequencies or rates. See Chapter 5 for more information.

**asynchronous digital signal hierarchy (ADH)** In time division multiplexing systems, to permit higher levels of multiplexing concentration, a multilevel TDM digital signal, DS"n", hierarchy has been developed. Figure 5.2 presents the U.S. TDM ADH and is explained in Chapter 5. Although every time division multiplexer must be pairwise synchronous with a connected demultiplexer, what justifies calling the current U.S. TDM hierarchy asynchronous is the fact that different pairs are not synchronous, but rather asynchronous.

**asynchronous transfer mode (ATM)** A broad-bandwidth, low-delay, packet-like (cell relay) switching and multiplexing technique. It is essentially connection oriented, although it is envisioned to support all services. ATM networks will accept or reject connections based on users' average and peak bandwidth requirements, providing flexible and efficient service for LAN-to-LAN, compressed video, and other applications that involve variable bit rate (VBR) traffic.

**ATM Adaptation Layer (AAL)** The ATM standards layer that allows multiple applications to have data converted to and from an ATM cell. A protocol used that translates higher-layer services into the size and format of an ATM cell.

**ATM Adaptation Layer 1 (AAL1)** Intended to map voice traffic onto CBR virtual circuits. The header contains clock recovery bits and sequence numbers, but no error control bits. Only one byte is required for AAL1 headers.

**ATM Adaptation Layer 2 (AAL2)** Maps video traffic onto VBR-RT virtual circuits, using an error-checking code for the whole cell. AAL2 headers are three bytes in length.

**ATM Adaptation Layer 3/4 (AAL 3/4)** Used for mapping data traffic onto VBR-NRT virtual circuits. AAL3 is used for connection oriented service and AAL4 for connectionless service. This layer provides error checking, sequencing, and identification of cells as parts of messages. The difference between this AAL 3/4 and AAL2 is that synchronism is not maintained across networks.

**ATM Adaptation Layer 5 (AAL5)** Intended for use with connection-oriented data, but does not insert header information into every cell. Information such as message length and an error-checking code for the whole message (not each cell) is appended to messages before fragmentation into cells. In North America, AAL5 has supplanted AAL3/4 as the adaptation layer of choice for data transmission.

**automatic call distribution (ACD)** A means for efficiently directing and managing large numbers of incoming calls to specific departments/terminals within an organization.

**automatic number identification** A carrier service that enables called-party subscriber equipment to display calling-line billing numbers.

**Automatic Retransmission reQuest (ARQ)** In systems using error-detecting codes, messages sent to data sources by receivers that detect errors in received codewords (ARQ).

**auto-sensing** A property of LANs where devices handshake with each other when originally connected to determine each other's capability and negotiate the highest-speed performance common operating mode.

**available bit rate (ABR)** An ATM service that provides only a minimum cell rate to the user. The user may transmit above this rate, but must respond to congestion messages from the network and control its rate.

**average delay** The average elapsed time between when a user enters a system and when service is completed.

**average delay time** The average time that a packet spends in a network (queuing time plus transmission time for the packet).

**backbone network** A transmission facility designed to interconnect often lower-speed distribution networks, channels, or clusters of dispersed terminals or devices.

**backbone wiring** In a premises distribution system, the cable connecting telecommunications closets and equipment rooms within a building, and/or between buildings in a campus. Backbone wiring is sometimes referred to as the *riser subsystem.*

**backplane** A physical area (component and/or printed circuit card) on which major switch fabric components are mounted and to which individual port cards are connected.

**balun** A small passive device that connects a balanced line, such as a twisted pair, to an unbalanced line, such as coaxial cable.

**bandwidth** A frequency range, usually specified by the number of hertz in a band or between upper and lower limiting frequencies. Alternatively, the frequency range that a device is capable of generating, handling, passing, or allowing. See Chapter 5 for additional explanation.

**baseband signals** Signals in the original format in which they were generated, unchanged by any modulation process. Analog voice signals generated by microphones and digital signals generated by PCs or other DTE, described in Chapter 3, are examples of baseband signals.

**Bell operating company (BOC)** The common term for one of 24 local exchange carrier telephone companies that were part of the Bell System prior to divestiture. All but two of the BOCs (Southern New England Telephone in Connecticut and Cincinnati Bell in Ohio) were owned and managed by one of seven (now four) regional Bell operating companies (RBOCs). Approximately 80 percent of America's local exchange users are served by the BOCs.

**bi-directional line switched ring (BLSR)** A SONET ring configuration that carries primary path traffic in the same manner as directly connected point-to-point fiber nodes; that is, separate fibers are used for each signal transmission direction. The important characteristic of BLSRs is that capacity required between two nodes (i.e., the timeslots dedicated to carrying traffic in both directions between those nodes) is available for reuse elsewhere on the ring.

**bi-directional WDM** In fiber optic cable-based transmission systems, the ability to transmit separate lightwave signals in both directions simultaneously over a single strand of fiber, with no or minimal mutual interference.

**binary coding** A coding scheme in which each element can assume only one of two possible states. In binary format, the number 00000001 is equivalent to the decimal system number 1. The binary equivalent to the decimal number 2 is 00000010. The binary equivalent to the decimal number 4 is 00000100. As can be seen, each movement of digits to the left corresponds to a multiplication by a factor of 2 rather than 10, as in the decimal system. Thus the binary number system is said to be of base 2 whereas the decimal system is said to be of base 10.

**binary digit** See *bit.*

**bipolar signals** T1 carrier systems carry 24 channels on two pairs of copper twisted-pair cable (one pair for each transmission direction). Signals are applied directly to cable pairs in bipolar format in which positive and negative pulses, always alternating, represent one binary signal state (for example the state corresponding to a binary bit value of "1"). Under the Alternate Mark Inversion (AMI) convention, pulses correspond to binary "1"s and the absence of pulses corresponds to binary "0" states. With Alternate Space Inversion (ASI), pulses correspond to binary "0"s.

**bit** The most fundamental and widely used form of digital signals are binary signals, in which one amplitude condition represents a binary digit 1, and another amplitude condition represents a binary digit 0.

Thus a binary digit, or bit is one of the members of a set of two in a numeration system that is based on two, and only two, possible different values or states.

**bit-error rate (BER)** The ratio of the number of bits received with errors to the total number of bits transmitted. BER and the average number of error-free seconds are the principal impairment measurements for digital channels.

**bit rate** The capacity characteristic of digital signals as defined by the number of bits (or bytes) per second that a channel will support. For example, a transmission facility that can support information exchange at the rate of 1 megabit per second (1 Mb/s or 1,000,000 bits per second) delivers the same quantity of information, i.e., throughput, as a 1-kilobit-per-second (kb/s or 1,000 bits per second) facility, but in only 1/1,000 of the time.

**bit stream** A time profile of a stream of bits in a binary digital signal.

**blocking** An event in queuing systems that occurs when users find all resources busy when they request service.

**blocking probability** The chance for encountering busy-signals or being denied service when all network resources are in use.

**Bluetooth** A code name for a personal area wireless networking project and standard, originally envisioned as producing a single, unified wireless replacement for cables used to interconnect desktop PCs, laptops, printers, facsimile machines, pagers, personal digital assistants (PDAs), and cell phones.

**bridges** In IEEE 802 local area network (LAN) standards, devices that connect LANs, or LAN segments, at the data link layer. Bridges provide the means to extend the LAN environment in physical extent, number of stations, performance, and reliability. Bridges perform three basic functions: frame (as opposed to packet) forwarding; learning of station addresses; and resolving of possible loops in the topology by invoking what is known as a spanning tree algorithm. Self-learning bridges construct tables of network addresses by "listening" to source address information contained in data signal frames. Other functions include: the ability to filter traffic to keep traffic originating and terminating in one network segment from leaving that segment; restricting specified traffic to one segment that might otherwise be routed to other segments; and collecting and storing network management and control information obtained via traffic monitoring. Bridges operate at the data-link protocol layer (Layer 2), and consequently pass packets without regard to higher-layer protocols. Because of this, bridges can only connect segments of the same LAN types.

**bridging taps** Points in subscriber loop outside plant at which multiple wiring runs toward customer premises are joined; they are installed to make loop reassignments easier.

**Broadband Integrated Services Digital Network (BISDN)** CCITT developed a BISDN umbrella standard, incorporating underlying standards for integrated digital network switching, multiplexing, and transmission facilities, which will be able to meet expanding voice, data, video, and other requirements well into the future. In one of the first draft CCITT documents, BISDN is simply defined as "a service requiring transmission channels capable of supporting rates greater than the primary rate." In the U.S. the primary rate for "narrowband" ISDN (as the current standard is sometimes referred to) is 1.544 Mb/s.

**brouter** A device that combines the functions of a bridge and a router.

**bursty traffic** Traffic characterized by short periods of high activity followed by long periods of relative inactivity. Bursty traffic must be specified by its peak data rate as well as its average data rate.

**bus** A transmission path or channel using a medium with one or more conductors such as in Ethernet, wherein all connected network nodes listen to all transmissions, selecting certain ones based on address identification.

**business applications** Unique aggregations of telecommunications services that satisfy particular enterprise needs.

**busy hour** A measurement interval used to quantify offered traffic when engineering shared resources. It should represent the hour-long interval during which peak traffic levels occur and in which required grades of service must be maintained.

**busy signal** A call progress signal in network signaling systems signifying that called parties are off-hook—that is engaged in conversation or otherwise using their telephones.

**byte** An 8-bit quantity of information also generally referred to as an octet or character.

**cable** A group of metallic conductors or optical fibers that are bound together, usually with a protective sheath, a strength member, and insulation between individual conductors/fibers and for the entire group.

**call handling** See *signaling*.

**caller ID** An optional carrier service that enables called-party subscriber equipment to display calling-party identification that may include caller names, numbers, and other data.

**capacitive reactance** For time-varying electrical signals, one of the factors that govern the relationship between electrical current and voltage. The opposition to the flow of alternating or time varying current through a capacitor; a device that has the capability of storing or holding an electrostatic charge, such capacity being measured in units called farads in honor of Michael Faraday, a British physicist (1791—1867). See *resistance* and *inductive reactance*.

**carrier** 1. A local (intra-LATA) or long-distance (inter-LATA) telecommunications service-providing organization. 2. A waveform, pulsed or continuous, which is modulated by another information-bearing waveform.

**carrier identification code** A code assigned by the North American Numbering Plan Administrator (NANPA) that permits users to access an interexchange carrier by dialing a special prefix followed by the code.

**carrier system** A system for transmitting one or more channels of information by processing and converting them to a form suitable for the transmission medium used. Carrier systems are classified as either analog or digital systems.

**carrierless amplitude and phase (CAP)** A modulation scheme introduced by AT&T, which is able to transmit from two to nine bits per Hz of analog bandwidth, used in some DSL equipment.

**Category 3, 4, and 5 UTP cable** In late 1991, the EIA/TIA issued Technical Systems Bulletin (TSB) 36, which defined transmission performance standards for unshielded twisted pair (UTP) cable up to 100 MHz. This major milestone galvanized the UTP cable industry, arming them with a potent weapon in their attempts to delay the inroads of fiber to the desktop. Overall, EIA/TIA has defined five levels or categories of cable performance and are developing new Category 6 and 7 specifications. Category 3, a cable quality met by most existing premises UTP installations in the United States and originally intended for voice service only, is capable of supporting IEEE 10BaseT LAN operations (10 Mb/s) within cable-run distance limitations of roughly less than 300 feet.

EIA/TIA's Category 5 standard, issued in 1995, imposes stricter performance-degrading parameter requirements and supports 100BaseT and higher speed operation (greater than 100 Mb/s). It is important to note that EIA/TIA, ISO, or other UTP performance standards have nothing to do with cable manufacturing processes or materials. It is incumbent upon manufacturers who make those design decisions to ensure product "category compliance," normally accomplished via certification testing. Thus, it is always advisable to include post-installation verification testing as a formal contract requirement in any premises distribution system (PDS) procurement.

**CCITT** The International Consultative Committee for Telephone and Telegraph, a consultative committee of the International Telecommunication Union (ITU) which recommends international standards for telephone and telegraph services and facilities to aid international connectivity and interoperability. Its functions were absorbed by the ITU Telecommunication Standardization Sector (ITU-T) in 1993.

**CCITT Signaling System No. 6 (SS6)** The initial common channel signaling system standard produced by CCITT, now supplanted by SS7.

**CCITT Signaling System No. 7 (SS7)** International common channel signaling system standards (recommendations) established by CCITT.

**cell relay** The process of transferring data in the form of fixed-length packets called cells. Cell relay is used in high-bandwidth, low-delay, packet-like switching and multiplexing techniques. The objective is to develop a single multiplexing/switching mechanism for dividing up usable capacity (bandwidth) in a manner that supports its allocation to both isochronous (e.g., voice and video traffic) as well as packet data communications services. Standards groups have debated the optimum cell size. Small cells favor low delay for isochronous applications but involve a higher header-to-user information overhead penalty than would be needed for most data applications. The current CCITT specification for BISDN is for a 53-byte cell that includes a 5-byte header and a 48-byte payload.

**cell site base stations** Fixed-location transmitter-receiver (transceiver) sites that communicate with mobile-subscriber station equipment.

**Cellular Digital Packet Data (CDPD)** An IP-compatible packet-switching data protocol designed to work over AMPS analog FDMA (800-MHz) facilities. The packet-switching and handling parts of the protocol can also be used in digital TDMA PCS (1900 MHz) systems.

**centi call seconds (CCS)** A unit of the average traffic intensity of a facility during a period of time, a CCS is 100 call seconds of traffic during one hour. Therefore a single traffic source, e.g., one call that generates traffic 100 percent of the time, produces 36 CCS of traffic per hour, i.e., 3600 seconds of traffic every 3600 seconds. An equivalent amount of traffic could also be generated by 10 sources that only generate traffic 10 percent of the time. That is, 10 sources of traffic generating 3.6 CCS each, contributes the same total traffic as a single 36 CCS traffic source. An alternative measure for traffic is *erlangs*, where 1 erlang equals 36 CCS.

**CENTRal EXchange (Centrex)** A LEC-provided switching service for business customers that permits station-to-station dialing, listed directory number service, direct inward dialing, and station number identification on outgoing calls. The switching functions are usually performed in a central office. Digital Centrex offers the advanced features of fourth-generation PBXs, without the need to purchase or lease equipment, and, in most cases, eliminates the need for customer provided floor space, electrical prime power, and heating, ventilation, and air conditioning.

**central office (CO)** A telephone company building in which network equipment such as switches is installed.

**central-office connection** In interexchange carrier tariffs, a rate element for connecting interoffice channels to other facilities such as LEC-provided local channels. A COC is required at each end of an interoffice channel.

**central processing unit** That part of a computer that performs the logic, computational, and decision-making functions. In telephony, a processor that controls and coordinates the processing of traffic in a switch.

**certificate authority (CA)** A trusted entity that verifies user identities and issues private/public key pairs to users.

**channel** A single communications path in a transmission medium connecting two or more points in a network, each path separated by some means; e.g., spatial or multiplex separation, such as frequency or time division multiplexing. Channel and circuit are often used interchangeably, however circuit can also describe a physical configuration of equipment that provides a network transmission capability for multiple channels. The characteristics of channels and circuits are determined by the network equipment and media used to support them. See *circuit.*

**channel rate** The capacity characteristic associated with digital signals is *channel rate* or *bit rate*; that is, the number of bits (or bytes) per second that a channel or circuit will support.

**channel service unit (CSU)** Channel Service Units (CSUs) and Data Service Units (DSUs) are required to connect digital customer premises equipment (CPE) to carrier networks. A CSU is network channel terminating equipment (NCTE) attached as CPE to telephone company digital circuits, protecting the network from harm. Other CSU functions include line conditioning and equalization, error control (e.g., bipolar signal violations), and the logical ability to respond to local and network loop-back circuit testing commands. see *data service units (DSU)*.

**chip rate** The maximum rate at which a PN code generator can change state. See *pseudo-noise code generator*.

**chromatic dispersion** A limiting property of most transmission media by which complex signals are distorted because various frequency components that make up signals are affected differently by a medium's propagation characteristics.

**circuit** Unidirectional (one-way) or bi-directional (two-way) paths between communicating points. A simplex circuit is a transmission path (capable of transmitting signals in one direction only, e.g., broadcast radio). A half-duplex circuit is a bi-directional transmission path (capable of transmitting signals in both directions, but only in one direction at a time, e.g., citizen band radio). A full-duplex circuit is a bi-directional transmission path (capable of transmitting signals in both directions simultaneously, e.g., telephone voice signal circuits). Channel and circuit are often used interchangeably; however, circuit can also describe a physical configuration of equipment that provides a network transmission capability for multiple channels. See *channel*.

**circuit switching** A process that establishes connections on demand and permits the exclusive use of those connections until released. Packet and message switching, primarily used in data communications networks, are alternative switching techniques.

**circuit usage time** The length of time circuits are actually tied up in telephone systems. This includes conversation time plus the time circuits are tied up during call setup attempts before the distant party answers or when the distant party does not answer or is busy.

**circuit-associated signaling** A technique that uses the same facility path for voice and signaling traffic. Historically this approach was selected to avoid the costs of separate channels for signaling and because the amount of traffic generated by signaling is small compared to voice, minimizing the chance for mutual interference. Circuit-associated signaling can be contrasted with some common-channel signaling systems which use completely separate packet-switched networks for signaling traffic.

**city-wide digital Centrex** A capability to serve multiple business locations within a single NXX (exchange code), using multiple LEC central office Centrex switches. Outside callers are unaware that multiple business locations are involved.

**clock** A device that generates a signal that provides a timing reference. Clocks are used to control functions such as setting sampling intervals, establishing signal rates (bps), and timing of the duration of signal elements, such as bit intervals. In a completely synchronous TDM network, all participating nodes must use timing provided by a single master network clock. This is not the case today in most TDM networks.

**coaxial cable** An insulated central conductor surrounded by a second cylindrical conductor clad with an insulating sheath. The outer conductor usually consists of copper tubing or copper braid.

**code division multiplexing/code division multiple access** In frequency division multiplexing, a transmission circuit's frequency spectrum is separated into subbands, each supporting single, full-time communications channels on a non-interfering basis. In TDM, a transmission facility is shared in "time" rather than "frequency.. Unlike either FDM or TDM, with code division multiplexing (CDM), individual channel-signals are modulated with special, orthogonal coding signals in such a way that multiple signals can be transmitted in the same frequency band and at the same time without significantly interfering with each other. To recover each transmitted signal individually, receivers must be equipped with an identical version of the modulating orthogonal signals.

**codec** A contraction from enCOder/DECoder; often used to describe devices that transform (encode) analog signals into digital signals for transmission through a network in digital format, and decode received digital signals, transforming them back to analog signals.

**coherent demodulation** Demodulation processes in which local versions of the carrier signal are ideally in perfect phase and frequency agreement with the carrier signal used in the modulation process. At a minimum, coherent demodulation produces output signal-to-noise ratios 3 dB better than non-coherent demodulation. This has the same impact as doubling transmitter power and, in some cases, may double traffic-handling capacity on otherwise identical transmission systems.

**collision domain** All nodes in a shared environment that can listen to and detect collisions.

**committed burst size ($B_c$)** In frame relay networks, the maximum number of bits that a user can transmit over a defined measurement interval before frames are marked as discard eligible. See *discard eligible, CIR, committed rate measurement interval.*

**committed information rate (CIR)** In frame relay networks, throughput guaranteed by service providers over a defined measurement interval, provided customers observe limitations on bursting above the CIR. See *committed rate measurement interval, committed burst size.*

**committed rate measurement interval ($T_c$)** In frame relay networks, the measurement interval used in determining the committed information rate. CIR is the committed burst size divided by the committed rate measurement interval.

**common management information protocol (CMIP)** The OSI protocol for network management. A structure for formatting messages and transmitting information between reporting devices (agents) and data collection programs, developed by the International Organization for Standardization and designated ISO/IEC 9596.

**common-channel signaling (CCS)** A signaling system developed for use between stored program-control digital switching systems, in which all the signaling information for one or more trunk groups is transmitted over a dedicated signaling channel, usually, but not always, over facilities separate from the user-traffic-bearing facilities.

**communications** The process of representing, transferring, interpreting, or processing information (data) among persons, places, or machines. Communications implies a sender, a receiver, and a transmission medium over which the information travels. The meaning assigned to the data must be recoverable without degradation.

**competitive local exchange carrier (CLEC)** Companies that offer alternative exchange service to an area in competition with incumbent local exchange carriers (ILECs).

**computer-to-PBX interface** Two switch-to-computer specifications advanced by the industry are known as the computer-to-PBX interface (CPI) and the digital multiplexed interface (DMI) specifications. Both specifications operate at DS1 rates of 1.544 Mb/s. A single CPI/DMI connection multiplexes into a single connection what would otherwise involve 24 separate switch-to-computer channels. Introduced by Northern Telecom Inc. (now Nortel Networks Corporation) and AT&T in the early 1990s, these interfaces have been eclipsed by more recent computer-industry driven computer telephone integration (CTI) initiatives to use modern computer hardware and software technologies to improve plain old telephone service (POTS).

**conductor** In electrical circuits, any material that readily permits a flow of electrons (electrical current) through itself. Analogously, optical fibers are sometimes said to conduct lightwaves and are also referred to as conductors. In twisted-pair cables, the electrical signal wave propagates from the sending end to the receiving end in the dielectric material (insulation) between the two conductors.

**connectionless data services** Services that transfer information among service subscribers without the need for end-to-end connection establishment procedures.

**connection-oriented** A term applied to data services that use separate procedures for connection establishment and end-to-end information transfer (connection establishment must take place prior to information transfer).

**constant bit rate (CBR)** An ATM service used to emulate dedicated circuits. The cell rate is constant over time and timing synchronization is maintained between transmitting and receiving devices.

**contract tariffs** Tariffs containing rates resulting from corporate contracts. Since these tariff rates derive from competitive processes, they are filed as tariffs without detailed cost justification.

**control channels** In wireless systems, control channels exchange various types of signals that allow base stations to sense the presence of powered-on mobile telephone sets, to page or notify mobile telephone sets of incoming calls, to transfer dialing instructions for outgoing calls, and accomplish handoff and numerous other call-monitoring and supervisory functions.

**convergence** The name applied to a concept for satisfying all voice, data, video, imagery, and other applications and all access, transport, and other service requirements (all-applications/all-services) over single telecommunications facilities.

**conversation time** The time during a telephone call that users actually engage in conversation. In telephone parlance, this is the elapsed time between distant-end answer (or called party going off-hook at the start of a call) and either party going on-hook at the end of a call.

**Corporation for Open Systems (COS)** A nonprofit organization composed of manufacturing, service, and user organizations in the computer-communications area. COS seeks to facilitate the development of the international, multivendor marketplace through the development, introduction, and verification of OSI and ISDN standards and by ensuring vendor equipment interoperability.

**cross-connect** 1. In a premises distribution system, equipment used to terminate and administer communications circuits. In a premises distribution system (PDS) wire cross-connect, jumper wires or patch cords are used to make circuit connections between horizontal and backbone wiring segments. 2. In transmission systems, a patch panel for connecting circuits.

**crosstalk** In unshielded twisted-pair cable, for instance, crosstalk is caused by undesired signal energy radiated from one UTP pair, coupling to and interfering with desired signals carried in adjacent copper loops in the same cable bundle.

**CSMA/CD** Carrier sense multiple access with collision detection; a local area network contention-based access-control protocol by which all devices attached to the network listen for transmissions in progress before attempting to transmit themselves and, if two or more begin transmission simultaneously, are able to detect the collision. In that case each backs off (defers) for a variable period of time (determined by a preset algorithm) before again attempting to transmit. Defined by the IEEE 802.3 standard.

**custom local area signaling service (CLASS)** A set of local calling enhancements to basic exchange telephone service. CLASS features use digital switching and signaling to provide automatic callback, automatic recall, selective call forwarding and call rejection, customer originated call trace, calling number identification, bulk calling number delivery, and other features.

**customer premises equipment (CPE)** Comprises inside wiring, station, switching, multiplexing, LAN, and other customer-owned equipment located on the customer premises, except possibly coin-operated telephones.

**cyclic redundancy check (CRC)** Transmitters employing this technique generate a unique cyclic redundancy check (CRC) code based on segments of user information. Using received data bits, a remote station employs the same algorithm to recalculate the CRC code. The remote station detects the CRC code sent by the transmitting station and compares it with the locally generated CRC code. Generally, if the two CRC codes match, there are no errors. In T1 carrier systems, this technique detects 98.4 percent of all possible bit-error patterns. What's more, the information is stored and provides a historical record of performance over time so that degradation can be detected before total line outages occur.

**data circuit terminating equipment (DCE)** A generic term for network-embedded devices that provide an attachment point for user devices.

**data compression** Techniques which remove redundancy in transmitted bit patterns to reduce transmission rates by 20 to 200 percent. For example, a modem designed to send and receive data at 1200 bps without data compression may be capable of supporting 2400 bps with data compression, using the same network analog voice-grade channel.

**data service unit (DSU)** Channel service units (CSUs) and data service units (DSUs) are required to connect digital customer premises equipment (CPE ) to carrier networks; a hardware device providing an interface between a digital line and a unit of data terminal equipment. DSUs provide transmit and receive control logic, synchronization, and timing recovery across data circuits. DSUs may also convert ordinary binary signals generated by CPE to special bipolar signals used in T1 carrier systems. See *channel service units (CSU)*.

**data terminal equipment (DTE)** Any device that can send data, receive data, or perform both functions. (Note: sometimes DTE implies digital terminal equipment, a type of CPE used with digital service.) See *CSU* and *DSU*.

**dedicated service** A service that assigns resources to users for exclusive use on a full time basis. The rates for these services are on a monthly basis and independent of actual quantities of user traffic carried by the service.

**dedicated-access WATS** A WATS service that employs dedicated-access circuits (trunks) between business premises and IXC POPs, bypassing LEC switched-access facilities.

**dense wavelength division multiplexing (DWDM)** A designation given to high-capacity WDM equipment. It is typically applied to equipment in which more than eight wavelength (channels) are supported on a single fiber optic strand.

**dial-1 WATS** A switched service where all calls are completed as MTS calls but customer billing is made on a bulk rather than call-by-call basis, with discounts proportional to traffic levels. Dial-1 WATS calls are provisioned using LEC switched access.

**differentiated services** A Layer 3 solution to the quality of service problem that uses IP's "type of service" field to carry information about IP packet service requirements. It relies on traffic conditioners at the edge of networks to indicate each packet's requirements based on the needs of applications.

**digital access and cross-connect system (DACS)** Switching/multiplex equipment that permits per-channel DS0 (64-kb/s) electronic cross-connection from one T1 transmission facility to another, directly from the constituent DS1 signals. DACS eliminates the need for demultiplexing DS1 signals into 24 DSO channels, switching them through a separate switch, and re-multiplexing back to DS1 formats at network switching and multiplexing nodes.

**digital carrier system** A carrier system for digital signals that uses regenerative versus linear repeaters and time division multiplexing.

**digital multiplexed interface** See *computer-to-PBX interface.*

**digital signal** A signal (electrical or otherwise) in which information is carried in a limited number of different (two or more) discrete states. The most fundamental and widely used form of digital signals are binary signals, in which one amplitude condition represents a binary digit 1, and another amplitude condition represents a binary digit 0. See *binary digit* or *bit.*

**digital signal processing (DSP)** Calculations performed by general purpose or specialized digital processors on digital information or in real time on digitized signals. Digital signal processing is used extensively in telecommunications for tasks such as echo cancellation, call progress monitoring, signal filtering, and voice processing, and for the compression of voice and video signals.

**digital speech interpolation (DSI)** To enhance efficiency, where transmission resources are scarce or expensive (transoceanic circuits for example), techniques such as Time Assigned Speech Interpolation (TASI), for analog voice signals, or DSI, for digital voice signals, are used to detect speech pauses and switch circuits to serve active speakers.

**digital subscriber line (DSL)** A generic name given to a class of digital services that enable users to connect data terminal equipment directly to LEC subscriber lines without voice channel modems, and are typically able to support data rates at 10, 100, or even 1,000 times the voice channel modem speeds. IDSL, HDSL, HDSL2, ADSL, RADSL,

VDSL types of DSL, and their distinguishing characteristics and applications are described in Chapter 5.

**digital subscriber signaling (DSS)** User interfaces with SS7 are addressed in Q.930/931, part of ITU's General Recommendations on Telephone Switching and Signaling, a specification for digital subscriber signaling systems. DSS permits telecommunications users to capitalize on economies, service enhancements, and flexibilities offered by integrated (voice and data) digital networks.

**digital termination system (DTS)** A system employing short-hop, point-to-point microwave radios in the 10.6 GHz band to provide up to DS1 rate service directly to customer premises. The FCC has allocated a special microwave band for DTS.

**digitizing** A process of converting signals in analog format to digital format. In the first step, the amplitude of an analog signal is *sampled.* That is, an analog signal's amplitude is measured or determined periodically at what is known as the *sampling rate.* Since outputs from analog signal samplers are typically still in analog format, they must be *quantized,* or assigned to one of a set of allowable incremental values. Finally, quantized values are converted to digital representations, usually in binary word format. This last step, or the entire process, is referred to as *digitizing.* See Chapter 3 for examples.

**direct inward dialing (DID)** PBX-to-central-office trunks that allow incoming calls to a PBX to ring specific stations without attendant assistance. DID greatly reduces the number of required console attendants, compared with systems in which all calls must be extended by console attendants.

**direct outward dialing (DOD)** PBX-to-central—office trunks that allow outgoing calls to be placed directly by PBX stations.

**discard eligible (DE)** In frame relay networks, a bit in frame headers, which is set if more bits than the committed burst size are transmitted over the committed rate measurement interval. The frames containing the excess bits are marked discard eligible and are carried by the network on a best-effort basis. These frames are discarded first when network congestion is encountered. See *committed burst size, committed rate measurement interval, CIR.*

**Discrete MultiTone (DMT)** A modulation technique selected by the American National Standards Institute (ANSI), European Telecom Standards Institute (ETSI), and the International Telecommunications Union (ITU) as a standard for ADSL. DMT's ANSI T1.413 standard divides the available bandwidth into 256 subchannels with subcarriers that can be modulated from zero to a maximum of 15 bits/second/Hz, permitting up to 60 kb/s data rates per subcarrier.

**dispersion** A limiting property of most transmission media by which complex signals are distorted (rectangular pulses become flattened or rounded) because various frequency components that make up signals are affected differently by a medium's propagation characteristics and paths. Dispersion limits the upper bit rate that a medium can support by distorting the signal waveforms to the extent that transitions from one information state to another cannot be reliably detected by receiving equipment, (e.g., logical 1 to logical 0 value changes).

**downstream** Traffic flowing from server or host equipment toward subscriber terminal equipment.

**drop cables** Collectively known as outside plant, local loops consist of LEC main, branch, distribution, and drop cables segments, which connect with inside wiring. When attached to inside premises wiring, the outside plant is able to establish central-office to on-premises telephones and other CPE device connections. Drop cables are those portions of outside plant that directly connect network interfaces located on customer premises.

**DS"N"** Digital signal hierarchy is a time division multiplexed hierarchy of standard digital signals used in telecommunications systems. DS1 level in the hierarchy corresponds to a 1.544-Mb/s TDM signal, which comprises 24 DS0 signals. DS0 refers to individual digital signals at channel rates of 64 kb/s. Four DS1 signals digitally multiplexed produce a DS2-level signal, containing 96 DS0 channels, and requiring a transmission medium that supports 6.312 Mb/s. A DS3-level signal results from the digital multiplexing of seven DS2 signals, supports 672 DS0 signals, and requires a 44.736-Mb/s transmission medium. Finally a DS4-level signal supports six DS3 level signals and 4,032 DS0 signals and requires a 274.176-Mb/s transmission medium. The DS hierarchy accounts for nonsynchronism in the multiplexing plan, hence the term "asynchronous digital hierarchy" and the use of overhead bits— note that bit rates at higher levels are not integer multiples of 64 kb/s.

**D-type channel bank** Channel termination equipment used for combining (multiplexing) individual analog channel signals on a time-division basis. D-type channel banks provide interfaces for "n" analog signal inputs. Each analog input signal is directed to a codec for encoding to PCM samples; a part of same T1 carrier systems.

**dual-band telephones** Wireless telephones able to operate in two radio frequency bands, often the 800- and 1900-MHz bands.

**dual-mode telephones** Wireless telephones that support both analog and digital operations.

**dual-tone multiple frequency (DTMF)** The generic name for the tone-based address signaling scheme used to signal from telephones to switching equipment, in which 10-decimal digits and two auxiliary characters are represented by selecting one frequency from the of the following group: 697, 770, 852, 941 Hz and one frequency from the 1209, 1336, and 1447 Hz group.

**dumb terminals** Terminals limited to low-speed operation; they do not incorporate local processing and transmit characters one at a time as they are typed by operators.

**E&M leads signaling** An interface, used for connections between switches and transmission systems and between transmission systems themselves. Signaling information is transferred across the interface via two-state voltage conditions on two leads, each with a ground return, separate from the leads used for message information. The message and signaling information are combined and separated by means appropriate to the transmission facility.

**earth-coverage satellite antenna** Satellite antennas with beam widths wide enough to illuminate the maximum area on the earth "visible" from a particular satellite.

**electrical signal** A signal consisting of an electrical current (i.e., a flow of electrons) that varies with time or space in accordance with specified parameters.

**electronic mail** A generic term for noninteractive communication of text, data, image, or voice messages between a sender and designated recipients using telecommunications.

**electronic switched network (ESN)** In private networks, a service that provides user organizations with a uniform numbering plan and numerous call-routing features. The electronic tandem switching functions are furnished by either PBX or Centrex switching equipment.

**end office (EO)** A LEC (BOC or an ITC) switching system within a LATA where local loops to customer stations are terminated for purposes of interconnection with each other and with trunks. CO (central office) and EO are often used interchangeably.

**enhanced SMR** Preserves dispatch-like specialized mobile-radio operations, but adds cellular mobile-telephone dialing features and ESMR-to-PSTN wireline network connectivity, allowing SMR operators to compete directly with cellular carriers.

**entrance facilities** In a premises distribution system, the point of interconnection between the building wiring system and external telecommunications facilities (LEC networks, other buildings, etc.). Bellcore defines the interface with LEC networks as end-user points of termination (POT).

**equal level far-end crosstalk (ELFEXT)** The interference resulting from a single remote transmitter, transmitting at a power level equal to that of the desired signal's transmitter power.

**equipment identity register (EIR)** In wireless systems, a database that contains electronic (equipment) serial numbers.

**equipment room** In a premises distribution system, special-purpose room(s), with access to the backbone wiring, for housing telecommunications, data processing, security, and alarm equipment.

**erbium-doped fiber amplifier (EDFA)** Provides in-line amplification of signals in optical form (no conversion of optical signals back to electrical signal formats is required) and can amplify several independent (wavelength division multiplexed) signals with virtually no mutual interference and negligible crosstalk. Whereas electrical signal amplifiers require electrical power, EDFA amplifiers consist of 10 to 100 meters of optical fiber with erbium-doped cores (erbium is a rare earth element), and optical "pumping" signals coupled to erbium-doped segments. Pumping causes erbium atoms to transfer energy through stimulated emission and in the process amplifies traffic signals. Besides being amazingly simple, EDFA amplifiers can typically be spaced 10 to 100 times as far apart as conventional repeaters.

**erlang** The international dimensionless unit of traffic demand, obtained by multiplying arrival rate by holding time, named after A. K. Erlang, a Danish mathematician acknowledged to be the father of queuing theory. Erlang's formulae, derived in 1917, remain the backbone of queuing performance evaluation today. Quantitatively, one erlang of traffic is equivalent to a single user who uses a single resource 100 percent of the time. See *centi call seconds.*

**Erlang-B formula** A formula to compute blocking probability in a lost-calls-cleared queuing system using the offered load and the number of circuits as input.

**Erlang-C formula** A formula to compute average delay in a lost-calls-delayed queuing system using the offered load, the number of circuits, and the line speed of the circuits as input.

**error-detecting codes** A process in which error checking bits are added to user data in transmitting equipment in a manner that allows receivers to determine whether errors have occurred in intervening switching or transmission networks. See *forward error detection and correction* and Chapter 3.

**Ethernet** The name coined by Robert Metcalf in 1973 to refer to CSMA/CD local area networks. Used today to refer to 10-Mb/s LANs. See *fast Ethernet, gigabit Ethernet.*

**exchange carrier** Any company, BOC or independent, which provides intra-LATA telecommunications within its franchised area. See *local exchange carrier.*

**exchange** The second group of three digits of a telephone number in North America. In historical Bell System parlance, N refers to a digit which can assume the values 2—9 and X is a digit which can assume values of 0—9. Using this convention, NXX has become a commonly used term referring to the exchange portion of a telephone number.

**extended superframe format (ESF)** An extension of the superframe format of T1 carrier systems from 12 to 24 frames and the use of framing bits for error checking; a facilities data link (FDL) as well as frame synchronization. See *superframe format.*

**extranet** An intranet to which access is provided via the Internet. Extranets capitalize on the Internet's global-in-extent public services to provide remote access to corporate intranets for employees (away from their offices), and customers—wherever in the world they may be located. Employee access normally involves some form of password and firewall security protection methods.

**facilities-based private-switched network services** A private network for which LECs and IXCs dedicate physical switching and transmission facilities for the exclusive use of a particular customer.

**facility data link (FDL)** A communications link between channel service units (CSUs) and telephone company monitoring devices.

**far-end crosstalk (FEXT)** Crosstalk interference at the far end from a transmitter (intended receiver locations) generated by other remote or distant transmitter signals. See *crosstalk.*

**fast Ethernet** A CSMA/CD LAN operating at 100 Mb/s. See *Ethernet.*

**fast packet** A term referring to a number of broadband switching and networking paradigms (e.g., frame relay, cell relay). Implicit is the assumption of an operating environment that includes reliable, digital, broadband, nearly error-free transmission systems.

**Federal Communications Commission (FCC)** A board of commissioners empowered by the U.S. Congress to regulate all interstate and international communications, as well as use of the radio frequency media.

**fiber optic modems (FOMs)** Devices that convert signals in electronic format to signals in lightwave formats and vice versa.

**fiber-distributed data interface (FDDI)** The American National Standards Institute (ANSI) standard X3T9.5 for a 100-Mb/s token ring using an optical fiber medium.

**Fibre Channel** ANSI-developed standard for high-speed (133 Mb/s-to-1Gbps) information exchange between mainframes, mass storage devices, workstations, and other computer peripherals. Designed to generate less than one error in $10^{12}$ bits.

**file transfer protocol (FTP)** An Internet protocol used to transfer files between two host systems.

**foreign exchange (FX)** A service that provides circuit(s) between a user station, a PBX, or a Centrex switch, and a central office other than the one that normally serves the caller.

**forward channels** In wireless systems, base station-to-mobile set transmission links

**forward error detection and correction** A process in which error-checking bits are added to user data in transmitting equipment in a manner that allows receivers not only to detect errors but to deduce what the transmitted code must have been; the receivers are thus able to find and correct errors which may have occurred in intervening switching or transmission networks, eliminating or greatly reducing the need for retransmission. Such codes are also described as forward error detection and correction (FEDAC) codes. See *error-detecting codes* and Chapter 3.

**four-wire** Circuits are classified as either two wire or four wire, regardless of whether they use fiber or metallic cable, terrestrial or satellite radio links, infrared or other optical transmission, in atmospheric or free-space media. A four-wire circuit uses two sets of one-way transmission paths, one for each direction of transmission. It may be two pairs (four wires) of metallic conductors or equivalent four-wire as in multichannel transmission systems.

**fractional T1** A service that IXCs and some LECs are now able to offer using flexible multiplexing and DACS equipment. In the past, T1 service was available only in integral 24 channel DS0 increments. Business users with lesser requirements often had to resort to more expensive single-channel data services, such as Digital Data Service (DDS), or purchase a full T1 service and pay for the unused capacity. With fractional T1, users can obtain service in DS0 increments at per-channel rates that are almost always less expensive than other alternatives.

**fragmentation** A process of breaking an upper-layer protocol segment in several pieces in order to fit into the segment size of the current-layer protocol.

**frame** The word "frame" has several distinct meanings when used in telecommunications contexts. Those that apply herein are:

1. In time division multiplexing systems (see Chapter 5), a sequence of time slots each containing a sample from one of the channels carried by the system. The frame is repeated at regular intervals (normally the sampling rate used in analog-to-digital conversion processes for signals being multiplexed) and each channel usually occupies the same sequence position in successive frames. In a time division multiplexed (TDM) system, a repetitive group of signals resulting from a signal sampling of all channels, including any additional signals for synchronizing and other required systems.[1]

2. In high-level data-link control (HDLC), the sequence of contiguous bits bracketed by and including opening and closing flag (01111110) sequences.[1] In data communications networks conforming to the seven-layer ISO model, the task of level 2, the data-link layer, is to take a raw transmission facility and transform it into a channel that appears to the network layer free of transmission errors. It accomplishes this by breaking input data up into data frames. Since layer 1 merely accepts and transmits a stream of bits without regard to meaning or structure, it is up to the data-link layer to create and recognize frame boundaries. The word "frame" is not the official ISO term for a unit exchanged by layer 2 processes. The ISO term is "physical-layer-service-data-unit".[2]

---

1 *IBM Vocabulary for Data Processing, Telecommunications, and Office Systems.* 7th Ed. 1981.

2 Andrew S. Tanenbaum, *Computer Networks.* Prentice-Hall 1981.

3. In graphics and desktop publishing applications, a rectangular area in which text or graphics can appear.[3]

4. In video and animation, a single image in a sequence of images; a single, complete picture in video or film recording.[4]

5. In HTML, a feature supported by most modern Web browsers than enables the Web author to divide the browser display area into two or more sections (frames). The contents of each frame are taken from different Web page.[3]

6. A steel bar framework used to support physical metallic wire or fiber cross-connections typically found in telephone closets or rooms and central offices. See Chapter 9.

**frame check sequence** The last part of a Layer 2 frame (usually four bytes) that ensures frame integrity.

**frame relay** A network interface protocol defined in ITU-T Recommendation I. 122 "Framework for additional packet mode bearer services," as a packet-mode service. In effect it combines the statistical multiplexing and port sharing of X.25 packet switching with the high speed and low delay of time division multiplexing and circuit switching. Unlike X.25, frame relay implements no layer 3 protocols and only the "core" layer 2 functions. It is a high-speed switching technology that achieves ten times the packet throughput of existing X.25 networks by eliminating two-thirds of the X.25 protocol complexity. The basic units of information transferred are variable length frames, using only two bytes for header information. Delay for frame relay is lower than for X.25, but it is variable and larger than that experienced in circuit-switched networks. This means that currently frame relay is not suitable for voice and video applications where excessive and variable delays are unacceptable.

**frequency** Acoustic waves and electrical signals might be made up of only a single tone, like a single note on a piano. In this case the signal waveform is made up of repeating identical "cycles" and is said to be of a single frequency, equal to the number of cycles that occur in one second of time. In communications, frequency was traditionally expressed in cycles per second, but is now expressed in hertz (Hz), still equal to one cycle per second. Thus, one thousand cycles per second is equal to one thousand hertz, or a kilohertz (kHz).

---

3 Webopedia at internet.com.
4 Harry Newton, *Newton's Telecom Dictionary.* 15th Edition. Milton Freeman, Inc. 1999.

**frequency division multiple access** Whenever wireline or fiber optic receiving equipment, satellites, or mobile base stations are designed to receive and handle multiple signals, they are said to exhibit multiple-access capabilities. Accordingly, linking frequency, time, and code multiplexing to multiple-access capabilities produces frequency division multiple access (FDMA), time division multiple access (TDMA), and code division multiple access (CDMA).

**frequency division multiplexing (FDM)** Divides the frequency bandwidth (spectrum) of a broadband transmission circuit into many subbands, each capable of supporting a single, full-time communications channel on a non-interfering basis with other multiplexed channels. FDM multiplexing is generally suitable for use with analog carrier transmission systems.

**frequency shift key (FSK)** A type of modulation in which the frequency of carrier signals is modulated by information-bearing signals. In binary FSK, modulators produce two different output frequencies, one corresponding to logical "1" input baseband signal states, and the other to logical "0s".

**frequency-hopping** A technique in which input signals are assigned a unique frequency-hopping pattern, as opposed to being assigned to fixed subbands. In frequency-hopping multiplexing or multiple access systems, each individual input signal (or user) is assigned a hopping pattern in a manner that prevents or at least minimizes interference with other user signals. Frequency hopping has two principal advantages. First, it minimizes the effects of narrow-band interference by the ratio of the total bandwidth over which frequency hopping occurs to the interference bandwidth. Secondly, it tends to "hide" transmissions from unauthorized eavesdropping and it makes it much more difficult, in military situations, for adversaries to intentionally jam or degrade friendly communications.

**full duplex** A full-duplex circuit is a bidirectional transmission path capable of capable of carrying transmitted signals in both directions simultaneously (e.g., telephone voice signals).

**functional areas** Subsets of information system capabilities that accomplish or support specified categories or subsets of information operations.

**G.Lite** The first standard adopted by the ITU for consumer-oriented digital subscriber line (DSL) service. See *digital subscriber line*.

**gateway** A server that permits client terminal/station access to otherwise incompatible communications networks and/or information systems. A gateway is usually a protocol-translating device that connects a local area network to an external network or any two networks that use different protocols and operating environments.

**General Packet Radio Service (GPRS)** The European standard digital cellular packet-switched data service.

**geosynchronous earth orbiting satellites (GEO)** Satellites in orbits that make them appear stationary from any position on earth. They are at an altitude of about 22,300 statute miles.

**gigabit Ethernet** A CSMA/CD LAN operating at 1000 Mb/s. See *Ethernet.*

**glare** A condition that can occur if loop-start signaling is used on two-way trunks; it can lead to conditions that take trunks out of service.

**grade of service (GOS)** An estimate of customer satisfaction with a particular aspect of service such as noise, echo, or blocking. For example, the noise grade of service is said to be 95 percent if, for a specified distribution of noise, 95 percent of the people judge the service to be good or better. In traffic networks, GOS defines the percentage of calls that receive no service (blocking) or poor service (long delays). GOS measures apply to all aspects of telecommunications networks. In many cases the literature equates GOS only with the probability of a blocked call. When used without further explanation, GOS generally refers to blocking probability.

**graded index** In fiber optic cables wherein the index of refraction, as one moves outward from the axis, is varied so that the speed of light is greatest at the outer extremity. For this case, axial light rays travel more slowly than those that bounce off core-clad boundaries. As a result, although direct-mode paths are physically shorter, multimode path delays tend to be more equal and the equivalent bandwidth or high bit-rate cable capacity greater than similarly constructed step index fibers.

**grooming** Using their flexible add/drop capabilities, DACS can be used to groom or segregate DS0 channels by type (e.g., voice, data, special services 4-wire/2-wire etc.), increasing the fill of T1 lines and enabling more-efficient utilization of resources and the satisfaction of individual customer needs.

**ground-start** A supervisory signal generated by certain coin-operated telephones and PBXs by connecting one side of the line to ground (i.e., a point in an electrical circuit connected to earth).

**guided media** Media that constrain electromagnetic or acoustic waves within boundaries established by their physical construction. Examples include paired metallic wire cable, coaxial cable, and fiber optic cable.

**half duplex** A half-duplex circuit is a bidirectional transmission path capable of transmitting signals in both directions, but only in one direction at a time (e.g., citizen band radio).

**header** Control information appended to a segment of user data for control, synchronization, routing, and sequencing of a transmitted data packet or frame.

**hertz (Hz)** Measurement that distinguishes electromagnetic wave-form energy; number of cycles, or complete waves, that pass a reference point per second; measurement of frequency, by which one hertz equals one cycle per second.

**high bit-rate DSL (HDSL)** The first generation HDSL employs 2B1Q signaling; four-level pulse amplitude modulation (PAM) occupies the 0 to 80,000-Hz part of the spectrum and is able to transmit up to 160 kb/s rates over 18,000-foot loops.

**highway** Another term for bus. See *bus*.

**holding time** In a queuing system, how long users tie up resources before releasing them.

**horizontal wiring** In a premises distribution system, the connection between the telecommunications outlet in work areas and the telecommunications closet.

**hub/concentrator** A network device that regenerates and forwards Ethernet frames. In LANs, wiring concentrator equipment used in hierarchical star physical wiring topologies. Those directly connected to terminals or other user devices are often referred to as local hubs or concentrators. Central hubs are those at the highest hierarchical level. Hubs often provide the means for interconnecting 10BaseT, coaxial, or fiber optic cable LAN segments. Intelligent hubs may implement multiport bridging and network management functions.

**hybrid telephone system** Incorporates both traditional KTS and PBX functions and features and may be of analog or digital design. A strict definition of a hybrid is difficult, since the mix of functions and features in a particular product is based solely on what the manufacturer believes will produce a competitive edge.

**impairments** Degradation caused by practical limitations of channels, (e.g., signal-level loss or attenuation, echo, various types of signal distortion, etc.) or interference induced from outside the channel (such as power-line hum or interference from heavy electrical machinery). The measurement of transmission impairments is an important aspect of predicting whether or not telecommunications systems will sustain the business applications they are intended to support. Signal-to-noise ratio, percent distortion, frequency response, and echo are measurements that define impairments most noticeable by users of analog voice systems.

**impedance** For time-varying electrical signals the relationship between electrical current and voltage is principally governed by three circuit parameters dubbed resistance, inductive reactance, and capacitive reactance. In these cases, opposition to the flow of alternating current (i.e., resistance and reactance) is called impedance and assigned the symbol "Z".

**in-band** Signaling that uses not only the same channel path as the voice traffic for signaling information, but also the same frequency range (band) used for voice messages. For example, DTMF addressing signals use the same frequency band as the voice signals.

**incumbent local exchange carriers** Defined as LECs providing local exchange services in an area on the date of enactment of the Telecommunications Act of 1996.

**independent telephone companies** Those exchange carriers in business prior to the divestiture of AT&T in 1984, that were not affiliated with AT&T or what is described as the Bell System. ITCs are not generally subject to the restrictions of the MFJ, although some of the larger ones are bound by separate consent decrees. Southern New England Telephone and Cincinnati Bell are generally considered ITCs from a regulatory point of view.

**index of refraction** Defined as the ratio of the speed of light in a vacuum to the speed of light in a material. At the boundary between two materials, for example air and water, differences in the index of refraction of the two materials cause incident lightwaves to bend or be reflected at the boundary.

**inductive reactance** For time-varying electrical signals, one of the factors that govern the relationship between electrical current and voltage. The opposition to the flow of alternating or time-varying current through a wire or a coil (inductor). An inductor is a device that has the capability of resisting changes in electromagnetic fields, such capability being measured in units called henrys in honor of Joseph Henry, an American physicist (1791–1878). See *resistance* and *capacitive reactance*.

**Industrial Scientific Medical (ISM)** Several unlicensed frequency bands in the radio spectrum (902-928 MHz; 2.4-2.483 and 5.725-5.875 GHz).

**information communications functional area** Capabilities to move or transfer information from one location to another.

**information exchange functional area** Capabilities to switch, direct, route, multiplex, or inverse multiplex information.

**Integrated Services Digital Network (ISDN)** A set of standards being developed by the ITU-T and various U.S. standards-setting organizations. The then-CCITT formal recommendations, adopted in October 1984, first defined ISDN as "…a network, in general evolving from a telephony integrated digital network, that provides end-to-end digital connectivity to support a wide range of services, including voice and non-voice, to which users will have access by a limited set of standard multipurpose user-network interfaces." The concept of user access to an existing integrated digital network (IDN) underlies the ISDN.

**interexchange carrier (IXC)** A company that provides telecommunications services between LATAs.

**interexchange services** Services provided by a carrier other than the user's local exchange carrier, usually to satisfy user requirements outside the first carrier's serving area.

**interfaces** Common boundary points between two systems or pieces of equipment.

**inter-LATA** Services, revenues, functions, etc., that relate to telecommunications originating in one LATA and terminating outside that LATA. An *interexchange carrier* (IXC) is a company which provides telecommunications services between LATAs.

**intermediate cross-connects** In a premises distribution system, cross-connects located in telecommunications closets.

**International Organziation for Standardization (ISO)** A worldwide federation of national standards bodies, representing 130 countries. Established in 1947, the mission of ISO is to promote the development of standardization.

**International Telecommunication Union (ITU)** Telecommunications standards date back to the 1880s with the founding of the International Telegraph Union (ITU) based in Geneva, Switzerland as the driving force to achieve interconnectability and interoperability among European telegraph systems—then mostly government appendages to railways or postal administrations. The ITU constituted a collegial process for establishing recommendations, which, while not binding on member telegraph administrations, were considered as treaties by foreign ministries. With the advent of trans-Atlantic cables, the U.S., through the State Department, became a signatory to ITU recommendations, an arrangement that persists to this day. In 1993, a new ITU—the International Telecommunication Union was reborn and reorganized to focus on the needs of modern telecommunications. See Appendix A.

**Internet** The Internet comprises an unstructured interconnection of thousands of separate networks on a worldwide basis. Chapters 2 and 7 chronicle the development of the packet-switched network concept, begun in 1969 by the Advanced Research Projects Agency. At the heart of the rapid growth of ARPA's original network, culminating in today's ubiquitous Internet, is a set of simple, yet powerful protocols (IP) that have stood the test of time to become the most popular solutions for internetworking ever devised.

**Internet Engineering Task Force** The key to Internet growth is the widespread adoption of simple, efficient protocols that can be used across many computer platforms. This feat is accomplished not by any controlling government authority, but by a largely volunteer community operating under a self-governing structure whose heart is the Internet Society and one of its components, the thirteen member Internet Architecture Board (IAB). One of the IAB's task forces, the Internet Engineering Task Force (IETF), coordinates the technical aspects of the Internet and produces numerous protocol standards, known as Requests for Comment (RFC).

**Internet Protocol (IP)** A connectionless protocol, at what we now refer to as Layer 3, for use in the Internet. As such, it is primarily concerned with delivering packages of bits from sources to destinations over interconnected systems of networks. See Chapter 7 for more details.

**Internet Service Provider (ISP)** A company or entity providing dedicated or dial-up access to the Internet and other services.

**Internet suite of protocols** A collection of computer-communication protocols originally developed under DARPA sponsorship, including the transmission control protocol/Internet protocol (TCP/IP).

**interoffice channel** A rate element in local exchange carrier tariffs, connecting two serving COs (more accurately, serving wire centers). The interoffice-channel rate element is mileage dependent, and wire center V&H coordinates are used to calculate mileage charges. In interexchange carrier tariffs, this rate element connects two POPs.

**interoffice transmission** Facilities used to connect LEC switching systems.

**interval** In telecommunications applications, interval may relate to the duration of phone calls or data communications session lengths, or to multiplexer bit or frame length duration.

**intra-LATA Services** Revenues, functions, etc., that relate to telecommunications originating and terminating within a single LATA (the domain of LECs).

**intranet** IP-based private network employing Internet technologies and standards and supporting user interfaces and capabilities nearly identical to those of the Internet.

**inverse multiplexing** A multiplexing technique that converts a single high-speed signal into several low-speed signals. This technique is used if high-speed circuits are not available for transmission of the input signal.

**IP Version 6 (IPv6)** A new version of Internet Protocol adopted to resolve long-term address and quality of service issues.

**IP-over-photons** A term used to describe the mapping of IP router packets directly to optical transmission structures, without the use of intervening ATM or SONET equipment.

**ISDN over digital subscriber line (IDSL)** A form of DSL that uses 2B1Q line coding on ISDN circuits to provide symmetrical service at up to 144 kb/s.

**isochronous** Generally refers to a class of traffic in which signal frequency or timing characteristics must be preserved if received information is to be intelligible or useful. Data traffic and e-mail, for example, exhibits no intrinsic timing or rate dependence. However, voice and video traffic are intimately tied to transmission timing and rate. Slow the delivery rate and at best altos may sound like bassos, and at worst in digital voice systems, become totally unintelligible.

**isochronous signals** Periodic signals in which the time interval that separates any two corresponding significant occurrences or level transitions is always equal to some unit interval or a multiple of that unit interval. For example, in digitized voice signals, ideally voice samples occur isochronously at precisely the sampling interval or frame rate. Packet data signals are not isochronous.

**jabber detector** A timer circuit that protects the LAN from a continuously transmitting terminal.

**key service unit (KSU)** In a key telephone system, a centrally located control unit that provides interfaces to LEC networks and contain common KTS control functions.

**key telephone system (KTS)** An arrangement of multiline telephones and associated equipment that permits the station user to depress buttons (keys) to access different central office or PBX lines, as well as to perform other functions. Typical functions include answering or placing a call on a selected line, putting a call on hold, using the intercom feature between phones at the same location, or activating a signal buzzer.

**KSU-less KTS** KTS systems in which all common control functions are implemented in printed circuit boards (PCBs) contained within each multiline telephone, and hence do not require separate or stand-alone KSUs.

**LDN** Listed directory number; generally an organization's main telephone number that appears in the telephone book.

**least-cost routing** A switch features that supports the automatic selection of preferred service options in response to offered traffic.

**light-emitting diodes (LEDs)** LEDs and light amplification by stimulated emission of radiation devices (lasers) are two solid-state or semiconductor conversion/transmitter technologies; that is, devices that convert signals in electrical formats to signals in light formats. LEDs and lasers are used in fiber optic cable-based transmission terminal equipment.

**line** See *loop transmission facilities*.

**line conditioning and equalization** In telephone networks the spacing and operation of amplifiers and filters so that gain provided by the amplifiers for each transmission frequency compensates for line signal-loss at the same frequency.

**line speed** Circuit transmission speed, usually measured in bits per second.

**loading coils** Inductors in outside plant, originally installed to optimize voice service on long subscriber-loop runs.

**local access and transport area (LATA)** A geographic area (called an exchange or exchange area in the MFJ) within each BOC's franchised area that has been established by a BOC in accordance with the provisions of the MFJ for the purpose of defining the territory within which a BOC may offer its telecommunications services. In 1989, there were 198 LATAs.

**local area network (LAN)** A premises high-speed (typically in the range of 10-to-1000 Mb/s) data communications system wherein all segments of the transmission medium (typically coaxial cable, twisted-pair, or optical fiber) are contained within an office or campus environment.

**local channel** A rate element in local exchange carrier tariffs. This element specifies charges for local loops from the customer's premises to serving wire centers. Local-channel charges are normally not mileage-dependent.

**local exchange carrier (LEC)** Any company that provides switched telecommunications services within a defined area. In this book, the term LEC is used when referring to either BOC or independent local exchange carriers.

**local exchange services** Services provided by the carrier serving the premises of the user.

**local loop** A transmission path between a user/customer's premises and LEC central office or some other service provider's premises.

**local multipoint distribution service (LMDS)** A regulatory designation for broadband fixed wireless systems that in the United States operate in licensed allocations in the 28 and 31 GHz bands.

**loop** short for *local loop*. See *loop transmission facilities*.

**loop signaling interface** One of three types of signaling interfaces in modern telephone systems that implement several different types of signaling associated with local subscriber loops.

**loop signaling** A method of signaling over direct current (dc) circuit paths that utilizes the metallic loop formed by the line or trunk conductors and terminating circuits.

**loop transmission facilities** Facilities connecting LEC switching or inter-office transmission systems to customer premises equipment throughout serving areas. A loop is a transmission path between a customer's premises and an LEC central office or some other service provider's premises. The most common form of loop, a pair of wires, is also called a *line*. A loop can be derived from digital loop carrier (DLC) systems, also referred to as *subscriber loop carrier* (SLC) systems.

**loop-start** In telephone use, going off-hook closes the switchhook switch and permits current to flow through the line from the battery to a current sensing device, both at the central office. *Loop-start* is a supervisory signal generated by a telephone or a PBX in response to completing the loop current path.

**lost-calls-cleared** A queuing discipline where users who encounter all resources busy leave without receiving service.

**lost-calls-delayed** A queuing discipline where users who encounter all resources busy join a queue and wait for service.

**low earth orbiting satellites (LEO)** Satellites orbiting hundreds of miles above the earth.

**main cross-connects** In a premises distribution system, cross-connects located in an equipment room.

**managed objects** Resources supervised and controlled by network management.

**management agent** Hardware and/or software in an object, which exchange management information with management stations.

**management and control** Refers to capabilities to plan, organize, design, optimize, engineer, implement, operate, monitor, provision, maintain, synchronize, supervise, manage, control, and administer systems, elements, processes, organizations, and events. Demonstrating the breadth of management and control (M&C) functionality, as lengthy as this list is, each item implies additional or subsidiary capabilities. For example, in telecommunications systems, the ability to "monitor" normally implies comprehensive performance assessment facilities to detect, isolate, report, and record network faults; to measure offered and refused (busy condition) traffic, to measure call completion times and call duration; and numerous other parameters critical to efficient operations.

**management information base (MIB)** Defines information about managed objects. Within MIBs, managed objects are described in terms of 1. object attributes and characteristics; 2. operations performed by or on objects; 3. notifications or reports objects can make; and 4. an object's behavior or response to operations performed on it.

**management process** An application process responsible for management activities.

**m-ary** In modulation systems, the number of discrete signal states that modulator output symbols may take. For example, a 4-ary phase shift key (PSK) modulator output produces carrier-signal signals in just four discrete phase positions (in the range of 0 to 360 degrees) relative to some fixed non-phase-varying reference carrier signal.

**material dispersion** A limiting property of most transmission media by which complex signals are distorted because various frequency components that make up signals are affected differently by a medium's propagation characteristics.

**maximum burst size (MBS)** In ATM networks, the maximum number of consecutive cells that may be transmitted at the peak cell rate. It is roughly equivalent to the burst size defined for frame relay virtual circuits. This parameter applies only to the VBR classes.

**media** See *transmission medium, guided media, unguided media.*

**media access control addresses** Frame addresses used with Layer 2 protocols. The term is usually applied to LAN protocols.

**message telecommunications service (MTS)** Standard (direct-distance dial [DDD]) or operator-assisted long-distance service.

**message units** Usage charges on local calls, sometimes depending on whether calls are within basic service areas or in expanded service areas.

**metropolitan area network (MAN)** A network in which communications occur within a geographic area of diameter up to 50 kilometers (about 30 miles) at selected data rates at or above 1 Mb/s consistent with public network transmission rates. Also, any network in which communications cross public rights of way and occur within a geographic area of diameter up to 50 kilometers, independent of data rates. See *wide area network.*

**microwave multipoint distribution systems (MMDS)** A 2 GHz mid-70s technology originally intended as a wireless alternative to cable TV systems. Currently being adapted for Internet and other data communications applications.

**microwave** In telecommunications, frequencies above 1 GHz.

**minimum annual commitment** The minimum revenue that a customer must pay annually to qualify for the rates contained in his special contract services agreement.

**mobile services switching center (MSC)** GSM terminology for U.S. mobile telephone switching office. See *MTSO.*

**mobile subscriber network** A network designed to support subscribers who are free to move; the network can deliver services to such subscribers while they are in motion or at rest.

**mobile telephone switching office (MTSO)** The principal component in a mobile subscriber network that provides switched access and transport services to subscribers. MTSOs typically provide interconnections to other telephone networks, accumulate calling information for billing purposes, control base-station operations such as subscriber hand off and other network management, and control performance monitoring and diagnostic capabilities.

**modems (MOdulator/DEModulators)** Devices that transform digital signals generated by data terminal equipment (DTE) to analog signal formats, suitable for transmission through the extensive, worldwide connectivity of public and private, switched (dial-up), and non-switched telephone voice networks.

**modification of final judgment (MFJ)** In 1983, a ruling issued by U.S. District Court Judge Harold Greene, which concluded the U. S. Justice Department's antitrust suit against AT&T by modification of an earlier (1956) consent decree's final judgment.

**modular RJ-XX connectors** Plastic plugs providing a simple means by which technically untrained customers can connect telephone devices—purchased at phone-center stores—to telephone networks (without a qualified telephone company technician residence visit).

**modulation** A process of varying certain parameters of a carrier signal—i.e., a signal suitable for modulation by an information signal—by means of another signal (the modulating or information-bearing signal). For instance, in AM broadcast radio, a radio frequency (RF) carrier signal (at a frequency assigned by the FCC to a particular station, e.g., 630 kHz) is amplitude modulated by an analog voice or music audio electrical signal. In this way, the information in the audio signal can be carried via radio-wave propagation from radio transmitters to radio receivers,where it is reproduced for listening.

**multiline telephone** A telephone that incorporates visual displays and switches (keys) that permit the station user to access more than one central office or other line and to perform other desired functions. Typical functions include answering or originating a call on a selected line, putting a call on hold, operating an intercom feature, a buzzer, etc. Displays can indicate busy, ringing, and message-waiting status.

**multimode fiber(s)** Wide fiber cores that allow lightwaves to enter at various angles and reflect off core-cladding boundaries as light propagates from transmitter to receiver, permitting lightwaves to travel over multiple paths from cable inputs to outputs. See *optical fiber(s)*, *single mode fiber(s)*.

**multipath interference** Occurs when signals, traveling over multiple paths of differing loss and length, arrive at receivers and either reinforce or diminish one another.

**multiple access** Whenever wireline or fiber optic receiving equipment, satellites, or mobile base stations are designed to receive and handle multiple signals, they are said to exhibit multiple access capabilities. Accordingly, linking frequency, time, and code multiplexing to multiple access capabilities produces frequency division multiple-access (FDMA), time division multiple access (TDMA), and code division multiple access (CDMA), respectively.

**multiple-frequency signaling (MF)** An interoffice address signaling method in which ten decimal digits and five auxiliary signals are each represented by selecting two of the following group of frequencies: 700, 900, 1100, 1300, 1500, and 1700 Hz.

**multiplexing** Techniques that combine a number of individual communications channels into a common frequency band or a common bit stream for transmission, usually over single circuits. At receiving terminals, demultiplexing equipment or processes separate and recover the individual channel components of multiplexed signals. Multiplexing makes more efficient use of transmission capacity to achieve a low per-channel cost. Four basic multiplexing methods used in telecommunications systems are frequency division multiplexing (FDM), wavelength division (WDM), time division multiplexing (TDM), and code division multiplexing (CDM), all of which are treated in Chapter 5. WDM, introduced in Chapter 4's fiber optic media discussion, is a lightwave version of FDM.

**multiprotocol label switching (MPLS)** An IP networking approach that solves the quality of service problem by assigning explicit connection-oriented paths to various classes of traffic, obviating the need for packet-by-packet processing and switching.

**national backbone service providers (NSPs)** Carriers providing high-speed nodes and transmission links for Intrnet service across the country, interconnecting major cities.

**near-end crosstalk (NEXT)** See first the Glossary definitions of crosstalk. At central offices where all local loops terminate, main feeder cables carry more wires than either branch feeder or distribution cables. Hence, crosstalk is greatest at central offices and interferes more with weak signals entering offices than with strong signals leaving them. This general phenomenon is referred to as near-end crosstalk.

**negative acknowledgement** Networks employing error-detecting codes typically also employ Automatic Retransmission reQuest (ARQ) messages. ARQ messages are sent to data sources by receivers that detect errors in received codewords. In many cases, receivers return acknowledgement (ACK) messages when no errors are detected, and negative acknowledgement (NAK) messages when errors are detected. Of course, such arrangements mean that in order that data sources may retransmit codewords arriving at receivers with errors, they must store transmitted data until ACK messages are received.

**network** A telecommunications network is a system of interconnected facilities designed to carry the traffic that results from a variety of telecommunications services. The network has two different but related aspects. In terms of its physical components, it is a facilities network. In terms of the variety of telecommunications services that it provides, it can support many traffic networks, each representing a particular interconnection of facilities.

**network access points (NAPs)** Network facilities by which private commercial backbone operators can exchange traffic with each other and become, by definition, part of the Internet.

**network channel terminating equipment (NCTE)** User-premises equipment used with digital transmission circuits to provide for digital signal processing and to protect the network from harmful signals.

**network control point (NCP)** In voice virtual private networks a centralized database that stores a subscriber's unique VPN definition. Highly sophisticated, this database screens every call and applies call-processing control in accordance with customer-defined requirements.

**network interface (NI)** The connection between a telephone company's network and CPE is formally called the *network point of termination* (POT) or, alternatively, the *network interface* (NI).

**network interface cards** Printed circuit cards plugged into workstations, PCs, servers, or other LAN DTEs to implement LAN protocols and provide medium interfaces.

**network operating system** Software that controls the execution of network programs and modules.

**network point of termination** See *network interface (NI)*.

**network services** Specified sets of information transfer capabilities furnished to users between telecommunications network points of termination. Network services categories include access and transport, public and private, and switched and non-switched.

**nonassociated CCS** A technique in which signaling information is routed through signal transfer points over completely separate facilities so that the ability to complete a call on an end-to-end basis can be determined prior to the commitment of trunk and switch resources, an approach that greatly reduces call setup time and enhances routing flexibility.

**North American Numbering Plan Administrator (NANPA)** An entity responsible for the assignment of area codes, exchange codes and other codes necessary to implement the North American numbering plan. This function was previously accomplished by Bellcore and the RBOCs, but the establishment of local exchange competition mandated an independent administrator.

**numbering plan area** The first three digits of a telephone number in North America (commonly known as the area code).

**numerical assignment module (NAM)** In wireless telephones, a device that relates a MIN (telephone number) and ESN (manufacturer assigned serial number) to a particular service provider.

**offered load** Traffic demand on a queuing system, computed by multiplying the average arrival rate by the average holding time.

**off-hook** A supervisory signal corresponding to a condition caused by lifting a handset of a telephone connected to a network or equivalent actions in speaker phones where lifting a handset is not required to place or answer calls.

**off-net** Locations accessed to or from a transport service using switched access are known as off-net locations. Calls to or from such locations are referred to as off-net calls.

**on-hook** A supervisory signal indicating an idle circuit. See *off-hook*.

**on-net** Locations using dedicated access to a transport service are known as on-net locations; calls between such locations are referred to as on-net calls.

**Open Systems Interconnection (OSI)** Standards for the exchange of information among systems that are "open" to one another by virtue of incorporating ISO standards. The OSI reference model segments communications functions into seven layers. Each layer relies on the next-lower layer to provide more primitive functions and, in turn, provides services to support the next higher layer.

**operating system control (OSC)** In Centrex systems, OSCs consists of standalone computers, generally powerful workstations, that not only connect with the CO switch for feature management but also control numerous other premises-related functions and may be located on customer premises.

**operating telephone company** Any Bell operating company or independent telephone company (termed exchange carrier in the MFJ) operating in North America.

**optical fiber(s)** Lightguides for electromagnetic waves in the infrared and visible light spectrum composed of concentric cylinders made of dielectric materials with different indices of refraction (i.e., velocity of propagation normalized to the velocity of light in free space). At the center is a core comprising the glass or plastic strand or fiber in which a lightwave travels. A low index of refraction cladding surrounds the core and is itself enclosed in a light-absorbing jacket that prevents interference among multi-fiber cables. Multi-fiber cable can be purchased with between 2 and 144 fibers.

**optical networking units (ONUs)** Fiber optic-based equipment, often installed in outside subscriber-loop plant closer to subscriber locations than central office wiring centers, to ameliorate line-loss problems or to increase the number of subscribers that can be served ("pair gain systems").

**originating access service** Services that enable subscribers and/or equipment to initiate (place) calls or information transfer transactions.

**orthogonal codes** Coding signals used in code-division multiple access systems, designed so that the cross product of different coding signals is "zero", whereas the cross product of the same encoding and decoding signals is "1". In practice, approaching this result requires a high degree of synchronism between transmitters and receivers and high-quality coding signals. By high-quality codes we mean a set of codes, which when multiplied together, yield a product that is very close to "0" for different codes, and close to "1" for the same code.

**orthogonal frequency division multiplexing (OFDM)** A form of discrete multitone modulation that splits high-rate data streams into a number of lower-rate streams transmitted simultaneously over a number of subcarriers.

**out-of-band signaling** Signaling using the same channel path as the voice traffic but in a frequency band outside that used for the voice traffic. In digital systems, out-of-band signaling may take the appearance of an allocated bit position or a dedicated channel or time slot.

**outside plant** Those portions of local loops that consist of LEC main, branch, distribution, and drop cables segments. When attached to inside premises wiring, the outside plant is able to establish central office-to-on-premises telephones and other CPE device connections.

**packet** A quantity of data transmitted and switched as a composite whole. A packet consists of "user data" (for example, some portion of the contents of an electronic mail message) and "control information," the latter including the "network address" of both originating (sending) and terminating (receiving) equipment.

**packet assembler-disassembler (PAD)** Equipment that segments user message information or data (which may be any number of bits or bytes) into packets of limited length.

**packet header** Control information appended to a segment of user data for synchronization, routing, and sequencing of a transmitted data packet.

**parabolic dishes** Antennas whose shape follows the mathematical formula for a parabola. Energy impinging anywhere on the surface of the antenna is directed to a single point, known as the *focus* of the antenna.

**parity bit** The American Standard Code for Information Interchange (ASCII) uses 8-bit bytes or words, 7-bit bytes for data encoding and an eighth parity bit for error detection. The parity bit is set at the data source so that the total number of binary 1s in a word is odd (odd parity) or even (even parity). Receivers (data sinks) check each received word for even or odd numbers of binary 1s. By knowing whether the data source is using even or odd parity encoding, the data sink is able to detect all single bit-per-byte errors.

**path overhead** In SDH systems, a header segment that provides communications between SPE assembly and disassembly points, that is nodes where user information is mapped to and from SPEs. See Chapter 5.

**PBX attendant features** Features enabling a console (switchboard) operator to answer external calls, extend them to PBX stations, serve as a call-coverage point, and assist users in placing external calls.

**peak cell rate (PCR)** In ATM networks, the maximum source traffic rate, in cells per second, that can be transported over PVCs. This parameter applies to both CBR and VBR classes (it is the only parameter applying to the CBR class).

**peripherals** In telephone systems, equipment such as telephones, modems, and other items of voice terminal equipment that act as the "eyes and ears" of the PBX, furnishing information needed to set up, connect, supervise, and tear down calls. More generally, equipment not integral to but working with any type of information system, computer, or telephone switch.

**permanent virtual circuit (PVC)** A virtual circuit resembling a leased line in that invariant logical numbers identifying PVCs are dedicated to a single user.

**personal area network (PAN)** A Bluetooth project concept where all the devices a person owns work seamlessly together. Under this perception, PANs can be created anywhere, in homes, offices, hotel rooms, and even automobiles.

**personal communications** Provides at least one human operator with direct terminal access and real-time or near-real time interactive communications with a remote human operator or an information system resource. Personal communications can refer to a broad range of services, systems, and equipment, e.g., facsimile machines, landline telephones, cellular telephone systems, and emerging personal communication system (PCS) adjuncts, and a variety of radio-based systems/devices including pagers, hand-held remote data entry terminals, and autonomous citizen band-like systems.

**personal communications service** The FCC defines PCS broadly as a family of mobile or portable radio services that can be used to provide service to individuals and businesses, and that can be integrated with a variety of competing networks. The term often encompasses a wide range of wireless mobile services, chiefly two-way paging and cellular-like calling services that are transmitted at lower power and typically higher frequencies (1850—1950 MHz) than analog cellular services.

**personal communications systems** Sometimes used interchangeably with personal communications service. A system able to render personal communications services.

**Personal Handyphone System (PHS)** A Japanese system for delivering personal communications services.

**per-trunk signaling** A signaling system approach wherein signals pertaining to a particular call are transmitted over the same trunk that carries the call.

**phase shift key (PSK)** In binary PSK modulation processes, baseband information signal 1s and 0s produce diametrically opposed—or phase reversed versions—of input carrier signals as modulator outputs. Since each carrier signal cycle is said to consist of 360 degrees (360°), binary

PSK modulators shift carrier phases by 180° for each baseband signal's transition from logical 1 to logical 0 states. In 4-ary PSK modulators, carriers are shifted among four possible phase alignments, relative to some fixed carrier phase, e.g., 0°, 90°, 180°, and 270°. M-ary PSK modulators shift carrier phases among m-possible phase alignments.

**photonics** A science and technology based on and concerned with the controlled flow of photons, or light particles.

**PIN diode** A photodiode made with an intrinsic undoped material between doped P and N layers and used as a lightwave detector. See *semiconductor.*

**plain old telephone service (POTS)** Telephony services considered "basic" decades ago; essentially a minimal service that allows subscribers to place and receive calls.

**plesiochronous** A condition among signals or signal components derived from different frequency or timing reference sources, but of nominally the same frequency, within some stated degree of precision. See Chapter 5 for more information.

**pointers** In SDH systems, pointers are overhead bytes that provide a simple means of dynamically and flexibly phase-aligning STS payloads, thereby permitting ease of dropping, inserting, and cross-connecting payloads in networks. Transmission signal wander and jitter are also readily minimized with pointers. See Chapter 5.

**point-of-presence (POP)** A physical location within a LATA, which an IXC establishes for the purpose of gaining access to LEC networks within the LATA using LEC-provided access services. An IXC may have more than one POP within a LATA and the POP may support public and private, switched, and non-switched services.

**point-to-point network** A network consisting of a dedicated connection between two pieces of equipment.

**Poisson distribution** A probability distribution where the inter-arrival times of users follows a negative-exponential distribution. This is the normal assumption in most queuing systems and is the most common occurrence in nature.

**port** A switched data service pricing element corresponding to the physical termination of an access circuit on a switch. The port speed usually matches the access circuit's bit rate.

**port oriented** An alternative to specifying maximum numbers of lines and trunks separately; in port-oriented PBXs and KTSs, lines and trunks can be intermixed to a maximum total capacity.

**power sum equal level far-end crosstalk (PSELFEXT)** Sums the total crosstalk interference power resulting from all adjacent cable pairs. See *crosstalk, equal level far-end crosstalk (ELFEXT),* and *crosstalk.*

**premises access services** Facilities underlying premises access services comprise all manner of terminals, PCs, and workstations designed directly to support operators; remotely controllable or accessible automation and data-processing equipment; a growing list of other data-network-compatible information origination/termination data termination equipment, and a premises distribution system.

**premises distribution systems** Wired, wireless, or hybrid transmission networks, inside buildings or among buildings on a campus. These networks connect desktop and other terminal equipment with common host equipment (e.g., switches, computers, and building automation systems), and with external telecommunications networks at network interfaces.

**private branch exchange (PBX)** A premises switching system, serving a commercial or government organization, and usually located on that organization's premises. PBXs provide telecommunications services on the premises or campus (e.g., internal calling and other services) and access to public and private telecommunications network services.

**private network** A network made up of circuits and, sometimes, switching equipment, for the exclusive use of one organization.

**processing gain** In CDMA systems, the spread-spectrum signal-to-noise level improvement factor. It is given by the ratio of the spread bandwidth (or chip rate) to the information signal bandwidth or bit rate.

**propagation** The movement or transmission of a wave in a medium or in free space.

**propagation loss** A loss of signal power level proportional to the square of the distance separating ratio transmitters and receivers. Factors other than square-law distance effects may also induce propagation loss.

**protocol** Strict procedures for the initiation, maintenance, and termination of data communications. Protocols define the syntax (arrangements, formats, and patterns of bits and bytes) and the semantics (system control, information context or meaning of patterns of bits or bytes) of exchanged data, as well as numerous other characteristics (data rates, timing, etc.).

**public-key cryptography** A cryptographic system using a pair of keys with the following property: messages encoded with one key can only be decoded with the other key. One key is held in secret within a particular user's equipment (a private key); the other key, a public key, is shared and can be stored in any other user's equipment.

**public-key infrastructure (PKI)** The entire process of issuing, using, and validating digital certificates and the entities involved in the process. The X.509 standards developed by the Internet Engineering Task Force (IETF) are widely accepted as the basis for such infrastructures.

**public switched telephone network (PSTN)** Denotes those portions of the LEC and IXC networks that provide public switched telephone network services.

**public utility commissions** State regulatory bodies that preside over state-level telecommunications regulation.

**pulse code modulation (PCM)** A modulation scheme involving conversion of a signal from analog to digital form by means of coding. See *modulation* and Chapter 3.

**quality of service (QoS)** In voice applications, normally encompasses a wide variety of parameters including loudness and signal-to-noise ratio. When describing packet-switched networks, QoS usually refers to the ability to meet isochronous traffic latency and delay and other requirements related to particular classes of service.

**quantizing noise** In any analog-to-digital conversion process, e.g., PCM, the difference between the converted binary value and the actual analog signal's amplitude. See *digitizing* and Chapter 3.

**queuing theory** The discipline of computing performance for a given traffic demand, number of resources, and probabilistic distribution of arrivals and service times.

**rake receivers** The ability to detect multiple received signals separately, and by appropriate signal-processing, combine or select only the combination that maximizes performance.

**rate adaptive DSL (RADSL)** DSL equipment that supports data throughput rates in accordance with actual and time-varying signal-to-noise ratio and other signal impairment conditions. Both DMT and CAP support such DSL operations. With DMT, adaptive rate adjustment steps may be as small as 32 kb/s, a very fine gradation for payload bandwidths in the order of 6—8 Mb/s. CAP provides 340-kb/s adaptive steps with the lowest rate alternative to no service at all being 640 kb/s.

**regional Bell operating company (RBOC)** One of seven regional companies created by the AT&T divestiture to assume ownership of the Bell operating companies. They were Ameritech, Bell Atlantic, Bell South, NYNEX, Pacific Telesis, Southwestern Bell, and US West. Mergers have reduced their number to four. Southwestern Bell, Pacific Telesis, and Ameritech became SBC Corporation; Bell Atlantic and NYNEX became Bell Atlantic (which merged with GTE Corporation to become Verizon). The term regional Bell holding company (RBHC) was originally used to refer to RBOCs, but fell into disuse.

**remote switch modules (RSMs)** In campus installations, some vendors place a central switch in one building and RSMs in other locations to concentrate traffic and simplify inter-building wiring.

**repeater** In digital transmission, equipment that receives a pulse train, amplifies it, re-times it, and then reconstructs the signal for retransmission. In IEEE 802 local area network (LAN) standards, a repeater is essentially two transceivers joined back to back and attached to two adjacent LAN segments. See *transceiver*.

**resistance** A measure of opposition to electrical current by materials or free space, when a potential difference (voltage) is applied between two points. According to Ohm's Law, when a constant voltage (non—time-varying) of one volt amplitude is applied to a resistor with a resistance value of one ohm, it produces a steady or constant current of one ampere.

**return loss** The amount of transmitter signal power reflected back to transmitters caused by mismatches between transmitter and cable network impedance.

**reverse channels** In wireless systems, mobile set-to-base station transmission links.

**ring protection switching** A mechanism used to detect failures in ring-topology systems and to restore service using an alternate path within the ring in 50 milliseconds.

**ring tripping** When called parties answer calls (go off-hook), ring tripping occurs immediately so that they do not hear 20-Hz "ringing" alerting signals.

**ringing signal** In telephone signaling systems, a 20-Hz alerting signal that causes called telephones to ring.

**roaming** A mobile customer's ability to use a mobile telephone outside the home service area of the operator with which the customer established mobile service.

**router** In IEEE 802 local area network (LAN) standards, devices that connect autonomous networks of like architecture at the network layer (layer 3). Unlike a bridge, which operates transparently to communicating end-terminals at the logical link layer (layer 2), a router reacts only to packets addressed to it by either a terminal or another router. Routers perform packet (as opposed to frame) routing and forwarding functions; they can select one of many potential paths based on transit delay, network congestion, or other criteria. How routers perform their functions is largely determined by the protocols implemented in the networks they interconnect. A router has all the attributes of a packet switch, but in addition has the ability to interconnect networks of differing design

**sampling** See *digitizing* and Chapter 3.

**satellite communications** Essentially uses radio, line-of-sight propagation from a transmitting earth terminal (usually ground-based, but sometimes shipborne or airborne) through free-space (the atmosphere and outer space) media to satellites, and back again to earthbound receiving terminals. In essence, satellites are equivalent to orbiting microwave repeaters. In addition to signal-repeating functions, some communications satellites provide signal-processing and switching capabilities.

**section overhead (SOH)** In SDH, STS-n frames are 125 microseconds long and consist of two parts, a section overhead (SOH) portion occupying the first three columns and all nine frame rows, and a synchronous payload envelope (SPE) into which user information is mapped. See Chapter 5.

**semiconductor** Semiconductors are materials midway between conductors and insulators. Semiconductors will only conduct electricity when a certain threshold voltage has been reached. Because the energy needed to "turn on" a semiconductor can be high, most semiconductors have their conductivity enhanced through a process called *doping*. Doping consists of adding an impurity that has either a surplus or shortage of electrons. Semiconductors with impurities that provide a surplus of electrons are called N-type semiconductors. Semiconductors with impurities that have a shortage of electrons are called P-type semiconductors. Together N-type and P-type semiconductors are the basic building blocks of nearly all solid-state electronic devices. Semiconductors have a resistance to electricity somewhere between a conductor (e.g. a copper wire) and an insulator (e.g. plastic). Hence the word "semi" conductor. Silicon and germanium are the two most commonly used semiconductor materials. The flow of current in a semi-conductor can be changed by light or the presence or absence of an electric or magnetic field.[1]

**server** Device that provides multiple connected workstations with various host-like services. In a network, equipment that makes available file, database, printing, facsimile, communications, or other services to client terminals/stations with access to the network. A gateway is a server that permits client terminal/station access to external communications networks and/or information systems.

**server farm** A configuration of servers used to provide access (usually Internet access) to applications for a large user base. Server farms typically contain as many as 50 multi-processor computers and hundreds of gigabytes of database storage, as an alternative to a single large mainframe computer.

**service level agreement (SLA)** A written agreement between a carrier and a customer that specifies delivered service performance levels to be monitored by the carrier and results to be provided to the customer, usually on a monthly basis. Carrier failure to meet performance levels results in credits to the customer.

**service management system (SMS)** In virtual private networks, a facility used to build and maintain a VPN database allowing customers to program specific functions for unique business applications. The SMS contains complete specifications of customer-defined private network specifications including location data, numbering plan, features, screening actions, authorization codes, calling privileges, etc. This information is downloaded (transmitted) to network control points (NCPs), which implement its instructions on a customer-by-customer basis.

**Service Management System/800** A national database system supporting toll-free services. LECs now access this database system to determine IXCs associated with particular 800 numbers.

**session** A connection between data terminals that allows them to communicate.

**Shannon's Law** In an historic 1948 paper, Claude Shannon proved that the maximum error-free bit rate that a bandwidth-limited channel can sustain is completely defined in terms of signal power-to-white noise power ratio, its bandwidth, and no other factors.

---

1 Harry Newton, *Newton's Telecom Dictionary*, Flatiron Publishing, 1998.

**shared service** A service that provides resources shared by many customers. Customers may request resources at any time, and pay in accordance with actual service usage.

**shielded twisted pair** Twisted copper paired wire cable with an outer metallic sheath surrounding insulated conductors. See *unshielded twisted pair.*

**sidetone** In telephone systems, the ability to hear oneself through telephone handsets.

**signal** Usually a time-dependent value attached to an energy-propagating phenomenon used to convey information. For example, an audio or sound signal in which the data are characterized in terms of loudness and pitch.

**signaling** The process of generating and exchanging information among components of a telecommunications system to establish, monitor, or release connections (call-handling functions) and to control related network and system operations (other functions).

**signaling interfaces** Interfaces are common boundary points between two systems or pieces of equipment. Interfaces exist between station equipment and transmission systems, between transmission systems and switching systems, and between transmission systems themselves, that are able to exchange signaling information.

**signaling system** From a top-level engineering perspective, all the signaling components (distributed among station equipment, switching, transmission, and other components of a telecommunications system) that comprise a signaling system.

**signaling system No. 7 (SS7)** International common-channel signaling system recommendations established by the CCITT.

**signal-to-noise ratio** In analog or digital systems, performance is ultimately limited or determined by the ratio of desired signal power to undesired naturally occurring or man-made noise or interference power with which receivers or signal regenerators must contend. Signal-to-noise ratios (SNRs) are most often stated in decibels, or dBs; that is, ten times the logarithm of the ratio of desired signal power to undesired noise power. If the mathematical term *logarithm* is unfamiliar, there is no need for concern. In practical systems SNRs of 10 dB are generally regarded as quite good (at 10 dB, signal power is ten times as large as noise power).

**simple mail transfer protocol (SMTP)** An Internet protocol used to send electronic mail messages from one host to another.

**simple network management protocol (SNMP)** The application protocol offering network management service in the Internet suite of protocols. A structure for formatting messages and transmitting information between reporting devices (agents) and data collection programs. Developed jointly by the Department of Defense, industry, and the academic community as part of the TCP/IP protocol suite; ratified as an Internet standard in Request for Comment (RFC) 1098.

**simplex** A simplex circuit is a transmission path capable of transmitting signals in one direction only, e.g., broadcast radio.

**single frequency signaling (SF)** A signaling method of conveying addressing and supervisory signals from one end of a trunk to the other, using the presence or absence of a single specified frequency, which exhibits significant susceptibility to fraud, and has all but vanished from the scene. A 2,600 Hz tone is commonly used.

**single-mode fiber(s)** Sufficiently small fiber strand core diameters that electromagnetic waves (lightwaves) are constrained to travel in only one transverse path from transmitter to receiver. This requires the utmost in angular alignment of light-emitting devices at points where light enters the fiber, and it results in higher transmitter/receiver costs than multimode fiber systems.

**socket** In TCP/IP systems, the combination of a port number and an Internet address.

**soft handoff** In cellular or satellite systems, a procedure whereby a subscriber signal in a "cell" is not handed off to another cell until that subscriber's signal is detected and received in the new cell while the original connection is still maintained.

**solid state** Any semiconductor device that controls electrons, electric fields, and magnetic fields in a solid material—and typically has no moving parts.

**SONET** An abbreviation for Synchronous Optical Network; SONET's basic design and operational concepts were formulated by R. J. Boehm and Y. C. Ching of Bellcore . By the end of 1994 the Bellcore work was submitted to ANSI's T1 Committee and in 1988 SONET interface standards, to be used in LEC/IXC optical networks, were approved. Like the ITU standards, SONET defines a hierarchy of synchronous transport signals (STS-n). SONET defines standard optical signals, a synchronous frame structure for multiplexing digital traffic, and operations procedures so that fiber optic transmission systems from different manufacturer/carriers can be interconnected.

**SONET ring service** Dedicated, high-capacity, self-healing network service designed to provide increased reliability, availability, and functionality via ring topologies among multiple customer locations and telco central offices.

**space division switch** A switch, which implements the switch matrix using a physical, electrical, spatial link. Where older space division switches used electro-mechanical mechanisms with metallic contacts, modern space-division switches are implemented electronically using integrated circuits. (Usually denoted by "S" in combined time and space division switches.)

**special access** Dedicated access provided by local exchange carriers to interexchange carriers for the purpose of originating or terminating inter-LATA traffic. Rates for special access are regulated by the FCC.

**special services** Any of a variety of LEC and IXC switched, non-switched, or special-rate services that are either separate from public telephone service or contribute to certain aspects of public telephone service. Examples include PBX tie trunks, foreign exchange (FX), and private-line services.

**spectral density** Specifies power levels for each frequency, or spectral component, of a signal.

**speed** A colloquial term, often used to connote the bit rate or throughput rate of switching, multiplexing, transmission, or other facilities or equipment used in digital communications.

**spot-beam satellite antenna** One that has a narrow beam width and covers only a small area of the earth's surface.

**spread spectrum** A modulation technique where the resulting modulated-signal bandwidth is much greater than the input-signal bandwidth.

**station equipment** A component of telecommunications systems such as a telephone or data terminal, generally located on the user's premises. Its function is to place calls and access services from networks, to transmit and receive user information (traffic), and to exchange control information with networks.

**station features** Station equipment capabilities that help individual telephone users to communicate with people and to access other information resources efficiently.

**station message detail recording (SMDR)** Capabilities, usually integral to telephone switches but possibly standalone, that detect and record calling and called numbers together with time, date, duration of calls, and other specified information.

**step index** An optical cable that exhibits a uniform refractive index across the core and a sharp decrease at the core-cladding boundary. Fiber cables can be manufactured to exhibit a constant index of refraction from core centers to core extremities, or treated to be variable. see Chapter 4 and *graded index.*

**stimulus/response** All signaling approaches can be classified as either stimulus/response or message based. As stored program control switches replace electromechanical switching systems, the trend is clearly toward SS7, message-based signaling.

**stored program control (SPC)** In today's digital circuit switches, connections through matrices and other switching system operations are controlled by a central computer, a switching system design attribute dubbed stored program control.

**structural dispersion** A type of dispersion in fiber cables attributable to unequal fiber-material structural alterations caused by different signal wavelength components. See *dispersion* and Chapter 4.

**subscriber identification module (SIM)** In GSM wireless telephones sets, a receptacle, into which small plastic SIM cards can be inserted. Each SIM smart card contains a microprocessor with eight kilobytes of memory that contains all the information needed to identify a particular subscriber and an algorithm used to encrypt that subscriber's voice or other information transmitted over the airwaves by GSM-conforming mobile sets.

**subscriber line charges** Fees mandated by the FCC to assure that costs for local-loop service are adequately provided for; they are charged to users monthly on a per-line basis.

**subscriber loop carrier (SLC)** Electronic equipment, often installed in outside subscriber-loop plant closer to subscriber locations than central-office wiring centers, to ameliorate line loss problems or to increase the number of subscribers that can be served ("pair gain systems").

**superframe format (SF)** Framing format (D3/D4—mode 3) The most widely used T1 carrier framing format in which the bipolar bit stream is organized into superframes, each consisting of 12 frames. To ensure timing, the signal must consist of at least one "1" bit in every 15 bits, and at least 3 "1" bits in every 24 bits. See *extended superframe.*

**supervisory signals** Signals used to indicate or control the states of circuits involved in a particular switched connection. A supervisory signal indicates to equipment, to an operator, or to a user that a particular state in the call has been reached and may simplify the need for action.

**suppressed carrier modulation** One reason why suppressed carrier modulation is used is that carrier signals contain no modulating or information-bearing signal content. Despite this, transmitting them "usurps" up to 50 percent of a transmitter's total power. By suppressing carriers, transmitter power that would have been dissipated in "informationless" carrier signals is used instead to increase information-bearing signal component signal-to-noise ratio.

**sustainable cell rate (SCR)** In ATM networks, the guaranteed average cell rate, in cells per second, of virtual circuits over time. It is roughly equivalent to the Committed Information Rate defined for frame relay virtual circuits. This parameter applies only to the VBR classes.

**switch matrix** A mechanism within switches that provides electrical paths between input and output signal termination points. Modern matrices are electronic and involve either time or space division switching. A time division switch employs a TDM process, in a time-slot interchange (TSI) arrangement. In space division, a physical, electrical, spatial link is established through the switch matrix. Whereas older space division switches used electro-mechanical mechanisms with metallic contacts, modern space-division switches are implemented electronically using solid state integrated circuits.

**switched access** Shared access provided by local exchange carriers to interexchange carriers for the purpose of originating or terminating inter-LATA traffic. Rates for switched access are regulated by the FCC.

**switched services** Public telecommunications services where users share transmission and switching facilities.

**switched virtual circuit (SVC)** A service that permits a user to establish virtual circuits between arbitrary network interface points on a call-by-call basis.

**switchhook** A plunger-activated switch built into telephone set cradles.

**switching** The process of connecting appropriate lines and/or trunks to form a desired communications path between two station sets, or more generally, any two arbitrary points in a telecommunications network. Included are all kinds of related functions, such as signaling, monitoring the status of circuits, translating address to routing instructions, alternate routing, testing circuits for busy conditions, and detecting and recording troubles.

**switching fabric** A relatively new expression in the telecommunications lexicon. In this context, "fabric" refers to the physical structure of a switch or network. It includes all the "interwoven" information-bearing components or channels that transport signals from input-to-output ports; a port is defined as a physical device interface. Switching fabrics do not encompass central control computers or adjunct components. The term fabric can apply to all input port-to-output port information-bearing physical network structures.

**switching systems** Systems that interconnect transmission facilities at various locations that route traffic through a network.

**synchronization** In communications networks the process of adjusting two or more signals so that corresponding signal characteristics or event time epochs or long and/or short term frequency, bit, byte, frame, or other periodic signal component rates correspond to one another in some defined or fixed relationship.

**synchronous** An adjective relating to a condition; perhaps the best definition is "a condition among signals or signal components derived from the same frequency or timing reference source and hence identical in frequency or rate." See Chapter 5 for more information.

**synchronous digital (signal) hierarchy (SDH)** Appearing to be an efficient means of transmission for optical networks, concepts for synchronous digital transmission first appeared in the early 1980s. On the international front, ITU's I.121 Broadband Integrated Services Digital Network (BISDN) (see Chapter 8) umbrella standard addresses switching, multiplexing, and transmission facilities able to meet expanding broadband, multimedia requirements. In particular, the G.707/708/709 recommendations define synchronous transport module (STM-n) signals as building blocks for a synchronous digital hierarchy (SDH). SDH is a digital transport structure that manages user information "payloads" and transports them through synchronous transmission networks.

**synchronous digital services** Digital private line services engineered to superior bit-error rate and error-free seconds specifications when compared to analog circuits with modems or T1-derived circuits.

**synchronous payload envelope (SPE)** In SDH, STS-n frames are 125 microseconds long and consists of two parts, a section overhead (SOH) portion occupying the first three columns and all nine frame rows;

and, a synchronous payload envelope (SPE) into which user information is mapped.

**synchronous transport module-n (STM-n)** G.707/708/709 recommendations define synchronous transport module signals as building blocks for a synchronous digital hierarchy (SDH). See *synchronous digital (signal) hierarchy (SDH)*.

**synchronous transport signals** A hierarchy of electrical synchronous transport signals (STS-n) defined in SONET specifications that correspond to SDH, STM-n signals and optical carrier, OC-n, optical signals.

**Systems Network Architecture (SNA)** IBM's proprietary description of the logical structure, formats, protocols, and operational sequences for transmitting information units and controlling network configuration and operation.

**T carrier** Introduced in 1962 within the old Bell System, T1 carrier systems have played a major role in the transition from analog to digital facilities. As one of the fastest-growing LEC and IXC facility segments in the early 1990s, T carrier became commonplace in large private networks, and, due to declining costs and wider service options, was increasingly found in moderate-sized networks, covering over 120 million miles in the North American continent. At the end of the decade (1990—2000), Bell Atlantic forecasted a 35 percent growth rate for T1 service. T carrier supports 24 full-duplex voice channels using just two pairs of unshielded twisted-pair (UTP) 19 AWG cable, constituting a low cost and reliable alternative to single, analog channel-per-twisted-pair voice frequency or other analog frequency division multiplexing (FDM) carrier systems.

**tandem switching system** A broad functional category describing systems that connect trunks to trunks and route traffic through a network.

**tariff** The published rate for a specific telecommunications service, equipment, or facility that constitutes a public contract between the user and the telecommunications supplier or carrier. Tariffed rates are established by and for telecommunications common carriers in a formal process in which carriers submit filings for government regulatory review, possible amendment, and approval.

**telecommunications** Any process that enables one or more users to pass to one or more other users information of any nature delivered in any usable form, by wire, radio, visual, or other electrical, electromagnetic, optical means. The word is derived from the Greek *tele,* "far off," and the Latin *communicare,* "to share."

**Telecommunications Act of 1996** On February 1, 1996, Congress passed the Telecommunications Act of 1996 (the "Act"), the first comprehensive rewrite of the Communications Act of 1934. The Act dramatically changed the ground rules for competition and regulation in virtually all the sectors of the communications industry. See Chapter 2 for details.

**telecommunications business applications** At the highest level, business applications are unique aggregations of telecommunications services that satisfy particular enterprise needs, for example medical/health care, hospitality, airline reservation, etc. Lower-level business applications, for example, station-to-station calling within a premises, are enterprise independent. For these situations telecommunications services correspond directly to generic business applications.

**telecommunications closet** In a premises distribution system, an area for connecting the horizontal and backbone wiring and for containing active or passive PDS equipment.

**telecommunications network** A system of interconnected facilities designed to carry traffic from a variety of telecommunications services. The network has two different but related aspects. In terms of its physical components, it is a facilities network. In terms of the variety of telecommunications services it provides, it can support a set of many traffic networks, each representing a particular interconnection of facilities.

**telecommunications network protocol (TELNET)** An Internet protocol that allows remote host access and terminal emulation.

**telecommunications service** A specified set of information transfer, and information transfer-supporting capabilities delivered to a group of users by a telecommunications system.

**terminal** A device, usually equipped with a keyboard, often with a display, capable of sending and receiving data over a communications link.

**terminating access service** Services that enable subscribers and/or equipment to accept (receive) calls or information transfer transactions.

**terrestrial microwave radio transmission systems** Systems consisting of at least two radio transmitter/receivers (transceivers) connected to high-gain antennas (directional antennas that concentrate electromagnetic or radio wave energy in narrow beams) focused in pairs on each other. The operation is point-to-point; that is, communications can be established between only two installations. This is contrasted to point-to-multipoint systems, such as broadcast radio or citizen band radios.

**thermal noise** In all electrical circuits an ever-present phenomenon is thermal noise, unwanted noise power that comes from electron, atom, and molecule movement. Sometimes referred to as *random* or *white noise*, in most communications channels or devices internally generated thermal noise produces a "flat" frequency spectrum. That is, in a band-limited channel (a channel with finite or limited-frequency bandwidth capability), the noise-power level at any frequency within the band, on average, is the same as the level at any other frequency in that band. Thermal noise gets its name from the fact that its noise level or power is proportional to temperature.

**time-assigned speech interpolation (TASI)** An analog technique used to transmit energy on an analog voice circuit only when a person is talking. See *digital speech interpolation*.

**time division multiple access (TDMA)** Whenever wireline or fiber optic receiving equipment, satellites, or mobile base stations are designed to receive and handle multiple signals, they are said to exhibit multiple access capabilities. Accordingly, linking frequency, time, and code multiplexing to multiple access capabilities produces frequency division multiple access (FDMA), time division multiple access (TDMA), and code division multiple access (CDMA).

**time division multiplexing (TDM)** A transmission facility shared in time (rather than frequency); i.e., signals from several sources share a single channel or bus by using the channel or bus in successive time slots. A discrete time slot or interval is assigned to each signal source.

**time division switch** A switch that implements the switch matrix using the TDM process, in a time-slot interchange (TSI) arrangement. (Usually denoted by T in combined time and space division switches.)

**time multiplexed switch (TMS)** In large switches employing both time and space division mechanisms, space division modules are referred to as time multiplexed switches (TMSs). A TMS is defined as an element of a time-division switching network that operates as a very high-speed space-division switch whose input-to-output paths can be changed for every time slot.

**time-slot interchange** See *time division switch*.

**token passing bus LAN** A LAN using a deterministic access mechanism and topology in which all stations actively attached to the bus "listen" for a broadcast token or supervisory frame. Stations wishing to transmit must receive the token before doing so; however the next logical station to transmit may not be the next physical station on the bus. Access is controlled by preassigned priority algorithms; defined by the IEEE 802.4 standard.

**token passing ring LAN** A LAN using a deterministic access mechanism and topology, in which a supervisory frame (or token) is passed from station to adjacent station sequentially. Stations wishing to transmit must wait for the "free" token to arrive before transmitting data. In a token ring LAN the start and end points of the medium are physically connected, leading to a ring topology; defined by the IEEE 802.5 standard.

**traffic** The flow of information within a network, among nodes, over links.

**traffic engineering** A branch of network design science that predicts and measures performance.

**traffic parameters** In ATM networks, parameters that relate to the characteristics of information being transported by virtual circuits. These include peak cell rate, sustainable cell rate, and maximum burst size.

**transceiver** A generic term describing a device that can both transmit and receive. In IEEE 802 LAN standards, a transceiver consists of a transmitter, receiver, power converter, and for CSMA/CD LANs, collision detector and jabber detector capabilities. The transmitter receives signals from an attached terminal's network interface card (NIC) and transmits them to the coaxial cable or other LAN medium. The receiver receives signals from the medium and transmits them via the transceiver cable and NIC to the attached terminal.

**transfer mode** A generic term for switching and multiplexing aspects of broadband integrated services digital networks (BISDN), adopted by CCITT Study Group XVIII.

**transmission control protocol/internet protocol (TCP/IP)** The transport layer and Internet layer, respectively, of the Internet suite of protocols. TCP corresponds to layer 4 of the OSI protocol stack; IP performs some of the functions of layer 3. It is a connectionless protocol used primarily to connect dissimilar networks.

**transmission facilities** Facilities providing communication paths that carry user and network control information between nodes in a network. In general, transmission facilities consist of a medium (e.g., free space, the atmosphere, copper, or fiber optic cable) and electronic equipment located at points along the medium. This equipment amplifies (analog systems) or regenerates (digital systems) signals, provides termination functions at the points transmission facilities connect to switching systems, and may provide the means to combine many separate sets of call information into a single "multiplexed" signal to enhance transmission efficiency.

**transmission impairments** Degradation caused by practical limitations of channels (e.g., signal-level loss due to attenuation, echo, various types of signal distortion, etc.), or interference induced from outside the channel (such as power-line hum or interference from heavy electrical machinery).

**transmission medium** Any material substance or "free space" (i.e., a vacuum) that can be, or is, used for the propagation of suitable signals, usually in the form of electromagnetic (including lightwaves) or acoustic waves, from one point to another; unguided in the case of free-space or gaseous media, or guided by a boundary of material substance.

**transmission system** A system consisting of a medium; that is, any material substance or "free space," (i.e., a vacuum) that can be used for the propagation of suitable signals from one point to another; and termination equipment needed to generate and receive signals compatible with the medium selected. From a top-level network engineering, total telecommunications system perspective, the aggregation of all facilities used to interconnect network nodes.

**transmission time** The time required to transmit a single packet once a circuit is available.

**transport** Switching, multiplexing, transmission, and any other facilities and related services that support information transfer between originating and terminating access services.

**trunk** A communication path connecting two switching systems. As an example, trunks are used to establish end-to-end connections between loops to customer station equipment.

**trunk group** A group of identical trunks within or between telephone company switching systems in which all paths are interchangeable. A trunk group can be searched in sequence until a free trunk is found.

**twisted pair** The most common type of transmission medium, consisting of two insulated copper wires twisted together. The twists or *lays* are varied in length to reduce the potential for interference between pairs. In cables greater than 25 pairs, the twisted pairs are grouped and bound together in a common cable sheath. See *unshielded twisted pair.*

**two-way trunk group** A group of central office trunks used for outgoing and incoming local and long distance calls.

**two-wire** A two-wire circuit uses a single transmission path for each direction of transmission. Two-wire circuits are normally used for local loops to subscribers (as opposed to four-wire circuits) due to the enormous wire and cable investment required to serve hundreds of millions of telephones. See *four-wire.*

**unguided media** Any medium in which boundary effects between "free space" and material substances are absent. The free-space medium may or may not include a gas or vapor. Unguided media including the earth's atmosphere and outer space support terrestrial and satellite radio and optical transmission.

**unidirectional path switched ring (UPSR)** A SONET ring configuration in which primary path traffic travels in only one direction on the ring. For full-duplex operation between two nodes, the forward traffic travels in one direction and the reverse traffic travels in the same direction (i.e., on the same fiber) around the other side of the ring.

**unidirectional WDM** In fiber optic cable-based transmission systems, the ability simultaneously to transmit separate lightwave signals but only in one direction over a single strand of fiber, with no or minimal mutual interference.

**Universal Mobile Telephone Service (UMTS)** A member of the International Mobile Telecommunications-2000 (IMT-2000) family of systems formerly known as the Future Public Land Mobile Telecommunications System (FPLMTS), an ITU project. With UMTS, single portable telephones and integrated telco networks may satisfy indoor, outdoor, pedestrian, and vehicular communications requirements.

**unshielded twisted pair (UTP)** Two wood-pulp or plastic-insulated copper conductors (wires), twisted together into pairs, capable of propagating electromagnetic waves. The twists, or lays, are varied in length to reduce the potential for signal interference between pairs, in multi-pair cables. Wire sizes range from 26 to 19 gauge (i.e., 0.016 to 0.036 inch in diameter) and are typically manufactured in cables of from 2 to 3,600 pairs. Shielded twisted-pair cable is similar to UTP, but the twisted pairs are surrounded by a cylindrical metallic conductor which is clad with an insulating sheath. See *cable, conductor.*

**unspecified bit rate (UBR)** An ATM service that is a best-effort service, with no guarantees of either throughput or cell loss.

**upstream** The direction of traffic flowing from subscriber terminal equipment toward server or host equipment.

**user datagram protocol (UDP)** An Internet protocol that provides connectionless transport services for applications not requiring TCP reliability. Its shorter header and lack of connection setup overhead make it more efficient when the amount of data to be transmitted is small.

**V&H coordinate system** The grid developed to permit accurate mileage determination in North America. Its origin is in Greenland and vertical (V) and horizontal (H) numerical coordinates increase toward the west and south.

**vampire taps** A means for penetrating coaxial cable cladding and outer metallic sheathing to make contact with its inner conductor.

**variable bit rate—non-real time (VBR-NRT)** An ATM service that allows users to send data at rates that vary with time. Networks can then invoke statistical multiplexing to conserve resources. Since applications using VBR-NRT are insensitive to cell-delay variation, synchronism between user devices is not maintained across the network.

**variable bit rate—real time (VBR-RT)** An ATM service that allows users to send data at a rate that varies with time. This service is similar to VBR-NRT except that synchronism is maintained across networks and the sustainable cell rate is maintained with less cell delay variation than in VBR-NRT service.

**very small aperture terminal (VSAT)** Earth terminals using small antennas (1.5—6 feet in diameter). This technology typically operates in the Ku band (11/14 GHz) and Ka band (20/30 GHz).

**VF transmission** Two broad categories of transmission systems are voice frequency (VF) transmission systems and carrier transmission systems. VF transmission supports the ubiquitous local loops connecting business and residential customer premises with serving central offices. VF transmission includes LEC main, branch, distribution, and drop cable subscriber-loop facilities and in the past was used for short interoffice connections.

**video conferencing** The real-time, usually two-way transmission of voice and images between two or more locations. Today, both voice and video analog signals are digitized by video codecs before transmission, which can involve wide bandwidths. To conserve bandwidth, some systems employ freeze frame, where a television screen is only "repainted" every few seconds. Codecs for higher-quality full-motion video attempt to minimize bandwidth requirements by taking advantage of intervals with relatively little motion (which require smaller bandwidths), and by trading off smooth-motion tracking and picture resolution.

**virtual private network (VPN) service** A switched service that uses a carrier's shared facilities to provide customers with the attributes of a private network. This includes using dedicated access on both ends of calls and the involvement of a database-controlled central processor in the routing of every call. Customers can define, change, and control network resources with flexibility the same as or greater than that afforded by facilities-based private networks.

**voice message processing** Instead of having to hang up after dialing to a ring-no answer or busy signal, voice message processing is the generic term used to describe capabilities that allow callers to hear telephones being answered, listen to prerecorded greetings (generally in the called party's voice), and hear instructions regarding how to record, review, change, or erase voice messages to be left for called parties. Later, called parties can retrieve messages and perform additional annotation, distribution, and delivery-verification functions. Also referred to as voice mail.

**Voice over IP (VoIP)** The transmitting of voice signals over IP-based packet-switched networks, in particular, the Internet.

**voltage** The amplitude of an electrical signal is measured in *volts* and referred to as voltage.

**waveforms** Amplitude (magnitude) versus time representation of signals; time-varying properties or shapes of electrical signals or other phenomena such as acoustic or sound waves. In the case of sound waves, how the compression and expansion of air would appear if sketched on a paper as a function of time. See Chapter 3 for example sketches.

**wavelength** A distance parameter associated with the motion of waves through or along a medium or free space. By *wave motion*, we mean the propagation of energy (or some kind of disturbance).

**wavelength division multiplexing (WDM)**  A type of multiplexing.  Since wavelength and frequency are directly related by propagation velocity in carrier media, wavelength division multiplexing could also be described as frequency division multiplexing. Industry practice, however, uses the term WDM for optical signal multiplexing. Compare with *FDM, TDM, CDM.* See Chapter 5.

**white noise** See *thermal noise.*

**wide area network (WAN)** A network providing services beyond the distance limitation of MANs. See *metropolitan area network (MAN).*

**wide area telecommunications services (WATS)** A service permitting customers to make (OUTWATS) or receive (INWATS) long-distance voice or data calls and to have them billed on a bulk rather than an individual call basis. The service is provided by means of special private access lines connected to WATS-equipped central offices. A single access line permits inward or outward service but not both.

**window size** The number of bytes of data receivers can accept and store in their buffers before sending acknowledgements.

**wink-start** A supervisory signal that consists of an off-hook followed by an on-hook signal, exchanged between two switching systems. The wink-start signal is generated by the called switch to indicate to the calling switch that it is ready to receive address signal digits.

**wire center** A location of one or more local switching systems and a point at which customer loops converge.

**wireless KTSs** KTS systems in which wired connections are only required between a single unit and LEC network interfaces. All other connections among handheld or desktop telephones are accomplished via radio links, that is, wireless means.

**wireless local loop** Local loop facilities employing unguided (usually radio) facilities in lieu of cable (historically copper twisted pair) to provision transmission paths between customer and service-provider premises. Ordinarily used to denote facilities intended for fixed subscriber equipment, but facilities supporting mobile subscribers may also be employed.

**wire-line** Originally companies that employ copper-wire subscriber-loop facilities were referred to as wireline carriers as opposed to telephone companies using radio loops. Since 50 percent of the cellular licenses were issued to wire-line carriers, the name is something of a misnomer. Usually, carriers using metallic or fiber cable to deliver services to subscribers are regarded as wire-line carriers.

**word length** The quantity of bits used to represent binary numbers.

**workgroup** In a LAN context, a group of people (or rather their workstations) attached to a single LAN segment.

**workstations** Input/output devices used by operators, which can process data independent of host processors and can be configured to exchange data with other workstations, host processors, or servers.

**World Administrative Radio Conference (WARC)** A part of the ITU that assigns both bandwidth and spectral location to specific wireless service classes.

# ACRONYM LIST

| | |
|---|---|
| AAL | ATM adaptation layer |
| ABR | available bit rate |
| ACD | automatic call distribution |
| ACF | access coordination fee |
| ACK | acknowledgment |
| ADC | analog-to-digital codec |
| ADH | asynchronous digital signal hierarchy |
| ADSL | asymmetric digital subscriber line |
| ALI | automatic location identification |
| AMI | alternate mark inversion |
| AMPS | Advanced Mobile Telephone Service (AMPS) |
| ANI | automatic number identification |
| ANSI | American National Standards Institute |
| ARPA | Advanced Research Project Agency |
| ARQ | Automatic Retransmission reQuest |
| ASB | asynchronous balanced mode |
| ASI | alternate space inversion |
| ASIC | application-specific integrated circuit |
| ASK | amplitude shift key |
| ASP | Application Service Provider |
| ATM | asynchronous transfer mode |
| AWG | American Wire Gauge |
| B8ZS | bipolar eight-zero substitution |
| $B_c$ | committed burst size |
| BECN | backward explicit congestion notification |
| BER | bit-error rate |
| BF | framing bit |
| $BF_t$ | terminal framing bit |
| BIPS | billion instructions per second |
| BISDN | broadband integrated services digital network |
| BLSR | bidirectional line switched ring |
| BOC | Bell-operating company |
| BRI | basic rate interface |
| BS | base station |
| BSC | base station controller |
| BTA | basic trading area |

| | |
|---|---|
| CA | certificate authority |
| CALC | customer access line charges |
| CAP | carrierless amplitude and phase |
| CAP | competitive access provider |
| CBR | constant bit rate/case based reasoning |
| CCIS | common-channel interoffice signaling |
| CCS | centi-call seconds |
| CCS | common-channel signaling |
| CCSA | common-control switching arrangement |
| CDM | code division multiplexing |
| CDMA | code division multiple access |
| CDPD | Cellular Digital Packet Data |
| CIC | carrier identification code |
| CIR | committed information rate |
| CLASS | custom local area signaling service |
| CLEC | competitive local exchange carrier |
| CMIP | Common Management Information Protocol |
| CMISE | Common Management Information Service Element |
| CMOL | CMIP over Logical Link Control |
| CMOT | Common Management Information Services over TCP/IP |
| CO | central office |
| COC | central office connection |
| CPI | computer-to-PBX interface |
| CPU | central processing unit |
| CRC | cyclic redundancy check |
| CSDS | circuit-switched data service |
| CSMA/CD | carrier sense multiple access w/collision detection |
| CSR | Centrex station rearrangement |
| CSU | channel service unit |
| CTIA | Cellular Telecommunications Industry Association |
| DACS | digital access and cross-connect system |
| DARPA | Defense Advanced Research Projects Agency |
| DBS | direct broadcast satellite |
| DCE | data circuit terminating equipment |
| DCP | digital communications protocol |
| DCS | digital cross-connect system |
| DDD | direct distance dialing |
| DDN | Defense Data Network |
| DE | discard eligible |
| DEMS | digital electronic message service |

| | |
|---|---|
| DID | direct inward dialing |
| diffserv | differentiated services |
| DLCI | data-link connection identifier |
| DMI | digital multiplexed interface |
| DMT | discrete multitone |
| DOD | direct outward dialing |
| DQDB | distributed queue dual bus |
| DSI | digital speech interpolation |
| DSL | digital subscriber line |
| DSP | digital signal processing, digital signal processor |
| DSS | digital subscriber service, digital subscriber signaling |
| DSS/BLF | direct station selection/busy lamp field |
| DSU | data service unit |
| DTE | data terminal equipment |
| DTMF | dual-tone multiple frequency |
| DTS | digital termination system |
| DWDM | dense wavelength division multiplexing |
| E-mail | electronic mail |
| EDFA | erbium-doped fiber amplifier |
| EF | expedited forwarding |
| EIA | Electronic Industries Association/ Electronic Industries Alliance |
| EIR | equipment identity register |
| EKTS | electronic key telephone system |
| ELFEXT | equal level far-end crosstalk |
| EMI | electromagnetic interference |
| ENFIA | exchange network facilities for interstate access |
| EO | end office |
| EPSCS | enhanced private-switched communications service |
| ESF | extended superframe |
| ESN | electronic switched network/electronic serial number |
| FCC | Federal Communications Commission |
| FCS | frame check sequence |
| FDDI | fiber-distributed data interface |
| FDL | facility data link |
| FDM | frequency division multiplexing |
| FDMA | frequency division multiple access |
| FECN | forward explicit congestion notification |
| FEDAC | forward error detection and correction |
| FEP | front-end processor |
| FEXT | far-end crosstalk |

| | |
|---|---|
| FIPS | Federal Information Processing Standard |
| FOM | fiber optic modem |
| FSK | frequency shift key |
| FSS | fully separated subsidiary |
| FTAM | file transfer access and management |
| FTP | file transfer protocol |
| FX | foreign exchange |
| GEO | geosynchronous earth orbiting satellite |
| GOS | grade of service |
| GOSIP | Government Open Systems Interconnection Profile |
| GPRS | General Packet Radio Service |
| GPS | Global Positioning System |
| GSM | Groupe Speciale Mobile, Global System for Mobile Communications |
| GUIs | graphical user interfaces |
| HDLC | high-level data-link control |
| HDSL | high bit-rate DSL |
| HLR | home location register |
| HVAC | heating, ventilation and air conditioning |
| I-MAC | isochronous media access controller |
| IAB | Internet Architecture Board |
| IC | information communications functional area |
| IDN | integrated digital network |
| IDSL | ISDN over digital subscriber line |
| IEC | International Electrotechnical Commission |
| IETF | Internet Engineering Task Force |
| ILEC | incumbent local exchange carrier |
| IMEI | international mobile station equipment identity |
| IMSI | international mobile subscriber identity |
| IMT-2000 | International Mobile Telecommunications for the year 2000 |
| IOC | interoffice channel |
| IP | Internet protocol |
| IPSec | IP Security |
| IPX | internetwork packet exchange |
| ISDN | integrated services digital network |
| ISM | Industrial, Scientific, and Medical |
| ISO | International Organization for Standardization |
| ISP | Internet service provider |
| ITC | independent telephone company |
| ITU | International Telecommunication Union |

| | |
|---|---|
| IVR | integrated voice response |
| IX | information exchange functional area |
| IXC | interexchange carrier |
| JPEG | Joint Photographic Experts Group |
| JTM | job transfer manipulation |
| KSU | key service unit |
| KTS | key telephone system |
| L2TP | Layer 2 Tunneling Protocol |
| LAN | local area network |
| LAP-B | Link Access Protocol-B |
| LASER | light amplification by stimulated emission of radiation |
| LATA | local access and transport area |
| LCR | least-cost routing |
| LDN | listed directory number |
| LEC | local exchange carrier |
| LED | light emitting diode |
| LEO | low earth orbiting satellite |
| LMDS | local multipoint distribution service |
| LORAN | Long Range Aid to Navigation |
| LSI | large-scale integrated circuit |
| M&C | management and control |
| MAC | media access control, minimum annual commitment, moves, adds, and changes |
| MACSTAR | multiple access customer station rearrangement |
| MAE | metropolitan area exchange |
| MAN | metropolitan area network |
| MAP | management application part |
| MBS | maximum burst size |
| MCU | mobile control unit |
| MF | multiple frequency |
| MFJ | Modification of Final Judgment |
| MFOTS | Military Fiber-Optic Transmission System |
| MHS | message handling system |
| MIB | management information base |
| MIN | mobile identification number |
| MIPS | million instructions per second |
| MMDS | microwave multipoint distribution systems |
| MPEG | Moving Pictures Experts Group |
| MPLS | multi-protocol label switching |
| MSC | mobile switching center |
| MSN | mobile subscriber networks |

| | |
|---|---|
| MSS | metropolitan switching system |
| MTA | major trading area |
| MTS | message telecommunications service |
| MTSO | mobile telephone switching office |
| NACK | negative acknowledgment |
| NAM | numerical assignment module |
| NANPA | North American Numbering Plan Administrator |
| NAP | network access point |
| NCP | network control point |
| NCSA | National Center for Supercomputer Applications |
| NCTE | network channel terminating equipment |
| NEXT | near-end crosstalk |
| NI | network interface |
| NIC | network interface card |
| NIST | National Institute of Standards and Technology |
| NOC | network operations center |
| NOS | network operating system |
| NPA | numbering plan area |
| NSEP | National Security and Emergency Preparedness |
| NSF | National Science Foundation |
| NSP | national backbone service provider |
| NT | network termination |
| NTSC | National Television System Committee |
| OA&M | operation administration and maintenance |
| OADM | optical add/drop multiplexer |
| OCC | other common carrier |
| OFDM | orthogonal frequency division multiplexing |
| OMC | operations and maintenance center |
| ONU | optical networking unit |
| OPX | off-premises extension |
| OSC | operating system control |
| OSI | open systems interconnection |
| OSS | operations support system |
| OXC | optical cross-connect |
| P-MAC | packet media access controller |
| PAD | packet assembler-disassembler |
| PAN | personal area network |
| PAR | positive acknowledgment with retransmission |
| PBX | private branch exchange |
| PCB | printed circuit board |
| PCM | pulse code modulation |

| | |
|---|---|
| PCMCIA | Personal Computer Memory Card International Association |
| PCR | peak cell rate |
| PCS | personal communications service/ personal communications system |
| PDA | personal digital assistant |
| PDS | premises distribution system |
| PHS | Personal Handyphone System |
| PHY | physical-layer protocol |
| PIC | primary interexchange carrier |
| PICC | presubscribed interexchange carrier charge |
| PKI | public-key infrastructure |
| PMD | physical-layer media dependent |
| PN | pseudo-noise (code generators) |
| POH | path overhead |
| POP | point-of-presence |
| POT | point of termination |
| POTS | plain old telephone service |
| PPSN | public packet switched network |
| PPTP | Point-to-Point Tunneling Protocol |
| PRI | primary rate interface |
| PSK | phase shift key |
| PSN | packet-switched network |
| PSPDN | packet-switched public data network |
| PSTN | public-switched telephone network |
| PTT | postal, telephone, and telegraph |
| PUC | public utility commission |
| PVC | permanent virtual circuit |
| QoS | quality of service |
| RADSL | rate adaptive DSL |
| RBHC | regional Bell holding company |
| RBOC | regional Bell operating company |
| RBS | robbed bit signaling |
| RDSS | radio determination satellite services |
| RF | radio frequency |
| RFC | request for comment |
| RFI | radio frequency interference |
| RFP | request for proposal |
| RSM | remote switch module |
| RSU | remote switching unit |
| SAFENET | Survivable Adaptable Fiber-optic Embedded Network |

| | |
|---|---|
| SCM | subcarrier multiplexed |
| SCP | service control point |
| SCR | sustainable cell rate |
| SDH | synchronous digital hierarchy |
| SDN | software defined network |
| SF | single frequency |
| SID | system identification number |
| SIM | subscriber identification module |
| SLA | service level agreement |
| SLC | subscriber line charges |
| SMDR | station message detail record |
| SMDS | switched multimegabit data service |
| SMI | Structure of Management Information |
| SMR | specialized mobile radio |
| SMS/800 | Service Management System/800 |
| SMT | station management technology |
| SMTP | Simple Mail Transfer Protocol |
| SNA | Systems Network Architecture |
| SNI | subscriber network interface |
| SNMP | Simple Network Management Protocol |
| SOH | section overhead |
| SONET | synchronous optical network |
| SPC | stored program control |
| SPE | synchronous payload envelope |
| SS | signaling system |
| SSN | switched service network |
| STM | synchronous transfer mode |
| STM-n | synchronous transport module "n" |
| STP | shielded twisted pair, signal transfer point |
| STS-n | synchronous transport signal "n" |
| SVC | switched virtual circuit |
| TA | terminal adapter |
| TASI | time assigned speech interpolation |
| $T_c$ | committed rate measurement interval |
| TCP | transmission control protocol |
| TDM | time division multiplexing |
| TDMA | time division multiple access |
| TELNET | Telecommunications Network |
| TIA | Telecommunications Industry Association |
| TMS | time multiplexed switch |
| TP | transaction processing |

| TSI | time-slot interchange |
|-----|------------------------|
| UBR | Unspecified Bit Rate |
| UDP | User Datagram Protocol |
| UIS | Universal Information Services |
| UMTS | Universal Mobile Telephone Service |
| UNI | user-network interface |
| UPSR | unidirectional path switched ring |
| USF | Universal Service Fund |
| UTP | unshielded twisted pair |
| VAD | value-added distributor |
| VAN | value-added network |
| VAR | value-added reseller |
| VBR | variable bit rate |
| VBR-NRT | variable bit rate—non-real time |
| VBR-RT | variable bit rate—real time |
| VCI | virtual circuit identifier |
| VCS | virtual circuit switch |
| VDSL | very high-speed DSL |
| VF | voice frequency |
| VLR | visitor location register |
| VoIP | Voice over IP |
| VPI | virtual path identifier |
| VPN | virtual private network |
| VRU | voice response unit |
| VSAT | very small aperture terminal |
| VSELP | vector sum excited linear predictive coding |
| VTNS | Virtual Telecommunications Network Service |
| WAN | wide area network |
| WARC | World Administrative Radio Conference |
| WATS | wide area telecommunications services |
| WCDMA | wideband code division multiple access |
| WDM | wavelength division modulation |
| WLAN | wireless local area network |
| WLL | wireless local loop |
| WPDS | wireless premises distribution system |

# BIBLIOGRAPHY

Bates, Regis J. *Broadband Telecommunications Handbook.* McGraw-Hill 2000.

*Bellcore Notes on the Networks*—1997. Special Report SR-2275 Issue 3, December 1997.

Black, Uyless. *Second Generation Mobile and Wireless Networks.* Englewood Cliffs, NJ: Prentice-Hall 1999.

Breyer, Robert and Riley, Sean. *Switched, Fast and Gigabit Ethernet* 3rd. ed. Macmillan Technical Publishing 1999.

Calhoun, George. *Wireless Access and the Local Telephone Network.* MA: Artech House 1992.

Cavanagh, James P. *Applying the Frame Relay Interface to Private Networks.* New York, IEEE Communications 1992.

Dicenet, G. *Design and Prospects for the ISDN.* MA: Artech House 1987.

Freeman, Roger L. *Telecommunication System Engineering.* 2nd ed. New York, John Wiley & Sons 1989.

Garg, Vijay K. and Wilkes, Joseph E. *Wireless and Personal Communications Systems.* Englewood Cliffs, NJ: Prentice-Hall 1996.

Goralski, Walter. *ADSL and DSL Technologies.* McGraw-Hill 1998.

Held, Gilbert. *Voice Over Data Networks.* McGraw-Hill 1998.

*IBM Vocabulary for Data Processing, Telecommunications, and Office Systems.* Seventh edition, 1981.

IEEE Standards for Local Area Networks: *Logical Link Control.* New York: Institute of Electrical and Electronics Engineers 1988.

IEEE Standards for Local Area Networks: *Supplements to Carrier Sense Multiple Access with Collision Detection (CSMA/CD) Access Method and Physical Layer Specifications.* New York: Institute of Electrical and Electronics Engineers 1987.

IEEE Standards for Local Area Networks: *Token Ring Access Method and Physical Layer Specifications.* New York: Institute of Electrical and Electronics Engineers 1988.

IEEE Standards for Local Area Networks: *Token-Passing Bus Access Method and Physical Layer Specifications.* New York: Institute of Electrical and Electronics Engineers 1988.

IEEE Standards for Local Area Networks: *Carrier Sense Multiple Access with Collision Detection (CSMA/CD) Access Method and Physical Layer Specifications.* New York: Institute of Electrical and Electronics Engineers 1988.

Lee, Byeong Gi, Kang, Minho and Lee, Jonghee. *Broadband Telecommunications Technology.* MA: Artech House 1993.

Lee, Jhong Sam and Miller, Leonard E. *CDMA Systems Engineering Handbook.* Boston: Artech House 1998.

Malamud, Carl. *STACKS—Interoperability in Today's Computer Networks.* Englewood Cliffs, NJ: Prentice-Hall 1992.

Martin, James. *Telecommunications and the Computer.* 3rd ed. Englewood Cliffs, NJ: Prentice-Hall 1990.

Miller, Mark A. P. E. *Internetworking, Second Edition.* New York: M&T Books 1995.

Miller, Mark. *Troubleshooting TCP/IP, Second Edition.* New York: M&T Books 1996.

Minzer, Steven E. *Broadband ISDN and Asynchronous Transfer Mode (ATM).* New York: IEEE Communications 1992.

Newton, Harry. *Newton's Telecom Dictionary. 15th edition.* New York: Miller Freeman 1999.

Noll, A. Michael. *Introduction to Telephones and Telephone Systems.* MA: Artech House 1986.

Project 802—*Local & Metropolitan Area Networks.* Institute of Electrical and Electronics Engineers, Inc. Unapproved Draft—Published for Comment Only.

Rey, R. F., ed. *Engineering and Operations in the Bell System, Second Edition.* NJ: AT&T Bell Laboratories 1984.

Rider, Michael J. *Protocols for ATM Access Networks.* New York: IEEE Network, January 1989.

Sapronov, Walter. *Telecommunications and the Law.* MD: Computer Science Press, 1988.

Scott Schauer. *Frame Relay: Designing Virtual Private Networks.* IL: Business Communications Review, 1991.

Stallings, William. *ISDN: An Introduction.* New York: MacMillan 1989.

Tanenbaum, Andrew S. *Computer Networks.* Englewood Cliffs, NJ: Prentice-Hall 1981.

Tedesco, Eleanor Hollis. *Telecommunications For Business*. MA: PWS-Kent 1990.

*Telecommunications Transmission Engineering, Third Edition*. Vol. 1: *Principles*. Bellcore 1990.

*Telecommunications Transmission Engineering, Third Edition*. Vol. 2: *Facilities*. Bellcore 1990.

*Telecommunications Transmission Engineering, Third Edition*. Vol. 3: *Networks and Services*. Bellcore 1990.

Thomas, Stephen. *IPng and the TCP/IP Protocols*. NewYork: John Wiley and Sons, 1996.

Weik, Martin H. *Communications Standard Dictionary*. New York: Van Nostrand Reinhold 1983.

# INDEX

Note: Boldface numbers indicate illustrations, italic t indicates a table.

## X

## Y

# ABOUT THE AUTHORS

**Joseph A. Pecar** is President of Joseph A. Pecar and Associates, an information systems engineering and integration firm he founded in 1983 after spending fifteen years with IBM, in a senior engineer/manager capacity. Mr. Pecar can be contacted at telecomfactbook@erols.com.

**David A. Garbin** is the Chief Engineer for Telecommunications and Networking at Mitretek Systems, a non-profit research and development organization that applies science and technology in the public interest. Mr. Garbin can be contacted at telecomfactbook@erols.com.